AF326926

PHYSICAL CHEMISTRY OF LEATHER MAKING

*We dance round in a ring and
suppose. But the secret sits
in the middle and knows.*

Robert Lee Frost

PHYSICAL CHEMISTRY OF LEATHER MAKING

by Krysztof Bienkiewicz

ROBERT E. KRIEGER PUBLISHING COMPANY
MALABAR, FLORIDA
1983

Original English Edition 1983
Revised and enlarged from Polish edition

Printed and Published by
ROBERT E. KRIEGER PUBLISHING COMPANY, INC.
KRIEGER DRIVE
MALABAR, FLORIDA 32950

Copyright © 1983 by Robert E. Krieger Publishing Co., Inc.

All rights reserved. No reproduction in any form of this book, in whole or in part (except for brief quotation in critical articles or reviews), may be made without written authorization from the publisher.

Printed in the United States of America

Library of Congress Cataloging in Publication Data

Bieńkiewicz, Krzysztof J.
  Physical chemistry of leather making.

Translation of Fizykochemia wyprawy skór.
  Includes index.
1. Tanning. 2. Chemistry, Physical and
theoretical. I. Title.
TS967.B4713   675'.23'015413   80-27191
ISBN 0-89874-304-4

FOREWORD

This volume deals primarily with the scientific background to the methods and processes by which raw hides and skins are converted to the various types of leather. Such a treatise is long overdue, for the last published works of a similar nature were: Thorstensen, 1969; Reed, 1966: O'Flaherty, Roddy, and Theis, 1962 (reprinted in 1976); Gustavson, 1956; McLaughlin and Theis, 1945. These are in the English language and the position in other tongues is little better, with a few exceptions. More recently there have been publications dealing with the more practical aspects of leather making. The authors are not able to digress unduly into the background science, and yet it is desirable to explain the reasons underlying the choice of methods. For this purpose an appropriate accompanying scientific textbook is required.

Over the last decade a great deal of new thought on leather making has been presented by the various national societies and institutes devoted to the dissemination of such scientific and technical knowledge. This information has been published in a score of journals in many languages. To conscientiously search through this world-wide literature is a formidable task and to collect, collate, and arrange this into rational textbook form is a praiseworthy effort, for which Dr. Bieńkiewicz is to be congratulated.

Significant in this updating of knowledge on "Leather Science" is the big expansion in methods of physical chemical investigation of the phenomena concerned. The first chapters of this book outline these techniques, viz., absorption spectroscopy in the visible, ultraviolet and infrared wave bands; X-ray diffraction; optical rotation; luminescence analysis; radio frequency spectroscopy; etc. Thermodynamic and kinetic theories are also discussed. The principles of these methods are given and their potential value to the "Leather Scientist" plus an account of the findings as related to the skin proteins and the effects of leathermaking processes on them.

Some of these results must stimulate new thoughts and revision of traditional theories in such important fields as depilation, tanning, and drying. Dr. Bieńkiewicz expounds on these subjects and introduces some new concepts. Of particular interest are attempts to explain the physical nature of leather, not only in terms of modified molecular structure, but relative to the macrophysics of fibrous structures—a field of more practical significance to the leather producer than detailed studies of amino acid composition.

The latter part of the book covers the technical application of this knowledge to leather-making processes. As such, it is not a practical handbook, but attempts to propound the scientific principles which underly, or should underly, the various processes. In a technology with at least a two thousand year history, one finds a great deal of practical dogma which is often at variance with scientific reasoning. Typical of such dogma are statements that "Good leather is made in the limeyard," "Slow drying gives a mellower leather," or "Sperm oil is the best for soft leathers."

These tenets may be well-founded on practical observation under a common set of process conditions in particular tanneries making a specific leather from specific hides or skins. However, the writer and many others have had the experience of transferring a leather-making process which operated successfully from one tannery to an apparently identical one in a different part of the world, only to find that the leather produced was not identical. Obviously some factors were not identical and the search for these can be a long and expensive operation. A sound knowledge of the basic science underlying the processes is essential in carrying out a systematic check of all the factors which could play a part. Thus local atmospheric conditions, e.g., temperate, tropical, arid, will affect not only the handle of the leather but at least the following processes: all piling, conditioning, and drying processes, soaking, liming, deliming, pickling, washing, and tanning.

The practical tanner may find in these chapters some control factors which he has overlooked whilst the scientific research worker may be more careful in producing data for the tanner until he has verified that all the factors relevant to his results are known and stated. Thus many technical problems may be clarified and occurrences minimized which lead to such phrases as "O.K. in theory, but it doesn't work in practice" or the converse "The process works (or doesn't) in practice, but we don't know why." In either case, the reliability of the process must be in question.

I appreciated collaboration with Dr. Bieńkiewicz for some months when he was seconded to my staff at the Turkish Leather Research and Training Institute, Pendik, Istanbul, as a United Nations Development Programme Expert, where it may be said that "East meets West" not only in the usual context, but in making me aware of the considerable amount of novel leather technical thought current in Eastern Europe of which I was relatively ignorant. Other English-speaking leather technologists may find this book stimulating for this reason.

J.H. Sharphouse B.Sc., etc.
One-time Head of Leather Technology, Northampton
One-time Project Manager for F.A.O.
Dericilik Arastirma ve Egitim Enstitusu
Istanbul

Preface

"There can be, however, no harm in an attempt to describe the physical chemist: he is a man who observes natural phenomena and records his observations, but maintains a judicious balance between the material changes themselves, the physical effects which accompany them and the mathematical machinery by means of which the whole can be most compactly condensed.

. . . The influence of heat, light and mechanical and electrical forces on the various states of matter forms the subject of thermochemistry, photochemistry, piezochemistry and electrochemistry, which rank among the most important subdivisions of physical chemistry." (Encyclopaedia Britannica.)

As we see, physical chemistry is a very broad field which is difficult to cover. To prepare this book was even a great task for myself. While seeking the explanation of processes and phenomena which can occur during leather making, I found that a great deal of them is well explained, many have explanations supposed but not proved, many have diverse, contradictory explanations, and still others could not find any explanation at all. I realize that there are many points too complex to allow exact elucidation since too many independent variable factors come into play. (This is particularly the case of the finishing processes.) But still I know that the tanner is fond of his material and would like to know what happens to it during its processing. So do I.

From my experience I know that the best way to approach a problem is to try to write down one's own thoughts about it in the simplest and clearest way. Then one can see the missing links in its understanding, or in one's knowledge or that of other people dealing with the same problems.

To understand the physico-chemical phenomena of leather making a certain knowledge of biochemistry is necessary. Usually—if at all—it is treated occasionally at chemistry departments, where leather technology is taught. I decided thus to include certain necessary items of biochemistry, especially as in this domain there is a real flood of information. It was not my intention—and this concerns other problems as well—to include too much detailed, specific information as that would lead to far too many points, and the reader might come to believe that science is too distant from practice, which would be only partly true. Several years ago I presented the result of my efforts to Polish readers. All to my satisfaction they have accepted it. Now a very well recognized American Publisher has offered me the opportunity to present my book to the English readership, and I wish to thank him for that.

There are still many points which could be approached here. I will not do that, but will rather let the book speak for itself. There is one thing that deserves explaining: I tried to eliminate the trade names of the preparations discussed, as that always leads to the problem of answering 'why mention is made of these and not those?' which I would be unable to answer.

I would like to express my gratitude to the people who helped me in this work. In the first place, I wish to thank Professor J. H. Sharphouse, who introduced me to the Publisher and convinced him to publish my book in English. He has also read the manuscript, given me some valuable advice and remarks, and wrote the section "oil tannage" by himself. I had the pleasure to work with him—unfortunately not for long enough.

I wish to acknowledge my friends, Dr. O. Rodziewicz from Technical High School in Radom and Drs. B. Felicjaniak and R. Radzynkiewicz—both from the Leather Research Institute in Lódź who discussed with me parts of the text and gave me their advice. Thanks are due to my Institute's librarians, Mrs. W. Janiszewska and Miss J. Granosik, who were always ready to hunt any source of literature I asked for, and to my co-worker Mrs. J. Pawlak, for help in corrections and in the drawing of formulas. All authors and publishers who gave me the permission to reproduce their results in the form of photographs, figures and tables, included in this book, are duly acknowledged.

Finally I wish to give my wife Lucy my affectionate thanks for her patience and continued help in the editing and correction of the book, despite her own professional duties quite different than mine.

K. J. Bieńkiewicz

Contents

1.

SKIN AS A COMPARTMENT CONNECTIVE TISSUE IN THE ORGANISM

Animal skin constitutes a barrier between the organism and the external world. That function of the skin is decisive for its histological and physiological properties. Skin is one of the principal adaptation organs of the organism, almost impermeable to water, aqueous solutions and pathogenic microorganisms. Nerve ends contained in it are receptors of touch (pressure), heat, cold and pain stimuli. From the physical point of view, skin reveals a specific viscoelastic behavior, permitting to avoid small injuries (because of its suppleness) and adaptation to change in shape and size of the body (in view of its stretching and shrinking ability). From the chemical point of view, the components of skin reveal a lower reactivity than the components of other organs, which can be regarded as an adaptational feature. The dye contained in the epidermis, (or in the hair coat) protecting the organism against the effects of electromagnetic radiation, can serve as an example.

In the present chapter the functioning of skin and the mechanisms involved of live organisms will be explained. As a result of man's intentional action, skin is transformed into another material having different, and from a certain point of view, more advantageous properties. It is that new material which we call leather. In order to know more about the properties of leather and to understand the changes taking place in the course of transforming skin (or hide) into leather we shall consider the structure of raw skin and the properties of its components. The purpose of processing skin is to preserve such properties of the raw material as strength, viscoelasticity, abrasion resistance, and to eliminate other properties, such as liability to putrefaction, or becoming calloused on drying. Some properties are improved, e.g., the thermal resistance and gas permeability.

1.1. Ontogenesis and histology of skin

Dermis and epidermis develop in the fetal life of mammals from two different germinal sheets, the first one from the external (ectoderma) and the second one from the internal (mesoderma). Initially epidermis is a single layer of cells. In later periods of fetal life the cells form two and then more layers. The developed epidermis, constituting 1 to 5% of skin thickness, consists of four or five layers, depending on the epidermis topography. Single layer cells are different in a way, which allows for their histological recognition. The thickness of the epidermis is usually reversely proportional to the thickness of the hair coat. Epidermis is removed in the leather manufacturing processes. It is, however, of great importance both from the technological and histological views.

The epidermis

Mammal epidermis consists, if we start from the internal side, i.e., from the basal membrane, of the following layers (strata): *basale, spinosum, granulosum, lucidum* and *corneum*.

The *stratum basale* consists of one layer of cylindric cells (keratinocytes) lying on the papillae of dermis, and is connected to it by the basal membrane (see below). This layer is also called the *Stratum germinativum*, as it is there that the main epidermis cell formation processes occur. According to old views all these processes concentrated there, but today we know, however, that it is not so.

In the human and animal basal layer of epidermis a dye is formed, called melanine. Melanines are the most common dyes of the animal world. They occur in skin, epidermis, feathers, eye and other parts of the body of vertebrate and invertebrated animals. From the point of view of tanning chemistry, it is most important to remove the skin coloring due to melanine and the remainings of hair containing this dyes. Our knowledge of the structure of those macromolecular substances is far from exact. A basic component of melanines is 5,6-dihydroxy indole (dihydroxybenzpyrrole)

Melanines are formed in melanocytes (cells belonging to the nerve system), interspersed between the slightly below the cells of the basal layer. In the Golgi apparatus of a melanocyte [1] egg-shaped melanosomes are formed. Their size is 100 × 50 nm. Melanosomes are surrounded by a protein basal substance probably containing tyrosinase, an enzyme, favoring quinone formation.

According to Heidemann [2] the formation of dihydroxybenzpyrrole probably proceeds as follows:

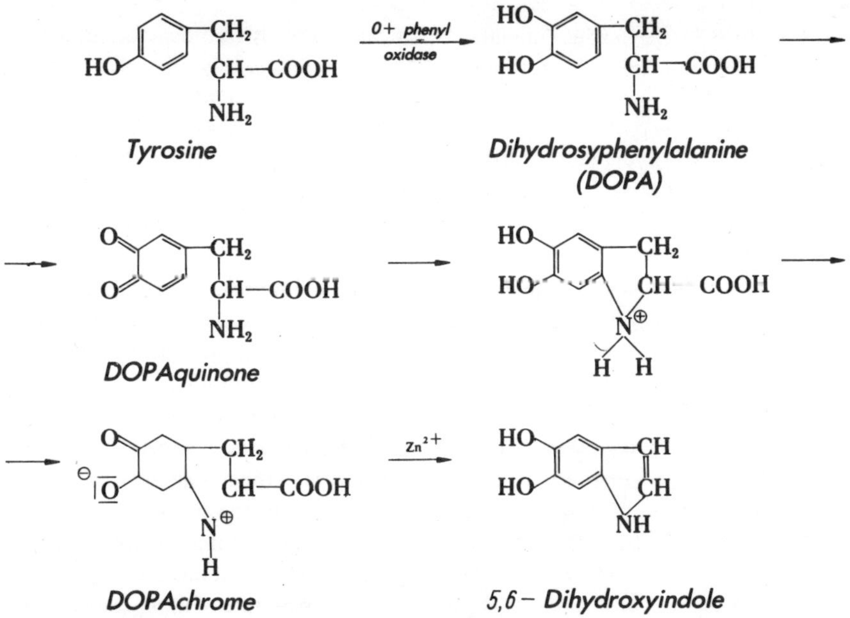

Tyrosine

Dihydrosyphenylalanine
(DOPA)

DOPAquinone

DOPAchrome

5,6 – Dihydroxyindole

The compounds, which are intermediates in this reaction have various lifetimes. The presence of DOPA chrome may be demonstrated by a spectroscopic method. The absorption maximum occurs at 475 nm, which corresponds to the color red. The isolated enzyme catalyzing this reaction is called catecholoxidase (E.C. 1.10.3.1. for explanation of enzyme classification see section 9.2). The majority of the intermediary products of the reaction have been isolated, or identified by the spectroscopic method, and the presence of certain functional groups and compounds has been detected using isotope tracers. How 5,6-dihydroxyindole polymerizes, is yet unknown. Investigations carried out using isotope tracers (T) and deuteration, and testing the products by the NMR method have shown that melanines are probably a mixture of various oxidized polymers. Bonds may occur on carbon 2, 4 and 7, e.g.:

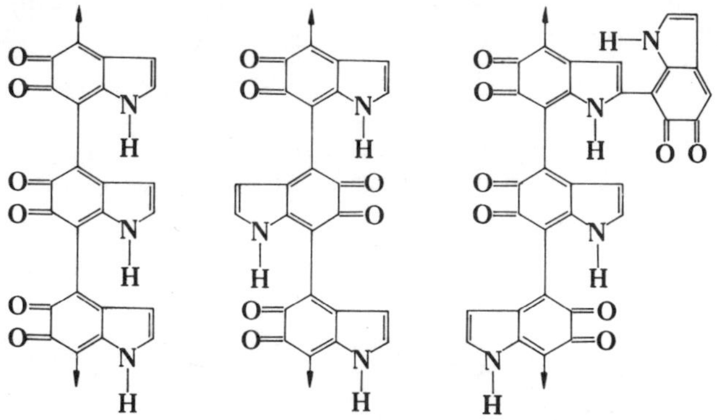

The melanines, occurring in natural media form protein complexes, melanoproteids, connected most probably through covalent bonds. It is difficult to classify proteins into those functionally bonded to melanines or just accompanying them. A new UV band has been observed, when the solutions of intermediate products in the reaction of melanine formation (5,6-indolquinone) are in contact with thiol groups. Condensed melanines do not give such reaction.

Stratum spinosum s.mucosum is thinner over papillae (extensions of dermis) and thicker between them; it consists of several layers of polyhedral cells, having a distinct nucleus and a lighter cytoplasma. These cells are connected by desmosomes (spina). Desmosomes are clearly visible in the intercellular spaces filled with lymph. Lymph plays an intermediary role in the nutrition of epidermal cells. It circulates among the cells as there are no blood vessels in the epidermis. Desmosomes are important elements of connection of the cells especially at sites where the cells are in close contact. They are visible under an electron microscope as intercellular slits of width up to 24 nm from which filaments of denser cytoplasm, tonofilaments, protrude. These slits are filled with a cementing material, probably glycosaminoglycans (see section 4.1). In *stratum spinosum*, especially in its deeper part, cell mitosis (partition) is still observed.

Stratum granulosum consists of 1–3 layers of fusiform (spindle form) cells. Those cells have weakly marked nuclei and are, except for the periphery of their cytoplasm, filled with keratohyaline grains. These grains are of irregular shape and size and are strongly basophilic. The importance of keratohyaline occurs in its participation in the keratinization process of epidermis. It was observed under an electron microscope that tonofibrils, or keratine fibers combine forming tight bundles, passing through keratohyaline grains. The latter substance does not contain thiol groups, but includes phospholipid protein complexes and nucleic acids. Some organella* are missing in the stratum granulosum cells (e.g., the Golgi apparatus), and they contain only few denatured mitochondria. A view exists that the metabolism of these cells is based on the anerobic action as is the case in fermentation processes (anoxybiosis).

Stratum lucidum consists of 2–3 layers of flat cells, filled with eleidine, a proteinous substance. They do not contain nuclei and have a very strong gloss; eosine stains them bright red. Stratum lucidum occurs only there, where the epidermis is callous, e.g., on human palms and feet, on dog and rabbit's paws, etc.

Stratum corneum is the external layer of epidermis. The cells are very flat and do not contain nuclei, but they are rich in keratin. Their cytoplasm consists of numerous tonofibrils immersed in a matrix substance, also called the intercellular substance or just matrix. In its lower layers the cells are connected by

---

*Basic information about cell ultrastructure, and biological meaning of terms, like organellae, Golgi apparatus, etc. can be found in new handbooks of biology.

desmosomes, and in the upper layer they are loose and gradually peel off in the form of microscopic flakes. Keratin, being a substance resistant to acids and enzymes, exerts an important protective action, especially when mechanic and chemical stimuli are involved.

Morphokinetics of the epidermis

Keratinocytes (epidermis cells) begin their life cycle in the basal (palisade) layer, and end it as horny flakes, deprived of nuclei which are rubbed off the surface. These flakes are formed by keratin and have a protective function. This process is shown schematically in Fig. 1.1. The basal cells lying on the basal membrane undergo mitosis. The cells lying over them, synthetize fibrous filaments about 8 nm thick and the matrix substance. These filaments form bundles surrounded by the matrix. In the differentiation process on the Golgi apparatus Oddland bodies—membrane coating granules—are formed. Those granules pass subsequently to the cell surface and disappear in the intercellular space. In a later development phase they spread on cell membranes. According to Parakkal and Alexander, they are a cellular binding agent and barrier for water [3].

Small, amorphic granules of keratohyalin form simultaneously with keratinosomes and grow gradually occupying much space in the cell. Cell organellae, after forming those differentiation products are resorbed and in this way the stratum corneum is formed. In this process cell membranes increase their thickness and the keratohyaline granules disappear; they are probably transformed

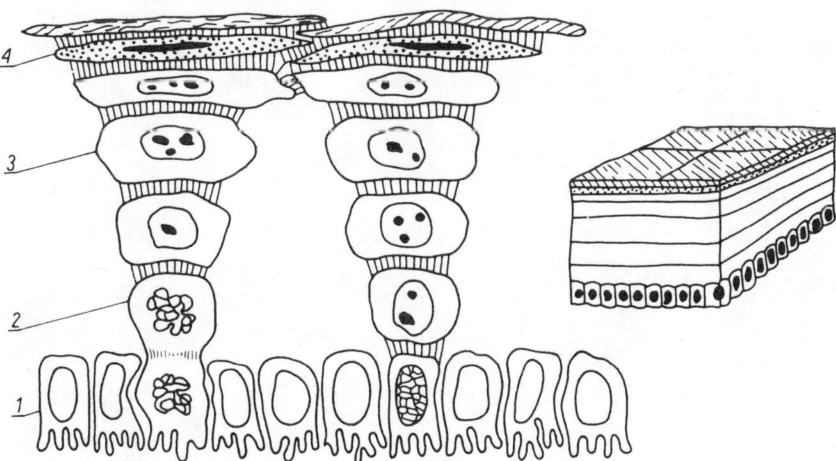

Fig. 1.1. Pathways of epidermis cells. Cylindric (palisade) cells, formed in the basal layer 1, which as proceeds upwards form polyhedral cells of *stratum spinosum* 2; then they become flat, filled with keratohyalin and form *stratum granulosum* 3. Finally they are transformed into *stratum corneum* 4.

into microfibriles in the matrix. Keratohyaline granules consist of histidine-rich protein. Their biological role in the keratinization process has not been elucidated so far. The completely keratinized cells are filled with filaments immersed in the matrix. In external layers they peel off.

The process of cell migration to the external side of the epidermis is accompanied by gradual dehydration. The water content in human basal layer is 60–70% and 10% in the horny (corneum) layer. The histological changes are accompanied by chemical ones, consisting in that the live protoplasm of basal layer transforms into keratohyalin or into eleidin, and eventually into keratin.

Epidermis has a very important function in the first contact of the organism with the surroundings. In man the epidermis cell lives from its formation to peeling off which is about one month. The duration of this process in animals has not yet been estimated. Continuous renovation of the epidermis is of great importance, and perturbations in this process may affect the whole organism.

Basal membrane

Between the epidermis and the dermis there is a layer of basal membrane (*membrana basalis*). Its existence has been considered doubtful by histologists even quite recently. Now it is known that in order to make it visible, appropriate staining has to be applied (PAS staining, consisting in applying periodic acid + leukofuchsin). The basal membrane, some tens of nm thick, separates the epidermis from the dermis. A really good picture of the basal membrane is obtained at × 20,000-40,000 magnification (electronic microscope, heavy metal

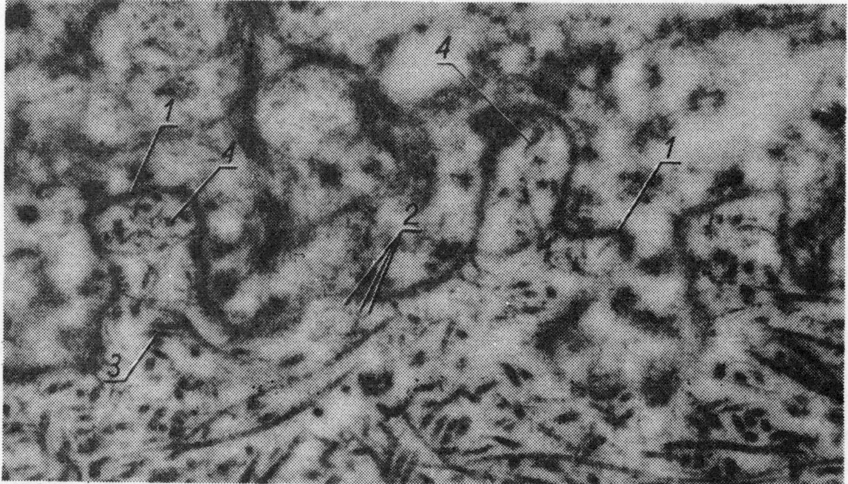

Fig. 1.2. Cross section of the layer between the epidermis and dermis. Cattle hide, magnification 25 000: 1 - basal membrane, 2 - connecting fibrils, 3 - desmosomes, 4- papillae, connecting epidermis to dermis in a mechanic way. Courtesy Dr. T. Stirtz.

shadowing). Stirtz has made a very good photograph (Fig. 1.2) where not only the basal membrane (1) is seen but also the fibrils (2) several nm thick penetrating into it and connecting the membrane with the dermis, as well as the desmosomes (3) and the appendices (papillae) (4), connecting it with the epidermis. The picture is about ⅔ of the basal layer cell.

The basal membrane contains collagens of several specific kinds [collagen IV A, B and probably C (see section 2.2)]. So far those collagens have not been exactly examined, neither from the chemical nor the physiological points of view, because of experimental difficulties in obtaining them in a pure form. Their characteristic feature is, however, the high hydroxylysine content of 25–60 residues per molecule and susceptibility to proteolytic enzyme attack. Many carbohydrate molecules are attached to their molecules, so it may be supposed that they are important for the transport of nutritional substances to the epidermis.

According to Stirtz [4] the basal membrane is destroyed during enzymatic (Arazym) unhairing. Other investigators, e.g., Reed [5] have shown that the basal membrane is the principal obstacle in the penetration of fats, emulsions used for finishing and fillers into the leather. For this reason, though the basal membrane is histologically a very small fraction of the skin, it is very important for leather processing to make sure that it has been completely removed, though this is still under discussion.

The epidermis and the basal membrane together constitute a thin, external layer of the skin. Keratin is the dominating protein in epidermis. It is a protein containing numerous disulfide bridges buffering the action of liming baths, dissolving in bases and therefore removed in the liming process. If we assume that its basal component is the collagen-glycosaminoglycan complex (see section 2.2), we will probably have to recognize that it is in principle removed in the liming process when the skin collagen network is disclosed.

Dermis *(corium)*

Dermis *(corium)* consists of two layers: the papillar, called thermostatic *(stratum papillare seu termostaticum)* and reticular *(stratum reticulare)*. The reticular layer passes without a clear limit into the subcutaneous tissue *(tela subcutanea)*. No distinct limitation between these two layers exists.

In the corium cells the fibril-forming ones—the fibroblasts—prevail. Beside them in the skin there occurs the intercellular substance (matrix) and two kinds of fibers: collagen and elastin. The third kind called reticulin, is now no more considered to be a specific protein. Quantitatively, the major part of the solid substance are collagen fibers—the basis of leather making. Both skin layers differ in thickness and in the way the collagen fibers are interwoven. In the papillar layer near to the grain they are thinner and more densely interwoven. Their thickness increases and the network density decreases in deeper layers. The surface of this network, remaining after removal of the epidermis and basal membrane, is called grain and has a pattern typical for animal species and sample location in the skin. In the cross section it is also typical. The surface or cross section pattern is best visible at a magnification $\times$ 4–8.

The name papillar layer is derived from the *papillae* covering the dermis and penetrating through to the epidermis (see Fig. 1.2). The name reticular layer is derived from the network of vessels which are better developed in it.

In corium there are cavities (hair pockets) lined with a special tissue (section 12.3). In those cavities lie hair bulbs. The part of the hair inside the skin is the growth zone. Hair formation, its aging and dying are discussed in section 3.1. From the histological and chemical points of view, hair constitutes a part of the epidermis. The structural protein of hair is keratin.

Skin is a connective tissue organ. Its functioning is possible due to the supplying of skin with blood and lymph through appropriate vessels and spaces. It also contains the endings of touch nerves (touch bodies of Meissner), nerve trunks as well as sebaceous, sweat and odor glands. Skin also contains muscles (hair muscles), fatty cells and slices of fat, mast cells and other components of lesser importance from the technological view point. All these components are either totally or partly destroyed in leather making. Description of their function is a matter of general histology or skin histology. The collagenous fiber network is the component of skin that persists through the whole process of leather making.

Depending on the animal species, race and individual properties, significant differences in skin thickness may occur, as well as in the epidermis to corium thickness ratio. Epidermis thickness in animals with a well-developed hair coat is about 1% of the total skin thickness, in animals with a poor hair coat (pig) it is about 5%. The relation between the particular skin layers changes as well, and the differences depend on the topography, age and sex of the animal. The thickness of papillar layers is about 50% in cattle fetuses, 35–40% in calves, and from 30% in cattle butts to 40% in bellies. In sheepskins this thickness may be as high as 70%, and in pigskins the histological differences between layers are not so pronounced, so it was presumed that they do not exist, which is, however, incorrect, since there are some differences and the layers are approximately of equal thickness. In horse hides these limits are very well expressed; almost no sweat glands can be observed in sheep and dog skins, etc.

All these differences were investigated some time ago. Now we just presume that the properties of animal skin can be very versatile. Today the general formulation of regularities controlling the mechanical properties of the collagen network in live organisms is considered to be much more interesting than, e.g., the topography of skin properties [6]. Our field of interest is the skin collagenous tissue and we will not discuss other kinds of connective tissues like arterial walls, bone, eye lens, etc. Furthermore, we will not discuss the details of skin topography, as this subject is covered by the many handbooks of leather technology.

1.2. Mechanical functions of skin

The mechanical functions of skin are very difficult to define. One of the obvious ones is to keep the body in shape and to protect deeper lying tissue. An injury to an animal arises from accidental contact with surrounding objects during locomotion; in this case the function of dermis and some epidermal components, principally hair, may be similar, i.e., essentially that of energy absorption and dispersion in as large a volume as possible [7]. If the speed of movement of an animal were constant, it could be expected that the energy to be absorbed per unit area of skin should be proportional to body weight, i.e., skin thickness would be proportional to the linear dimension of the animal. In fact, the skin thickness is roughly proportional to the square root of body weight. Another feature of skin is its resistance to piercing, e.g., against predator attack.

In some body parts skin is permanently lightly stressed, in some others it is loose, whereas in definite places it strictly adheres. The tearing of skin requires great force. Skin extends under prolonged stress, it yields under pressure forming a thicker ring around the area where pressure is applied. This is why clamping of samples in dynamometer jaws should be avoided when testing raw hide, and instead skin rings are used, which are obtained from bodies of small animals, or from limbs of bigger ones. Skin may be stretched by 10–50% (depending on the age and kind of animal), when the stress is maintained for several seconds. Skin breaks if the load applied is greater than the breaking strength. If the stress is applied for some time an irreversible plastic component accompanies the elastic one. Human skin is usually under a load of about 20 g/mm^2 (kPa) and it is extended by about 10–20% [8].

The knowledge of mechanical properties of skin is based on the investigation of its features (as of a solid), on the analysis of the tissue microstructure (collagen fibril systems included), and on comparison of the skin tested with other materials.

A collagen structure unit (notion usually related to the mechanical function) is a fibrile consisting of 'bundle' of molecules, in parallel order, of cylindric shape of approximately uniform diameter, which varies in dependence on the degree of tissue development and on topography. The fibril diameter may range from 10 to 200 nm. Its length is usually not given in the literature. We do not know much about fibril endings, sometime under a microscope we observe thinning endings which are considered to be the natural end of the fibril. In the range of small stresses, skin obeys Hooke's law, which in differential form is expressed by:

$$\frac{dT}{dS} = X \text{ (Young's elasticity modulus)}$$

Application of this law to connective tissue requires the assumption that the stressing load acts only for a short time. The load (force) is proportional to the stress when it does not change the structure of the tissue. Slipping T under constant load is a process which depends on the viscous resistance. This relationship can be linear if one assumes that slipping does not change the structure of the tissue significantly:

$$\frac{dS}{dt} = \mu T$$

where S is the deformation, T the stress, and μ the viscous extensibility constant. Slipping as well as elastic elongation can be determined using the same methods at constant load.

The basic mechanical function of such structures is their resistance to extending them along their axis. The mechanical properties of tissue in which these structures occur are determined chiefly by their orientation and arrangement. Arrangement of collagen fibrils closest to parallel is observed in tendons through which the muscles act on the skeleton (transmission of stretching force in one direction). Tendons constitute almost unstretchable (less than 5% under usually occurring forces), almost ideal elastic connections of great modul, about 10^9N per square meter. Layer systems are also known of varying orientation, e.g., crossing layers of collagen in the cornea. This layer structure may also occur as well in skins of fish and of reptiles.

Collagen fibrils of mammal skin form a network which can be apparently considered as random. However, a detailed investigation carried out so as to assure a high level of statistical confidence, has shown regularities in the network related to the function of skin as the external shield of the organism and at the same time 'bag' in which it is contained.

Fundamental for understanding the behavior of the network of fibrils, decisive for the physical functions of skin, is to imagine that network as a net made of yarn with no ends. Three steps can be distinguished when such a net is subjected to stress (Fig. 1.3). The first step is the straightening (stretching) of the net in

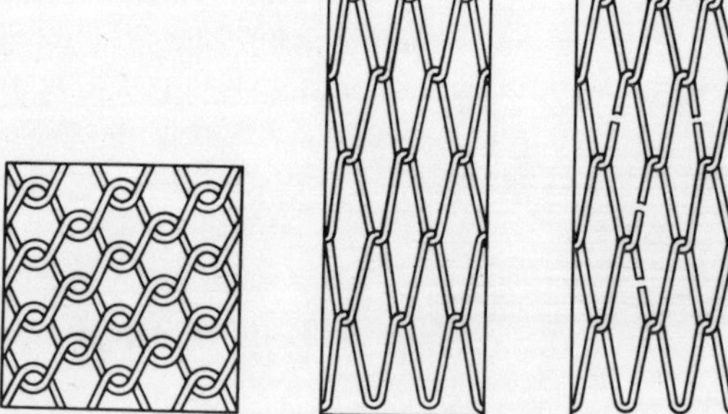

Fig. 1.3. Three steps in the stretching of a fiber net. Scheme of fully ordered network.

the direction of the force applied, when the action of small force gives great elongation, the fibers order themselves with respect to each other and will stretch slightly. Lying in the intercellular spaces filled with the matrix the fibers slip easily, but they recover very slowly after the force is removed. In skin no specific mechanisms of recovery is known. So it is mainly subject to plastic elongation, including, however, an elastic component. When the fibers contact each other at various angles, they meet resistance of cell membranes and other elastic elements. Shifting of fibers with respect to each other can be revealed by the model shown in Fig. 1.4 in the form of two ensembles of stiff rods immersed in a viscous liquid. Every ensemble is fixed on one (opposite) side. Such slipping occurs in intercellular space; participating in it are fibrils crosslinked by various bonds. It is difficult to assume that slipping occurs inside the fibril, as then crosslinking bonds would be destroyed and thus the properties of a molecule, e.g., its solubility would change. In the same way, we can arrive at the compression model for skin. It may be pictured as a pile of stiff plates immersed in a liquid. During compression the liquid is squeezed out from the spaces between the plates.

The second step is the elongation of the fibers themselves. A small elongation already requires a great force for overcoming intramolecular bonds, due to which collagen helices are formed. That force has to also overcome intermolecular and hydrogen bonds. In this step the structure of the collagen molecule changes and so do the supramolecular structures. In this phase elastic deformation is the main process, and no structures are yet broken, but they are disturbed in a way comparable to the stretching of the dynamometer spring.

The third step is where the break takes place. The fibers break, the slope of the stress-strain curve decreases, the plastic deformation part increases. Finally disruption occurs and the material is destroyed.

The described process of skin stretching is of basic importance for its characteristics. General quantitative data cannot be given as they vary within broad limits. The sigmoidal shape of the Grassmann curve, varying with the technological operations carried out on the skin, characterises well the raw material as

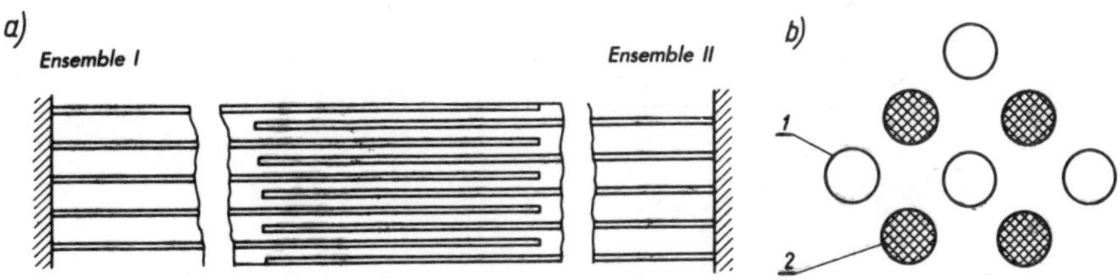

Fig. 1.4. Distribution of stiff rods, fixed at one end: (a) model of slipping of fibers in matrix substance, (b) distribution of unordered rods; 1 - rods of ensemble I, 2 - rods of ensemble II.

well as leather in all processing stages, primarily by normalizing the curves, which makes possible comparison of the materials tested, irrespective of the absolute force necessary for breaking the sample. In Figs. 1.5 and 1.6 some examples are given illustrating how the curves are constructed and how their shape is influenced by liming, bating and tanning.

The curves are plotted by laying off on the ordinate where the force is applied, expressed as fraction of breaking load (in percent), and the elongation, in percent of initial sample, on the abscissa. In the upper right corner of every plot, the absolute value of the breaking force in kg/mm^2 (kP) is given.

The difference between fresh and dried hide (Fig. 1.5 a and b) is due to the cementing of collagen fibers by the matrix components, which results in the disappearing of the plastic component. In the first elongation phase the slope of the elongation curve is greater because the fibers cannot be shifted one against another. Thus Sample b is much more resistant than Sample a. The curve in Fig. 1.5 c is not so steep as that in Fig. 1.5 a, so the elongation at rupture is much greater in the former case, and due to the removing of non-collagenous components of hide during liming, the system becomes looser. The curve in Fig. 1.5 d is typical for bated hide, when the structure is still looser. In the case of boxcalf, which is chrome tanned leather, the curve in Fig. 1.6 still keeps its sigmoidal shape; the elongation at break is smaller and the upper part of the curve is steeper. This points to a greater contribution of the elastic deformation component due to the crosslinking of the material by means of the chromium tanning agent. Curves d and e characterize sole leathers of vegetable and chrome

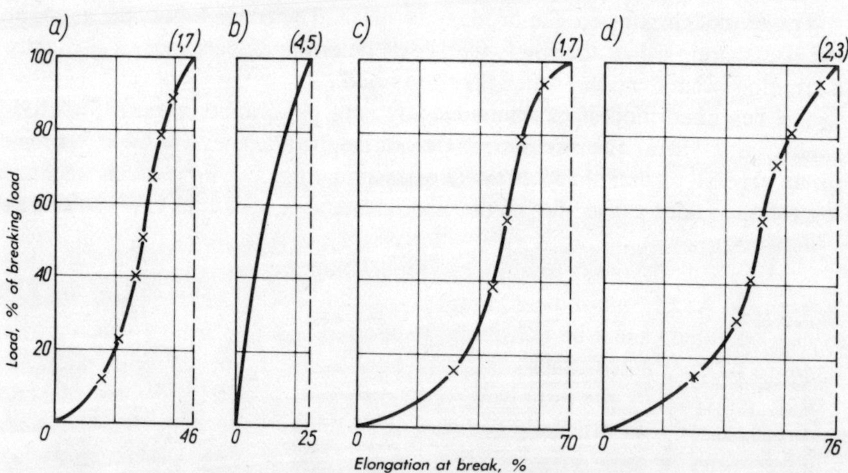

Fig. 1.5. Grassmann curves for raw hides and hides during processing: (a) raw, wet hide, (b) dry hide, (c) limed, wet hide, (d) bated, wet hide (in the right upper corner values are given of breaking load in MPa., kg/mm^2).

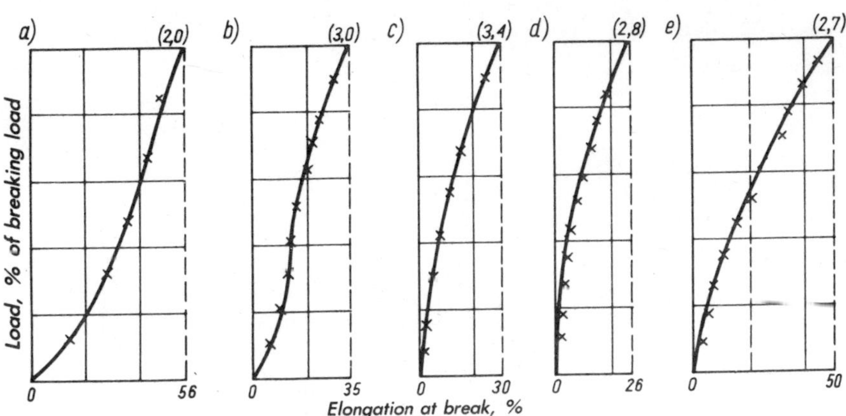

Fig. 1.6. Grassman curves for leather: (a) box calf, (b) bridle, (c) belting, (d) sole, vegetable-tanned, (e) sole, chrome-tanned.

tanning. The initial sector of those curves has a high slope, and due to tanning some reversion of the curve occurs. That is because the fibers stick together in such a way as that which has been observed in dried hide; the liming of sole leathers is short and weak, and they are not bated. Another factor influencing the shape of the curves is the filling of the interfibrillar spaces with molecules of the vegetable tanning agents. Besides, sole leathers are roller pressed, the reason for which those internal spaces are reduced, and so as is the relative mobility of the fibers. The contents of fats and water in leather are significant because they reduce the friction between the fibers (see section 4.2 and chapter 5).

When characterizing leather, the layer and directional properties should be also considered. The strength of the fiber network increases from the leather surface into its depth, and attains a maximum value in the layer adhering to the subcutane tissue. The above is related to the fiber thickness and density of the network.

Many scientists have investigated the topographic distribution of animal skin strength. As a result of the differences between species and even individuals the data obtained cannot be transferred to another place or time, if detailed values are to be considered. Such data have to be adapted, which is equivalent, as a rule, to repeating the test. However, some general ideas about the prevailing directions of fiber orientation can be found in Maeser's work [9]. That author gave values of some strength factors in the form of ellipses in which the axes ratio is the measure of the indexes ratio, and the inclination angle of larger axis with respect to the skin axis shows the direction in which the value in question has its maximum. In Fig. 1.7 the distribution and shape of ellipses of tensile

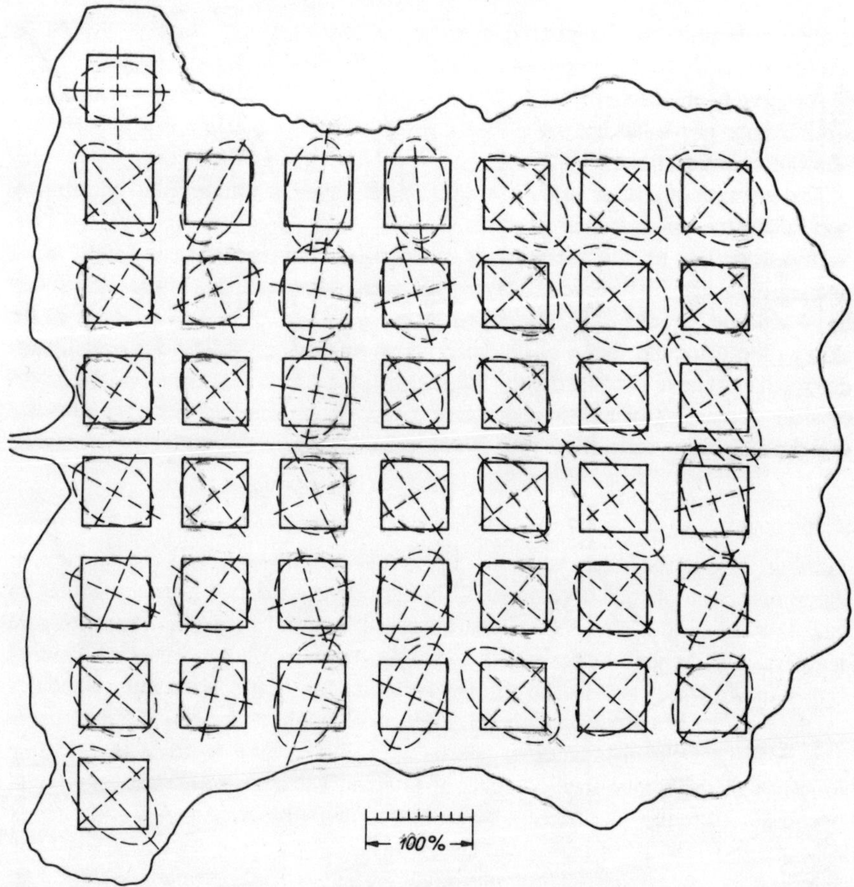

Fig. 1.7. Tensile strength ellipses of leather (cattle). According to [9]

strength of leather from cattle hide are shown. The data of Maeser are based on the testing of 21 pieces of whole chromium tanned kips. He collected 44 samples from each piece and tested them on an Instron dynamometer. Analyzing the Maeser ellipses one can see a significant difference when looking at the position of sample. It is also important from which side of the hide the sample has been taken. The asymmetry of hides, recently tested and confirmed by van Vlimmeren who used modern methods, is related to the position of various organs inside the body [10].

Considering the mechanical properties of live animal skin, one has to take into account its whole thickness in relation to the epidermis. It is generally believed that the role of the epidermis can be neglected, except for some very slight stresses. However, this opinion is not justified as the role of this layer is different: it protects the deeper layers from the action of shearing forces, in which this action would result in superficial abrasion.

As already mentioned, the hair coat is usually of protective importance. A similar role is played by the external layer of corium, consisting of soft and flexible collagen fibers.

The strength of leather as a whole does not decrease in a statistically significant way when the epidermis is removed.

Investigations on the variation of mechanical characteristics of hides, skins and leathers with differences of layer structures, have been discussed extensively by Ward and Brooks [11]. Referring to the direction of collagen fibers in the skin as the principal factor of its mechanical resistance, Shestakova and Shneiderovich [12] have proposed a rheologic characteristic of skin based on the model of a net consisting of elastic and plastic elements (spring and piston) connected in a rhomboidal system (Fig. 1.8). That model is different from that of the elastic

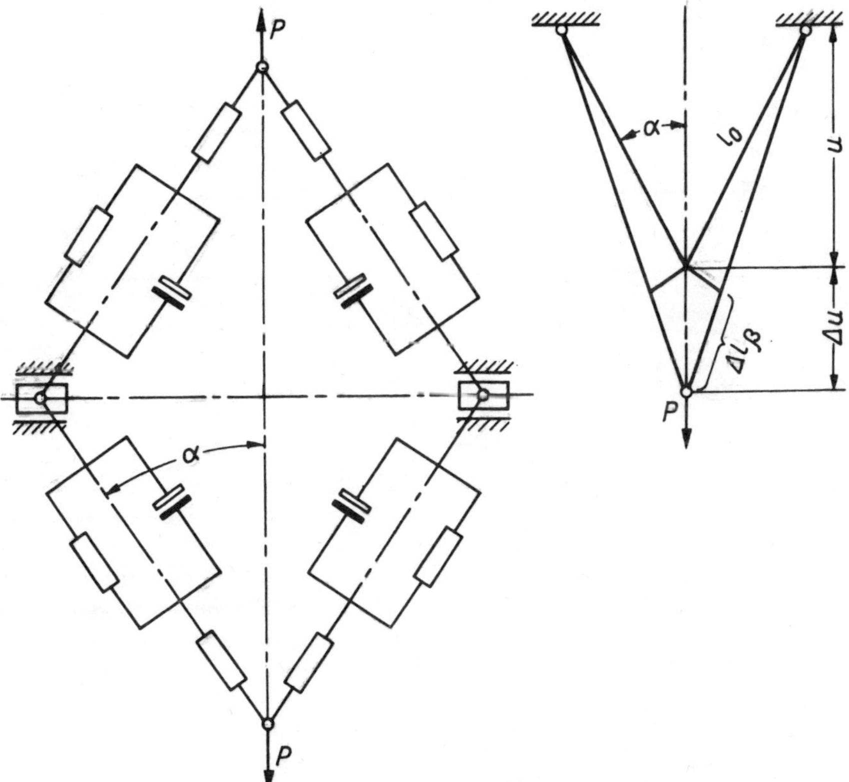

Fig. 1.8. Rheologic Burgers model proposed for collagen fibers. P-direction of force u applied, Δl - change of the initial length l_0 due to the action of force u + Δu.

body in that deformation in it is time-dependent. This system is known as the Burgers, or Kelvin-Voigts model. Its elements are a spring and a piston, which are in a parallel connection. An additional spring is included into the system as is in the model of plastified polivinyl chloride. All four unit elements are connected by joints. In this model, according to the theory of viscoelastic bodies, elastic deformations are delayed due to the resistance of the piston (damper). If the force applied is constant, the elongation rate will decrease steadily, since increasing stress is consumed to move spring and decreasing stress acts on the piston. Thus the elongation rate becomes zero, when the force applied is fully equivalent to the spring stress. Delayed elastic deformation, or creeping, is illustrated in Fig. 1.9 a. The load has been applied at point A, and removed at

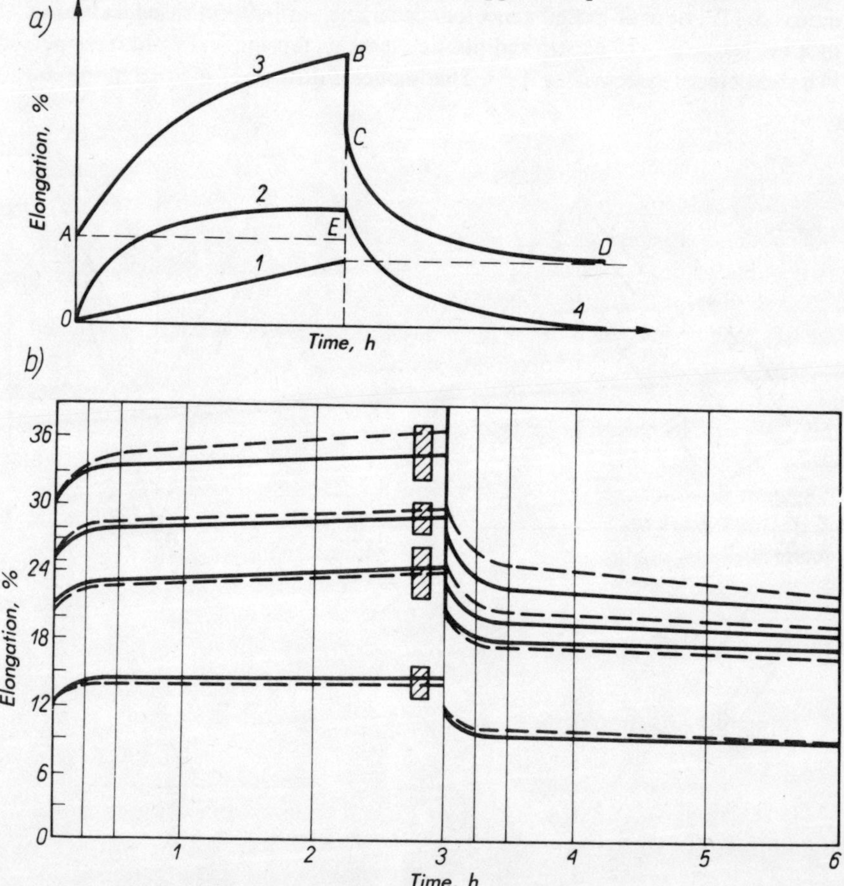

Fig. 1.9. Creep and reverse creep curves: (a) typical for PVC, (b) for chrome side, according to Shestakova. Note the similarity of the shape of curves in both figures. 1 - curve of viscous flow (corresponding to Newtonian liquid), 2 - component of momentary elastic deformation, 3 - component of delayed elastic deformation, 4 - curve of delayed reversion, CD curve of delayed back-creep.

point B. The creep curve (3) is thus sum of components (1) and (2). The inclination of the curve after the elapse of a suitable amount of time becomes equal to zero. The creep curve then has the same shape as the viscous flow component. Removal of the load at that very moment results in a recovery of elastic deformation along the BC line (= OA). With time recovery, delayed elastic deformation takes place. Viscous flow is a permanent remainder. Curve (4), illustrating delayed elastic deformation, is a mirror reflection of curve (2). Such a creep curve is typical for the generalized Burgers (or Kelvin-Voigts) model of viscoelastic body. In Fig. 1.9 b., the curves for chrome sides are shown in an analogous presentation. In the work quoted [12], several simplifying assumptions have been made without which mathematical solution of the problem is impossible in view of the numerous unknown values. The most important of these are:

(1) The non-linear deformation is a geometric configuration change of the rhom-
    boidal elements.
(2) The properties of fiber bundles are linear-elastic and linear-plastic.
(3) The angle of network configuration depends on the deformation only, and
    all the fiber bundles have the same elastic properties.

Based on these assumptions, the geometry of the model and the data from investigations of properties of particular kinds of leather, the authors presume that the creeping of leather, i.e., the processes of delayed action and relaxation of stresses, can be presented in the form of creeping curves only if it is assumed that rheological parameters of stresses and deformations are non-linear. This non-linearity can be considered to be the result of changes in the network geometry, linear viscoelasticity of the fibers forming the system in question being assumed.

The main directions of the collagen fiber bundles are shown in Fig. 1.10. More detailed data concerning the structure of the network with dependence on the characteristics of the animal, processing applied and other features, can be found in the atlas, "Hides, Skins and Leather under the Microscope", in the papers by Morgan [13] and Ward and Brooks. Raabe [14] dealt extensively with the problems of leather strength and of hides and skins.

1.3. Versatility of mechanical properties of hides and skins

Aging of an organism influences the mechanical properties of the skin. It is difficult to explain to what degree these changes result from changes in the collagen fibers due to the increasing degree of crosslinking, and to what degree from changes of collagen content in skin take place. In the course of aging, the thickness of fibers changes as does the packing density of collagen in skin. The amount of collagen in skin increases with age, but it can also decrease because of illness, senility, sex, nutrition, work, motion and many other factors.

Few papers that have been written on leather making deal with viscoelastic

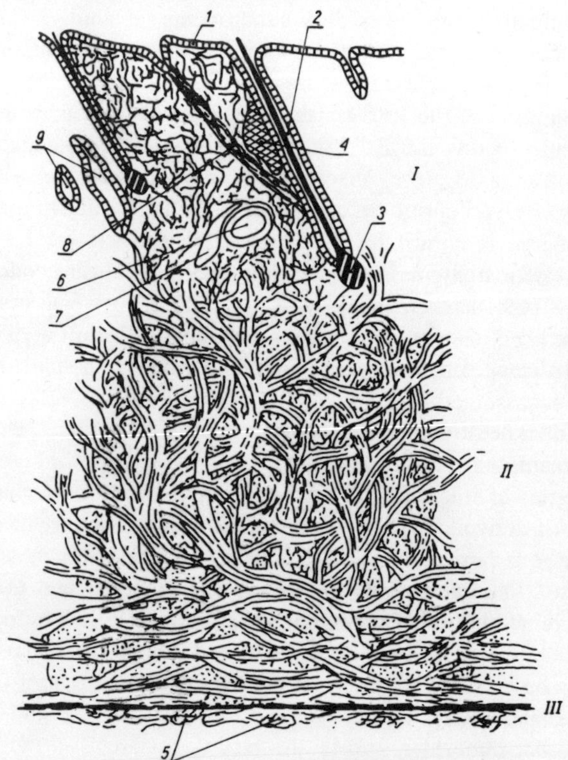

Fig. 1.10. Schematic cross section of the cattle hide. The papillar layer, histological not separated, reaches to the bottoms of hair bulbs. I - grain, II - corium, III - muscle tissue. 1 - epidermis, 2 - hair follicle, 3 - hair bulb, 4 - sebaceous gland, 5 - fat, 6 - artery, 7 - vein, 8 - hair muscle, 9 - gland.

properties of raw skins and hides, but deal rather more with those of tanned ones or with those of the material during its processing. Thus the remarks below concern leather properties and are closely connected to part 3.

Measurements cannot be taken on a sample of partial thickness (split), because during splitting, the part of shaving and buffing the fibers are cut and thus the strength of the skin changes. It is relatively easy to measure the surface mechanical resistance of a strip taken through the skin thickness, for the width of the strip has to be large enough to permit neglecting the decrease of its strength due to vertical cross section of the fiber bundles.

It may be difficult to take a sample of a definite shape from fresh skin in view of its unsatisfactory internal stiffness. Only in some cases can one cut samples of stabile size, e.g., from the neck of a hippopotamus, from seahorse hide or from a rat's tail. In these cases the hide has the structure of an unstretchable plate. According to the present author's experiments such hide is about 2.5 times stronger than normal hide, but its elongation is almost nonexistent (unpublished results).

Mitton [15] proposed to express the tear strength of leather by the linear equation

$$a = k (x - A)$$

where y is the tear strength and x the hide thickness, A is a constant specific for the given kind of leather (for sides $A = 0.3$ mm, for necks of vegetable or combined tanning $A = 0.8$ mm), k is a conversion factor indicating whether the sample is 'strong' or 'weak' in the given group.

It can also be assumed that the tear strength of leather is proportional to the square of its thickness: $y = kx^2$.

A typical example from Mitton's paper is shown in Fig. 1.11, where the tear strength of a grain split is shown and it is quantitatively described by square and linear equations derived by regression analysis of 42 samples. Mitton's investigations have been carried out by using the IUP/8 method on samples taken from standard sites. It is almost correct to assume that the mechanical functions of skin are determined by the organism itself, e.g., good sole leather comes from

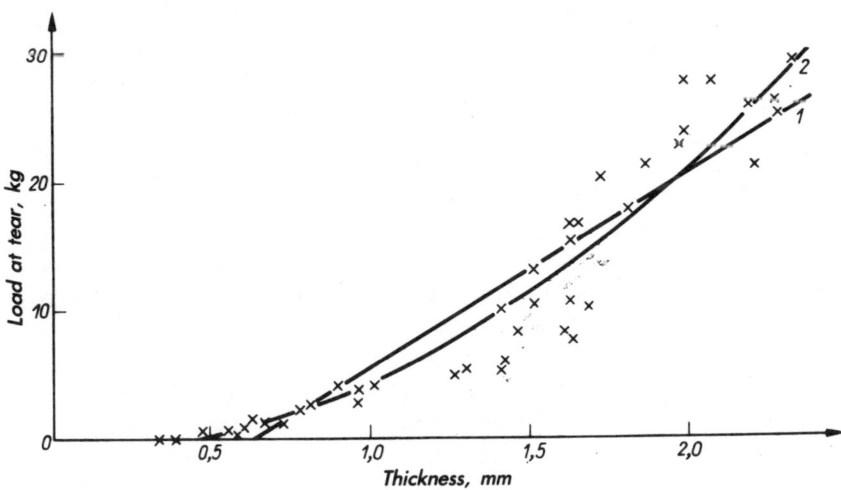

Fig. 1.11. Leather tear strength, and proposals for its recording by: 1 - linear equation, 2 - square equation. According to [15]

a specific area of the butt. The tanners can define this area by just touching it. A very convincing example of this is the 'shells' of horse hide. They are the parts of a horse butt located on either side of the backbone, and are composed specially of fine, closely packed fibers. There are less sweat glands in the shells, so they form an almost continuous network. The biological sense of these parts of horse hide, being of special value for tannery, has not yet been elucidated.

The investigations described above give a general view of the versatility of skin and leather mechanical properties. Recently, this versatility has been tested in detail, i.e., the influence of external factors on collagen fiber strength was examined, particularly of these, which can be applied to the tanning processes. Shimenovich and Mikhailov [16] have shown the decrease of collagen fiber strength under the influence of acids, alkalis and various salts, such as calcium and sodium chlorides, and enzymes which are used in tannery. The fibers tested were isolated from cattle hide. These changes are reversible, or almost reversible and the initial strength of fibers is recovered after the acting factors are removed. When these agents are used as mixtures, some of them do not decrease the fiber strength, e.g., 0.01 m H_2SO_4 in 0.05 m Na_2SO_4 makes the tested fibers about 10% stronger than the control sample. According to the quoted authors' conclusion based on previous X-ray investigation of Zajdes [17] as well as on their own experiments, no significant changes result in collagen structure after such treatments are given. However, as observed recently by Rajaram et al. [18], the length of collagen fibers influences their mechanical properties, such as stress-strain behavior, hysteresis, work recovery and stress relaxation. The quoted authors explain this by internal friction and an average length of fibrils. The stress and strain at break for the fibrils of different length can be represented by the following empirical equation:

$$\ln y = -4.11 + 2.04 \ln x + 0.73 \ln l$$

where y is the stress of break, x the strain of break and l is the fiber length. From this equation one can evaluate the average fiber length as equal to 0.45 mm. This gives us an idea of the linear dimensions of the fundamental filbrillar unit. This dependence has not been considered in the work previously quoted [16].

A very detailed description of viscoelastic properties of collagen after its drastic liming (several weeks), as is the case when preparing material for sausage castings, has been made by Branicz [19]. When testing the collagen prepared this way at various pH and water content levels he concluded that the visco-elasticity of collagen matter is non-linear and proposed some experimental equations for its definition. According to him, the coefficient of apparent viscosity reaches its maximum at pH 3.0–3.5 and increases slightly again, without reaching maximum at pH 6–8 (see Fig. 1.12). These observations can be transferred to

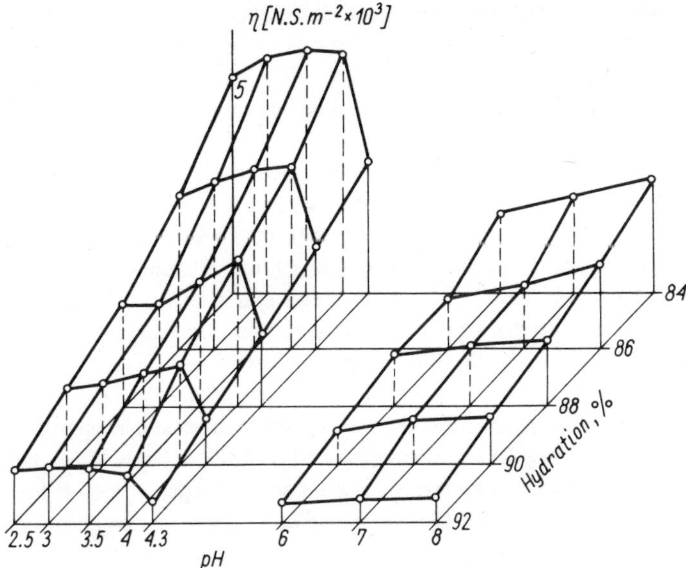

Fig. 1.12. Variations of apparent viscosity of the collagen matrix. Assumed shearing rate $\dot{\gamma}_{sc} = 0.4$ s⁻¹. Courtesy Dr. M. Branicz.

the limed and pickled pelt, and explain its properties. However, it should be kept in mind that in pelt the internal structure (crosslinking bonds) are destroyed to a lower extent than in the material tested. The creeping curve is very similar to that for PVC, and seems to indicate similar viscoelastic properties (Fig. 1.13).

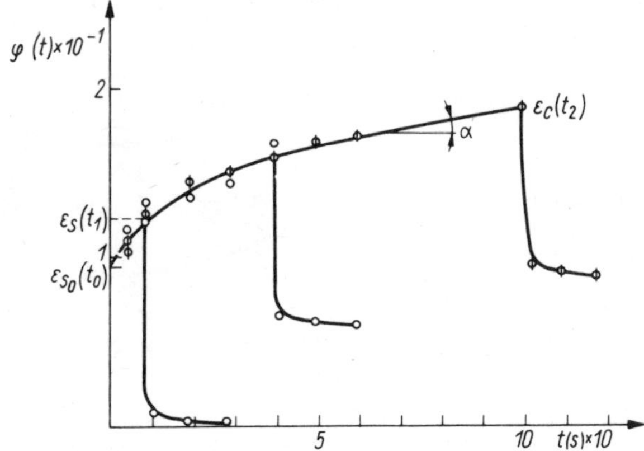

Fig. 1.13. Creeping curve of the collagen matrix. ε - shearing deformation at time t, γ - creeping function of a viscoelastic body per time unit. Courtesy Dr. M. Branicz.

Branicz considers collagen under shearing conditions as non-Newtonian liquid (see ch. 19) and describes its flow as being in accordance with Ostwald-de Wael equation

$$\tau = k\gamma_{sh}^{n}$$

where τ is the flow, $\dot{\gamma}$ the apparent viscosity, and coefficients n and k are pH and hydratation-dependent. These values can be calculated for pH ranging from 2.5 to 4.3 and 6.0 to 8.0, and are valid for limited shearing speed (0.026-0.226 $rad.s^{-1}$) in the Couett rotational viscometer. For the technologist this provides a certain explanation as to why when liming and pickling at too great a drum revolution speed, can damage to the processed leather be made if the limits of viscoelastic flow are exceeded, especially in short float, where hydration may be limited.

Correlation between abrasion resistance and viscoelastic properties of hides tanned with various agents, was investigated by Katrich et al. [20]. Testing the deformation under pressure, they found its strong dependence on the kind of tanning applied (see Fig. 1.14) and a close correlation of this dependence with

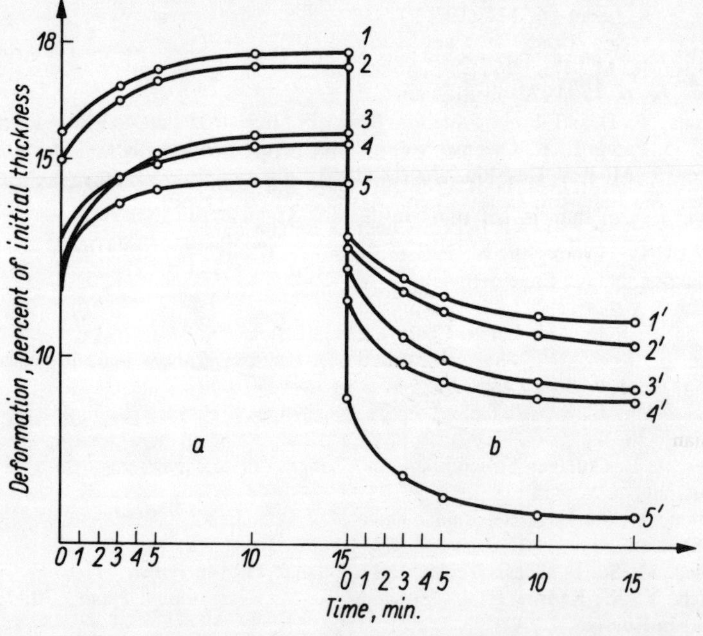

Fig. 1.14. Changes in relation deformation of heavy leathers (hydrated) (a) under pressure, (b) after pressure is removed in dependence on tanning technology. 1 and 1' - chrome tanned, syntan and vegetable retanned, 2 and 2' - chrome tanned, syntan retanned, 3 and 3' - chrome tanned, zirconium and syntan retanned, 4 and 4' - chrome tanned, zirconium, titanium and syntan retanned, 5 and 5' - chrome tanned, aluminium and syntan retanned. According to [20]

abrasion resistance. As can be seen from the figure, the chrome-syntan vegetable tanned leather is most resistant against pressure than compared to others. The authors were interested in practical results and did not give a theoretical explanation of the observed fact; it can be concluded, however, that none of the tanning technologies investigated can change the basic, viscoelastic behavior of leather, which remains in accordance with the Burgers model. Of course the skin component responsible for this property is collagen, which in living animals consists of fiber bundles immersed in the matrix and in leather of the same fibers containing some unnatural additional crosslinks, which includes a certain amount of tanning agents, and is immersed in substances added or not removed, i.e., water and fat.

REFERENCES

1. De Roberts, E. D. P., Nowinski, W. W., Caez, F. A. Cell Biology, Saunders Philadelphia 1970
   Kulonen, E., Pikkarainen, J. Biology of Fibroblast, Acad. Press London 1973
2. Heidemann, E. Leder *22*, 41 (1971)
3. Parakkal, P. F., Alexander, N. J. Keratinisation, Acad. Press N.Y. 1972
4. Stirtz, T. S. Leder *16*, 177 (1965)
5. Reed, R. J. Soc. Leath. Tr. Chem., *37*, 75 (1953)
6. Zurabian, K. M., Dumnov, V. S., Mateckene, N.Y., Erdynee, L.L. Kozh. Obuv. Prom., *14*, 6, 17 (1972) in Russian
7. Harkness, R. D. in Fibrous Proteins—Scientific, Industrial and Medical Aspects ed. D. A. D. Parry, L. K. Creamer vol. 1, Acad. Press London 1979
8. Kenedi, R. M. Res. Develop., *10*, 18 (1963)
9. Maeser, M. J. J. Am. Leath. Chem. Assoc., *55*, 501 (1960)
10. van Vlimmeren, P. J. XII IULCS Congress Prague 1971
11. Ward, A. G., Brooks, F. W. J. Soc. Leath. Tr. Chem., *49*, 312 (1965)
12. Shestakova, N. A., Shneiderovich, R. M. Kozh. Obuv. Prom., *14*, 6, 20 (1972) in Russian
13. Morgan, F. R. J. Am. Leath. Chem. Assoc., *55*, 4 (1960)
14. Raabe, E., Kornas, A. Physical Properties of Leather, Warsaw 1965 in Polish
15. Mitton, R. G. J. Soc. Leath. Tr. Chem., *48*, 195 (1964)
16. Shimenovich, B. S., Mikhailov, A. M. Kozh. Obuv. Prom., *18*, 7, 45 (1976) in Russian
17. Zaides, A. L. Collagen Structure and its Changes due to Treatments, Moscow 1966 in Russian
18. Rajaram, A., Sanjeevi, R., Ramanathan, N. J. Am. Leath. Chem. Assoc., *73*, 287 (1978)
19. Branicz, M. Sc. D. Thesis Techn. Univ. Gdansk 1978 in Polish
20. Katrich, V. N., Kedrin, E. A., Zurabian, K. M. Kozh. Obuv. Prom., *20*, 12, 29 (1978) in Russian

2.

COLLAGEN MACROMOLECULE: STRUCTURE AND PROPERTIES

2.1. Some general properties of biopolymers

Biopolymers fulfill numerous physiological functions among which the main are: maintaining the structure of cells and tissue (keratin, collagen), catalysis (enzymes) nutrition (glycogen), control of osmotic conditions (glycosaminoglycans, phospholipids), hormonal control (protein hormons), immunological reactions (antibodies, haptens) and storage of genetic information (nucleic acids).

The shape of the polymer molecule is of primary importance of its properties: amylose, which is a linear polysaccharide, yields resilient gels, and branched amylopectin—viscous solutions only. A linear polymer, which allows great freedom of rotation around its valence bonds, assumes the shape of a random coil. This is the most probable shape of the macromolecule, where its enthropy reaches maximum. Stretching of such a molecule may take place only after a decrease of entropy, i.e., after an increase of the intramolecular energy.

No real valence bonds have full rotational freedom. Even polymethylene allows for privileged atom position in the chain. The potential energy of every bond passes through three maxima and three minima in each rotation about the valence bond. A diagram of the changes of energy levels of the single carbon-carbon bond is shown in Fig. 2.1. As one can see, the trans-configuration, which is of lower energy, is more stable than the cis one. If in a polymer chain double bonds occur, or if the main chain has 'heavy' side chains, the limitations in rotation about the valence bond are very significant.

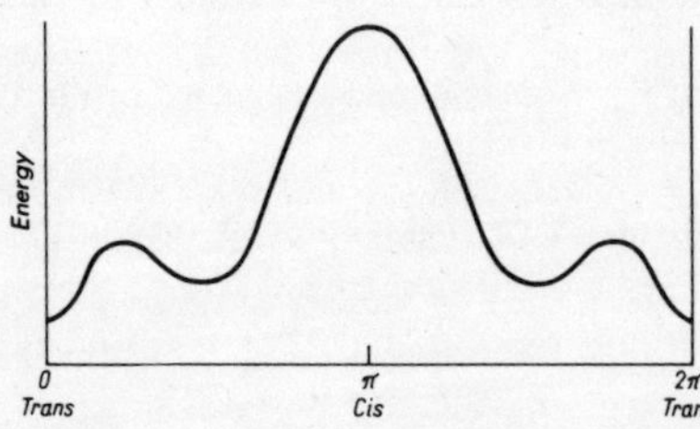

Fig. 2.1. Energy level changes occurring during rotation around the carbon-carbon bonds between two adjacent chain segments.

The mean distance $\bar{r}$ between the chain ends is found in the expression:

$$\bar{r}^2 = nl^2$$

where l is the length of each link (segment) in the chain and n the number of bonds in the molecule. The rotation limitation is equivalent to a decrease of n and increase of l. The value of l increases for each link quicker than the number of bonds in the molecule decreases. If we therefore assume the molecular weight to be constant, the value of $\bar{r}$ will increase: the random coil opens, extends, and assumes the form of a rod. The random coil may be the subject of many modifications, for example, a solvent molecule may attach itself to the polymer, making the coil swell, then the equation quoted above assumes the form of inequality:

$$\bar{r}^2 > nl^2$$

A solvent that causes the shrinking of the coil is a 'bad' solvent, whereas an 'ideal' solvent leaves the coil unchanged. Among known biopolymers the caoutchouc polymer has a shape close to that of the theoretical coil. It consists of 500–5000 isoprene units, polymerized in the cis-configuration. Stiffness of the linear polymer chain due to limitations in rotational freedom leads to departure from the coil structure; however, the latter may be due to many other reasons. Intermolecular interactions may be a result among many others, to electrostatic forces, hydrogen and hydrophobic bonds. A result of these interactions is the formation of very specific, ordered structures such as proteins and nucleic acids. Interaction between parts of the flexible chain may occur without formation of a specific structure. This may occur as a result of denaturation of proteins and nucleic acids.

2.2. Collagen

Collagen is one of the fibrillar components of connective tissue in animals. Its macromolecules are arranged parallel and close to one another. Bundles of linear polymers are formed as a result of aggregation of the chains constitute fibrils, by which multiple interconnection form fibers. In the living organism collagen penetrates connective tissue, e.g., corium, as fairly thick fiber bundles, in the form of a random, spatial network, where apparently no fiber ends are present.

The formation and turnover of collagen are the functions of connective tissue. Greatest collagen clusters are formed there, where that tissue is substantial for the given organ (skin, bones, tendon, eyes, etc.). A survey of kinds of connective tissues has been made, e.g., by Reich [1].

Collagen, being a characteristic extracellular component of connective tissue, has a dominant position in the molecular structure of higher animals. It is the main factor transporting the internal and external forces, acting on the organism. Among skin proteins it is quantitatively first, accounting for 60–80% of its dry matter, and 20–30% of total body protein. Corium of animals is used for production of leather, while from collagen gelatines for photographic, nutritional and technical use as well as glue are produced.

A number of monographs have been dedicated to collagen, and its biological [2], biochemical [3], structural [4] and technological [5] features.

The technology of leather making is, in the broadest sense, a series of operations which aim at isolating collagen by removing noncollagenous components of skin, and then at making it resistant to physical, chemical and biological factors. The first part of those processes is performed in the tannery beamhouse, the second in the tanning and finishing departments. This traditional division has its scientific justification. It is not easy to make a good choice of information useful for a tanning chemist. The present author tried to collect data published by the best-known teams of specialists who were known to concentrate on collagen investigations. It has been assumed that the following information is essential:

(1) Collagen biosynthesis, structure and amino acid composition with the present views on amino acid sequences and reactions of functional groups, which occur in side chains and lead to the formation of crosslinking bonds.

(2) Bonds occur in collagen and their properties. The bonds, occurring in collagen strongly affect its behavior in leather processing, its resistance to external factors and the structure of the collagen molecule.

(3) Spatial structure of the collagen molecule, essential for its physical properties as well as the present knowledge about chain conformation.

(4) Basic data concerning supramolecular structure of collagen and testing that structure with an electron microscope.

Biosynthesis and structure of collagen

The collagen molecule as a whole consists of three peptide chains containing about 1000 amino acid residues in each. These chains are stranded in a triple helix and interconnected by a system of crosslinking bonds, perpendicular and parallel to the molecules, forming intramolecular and intermolecular bridges.

Collagen biosynthesis takes place in special cells called fibroblasts. Closer consideration of that problem, however, exceeds the scope of this book and the reader is referred to many excellent reviews and monographs (e.g., [2]), and general aspects of protein biosynthesis are to be found in [6]. The peculiarities of collagen biosynthesis are connected with its unique structure. Here it will be only mentioned that it consists of two steps of which the first one is a sequence of amino acids is produced by α-amino peptide linkage yielding the so-called protocollagen which is a product of a structural gene action; according to our present knowledge, there are at least nine different structure genes specific for the particular chains in the triple helices of various types [7]. In the second one the post translational modifications result from the action of at least seven different enzymes. These modifications include conversion of proline residues to 4-hydroxyproline and to 3-hydroxyproline, conversion of lysine residues to hydroxylysine and glycosylation of hydroxylysine, oxidative deamination of lysine and hydroxylysine (which yields the crosslink-precursor aldehydes, i.e., allysine and hydroxyallysine) and proteolytic cleavage of procollagen to collagen (procollagen is hydroxylated, glycosylated protocollagen still attached to a specific 'information' peptide containing about 300 amino acids more than the matured collagen). When leaving the fibroblast, this peptide is split off by an extracellular enzyme as it does not have the specific collagen triple-helix structure that is necessary, and therefore it is accessible to many proteolytic enzymes. The remaining collagen, being insensitive to proteolytic enzymes except collagenase, starts forming fibrils by staggering its molecules parallel to each other in the intercellular space.

An instructive study of collagen biosynthesis, as well as of the significance of its particular stages, is that given by Nimni [7a], from which the Fig. 2.2 is

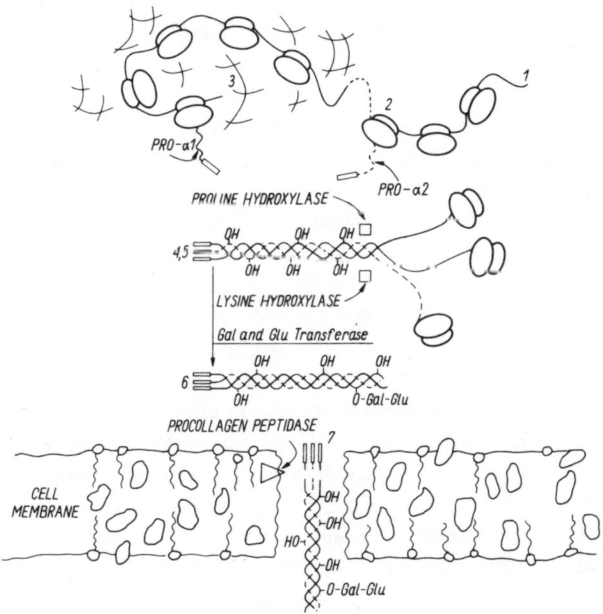

Fig. 2.2. Intracellular events leading to the completion of a collagen molecule. 1 Synthesis of specific mRNA for collagen; 2 translation of message by ribosomes; 3 clustering of polysomes and association with endoplasmic reticulum; 4 recognition and alignment of α chains aided by registration peptides; 5 coiling of chains and hydroxylation of specific proline and lysine residues; 6 glycosylation of specific hydroxylysine residues; 7 extrusion of completed molecule. According to [7a]

taken, illustrating the pathways of biosynthesis. More details about action of particular intra and extracellular enzymes acting in this process may be found in a quoted symposium report.

It is generally accepted to define the single chain as α-chain of collagen. Each collagen molecule consists of three α-chains, usually identical.* The only known exception is type I collagen consisting of two identical chains ($\alpha1$) and one different chain ($\alpha2$) which is denoted as $[\alpha1(I)]_2(\alpha2)$. This type, characteristic for mature skin, is the most important and thus most investigated component. It is the only heteropolymer among collagens [8]. Index I is used because the chains in the particular collagen types differ slightly in their amino acid composition.

Collagen specific for cartilage, type II, has three identical chains, and is designated as $[\alpha1(II)]_3$. Collagen type III, a minor but important constituent of skin, is defined as $[\alpha1(III)]_3$. It is to be found in most connective tissues, and is always associated with collagen I. The relative proportions of I and III collagens vary with age: collagen III occurs in a higher ratio in skin and blood vessels of fetuses [9], so sometimes it is called fetal collagen. Its content in human skin varies from 11 to 19% of total collagen [10], in a fetus in the early stages the content of collagen III is still higher. Its specific feature is that it contains an intrachain disulfide crosslink (two cysteine residues per 1000 amino acid residues) [11].

Collagen of the basement membrane is usually denoted as collagen IV (sometimes as BM collagen). According to more recent communications there are at least three, perhaps four, collagen types in the basement membrane (IV, A, B, and probably C) [12]. It is very difficult to separate the basement membrane collagens because they occur in very small amounts in skin, and probably are susceptible to attack by proteolytic enzymes other than collagenase. Those collagens contain much more hydroxylysine residues than other collagens. Their specific composition may perhaps be explained by their biological role, as mentioned above. For the tanning process it is important to know whether the basal membrane, forming a poorly permeable shield on the surface of the skin, is removed in the pretanning processes or not. This point will be discussed later (p. 282). Separation of collagens I–III can be performed in the preparative technique, e.g., by fractional salt precipitation: in cold neutral (buffered) solution type III precipitates from about 1.5molar NaCl, type I from about 2.2molar NaCl, and type II from about 4.4molar NaCl [13, 14]. Other separation techniques, as well as those for separating α-chains, are based on careful protease digestion and chromatographic processes, using CM-cellulose and molecular sieves. The differences between the particular α-chains in human skin are presented in Table 2.1.

---

*It must be emphasized, that α-collagen chain has only incidentally the same definition as α-helix of other proteins. Collagen α-chain (not helix!) is but a denotation of macromolecule unit, whereas in other proteins α-helix is a definite structure, quite different than this of collagen. This structure is shortly explained in section 3.1.

Earlier investigations of collagen, dissolved and precipitated in various ways, lead to the conclusion that not three but also two α-chain aggregates can be formed. The latter were referred to as β-molecules and denoted by $\beta 1, 2$ or $\beta 1, 1$ or $\beta 2,2$, depending on the kind of α-chains involved. Today we know that these are artificial formations, having no reflection in nature.

In 1952 the collagen structure was still classified by Lindestrom-Lang into structures of the first, second, third, fourth and fifth order. The first-order structure applies to the sequence of amino acids in polypeptide chains. The second-order structure relates to the conformation of polypeptide chain, the third-order refers to the position of a peptide chain in a triple helix. The fourth-order (quaternary structure) is the supramolecular structure of fibril formation. Finally, the fifth-order structure is that in which fibers can be seen at medium magnification under the microscope. This detailed classification is not, however, adequate to contemporary knowledge as it involves some inconsequences, for instance disulfide bonds are believed by some authors to belong to the first-order structure, while others assign them to the third order. Other crosslinking bonds present the same problem. It has not been established whether, e.g., the pleated sheet structure (see p. 74) is of second order, as the helix is, or of third order. It is not known either whether the third-order structure should include only the

Table 2.1

Amino acid compositions of human collagen α-chains*
(after Glanville and Kühn [ref. 7 ch. 1]

Amino acid	$\alpha 1(I)$	$\alpha 2$	$\alpha 1(II)$	$\alpha 1(III)$	$\alpha A(IV)$	$\alpha B(IV)$	$\alpha A(V)$	$\alpha B(V)$
Hydroxyproline	102	87	99	127	146	127	107	108
Aspartic acid	43	46	41	43	48	52	55	51
Threonine	17	19	20	13	18	28	27	22
Serine	39	36	26	39	34	26	34	26
Glutamic acid	74	69	88	72	77	62	90	99
Proline	139	123	119	109	79	69	92	119
Glycine	352	350	326	354	318	313	318	320
Alanine	121	113	98	97	31	49	59	46
Valine	21	34	18	14	30	24	30	19
Cysteine	—	—	—	2	—	—	—	—
Methionine	7	5	9	8	15	12	10	7
Isoleucine	7	15	9	13	34	42	18	20
Leucine	21	33	25	22	54	59	37	40
Tyrosine	2	4	1	3	5	6	—	1
Phenylalanine	13	12	13	8	24	37	12	12
Hydroxylysine	5	9	14	5	61	42	25	36
Histidine	2	11	2	6	6	8	10	7
Lysine	31	23	22	30	5	6	18	19
Arginine	51	53	50	47	20	45	57	48

*Compositions calculated as amino acid residues per 1000, rounded off to the nearest whole number.

distribution of both lower orders, or the factors giving shape and stability to the structure, e.g., angles between the bonds. Wetlaufer [15] proposed to use the notion of *'chain sequence'* for determining the order of acids in the amino acids in the chain and *'chain conformation'* for defining the arrangement of the main and side chains in space.

The notion of configuration should traditionally be reserved for optic asymmetry centers in molecules, so we speak of configuration of the amino acid residue, and the peptide *'chain conformation'* has been defined above. The sequence and conformation of a chain are factors that may suffice to describe the position of an amino acid residue or bond. Instead of the term 'quaternary structure' Wetlaufer proposed the term *'sub-unit array'* or distribution of sub-units in space, taking account of the interaction between those sub-units. The value of the term *'sub-unit array'* is, however, dubious when applied to collagen because one can first assume as sub-unit a fragment of a molecule, and secondly, we cannot apply this expression to fibrillar structures, i.e., to structures in which a network of fully developed and distributed in a specific manner fibers occurs. It seems that the J. Polym. Sci. definition of *'supramolecular structure'* is more appropriate. The *'short-range'* or *'long-range organization'* proposed by Veis [16] has not been widely accepted. Nevertheless, the *'short-range organization'* is important for technologists because of steric relations and accessibility of the particular collagen functional groups, participating in the leather making process. *'Short-range organization'* covers the sequence and structure of the peptide chain as well as the structure of the triple helix' coiled coil'. Thus the *'long-range organization'* covers the structure ranges as investigated with the electron microscope and low-angle diffraction of X-rays.

In the present work the Wetlaufer nomenclature is used, as is in some biochemistry handbooks [17, 18].

Amino acid sequence and protein structure range

The amino acid sequence is a typical feature of protein, determining its structure as a whole. Collagen, apart from its origin, contains 19 amino acids, among which are two that do not occur in other proteins, i.e., hydroxyproline and hydroxylysine. Besides, collagen contains more glycine than most other proteins, but it does not contain cysteine, cystine (with the exception of collagen III) and tryptophan.

According to Mikhailov [5], the typical features of collagen are:

(1) The number of glycine residues amounts to ⅓ of all amino acids residues.

(2) The number of iminoacids residues is ⅕ of all amino acid residues in mammals and birds*.

*The name iminoacid is currently used in biochemistry though it is not quite correct since those compounds are derivatives of pyrrolidine not imines. The systematic name of proline is: pyrrolidine-α-carboxylic acid, and that of hydroxyproline: β-hydroxypyrrolidine-α-carboxylic acid. However, basing on the known handbook of organic chemistry by Roberts-Caserio the name of iminoacids has been used throughout this book.

(3) The presence of two specific hydroxyaminoacids: hydroxyproline, hydroxylysine (hydroxyproline may occur in other proteins but in uncomparably smaller amounts).

(4) The presence of a certain amount of aldehyde groups (participating in cross-linking bonds, as suggested by the present author).

(5) The presence of hexoses, bound to protein side chains.

(6) The occurrence of characteristic hydrophilic and hydrophobic space groupings in a chain.

(7) The average molecular weight of one residue of 90.7.

(8) The number of aminoacids in a chain amounting to about 1000 on the average.

(9) The average molecular weight of one chain amounting to about 90,000.

The composition of collagen as investigated many times by various analytical techniques varies in a small range. The triangle shown in Fig. 2.3 is divided into parts proportional to the amino acids content, and also indicates the pro-

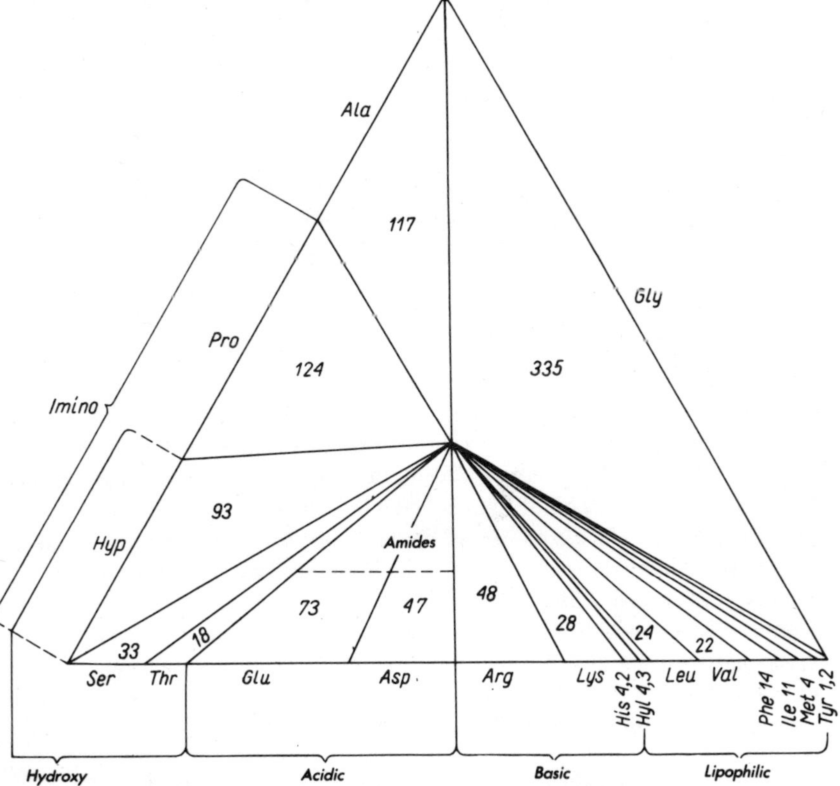

Fig. 2.3. Occurrence of particular amino acid residues and their groups in collagen of higher animals. Figures give the numbers of residues per 1000 AA residues. [4]

portion of amino acid groups: imino acids (Pro, Hyp), hydroxyaminoacids (Hyp, Ser, Thr), acidic amino acids (Glu, Asp and their amides) and basic amino acids (Arg, Lys, His, Hyl). The numbers refer to cattle hide collagen Hylys belonging to three groups: hydroxy, basic, and lipophilic (Leu, Val, Phe, Ileu, Met, Tyr). Two amino acids (Gly and Ala) do not belong to those groups.

In collagen 3-hydroxyproline occurs in amount of about 1% of the total amount of hydroxyproline; this was discovered in 1961 by Ogle et al. [19]. It was supposed to be an 'artifact' of genetic information. This way is, however, incorrect as it occurs in various chains in different amounts: collagens I and III contain 1 residue per α-chain, collagen II contains 2, collagen IV 10-15 residues per α-chain [20]. As it has been recently shown [21], two different enzymes in fibroblast catalyse proline hydroxylation: prolyl-4-hydroxylase and prolyl-3-hydroxylase, they have both been isolated and characterized. Hence, although the role and significance of 3-hydroxyproline has not yet been made clear, then it must be considered as a regular collagen component.

Among the aminoacids mentioned above, glycine, alanine, valine, leucine, isoleucine, phenylalanine and proline have nonpolar side chains. Serine, threonine, tyrosine, methionine and both hydroxyprolines are polar aminoacids with non-ionizing groups in the side chains. Glutamic and aspartic acids, as well as lysine, hydroxylysine, arginine and histidine are amino acids with ionizing side chain functional groups.

In Table 2.2 the properties of amino acids most common in proteins are compared.

To find the amino acid composition of a protein one has to isolate it as a pure substance, subject it to complete hydrolysis and separate the products of hydrolysis. Hydrolysis of collagen is usually performed in an acid medium, when part of the hydroxyamino acids undergoes decomposition. In order to obtain the most accurate data it is recommended to conduct acid hydrolysis at different times and then to calculate the hydroxyamino acid content by extrapolation to zero time. The data of Rozycka [22] shows that hydrolysis with 25% HCl 34 h at 140°C does not lead to hydroxyproline decomposition. Later this was confirmed by the work of a group of laboratories under the direction of Heidemann [23], based on the method of Langerwerf [24] (hydrolysis conditions 6% HCl, 110-130°C, 60 h). The great improvement of hydrolysate separation techniques, a.o. by introducing automatic amino acid analyzers, resulted in an almost full knowledge of amino acid composition of collagens of different origin. Investigations are now under way of the exact composition of the particular collagen chains and products of their partial hydrolysis (enzymatic, cyanogen bromide, etc.). In Table 2.3 the amino acid contents of some collagens of various origin are presented according to 1967 data. However, no changes have been reported since.

Pikkarainen and Kulonen [25] carried out a comparative study of collagens

Table 2.2

Amino acid residues and their characteristics

Residue	*Probable function in proteins and enzymes*
Arginyl	hydrophilic; electrostatic action.
Lysyl	" " " : attaching of prosthetic groups and cofactors in amide binding; participation in Schiffs bare formation: ligand in metal complexes.
Histydyl	hydrophilic or hydrophobic, depending on ionization; electrostatic action; proton transfer; ligands in metal complexes; hydrogen bonds, acceptor in transfer reactions.
Glutamyl } Aspartyl	hydrophilic; electrostatic action; ligands in metal complexes covalent bonds in esters or amides by carbonyl $=O$.
Glutaminyl	hydrophilic; hydrogen bonds.
Asparaginyl	" " "
Seryl	nucleophilic; hydrogen bonds; covalent OH bonds in esters.
Threonyl	nucleophilic; hydrogen bonds; covalent OH bonds in esters, hydrophobic.
Glycyl	lack of side chain makes possible bending of the chain and formation of crosslinking hydrogen bonds.
Alanyl } Valyl } Leucyl	hydrophobic interactions; determinants of stereospecifity and conformation, e.g., multiple alanyl residues favor helix formation, valyl and izoleucyl prevent it.
Tyrozyl	hydrophobic; hydrogen bonds, proton transfer; electrostatic action at high pH; ligand in metal complexes
Tryptophanyl	hydrophobic-hydrogen bonds.
Cysteyl	nucleophilic acceptor of acyl groups, hydrogen bonds; ligand in metal complexes.
Cystyl	crosslinking bonds with disulfide bridges.
Methionyl	hydrophobic; hydrogen bonds at S; ligand in metal complexes.
Prolyl	hydrophobic: breaking of structures of α helices; formation of specific conformations.
Hydroxyprolyl	hydrophilic; otherwise like prolyl.
Hydroxylysyl	like lysyl; binding sugars through-OH groups.

of various origin, analyzing them from the point of view of the evolution of species. They observed that with the evolution of animals the amount of iminoacids increases. This occurs at the cost of serine and threonine. From the genetic code it follows (see, e.g., [6]) that they differ only by the first letter of the code: U-serine, A-threonine, C-iminoacids, G-alanine. The number of hydroxyl group is constant due to proline hydroxylation. There is a clear difference in the content of acidic aminoacids in vertebrate and invertebrate collagen: vertebrate collagen contains 100-120 residues of glutamic and aspartic acid residues, whereas invertebrate collagen contains about 150 residues. In warmblooded (homeothermic) vertebrates (mammals, birds) there are about 50 serine and

Table 2.3

Average amino acid composition of collagen in some mammals; number of residues per 1000 amino acid residues

Amino acid	Wallaby tendon	Human skin	Rabbit skin	Rat tail tendon	Pig skin	Pig tendon	Sheep tendon	Cattle hide	Calf skin	Cattle tendon
Alanine	112.5	114.4	101.8	99.3	110.8	94.0	99.9	105.0	112.0	97.8
Glycine	320.0	324.4	307.4	351.0	326.0	341.0	327.0	334.0	320.0	336.5
Valine	23.2	24.5	21.5	22.5	21.9	22.2	25.1	19.0	20.0	21.5
Leucine	26.3	28.8	23.2	22.2	23.7	25.0	25.4	25.0	25.0	27.3
Isoleucine	8.9	10.4	15.0	13.2	9.6	12.5	12.7	11.0	11.0	14.5
Proline	119.1	125.1	141.9	123.0	130.4	119.7	120.0	129.0	138.0	144.2
Phenylalanine	16.0	12.6	12.3	14.3	14.4	14.0	13.6	13.0	13.0	15.3
Tyrosine	4.1	3.5	2.0	5.4	3.2	5.1	4.8	4.7	2.6	4.8
Serine	39.0	36.9	40.1	27.8	36.5	28.4	27.9	38.0	36.0	29.5
Threonine	20.1	18.3	19.7	19.1	17.1	18.6	20.6	17.0	18.0	18.9
Cystine	—	—	0.2	—	—	—	—	—	1.0	—
Methionine	6.6	7.0	8.6	5.8	5.4	5.8	5.7	6.6	4.3	3.6
Arginine	51.1	49.0	45.3	46.5	48.2	49.7	49.9	48.0	50.0	45.4
Histidine	5.1	5.4	5.5	3.3	6.0	4.1	4.2	4.6	5.0	6.5
Lysine	24.6	26.6	27.3	35.6	26.2	34.2	35.5	25.0	27.0	22.4
Aspartic acid	49.3	47.2	50.2	47.1	46.8	48.5	49.1	48.0	45.0	48.0
Glutamic acid	73.0	77.7	68.7	73.7	72.0	77.8	76.4	72.0	72.0	71.4
Hydroxyproline	92.8	90.9	103.6	90.4	95.5	99.2	102.4	92.0	94.0	83.4
Hydroxylysine	8.0	5.9	4.9	—	5.9	—	—	6.8	7.4	9.3

threonine residues per 1000 residues, whereas in coldblooded (heterothermic) vertebrates and invertebrates 70-100 residues occur. The sum of serine, threonine, iminoacids and alanine content (i.e., of the aminoacids being genetically related) is constant in vertebrate collagen, so as is the content of nonpolar amino acids, however, both these values are higher in vertebrates than in invertebrates. These generalizations do not apply to rainworm collagen. Further differences relate to the structure of higher orders. This evaluation shows the development trends in collagen formation in living organisms. Collagen formation is generally in accordance with the development of the organism as a whole and it adapts to the increasing life level. The collagen functions other than tissue-structure supporting, not yet fully understood, are in higher organisms much more complicated and variegated. It participates in immunological reactions, in wound healing, in preventing malignant tumors, in cell shaping, etc. The collagen molecule contains important genetic, extra-gene information. Thus it cannot be believed any more to be an inert structural protein; this opinion was still valid until about 10 years ago. Collagen significance in biology has been recently reviewed [26].

In Fig. 2.4 an evolution scheme of the α-chain is given according to Pikkarainen and Kulonen which illustrates the differences in collagen. Those differences depend on the 'genetic age' of collagen.

The collagen molecule, consisting of three chains, contains about 3000 amino acid residues. It has about 3000 Å in length and 14 Å in diameter, so it has the shape of a rod whose length is 200 times larger than its diameter. If it had the thickness of a pencil, it would have the length of 1.5 m. This rod is reinforced by crosslinking bonds.

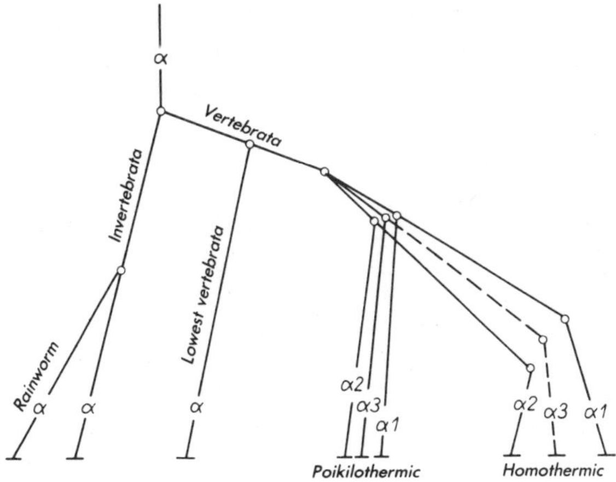

Fig. 2.4. Collagen evolution from the point of view amino acid content of the α-chain. According to [25]. The occurrence of $\alpha3$ chain has not been confirmed recently.

Bonds in collagen and intermolecular interactions

Many kinds of bonds of various energy occur in collagen. Most important are the covalent bonds formed as a result of interaction between the functional groups inside the amino acid chains in the collagen molecule. In that case two atoms are bonded which have an equal or similar affinity to electrons. They are bonded as a result of the tendency of both participants to attain a stable electron shell. Part of the electrons of the outer shell becomes common on bond formation; the configuration is stable if both atoms give each one electron to the common pair, the formed system is of lower energy, and the amount of energy evolved is the measure of the bond stability. An example of such bonds are the peptide bonds and bonds between collagen side chains. A specific kind of covalent bond is the co-ordination bond, characterized by its formation at the cost of an electron pair of one of the atoms. The co-ordination bond is also formed when a free electron pair of an atom is transferred to the octet hole of another atom, supplementing its outer shell. A semipolar bond is formed when one electron, e.g., of nitrogen atom, passes formally to the oxygen atom.

The co-ordination bond principle has been used to explain the structure of covalent complexes with overlapping electron clouds. Such complexes are formed due to co-ordination of weak electronegative ligands by weak electropositive central ions with high ionization energy. The number of co-ordination bonds, identical in this case to the co-ordination number of the central atom, is defined by the number of nonoccupied orbitals of low energy level able to intercept electrons of the partner, which are necessary to form the co-ordination bond.

Beside bonds, defined as intramolecular, which affect the shape and location of electron orbitals, intermolecular interactions occur between systems (molecules, atoms) in which the outer electron orbitals are already occupied. Intermolecular interactions are of importance for biological solutions and systems such as proteins and nucleic acids. They are of an electrostatic, induction, dispersion, resonance and donor-acceptor type character.

The energy of electrostatic interaction is the energy of the Coulomb interaction of two undisturbed electric charges, defined by the square of their wave function ψ, or undisturbed wave function of the ground state. That energy, like, e.g., ion-dipole or dipole-dipole interaction, depends on the dipole orientation.

Induction interaction occurs when one or both interacting systems are excited. The energy of that interaction is calculated by taking into consideration the action of the charge, dipole and quadrupole of one molecule with the induced dipole of another.

The energy of dispersion interaction is a time average of interaction effects on induced dipoles: as a result of fluctuation of the electron charge in one of the molecules, a temporary dipole is formed whose field induces a dipole in the other molecule. The dispersion interaction energy is referred to as van der Waals.

The energy of resonance interaction may be the result of interaction of two identical atoms, one of which is in the excited state. This energy decreases proportionally to R^{-3} (where R is the distance between the atoms), whereas the energy of dispersion interaction decreases proportionally to R^{-6}, the latter decrease being much quicker. The interaction energy may thus increase significantly, if one of the systems is excited. Donor-acceptor interaction is often defined as charge transfer interaction. In the case where no chemical bond is formed between two systems each of them keeps its electrons. One of them is an electron donor if its ionization potential is low. The other is an acceptor, if its electron affinity is high. Transfer of the electron from the first system to second one then takes place with a small energy loss, compensated by a stronger attraction. The donor-acceptor interaction strength can be predicted if the ionizing potential of system A (in approximation the energy of the highest occupied orbital in the ground state) and the electron affinity of system B (approximately the energy of the lowest empty orbital) are known. These kinds of bonds can either occur in the native collagen molecule or can be the result of the leather making process.

Peptide bonds. Amino acids, combining into a peptide chain by splitting off carboxyl OH from one and the proton from the α-amino group of the other, form peptides:

$$
\begin{array}{ccc}
& R_1 & & R_2 \\
& | & & | \\
NH_2-C-COOH & + & NH_2-C-COOH & \longrightarrow \\
& | & & | \\
& H & & H
\end{array}
$$

$$
\begin{array}{cc}
R_1 & R_2 \\
| & | \\
\longrightarrow \quad NH_2-C-CO-NH-C-COOH & + \quad H_2O \\
| & | \\
H & H
\end{array}
$$

The peptide bond, connecting two amino acid residues, is decisive for the biological properties of proteins. The geometry of the peptide bond is shown in Fig. 2.5. That bond is planar, where the oxygen atoms at the carbon and the hydrogen atoms at the nitrogen are in the trans-configuration. The distance between C and N atoms is 1.32 Å, and is smaller than that found between nitrogen and α-carbon, which equals 1.47Å. This points to a conjugation between the

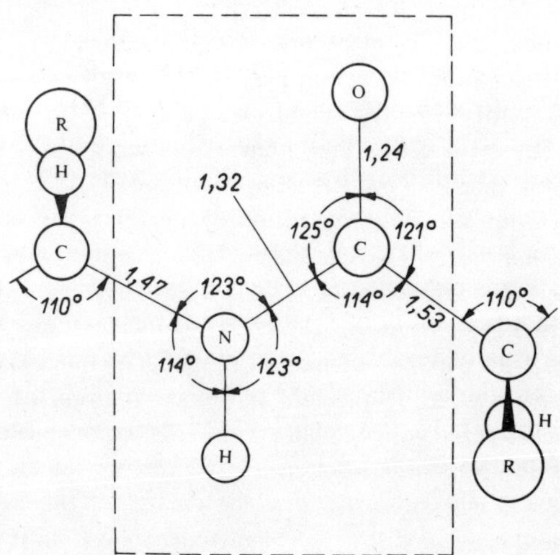

Fig. 2.5. Geometry of the peptide bond.

electron pair π of the C = O bond and the free electron pair of the nitrogen atom. The occurrence of delocalization (transfer) of electrons in the peptide bond is shown by the two mesomeric formulas (covalent /I/ and ionic /II/) given below:

$$R_1-\overset{\overset{\textstyle O}{\|}}{\underset{\underset{\textstyle H}{|}}{C}}-\overset{\cdot\cdot}{N}-R_2 \;\rightleftharpoons\; R_1-\overset{\overset{\textstyle \cdot\cdot\overset{\ominus}{O}}{|}}{C}=\overset{\oplus}{\underset{\underset{\textstyle H}{|}}{N}}-R_2$$

$$\text{I} \qquad\qquad\qquad \text{II}$$

Polycentric electron orbitals occurring in such a system make the calculation of bond energy very difficult.

The difference between bonds with delocalized electrons and those with non-delocalized ones lies in their additivity. The former have non-additive properties while those with non-delocalized electrons do.

The possibility to describe the multi-atom molecule, using localized molecular orbitals, implies that the properties of such a molecule are constant and are liable to insignificant changes caused only by the transfer of electrons from one molecule to another. This should not be expected in the case of a molecule with conjugated bonds. Some properties, which may be related to localized electrons in the bond, are constant in the whole group of non-conjugated molecules.

Among them are: bond energy and length, vibration frequency, force constants and dipole moments. The peptide bond assumes 40% of the character of a double bond due to electron delocalization of the C-N part (60% of structure I). The occurrence of the trans-configuration in the neighborhood of the C-N bond may be ascribed to greater stability of that bond as compared with the cis-configuration. The difference in stability of those bonds is about 2 kcal/mole (8.4kJ/mole). Single bonds of typical length (1.53 Å for C-C and 1.47 for C-N) occur between the atoms participating in the peptide bonds and the neighboring α-atoms. Therefore a possibility of rotation about these bonds does exist. The protein molecule is usually flexible and in solution it may exist as a random coil. However, in collagen additional bonds are making the molecule more rigid, thus its transition from triple helix to random coil is connected with the change in thermodynamic properties of the molecule.

As we see from the above characteristics of the peptide bond, it has to be treated as a whole and not as keto-imide bond (a previously used name), as it has properties neither of the keto nor of the imino group. The atom group forming the peptide bond has a definite energy level and space arrangement. The results of calculations made by Suard et al. [27] and material later supplemented by the same authors have lead to the conclusion that the interaction between groups contributing to the formation of two adjacent peptide bonds is insignificant. The influence of the side chains on the shape and energy of the peptide bonds is insignificantly small as well.

Imino acids account for approximately 20% of the components of the collagen peptide chain. The question arises: What are the properties of the peptide bond in which there is no hydrogen atom at nitrogen where the C-N bond has an extra 'shackle'? Exact calculation of the angles and distances between the atoms carried out for crystalline tripeptide L-leucyl-L-prolyglycine, account being taken of about 1700 reflexes in an X-ray diffraction spectrum (Fig. 2.6), has shown that this bond is planar as well, and that the angles and interatomic distances are very much like those found in other peptides. The dihedral angle between the peptide bond and the pyrrolidine ring is 7°. The length of the C-N bond is equal to 1.34 Å, which is a usual value for the peptide bond. Thus, the bond is as much of the character of a double bond as is other peptide bonds. Twisting of the chain by an angle of 120° occurs at the place of attachment of the pyrrolidine ring. This is due to the geometry of the ring and is essential for the unique shape of the collagen helix.

Hydrogen bond. Apart from the peptide bonds, others are present in collagen which stabilize the peptide chains, the most important being the hydrogen bond. It is formed as a result of a molecular interaction of donor-acceptor type, where a Brönsted acid A-H is bonded to a base B, i.e., one covalently bound hydrogen atom forms a second bond with another atom. A hydrogen bond is oriented in space and stronger than a bond of the van der Waals type. It is much weaker,

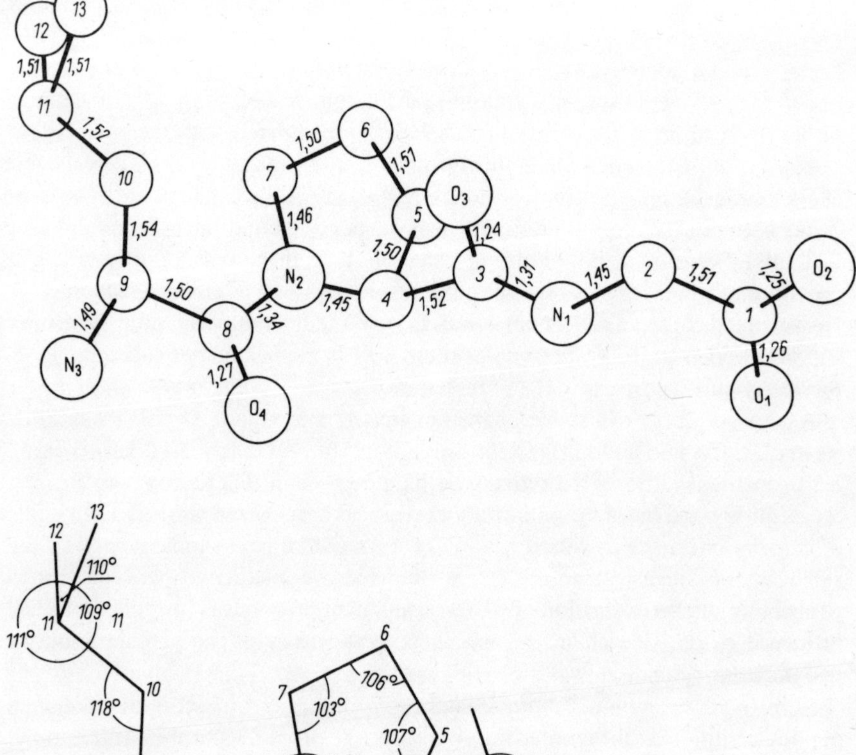

Fig. 2.6. Geometry of the tripeptide L-leucyl-L-prolyl glycine, distances and angles between the atoms. Carbon atoms are identified by numbers. Such groupings occur frequently in the collagen amino acid sequence.

however, than the usual chemical bonds and can be easily split by an increase in temperature or even by the action of an inert solvent. Thus, it plays an important role in biological systems including water; the conformation of biopolymers containing donors and acceptors, for instance, is frequently determined by the formation of splitting of hydrogen bonds.

Hydrogen bonds may be classified according to the nature of the atoms A and B, the structure of their entities, the strength, and the type of interaction.

A hydrogen bond can occur between atoms whose electronegativity is greater than that of hydrogen. Oxygen, nitrogen and halogen (X) atoms are the most frequent partners and may give rise to strong hydrogen bonds, in particular O-H . . . O, O-H . . . N, N-H . . . N and X-H . . . X. Although carbon, phosphorus and sulfur are usually involved in weak bonds, some relatively strong C-H . . . O and N-H . . . S bonds have been observed [28]. A hydrogen bond is called intramolecular when it is formed between groups within a molecule and intermolecular when it involves two or more molecules. Thus, the intermolecular hydrogen bond between the molecules of the same substance leads to self-association.

In order to classify hydrogen bonds according to their strength, we must first recall that the basic group B in the A-H . . . B system represents a supplementary attraction potential for the proton. The potential energy function of the free AH group is thus modified by the potential of the atom B. The proton shifts toward B and the equilibrium A-H distance increases while the A . . . B distance decreases to a value less than that of the van der Waals radii of the A and B atoms.

In addition to the enthalpy ΔH we have two main criteria for estimating the strength of the hydrogen bond: the A-H stretching frequency (in the spectroscopic band characterizing the bond, see chapter 7) or its relative shift $\frac{\nu_o - \nu}{\nu_o}$ (where ν_0 is the frequency of the 'free' A-H group) and the distances (R) A-H and A . . . B, which may be estimated, e.g., by the X-ray diffraction method. According to these criteria, hydrogen bonds may be regarded as weak, intermediate and strong. For the OH . . . O bonds which have been most investigated this approximate classification is as follows:

Hydrogen bond	$\frac{\Delta\nu}{\nu_0}$ %	R O . . . O Å	ΔH kcal/mole	ΔH kJ mole
weak	12	2.7	5	21
intermediate	12-22	2.7-2.6	6-8	25-33
strong	25-83	2.6-2.4	8	33

This classification can be presented as in Fig. 2.7, since the system can form a second potential well which is demonstrated schematically. The type of interaction has its reflection in that figure. If the AH . . . B has a potential curve similar to that in the figure, the bond is strong or moderate. For the $A^- . . . HB^+$ system well II is deeper than well I. Finally, the potential curve may be symmetric. When the potential barrier is small or equal to zero a 'hesitating proton' is involved. Thus we distinguish: an asymmetric double minimum, a symmetric double minimum and a symmetric single minimum with RA-H $= \frac{1}{2}$ RA . . . B (then usually A = B).

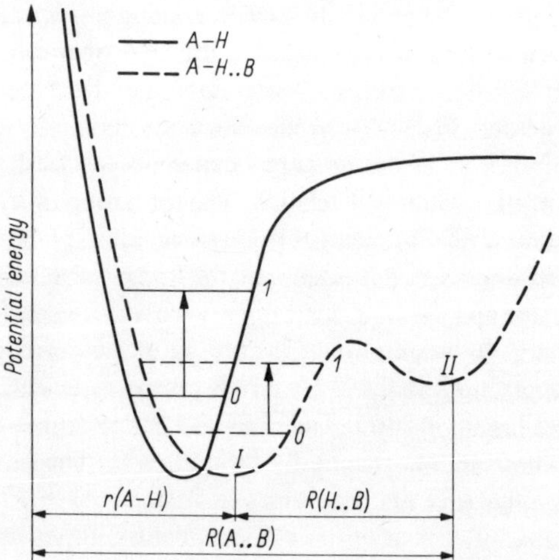

Fig. 2.7. Qualitative potential curves for the free (—) and hydrogen bonded (- - -) AH group.

The length of hydrogen bonds in collagen is approximately 3.0 Å. Among the bonds occurring in collagen the most common is

$$C = O \ldots H - N$$

According to the Rich-Crick's structure model II, this bond connects peptide groups between the oxygen and hydrogen atoms. From analysis of the collagen infrared spectrum, as well as from its circular dichroism spectrum, we see that collagen contains other hydrogen bonds, for instance of the type:

$$C - H \ldots O = C, \quad -N-H \ldots N -$$

or that in which oxygen of the water molecule participates. These weak bonds strongly affect the stabilization of the collagen molecule, and form a dense intramolecular and intermolecular network. They control the solubility and thermal stability of the molecule, influencing its shape and denaturation by precipitation from solution. Thermodynamic calculations suggest another reason of molecule stabilization as the total energy of hydrogen bonds is low. However, this reason has not yet been detected.

Hydrophobic bonds. Hydrocarbons are water-insoluble, and water does not adhere to their surface. Therefore they are defined as hydrophobic substances. Introduction of -COOH, -OH, or -SO_3H groups makes hydrocarbons hydrophilic in the modified part of the molecule. The presence of hydrophilic and hydrophobic groups in one molecule makes it polar. Dipole moments, e.g., in fatty acids, are almost independent of the length of carbon chain.

Earlier, the hydrophobic bond was believed to be formed as a result of action of van der Waals forces between hydrocarbon residues. It was shown, however, that the van der Waals forces account for only 45% of the formation of these bonds (in aqueous solution), and the remaining part of the energy comes from changes in the structure of water surrounding both residues.

The hydrophobic bonds in the collagen molecule are weak and connect hydrophobic groups of the protein side chains. The formation of hydrophobic bonds has been clearly shown by Sheraga [29]*. Hydrophobicity does not necessarily account for the repellence between water and the hydrocarbon chain. Some forces exist that attract water and the hydrocarbon molecules, weaker than the attractive forces between water molecules.

Hydrophobic bonds are formed if the groups of the hydrocarbon type, like the groups in the side chains of valine, leucine and phenylanine, occur very close one to another. They are the result of the action of forces similar to those attracting the molecules (not ions) in crystals. Water is expelled from the region where such a bond is formed. Attraction forces between hydrocarbon chains are due to induced dipoles, since the aliphatic hydrocarbons do not have any permanent dipole moments.

If the flexible molecules containing hydrophilic as well as hydrophobic centers are dissolved in water, they have the tendency to stay in such an order that their hydrophobic centers are directed in towards the molecule center. Such behavior allows to conclude as to the structure of a great, flexible molecule in an aqueous solution. When apolar groups approach each other until they come into direct contact, the amount of water molecules contacting them decreases (Fig. 2.8). This is a process from which the point of energy is the reverse of dissolving.

If 'ice-like' regions occur in solution (see chapter 5) and the hydrocarbon residues are present in the solution, removal of a part of the hydrocarbons is accompanied by 'melting' of the ice-like structures.

The sign of the thermodynamic parameters of hydrophobic bond formation

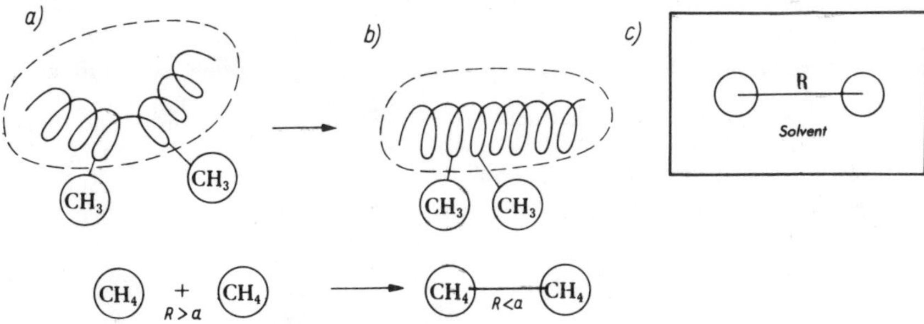

Fig. 2.8. Various phases of hydrophobic bond formation: (1) change of biopolymer conformation, allowing to form a bond, (b) dimerization of two simple groups. Such a pair forms a dimer, if the distance R is smaller than a definite parameter a. In Fig. (c) the conditions are shown, allowing to estimate the free energy of the two-molecule system in a solvent (water) as a function of distance between molecules.

*Cf. a new monograph 'Hydrophobic Interactions' A. Ben Neim, Plenum Press 1980.

is opposite to those of the dissolving process. Thus the factors described above, as well as the van der Waals forces acting between the hydrocarbon chains, contribute to the formation of hydrophobic bonds in aqueous solutions.

Crosslinking bonds and collagen aging. The knowledge of the character and properties of crosslinking bonds is of great importance to the tanning chemistry. The splitting of those bonds increases the solubility of collagen, which decreases the shrinkage temperature and influences many other properties. Increase in the amount of those bonds, which is equivalent to tanning, has an opposite effect. The splitting of the crosslinking bonds, due to, e.g., oxidation, produces typical destructs which pass on to the solution.

According to Siegel [30] the change from the soluble to the insoluble fibrous state is an essential biological requirement for crosslinking. Although many kinds of crosslinking reducible covalent bonds were reported to have been isolated and characterized, only two of them will be discussed here, those which are best known and have been recently characterized. The residues of dehydro-hydroxylysino-norleucine (I)

$$\underset{\underset{\displaystyle NH_2}{|}}{\overset{\overset{\displaystyle COOH}{|}}{CH}} - CH_2 - CH_2 - \overset{\overset{\displaystyle OH}{|}}{CH} - CH_2 - N = CH - CH_2 - CH_2 - CH_2 - \underset{\underset{\displaystyle NH_2}{|}}{\overset{\overset{\displaystyle COOH}{|}}{CH}}$$

and hydroxylysine − 5-keto-norleucine (II)

$$\underset{\underset{\displaystyle NH_2}{|}}{\overset{\overset{\displaystyle COOH}{|}}{CH}} - CH_2 - CH_2 - \overset{\overset{\displaystyle OH}{|}}{CH} - CH_2 - NH - CH_2 - CH_2 - \overset{\overset{\displaystyle O}{\|}}{C} - CH_2 - CH_2 - \underset{\underset{\displaystyle NH_2}{|}}{\overset{\overset{\displaystyle COOH}{|}}{CH}}$$

are typical components of such bonds. The first of these residues occurs mostly in skin, the second in cartilage. The enzyme, catalyzing the crosslinking reaction in question, lysil oxidase, has been isolated and characterized. It catalyzes the formation of lysine aldehydes (α-amino adipic acid -δ-semialdehyde) from lysine in the amino and carboxy terminal nonhelical regions (see below) of the α-chains at positions 9 and 1047 in the sequence, respectively. In the native fibril the carboxy terminal lysine aldehyde aligns, with the hydroxylysine in position 103 and the amino terminal lysine aldehyde, with the hydroxylysine residue in position 946 [31].

Considering the optimal interactions of the adjacent $\alpha 1(I)$ chains Piez find that the molecules align with an axial stagger of 233 residues [32] which is consistent with the well-established quarter-stagger hypothesis. Computer-aided studies utilizing the known amino acid sequence of both the $\alpha 1(I)$ and $\alpha 2$ chains have supported this finding [33]. The labile aldimine I may form a stable ketoimine spontaneously through *in vivo* Amadori rearrangement when both lysines are hydroxylated.

At present the best insight into the general structure of collagen crosslinking is provided by lathyrogens—selective enzyme inhibitors. These components prevent collagen and elastin from crosslinking thereby altering the tensile properties of the tissues containing those two proteins. The phenomenon of lathyrism can be explained as follows: lysine and hydroxylysine residues contained in the molecule are converted into the α-amino (or γ-hydroxy) adipic acid-δ-semi-aldehyde residue by lysil oxidase.

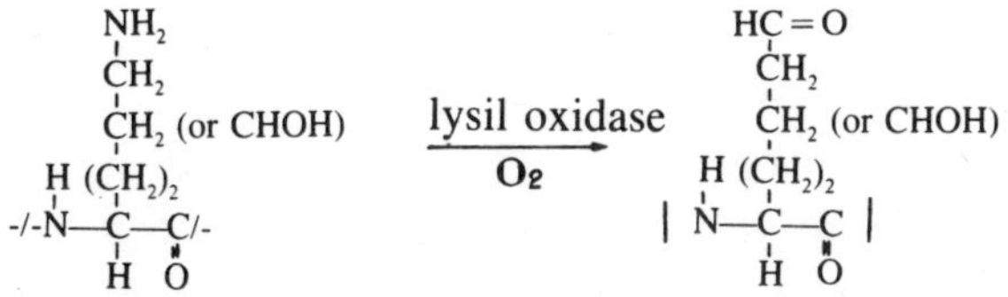

$$
\begin{array}{ccc}
\begin{array}{l}
NH_2 \\
| \\
CH_2 \\
| \\
CH_2 \ (or \ CHOH) \\
| \\
H \ (CH_2)_2 \\
| \quad | \\
\text{-/-N—C—C/-} \\
| \quad | \\
H \quad O
\end{array}
&
\xrightarrow[O_2]{\text{lysil oxidase}}
&
\begin{array}{l}
HC = O \\
| \\
CH_2 \\
| \\
CH_2 \ (or \ CHOH) \\
| \\
H \ (CH_2)_2 \\
| \quad | \\
\text{N—C—C} \\
| \quad | \\
H \quad O
\end{array}
\end{array}
$$

Allysine or hydroxyallysine

The lathyrogen, such as β-aminopropionitrile, specifically inhibits lysil oxidase activity by irreversibly combining with the enzyme [34]. When lysil oxidase is made inactive, the collagen molecules do not undergo enzymatic modification which is necessary for crosslinking. They self-assemble (this occurs with the presence of another mechanism), but cross-linking does not take place.

The reducible crosslinks are stable under physiological conditions and a non-additive modification, such as reduction or oxidation, would be unlikely to enhance their stability or function. Recently, many reports have appeared describing stable crosslinks. However, apart from the original reports very little additional evidence has come forth to support any one of them. Therefore, the real existence of these structures should be considered with caution.

Several compounds have been isolated and identified. They point to a relationship between collagen and elastin in natural conditions which is inferred from the fact that lathyrogenes do interfere with the crosslinking of both proteins (see chapter 4). Among the compounds isolated in addition to the already mentioned lysinonorleucine, hydroxylysinonorleucine and dihydroxylysinonorleucine, we have: hydroxymerodesmosine (I), aldol histidine (II) and histidino-hydroxymero-desmosine (III).

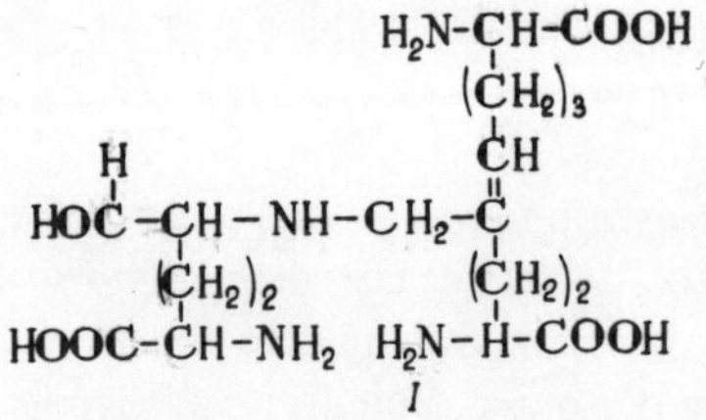

$$
\begin{array}{c}
H_2N-CH-COOH \\
| \\
(CH_2)_3 \\
| \\
CH \\
\| \\
\end{array}
$$

$$
\begin{array}{cc}
H & \\
| & \\
HOC-CH-NH-CH_2-C & \\
| & | \\
(CH_2)_2 & (CH_2)_2 \\
| & | \\
HOOC-CH-NH_2 & H_2N-H-COOH
\end{array}
$$

$$I$$

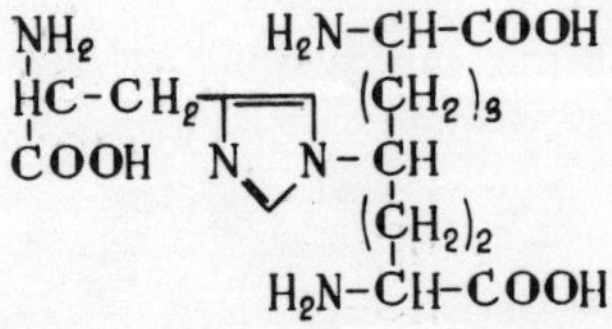

$$
\begin{array}{cc}
NH_2 & H_2N-CH-COOH \\
| & | \\
HC-CH_2 & (CH_2)_3 \\
| & | \\
COOH \quad N \quad N-CH \\
& | \\
& (CH_2)_2 \\
& | \\
& H_2N-CH-COOH
\end{array}
$$

$$II$$

$$
\begin{array}{cc}
NH_2 & H_2N-CH-COOH \\
| & | \\
CH & (CH_2)_3 \\
| & | \\
HOOC \quad N \quad N-CH \\
HOHC-CH_2-NH-CH_2 \quad CH \\
| & | \\
(CH_2)_2 & (CH_2)_2 \\
| & | \\
H_2N-CH-COOH & \\
& H_2N-CH-COOH
\end{array}
$$

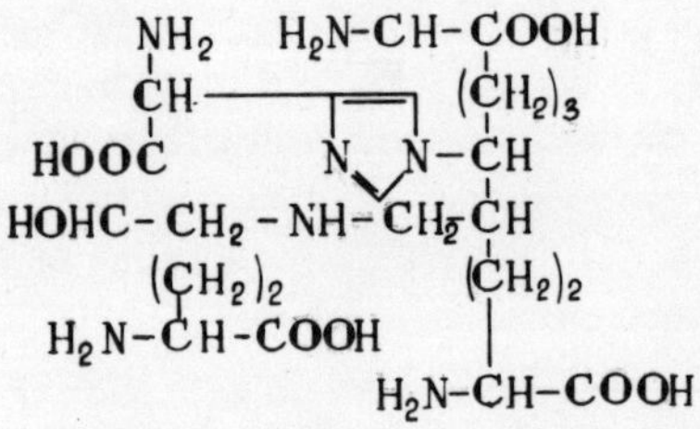

$$III$$

According to Housley et al. [35], the reaction between allysine, hydroxyally-sine and histidine leads to a crosslinking bond in the following reaction:

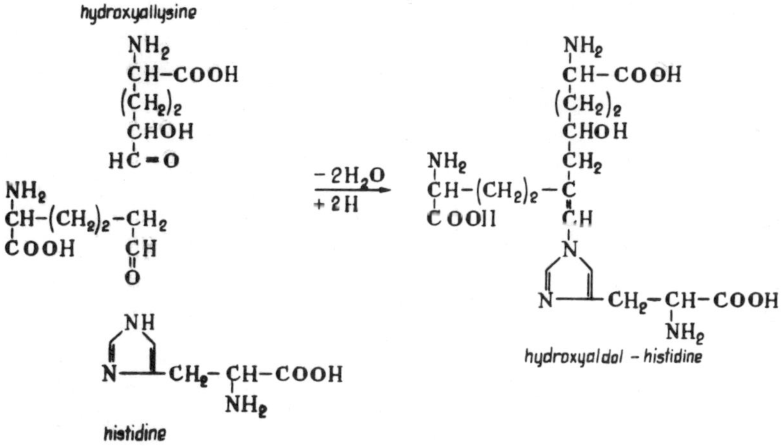

hydroxyallysine

$$NH_2$$
$$CH-COOH$$
$$(CH_2)_2$$
$$CHOH$$
$$HC=O$$

$$NH_2$$
$$CH-(CH_2)_2-CH_2$$
$$COOH \qquad CH$$
$$\parallel$$
$$O$$

$$NH$$
$$N \longrightarrow CH_2- CH-COOH$$
$$NH_2$$

histidine

$$\xrightarrow[+2H]{-2H_2O}$$

$$NH_2$$
$$CH-COOH$$
$$(CH_2)_2$$
$$CHOH$$
$$CH_2$$

$$NH_2$$
$$CH-(CH_2)_2-C$$
$$COOH \qquad CH$$
$$N$$

$$N \longrightarrow CH_2-CH-COOH$$
$$NH_2$$

hydroxyaldol – histidine

Hydroxyaldolhistidine has been isolated from connective tissue, although the site in the molecule (or molecules) where this crosslink occurs has not yet been localized.

The previously supposed existence of an ester-type bond, *via* hexose residue, probably derives from the fact that saccharide units have been found in collagen, which are attached to hydroxylysine by glycosidic linkage in the helical region of the molecule, either as galactosyl-hydroxylysine or as glucosylgalactosyl-

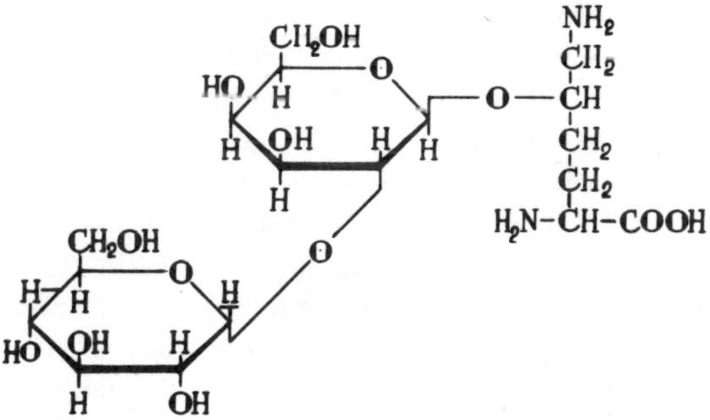

2-O-α-D - glucosyl-O-β-D-galactosylohydroxylysine

hydroxylysine. Type I and type III collagens contain about 0.4% of carbohydrates, and type II contains about 4%. The major sites of glycosylation are those which are involved in the intramolecular crosslink [36]. To date, no experimental evidence has been made available that would demonstrate the function of these carbohydrates. It has been thought that they may regulate the formation of crosslinks and aggregation of collagen molecules into the quarter stagger arrangement. The aspects of crosslinking are closely related to molecule aging. This is, however, a process also involving some genetic problems such as the change in the collagen I to collagen III ratio in skin, already discussed and occurring in tissue as a result of organism growth. In early postnatal tissues the amount of reducible crosslinks is high and decreases as the physical maturity progresses. The stable crosslinks replacing the reducible ones have not yet been determined with full certainty: possible crosslinks of that kind are hydroxyaldol-histidine [35] derived from histidine, hydroxylysine and lysine aldehydes and pyridinoline [37] from two hydroxylysine aldehydes and one lysine aldehyde.

Alterations of the physical and chemical properties of collagen fibers due to aging are very distinct. The fibers become increasingly insoluble, their ability to swell in acid solution decreases and so does the susceptibility to enzyme attack, whereas their mechanical strength and stiffness increases. The stiffness increases through the whole lifetime, creating brittleness which results in the decrease of tensile strength [38]. When artificially introduced crosslinks give rise to more than the optimum number of crosslinks, the connective tissue becomes brittle. The first possible reason for these phenomena is the conversion of aldimine crosslink to stable non-reducible forms. However, the stabilization of those crosslinks would not suffice, so it is thought that either the amino acid residues from the adjacent collagen molecules become involved in the formation of multivalent addition products, or additional divalent crosslinks are formed in new places. Another possibility is oxidation; oxidized collagen has a higher stability as compared with the non-oxidised one [38]. According to Light and Bailey [39], one can imagine a two-stage process, i.e., addition and then oxidation of the multiple bonds. The slow aging process in the collagen matrix may be controlled by the alignment of reducible crosslinks with other reactive residues and by the presence of oxygen in the extracellular fluid. This approach is supported by the observation that on aging, the carboxyterminal part of the molecule becomes involved in an aggregate of high molecular weight. The quoted authors predict a network of stable crosslinks bridging from fibril to fibril and originating from divalent reducible bonds.

Amino acid sequences in collagen

The amino acid sequence in the peptide chain, specific for each protein, determines the spatial shape of this chain and its biological function. The unique shape and properties of the collagen molecule are due to its amino acid composition and sequence.

Investigation of the amino acid sequence in the protein molecule is difficult and time-consuming. The first step is usually to obtain identifiable chain destructs and to estimate their sequence in the whole chain. This is done by partial hydrolysis. After identification and localizing the segments, the latter are examined by the Edman's step-by-step degradation in which every amino acid is separately split off and identified. This reaction has now been automated and can be carried out on chain segments of up to 60 amino acids in one run.

The amino acid composition of collagen was investigated in the years 1953–56, and the occurrence of tripeptides then became recognized, each of which starts with glycine and having other amino or imino acids in the two successive sites. However, no satisfactory information about the amino acid sequence could be obtained.

Partial hydrolysis is carried out in a chemical or enzymatic way. Preference is given to two chemicals, i.e., cyanogen bromide and hydroxylamine, and to enzymes such as collagenase, trypsin, chymotrypsine, pronase, etc. The obtained peptides, containing from several to several hundred amino acids, were then separated by chromatographic methods and identified by, e.g., electron microscope method (see p. 61). Collagen contains small amounts of methionine. Cyanogen bromide splits the peptide bonds, in which methionine carboxyl is participating, and a tetrahydrofurane ring is formed. Thus it can be concluded that the peptide bond in which

methionine carboxyl participates is formally the same as every peptide bond, but it can be changed by the adjacent sulfur atom. Methionine in reaction to cyanogen bromide eventually becomes homoserine—an amino acid with one CH_2 group more than serine:

$$H_3\overset{\oplus}{N}-\underset{\underset{\underset{CH_3}{|}}{\underset{\underset{S}{|}}{\underset{\underset{CH_2}{|}}{\underset{CH_2}{|}}}}{\overset{\overset{\overset{O}{||}}{C-O^{\ominus}}}{C}}-H \longrightarrow H_3\overset{\oplus}{N}-\underset{\underset{\underset{OH}{|}}{\underset{CH_2}{|}}{\underset{CH_2}{|}}}{\overset{\overset{\overset{O}{||}}{C-O^{\ominus}}}{C}}-H$$

Studies under way in several centers, among which the Max Planck Institute for Biochemistry in Munich is perhaps the most active, lead to complete recognition of amino acid sequences in $\alpha 1$ and $\alpha 2$ chains of calf and rat skin, and to almost complete recognition of the pig skin collagen amino acid sequence. Cyanogen bromide splits the $\alpha 1(I)$ chain into eight peptides of varying length; the $\alpha 2$ chain splits into five peptides. These peptides are denoted as CB peptides. Hydroxylamine splits the peptide bond between asparagine and glycine. This bond occurs in the $\alpha 1(I)$ chain five times, so six peptides are obtained. Occurrence of these bonds at sites other than those split by cyanogen bromide allows us to compare the positions of peptides due to cross-striation, observable under an electron microscope after staining.

Limited collagenase action leads to digestion of the molecule from the ends towards the center, and permits an isolation of a series of helical fragments. The fragments thus liberated were further separated by chromatography on CM-cellulose. Those containing methionine can be further cleaved by cyanogen bromide.

Today new details about amino acid sequences in collagen are brought in almost every issue of biochemical journals. It should be emphasized that collagen is at present the greatest protein molecule of a known sequence. Details regarding this sequence are given in monographs devoted to collagen biochemistry [2] or in surveys, e.g., [32]. By generalizing, we can describe the discussed sequence as follows:

1. The collagen α-chain consists of a central helical part containing 1011–1047 amino acid residues of which every third must be glycine, and of two non-helical end parts in which glycine occurs only in several irregular positions (the end parts were previously called *telopeptides*). The length of the end parts amounts to 6—25 amino acid residues.

2. The helical part contains about 20% iminoacids in the second or third positions, if we divide the molecule into tripeptides, each of which starts with glycine (-G-X-Y). In mammal collagens about ⅔ of the iminoacids are hydroxylated and are always in the Y-position (4-hydroxyproline). The only exception is 3-hydroxyproline which occurs in the X position however once or twice in the chain only.

3. The non-helical extensions are relatively rich in hydrophobic amino acids and

contain a lysine residue which can be enzymatically oxidized and serves as a functional group for the formation of intra- and intermolecular crosslinks. The amino terminal extensions (16–17 residues in α1 and 9–11 in α2) all start with a pyroglutamic acid residue probably formed during the enzymatic removal of the procollagen peptides. The carboxyterminal extensions (25 residues in α1 and 6–8 residues in α2) have not yet been thoroughly investigated, however, it appears that the same general comments can be applied here.

4. Hydroxylysine is the amino acid occurring exclusively in collagen. It is the only amino acid glycosylated at several sites, but not every residue in chain. Lysine, like proline, is hydroxylated only when it is in the Y-position.

5. The average content of proline plus hydroxyproline is equal throughout the chain, except for the C-terminal, which terminates with five consecutive tripeptides Gly-Pro-Hyp. This suggests an exceptional stability of the C-terminal helical region of the molecule.

Throughout, the chain regions occur with an imino acid content lower than average, and thus they should be susceptible to proteolytic attack. Investigations of the effect of proteases on collagen did not reveal the presence of such spots in the rigid rod of the collagen molecule. No position in the central part of the molecule is susceptible to proteolytic attack of pronase, pepsin or trypsin.

The structure of the collagen molecule, which will be described now, is based on the above findings. Nothing is known as yet about sites where artificial crosslinks appear as a result of tanning.*

Conformation of the collagen chain

The collagen molecule has been investigated by X-ray diffractometry since 1938, when the first attempts at the interpretation of the diffractograms were made. The Pauling-Corey's proposal of helix structure and the Fourier's analysis of the diffractograms suggested by Cochran deserve special mention. (The latter analysis is still being developed, and important papers have been devoted to it in 1979.) Recognition of the helical collagen structure comes from diffractogram analysis as well as from structural calculations.

Calculation of the structure requires a mathematical and crystallographic background. As the latter cannot be presented here, the reader is referred to Vol. 1 of *Collagen*, edited by Ramachandran [4]. Only a brief information is given here. The repetitive sequence in the collagen molecule has a helical conformation, so the repetitive segment is called the helical region.

The discontinuous helix consists of an infinite set of points, lying on a screw line and separated by a constant axial translation. According to the IUPAC-IUB, Commission on Biochemical Nomenclature (1970), this translation is denoted as h (unit height) and the angular separation t is termed the unit twist, and where

*See note p.66.

r_o is the radius of the helix. The whole pitch P is given by the expression:

$$P = 2\pi\frac{h}{t}$$

$\frac{P}{h}$ may be expressed as the rational fraction $\frac{n}{v}$, which means that the discontinuous helix has n points in v turns. The number of points N per turn is found from the expression:

$$N = \frac{2\pi}{r} = \frac{P}{n} = \frac{n}{v}$$

N being negative for the left-hand helix. The above values have been estimated repeatedly by many authors, starting with the calculation of the amino acid sequence and/or by X-ray diffractograms. The recent results, obtained in 1979 by Fraser et al. [40], are:

$$h = 2.98 \text{ Å}$$
$$|t| = 107°$$
$$|N| = 3.36$$

It should be noted that these results are obtained for the stretched rat tail tendon at 75% relative humidity. They can vary slightly, however, e.g., Ramachandran [4] obtained for unstretched collagen: h = 2.91, |t| = 111°, |N| = 3.25. Up until today, for technical reasons, no satisfactory estimate has been found for native hide collagen, though the obtained data are sure to be close to the above estimates. (Fig. 2.9)

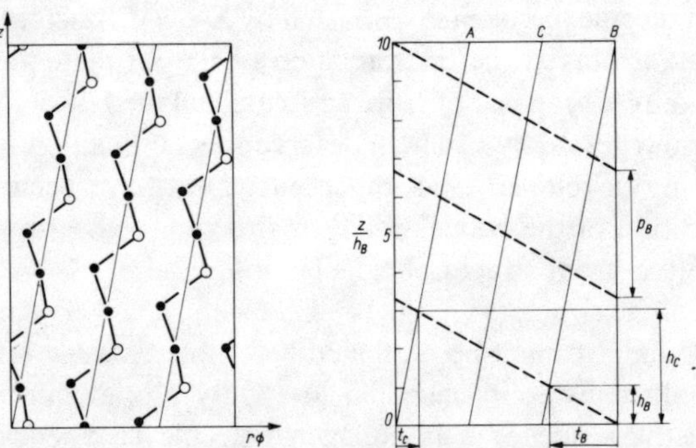

Fig. 2.9. Radial projections of the helical conformation of the collagen molecule. In the left-hand projection, the positions of the α-carbon atoms are shown. The individual chains have a left-handed screw sense and the three chains have a right-handed rope twist. In the right-hand projection, the dots represent equivalent points in each (Gly-X-Y) unit and the basic helix (broken line) and individual chain helices (full lines) are identified. The chains are identified by the letters A, B and C.

In this figure Z, r and φ are cylindrical polar coordinates, P is the z displacement per complete turn of the helix (pitch), t angular separation and h unit height of translation. According to [7 ch. 1] with permission

The values given above are very close to those observed in the synthetic polytripeptide (Gly-Pro-Pro)$_n$, viz. h = 2.87 Å, |t| = 108° and |N| = 3.33. So the influence of proline rings on collagen structure has been experimentally confirmed. The non-integer number of residues in one turn could not be explained until the Ramachandran and Kartha's [41] suggestion was accepted which states that the molecule has the form of a three-strand rope in which the individual chains have a left-hand helical conformation and the three chains are twisted around a common axis with a right-hand rope twist (Fig. 2.10). In this model,

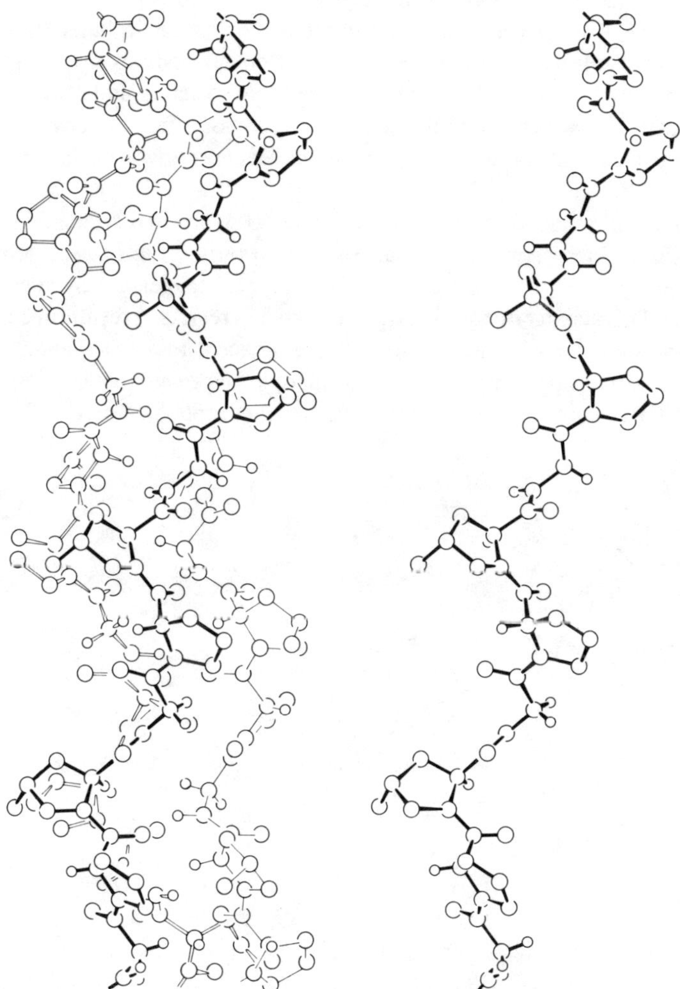

Fig. 2.10. Model for the structure of the collagen molecule obtained by refinement of the Rich-Crick structure II using the linked-atom least-squares refinement procedure. For clarity, hydrogen atoms not attached to the main chain have been omitted. The complete model is shown on the left and a single strand on the right. According to [7 ch. 1] with permission

two hydrogen bonds per tripeptide unit have been accepted, whereas in the structure, proposed by Rich and Crick [1, 4] (Rich-Crick's structure II), only one was formed. This was a matter of discussion for more than 20 years since it was very difficult to find an experimental proof for one of the two proposed concepts. Instead, a compromise proposal was forwarded by Ramachandran and Chandrasekharan [3] who suggest that ''collagen has a one bonded structure which contains water bridges'' (Fig. 2.11). Recently, Rich-Crick's structure II was found to be the most acceptable one, which has recently been slightly corrected by the least-squares method (Fraser et al., l.c. 1979). If a model with $t = 108°$, $N = -10/3$ is accepted, the pitch of the chain helix P_c is exactly 30 unit heights of the basic helix (86 Å long). Some deviations from strict helical symmetry may be expected in view of the varying structure and character of the side chains. The water bound to the chains does not affect the chain symmetry if it is accepted that more than one water molecule is involved in a bridge, as shown in Fig. 2.11.

Evolution and analysis of the sequence in the $\alpha 1(I)$ and in $\alpha 2$ -chains connected with electron microscopic investigations carried out by Fietzek and Kühn [36] have given a strong support of the axial quarter-stagger arrangement of the molecule. This stagger of 650 Å (distance D = 233 residues) results in a maximal interaction of polar sequence regions of the adjacent molecules and, likewise, of regions of hydrophobic residues. Thus it could be proved that such an ordered aggregation is controlled by electrostatic and hydrophobic forces. The quoted

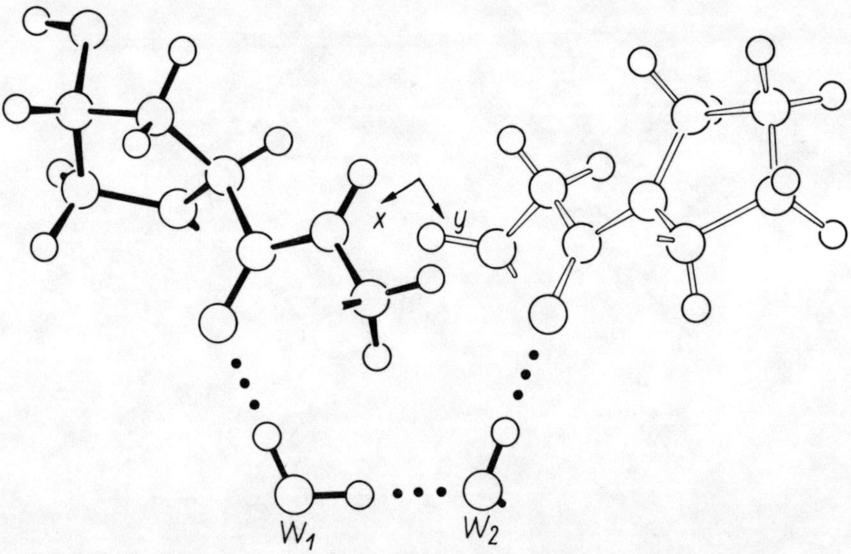

Fig. 2.11. Possible location of fixed water molecules forming a bridge between the C = O groups of the y residue and the glycine residue. According to [7 ch. 1] with permission

investigations were supported and continued by Hofmann et al. [42] who extended the calculations to encompass the whole $[(\alpha 1/I)]_2(\alpha 2)$ molecule, and the molecular packing of type I collagen. According to the quoted authors, the best fit with the highest interaction scores was found again at a triple-helical pitch of the 30 residues. The azimuthal angle of 90–108° between these interactions taking place on edges, suggested a five-stranded left-hand microfibril as being the lowest packing unit.

Aggregation of the molecules into fibrils is a process of self-assembly controlled by the amino acid sequence of the molecule. The controlling function is predominantly attributed to the changed polar amino acid residues of lysine, arginine, glutamic and aspartic acids. Hydrophobic amino acids, such as valine, leucine, methionine, phenylalanine and tyrosine, might also be expected to participate in the molecule aggregation. Collagen molecules are soluble in cold neutral salt solutions, since the electrostatic forces are not sufficient enough under these conditions to cause aggregation. Upon heating to 37°C hydrophobic interaction increases, resulting in aggregation of the molecules into fibrils.

Considering the collagen structure in these terms, the problem arises as to the number of and positioning of hydrogen bonds as well as that of the location of water molecules. The hydrophobic clusters do not participate in this, so the number of water hydrogen bonds and water molecules at the hydrophilic sites must be increased and exceed the previously accepted number, i.e., two per tripeptide. However, this problem needs further explanation. To date, the described collagen structure fulfills the stereochemical requirements as well as the experimental ones resulting from X-ray diffraction, circular dichroism and infrared spectroscopy. The question of how the triple helices are aligned will be dealt with in the section concerned with supramolecular structure. The stability and other properties of the triple helix were investigated in terms of molecule thermodynamics.

Many authors have approached the question of energetics of the collagen molecule through investigation of its thermal stability and denaturation thermodynamics (cf. [43]). However, the results obtained have lead to divergent conclusions. A recent paper by Privalov et al. [44] is based on a calorimetric study, and the use of isotope tracers and infrared determination of the H $\rightarrow$ D exchange. The aim of the quoted work had been the estimation of the enthalpy ΔH and entropy ΔS of the structure disruption. The structure stability is determined by the Gibbs energy function:

$$\Delta G(T) = \Delta H(T) - T\Delta S(T)$$

The quoted authors observed a dependence of the denaturation enthalpy on the temperature, as it was shown for globular proteins. It is generally assumed that this dependence is a result of the contribution of hydrophobic interactions to the stabilization of structure. The enthalpy dependence on temperature can be measured, since according to the Kirchhoff law

$$\frac{d\Delta H}{dT} = \Delta C_p$$

By investigating the enthalpies and entropies of various collagens (fish, frog, rat tail tendon) the quoted authors were able to demonstrate that the values obtained were not due to temperature dependences of those functions, but rather to the structural differences of the considered molecular systems resulting from their different chemical compositions, i.e., on the Pro-Hyp ratio. The transition enthalpy and Gibbs energy of stabilization were found from the relations

$$\Delta S_{tr} = \frac{\Delta H_{tr}}{T_{tr}}$$

$$\Delta G_{Tr} = \Delta H_N \left(\frac{T_{tr} - T}{T_{tr}} \right)$$

The best correlation was obtained between the total imino acid content and the enthalpy of collagen transition and the much poorer correlation for Gibbs energy. Thus the quoted authors concluded that this was due to the crosslinking influencing the unfolding entropy only, but not the enthalpy. This cannot be explained by the van der Waals interactions of molecules, since these links cannot change greatly with iminoacid content. No correlation has been detected between the enthalpy and the non-polar groups content.

Two kinds of exchanging hydrogens were found in collagen—fast and slow exchanging ones (below and over 1 hour)—and this was done by IR and isotope tracers investigations. The slow exchangeable hydrogens (participating in hydrogen bonds) extrapolated to zero-time gave the value of 1.7 ± 0.1 per tripeptide unit. Thus the quoted authors concluded that the amount of stable hydrogen bonds was the same for all collagens tested and that all these bonds were connected with the NH and CO peptide groups. Those authors further classified the slow-exchanging hydrogens into two groups and proposed a two-stage denaturation model, first, of micro-unfolding and second, of macro-unfolding. (Micro-unfolding is believed to be a non-denaturating process occurring in a tripeptide unit.) The denaturation process involves over 30 residues. The micro-process has a Gibbs energy of the order of 7-11 kJ/mole, whereas the macro-process—an energy of 200—400 kJ/mole. The total values for ΔH were found to be 4000—6500 kJ/mole and ΔS 14—21 kJ/mole. The effect of water on stabilizing the collagen structure has not yet been elucidated, however, it has been accepted as the only means of explaining the surprisingly high observed values of collagen denaturation enthalpy and entropy.

The relation between thermal stability of the collagen structure and its functioning in the organism was recently discussed by Tiktopulo et al. [45]. According to the authors, the natural selection is secured by correlation between the conformational mobility level of macromolecules and the thermal ecology of the organism. In their argumentation the authors assume that the correlation between thermal ecology and thermal stability is but of secondary significance.

The term 'mobility' is more and more often used in the literature. Its source is, however, undefined and lacks quantitative treatment. Frequently, hydrogen exchange has been accepted as one of the factors in the mobility of the protein structure. In fact, for this exchange, an opening of the structure is necessary; thus, it may be related to the thermodynamic characteristic.

According to the present author's opinion, the mobility of the macromolecular structure unit includes another factor, i.e., the motions of side chains due to the influence of its surroundings. Perhaps the example of the swimmer's hair spreading in water may best illustrate this phenomenon. If this motion facilitates the exchange of hydrogen it is of minor importance as that which is compared with the possibility of contact with other molecules present in the solution. This kind of mobility, demonstrated e.g., by Torchia and Vander Hart [46] in the ^{13}C NMR experiment is surely important in the processes in which contact of the side chain polar groups with surrounding molecules is essential, like in crosslinking through tanning with chromium salts (cf., section 7.7).

The data of Privalov et al. [44], though supported by convincing experiments, are in disagreement with some other ones in which the significance of hydroxyproline alone for collagen structure stabilization is emphasized. It seems impossible to answer this question without excluding Hypro hydroxyls due to their blockage or decomposition without damaging the molecule itself and without engaging NH_2-groups in the side chains in the process. As of yet we do not know such a reaction. The significance of Hyp will be considered again in Chapter 10.

The influence of additional crosslinking factors (chromium salts, formaldehyde, 1,5-difluoro-2,4-dinitrobenzene and its derivatives) on the thermodynamic properties of collagen characterising the helix stability has not yet been investigated. However, from this work one can conclude that collagen tanning should give a drastic change of the entropy of the system due to the action of crosslinking agents, since it is already known from previous papers that the ΔH values are not significantly changed by tanning, whereas T_S does change.

Considering that NH_3 liberation during liming Heidemann and Deselnicu [47] have come to the conclusion that asparagine does participate in the reaction, and that a ring is temporarily formed as it is in succinimide:

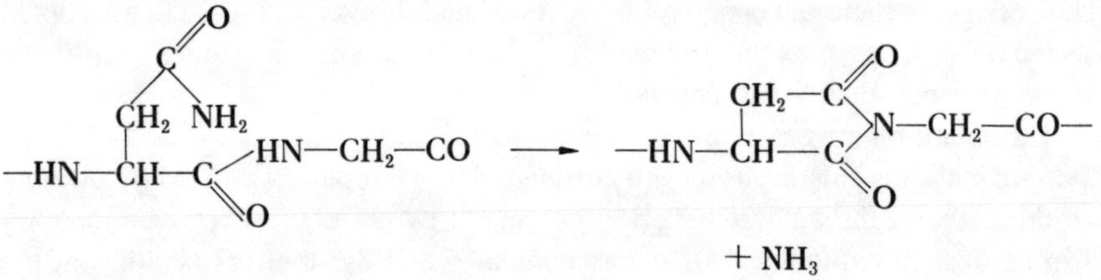

$$CH_2\!-\!C(\!=\!O)\cdots \quad -HN\!-\!CH\!-\!C(\!=\!O) \cdot HN\!-\!CH_2\!-\!CO\!- \longrightarrow -HN\!-\!CH\!-\!C \underset{\displaystyle O}{\overset{\displaystyle O}{\big\langle}} N\!-\!CH_2\!-\!CO\!- \;+\; NH_3$$

This ring is split during further hydrolysis into α-and β-peptides:

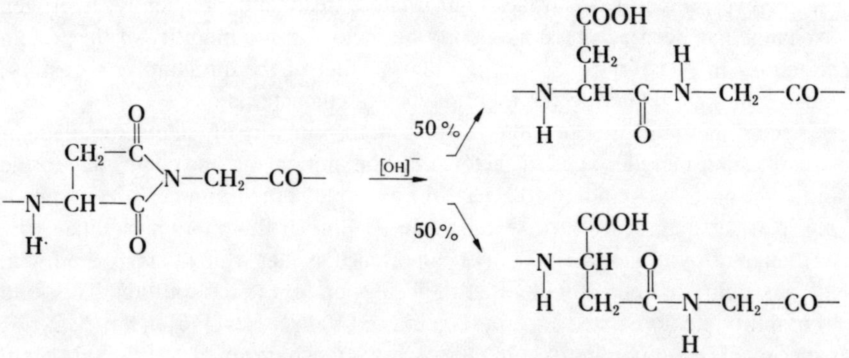

$$\underset{\overset{|}{H\cdot}}{\overset{}{-N}}\!-\!CH\!-\!C\!\Big\langle^{O}_{\;} \;N\!-\!CH_2\!-\!CO\!-\;\;\xrightarrow{[OH]^-}\;\;$$

50% →

$$-N\!-\!CH\!-\!C\!-\!N\!-\!CH_2\!-\!CO\!-$$
(COOH, CH$_2$, H, O substituents)

50% ↘

$$-N\!-\!CH\!-\!CH_2\!-\!C\!-\!N\!-\!CH_2\!-\!CO\!-$$
(COOH, H substituents)

The quoted authors have shown that in the industrial process, 1-2 molecules of ammonia are split from among 1000 amino acid residues in the collagen molecule. The importance of the reaction described above lies in the formation of a β-peptide together with the α-peptide (i.e. the peptide group is attached to the β-carbon), thus a change in the conformation of the peptide bond occurs at the reaction site. Such a change results in the extension of one of the peptide bonds, and thus in the distortion of the structure, as the chain is immobilized in a greater aggregate (Fig. 2.12).

In his paper on thermal stability of collagen, Heidemann [48] emphasized that the denaturation temperature of collagen is somewhat higher than the temperature of the body. It seems that this difference cannot be too great as that would hinder the metabolism. An example speaking in favor of this view is the denaturation temperature of collagen from liver trematode, causing distomiasis (fluke disease). Its T_D is dependent on the temperature of the host organism. Thus the denaturation temperature may be independent of the degree of genetic development, and be the result of adaptation to environmental conditions. It should be recalled here that we speak of denaturation temperature (or melting temperature, which in this case is not a quite exact expression) T_D or T_m, if the collagen solution is investigated. Shrinkage temperature T_S is the point of thermal transition in the solid state of the collagenous material. As a rule both changes are irreversible, unless it is indicated otherwise and the method of progression is defined.

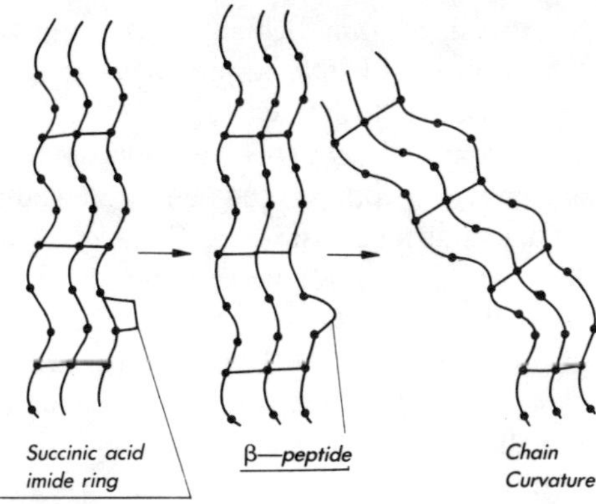

Fig. 2.12. Conformation change of the peptide chain giving distortion of the structure. According to [47]

Supramolecular structure and properties of collagen aggregates (fibers)

The arrangement of triple helices into greater aggregates was extensively studied by X-ray diffraction under low and medium angles and using an electron microscope. The interpretation of X-ray reflexes and of cross-striation of fibers treated with heavy metal salts as observed under × 30-300,000 magnification, was discussed. The conclusions resulting from the research on amino acid sequences as well as from dissolving, reprecipitation and repeated observation under an electron microscope, led to the view that the bundles of collagen molecules were ordered longitudinally with a regular quarter-staggered overlapping. The ends of the chains were separated by an interval of about 40 Å. The repeating distance in the electron microscope striation (D) was reported to be 640-670 Å, which corresponds to 233–234 residues. Until very recently it was believed that the triple chains formed a 'supercoil' consisting of 4 or 5 triple helices; Hulmes and Miller [49] again thoroughly analyzing the X-ray data collected by previous investigators came to the conclusion that such 'ropes' do not exist, and prepared a new model based on a quasi-hexagonal molecular packing without microfibrillar sub-structures and, hence, having the character of a molecular crystal [50]. According to these authors such sub-structures did not result from the X-ray data. Thus the supramolecular structure of an ensamble of collagen fibrils may be considered as crystalline, but only in terms of certain features. In section 7.1 the difference between a regular crystal and a crystalline (or quasi-crystalline) structure of a biopolymer will be explained.

In the organism collagen is a solid phase, dissolving without hydrolysis. The difference between the soluble (i.e., a biologically young) and insoluble (mature) form lies in number and kind to the crosslinking bonds it contains, as discussed above (see p. 48). Some collagens, particularly those from young animals and fetuses, may be extracted in their native form. The purpose of such a process is to obtain pure collagen in a strictly defined way. Dissolution, when the process parameters (temperature, ionic strength, pH, reaction time) are constant, leads to a viscous solution from which collagen can be precipitated by changing these parameters, e.g., after dialyzing it against water. Collagen solutions are unstable; after some time, the fibrils of native type forming aggregates do precipitate spontaneously. Even small changes in the precipitation conditions lead to products markedly differing in their properties. This allows us to characterize collagen by testing the particles precipitated from solution. A condition of such a characteristic is to test whether crosslinking or other nonpeptide bonds have been formed during preparation.

Multipurpose investigations have been carried out on collagen solubility, transformation of the insoluble into the soluble form and on the properties of the solutions. These investigations have for aim, e.g., structure and sequence investigations, and the obtaining of 'renatured' collagen fibers. Investigations on collagen solubility have been started in the Soviet Union in 1947 by Orekhovich and Tustanovskii. These authors have observed that collagen dissolved in buffers, e.g., citrate buffers at pH 3.0-4.5, precipitates in the form of fibrils when dialyzed against water. This preparation has been given the name of procollagen (it should not be confused with procollagen, discovered later in collagen biosynthesis) which suggests its being a collagen precursor. Gross et al. [51] have introduced the term tropocollagen. According to later papers tropocollagen is just a collagen molecule monomer. Now this expression is also related to other soluble collagen forms. It is recommended to use more general terms, such as soluble collagen or fibrous collagen. Among those well characterized forms we have acid soluble collagen and neutral salt soluble collagen.

Acid soluble collagen is usually obtained from disintegrated tissue. To remove protein and polysaccharide impurities, cold extraction with 0.3 molar disodium phosphate, 0.5 molar sodium acetate of 10% NaCl, is applied. Usually 0.5 molar acetic acid is used for collagen solubilization. The yield of acid soluble collagen increases with the decrease of the extraction process pH—probably due to the increase of swelling. Neutral salt soluble collagen is obtained at pH $\approx$ 9. Based on isotope-tracers investigations it was found that this is a specific collagen form. The yield of soluble collagens not only changes with the process conditions but also with the state of live tissue, animal age, etc. The purity of these collagens is determined by:

1. Checking the degree of removal of impurities (presence or absence of noncollagenous functional groups),
2. Determining the known collagen components such as hydroxyproline or glycine.

Collagen is precipitated from a solution of pH $\approx$ 7 at ionic strength 0.25, and from an acidic solution at a low ionic strength. Solubilization is also temperature dependent: Collagen fibrils precipitate from solutions heated to a temperature normal for the organism. In collagen purification various combinations of pH, temperature and ionic strength are applied. The obtaining of standardized purified collagen is described in detail by Rubin et al. [52].

According to Williams et al. [53] it is optimal for collagen fibril formation that the solvent contains 30 mmole phosphate, 30 mmole N [tris(hydroxymethyl-methyl-2-amino)] ethanesulfonic acid and NaCl to give an ionic strength of 0.225 at pH   7.3, the temperature ranges between 20 and 30°C and collagen concentration between 0.02 and 0.5 mg/l. Addition of phosphate is required to obtain fibrils of a native order.

It is difficult to find an experimental method that has not been used for obtaining information about collagen. In order to investigate supramolecular structures consisting of parallel ordered fibers, it is most advantageous to use an electron microscope. The band patterns visible after appropriate staining are characteristic, and are of great value for identification purposes. Observed for the first time in 1942, they became the basic tool of experimentation on the degree of structure ordering in collagen. Properly stained (contrasted), collagen fibrils will show a regular and reproducible band pattern. When the electron microscope experiment is conducted by the 'replique' technique, one can see that the band pattern consists of two macroregions differing in fibril thickness by about 15%. A detailed checking allows us to detect up to 16 bonds, repeated periodically, in dependence on the resolving power of the apparatus and on the applied staining method. In this method 'negative' staining is applied, i.e., in a medium of pH = 8, using substances that are not tanning agents for the sample but contain heavy metals, e.g., phosphotungstates. In such a case a similar reaction to the formation of a photographic negative occurs. Compounds containing heavy metals penetrate the looser regions easier than the more compact ones. The looser ones are stained deeper, as they contain more heavy metals, and thus they are less transparent for the electron beam. The positive (electron) staining is based on the tanning action of heavy metal salts on collagen. Among these salts we have Cr^{3+} salts, uranyl acetate, osmium tetroxide and phospho-tungstic acid at pH $\approx$ 5. Cross-striation makes it possible to identify the fibrils, as any change in their structure changes the band pattern in electron microscope picture, so it is considered to be the basic criterion of comparison. Those fibrils that have a typical band pattern are considered to be native. Definite operations done on collagen result in appropriate changes in the fibril band pattern. The identity period of this pattern is about 650 Å, and more precisely, it varies from 640 to 700 Å. This value is in accordance with the data obtained using low-angle X-ray diffractometers. It is specific for native collagen.

The occurrence of the band pattern is explained by the overweight of positively charged side chains (arginine, lysine, hydroxylysine, histidine) in some regions of the molecule. In those regions ions of opposite charge are adsorbed, e.g., the arginine residue combines more phosphotungstic acid than the anionic chrome complexes. In regions consisting of polar amino acids where negative charges prevail, compounds of cationic character are bound easier, making the fibrils stain darker at this site. Distribution of charges in collagen was tested by Borasky [54], who used cationic and anionic chromium complexes; he proposed identification of the regions (Fig. 2.13). Similar results were later obtained. Recently McLachlan [55] has investigated the dependence of many features of the molecule on its periodicity. In his work he availed himself of the knowledge of the amino acid sequence and the calculated correlations.

The apolar parts of the molecule have a linear distribution. They are located in groups, usually separated by D/5, D/6 and D/11 periods. The acidic and basic residues have a periodicity of D/6. These periodicities, pointed out by McLachlan [55] and Hulmes et al. [56], have confirmed the previous finding (of Hulmes et al. [32]) that a significant number of charged residues occur as base-acid pairs. This may be an important factor for skin tanning, and we are still awaiting the results. Instructive pictures have been taken under an ultramicroscope that concern details of the cross-striation pattern (Bruns [57]).

Taking into consideration that in the range of 640 Å, 16 dark bands are observed with the total number of sectors being 32, their average length is 20 Å. As it follows from collagen X-ray diffractograms, a 2.9 Å chain length is occupied by one amino acid, thus on the average we can say that one sector consists of seven amino acids. Such groupings give band components. Densitometric examination of the photographs of native fibrils leads to the conclusion that overlapping occurs due to which the neighboring elements are shifted by ¼ D.

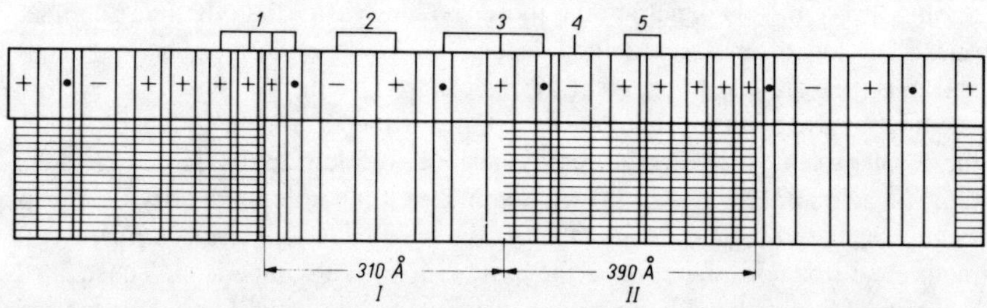

Fig. 2.13. Distribution of negatively and positively charged regions. Then on striated regions I are the regions of parallel fibrils. Striated regions II correspond to fiber overlapping. The figures correspond to bonds occurring when samples are treated with 1 - 2 × 10^{-2} m Mo_2Ac_2 (equilibrium staining), 2 - 1 × 10^{-5} m UO_2Ac_2 (cationic staining), 3 - 2 × 10^{-2} m UO_2Ac_2 (cationic-anionic staining, which occurs in the electroforesis-treated sample at the anode), 4 - 2 × 10^{-2} m UO_2Ac_2 (equilibrium staining at the cathode), 5 - 0.5% phosphotungstic acid, pH 7.5 (negative staining). Equilibrium staining-means that neutral $UO_2(Ac)_2$ complexes are used. The nonpolar regions remain then unstained. According to [54]

Intermolecular bonds may be classified into:

(1) Side-to-side bond, connecting the overlapping molecules shifted by ¼ of their length (quarter staggered). This bond is easily split by pepsin.

(2) Head-to-tail bond in which two different molecule ends participate. This bond is fairly resistant to pepsin and lies also in the molecules overlapping range.

(3) End-to-end bond, connecting two molecules by their (identical or similar) ends, located in their last residues and are somewhat pepsin-sensitive.

Zimmermann et al. [58] have investigated these bonds, finding proof of their existence. Only the molecules connected with the third way described can be extracted from collagen. Mild pepsin action splits bonds of the first and third type, leaving only these of the 2nd type. The quoted investigators have shown that all three kinds of bonds are really present. In Fig. 2.14 a two-dimensional picture of helical, crosslinked regions in the collagen molecule shows results from quarter-staggering. The four helical regions, probably connected with the C-end nonhelical region, have been denoted by C_1 to C_4, whereas the regions with probable crosslinking to the N-end helical region have been denoted by N_1 to N_4.

Within the microfibril only head-to-tail bonds are possible. Between fibrils side-to-side bonds and head-to-tail bonds can occur. Probably, the N-terminal nonhelical region is involved predominantly in the formation of head-to-tail bonds, whereas the longer C-terminal nonhelical region is supposed to establish interfibrillar bonds. This region, due to the aggregation of molecules into fibrils, will always be surrounded by clusters of hydrophobic residues present at the adjacent molecules, since it is hydrophobic itself.

Collagen fibrils can be precipitated in a way that allows us to obtain preparations with other, but strictly determined, regions of bright and dark bands in

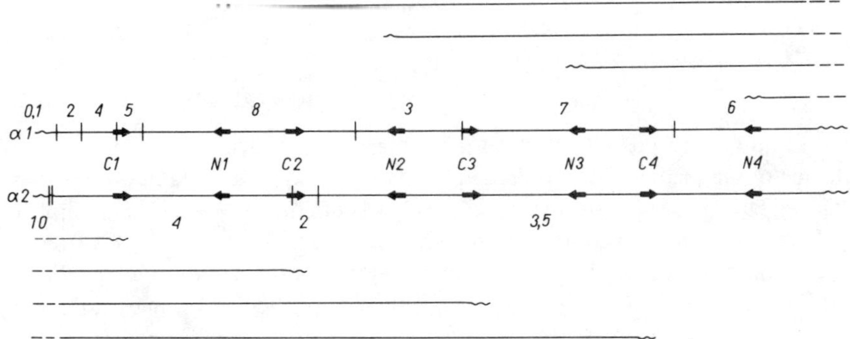

Fig. 2.14. Two-dimensional picture of helical, crosslinked structures obtained by overlapping of fibrils, shifted by ¼. The figures relate to peptides obtained from the α1 and α2 chain by cyanogen bromide action.

the electron microscope picture. Such experiments are important from the points of view of recovery (reconstitution) of collagen, and of its supramolecular structure as well.

Observing the supramolecular structure of the various shapes of collagen obtained by dissolving and precipitation, Mikhailov [5] advanced a theory which states that: The surface of collagen fibrils differs from their internal parts in regards to the glycosaminoglycan binding. This is almost clearly visible in the narrowed end fragments of the molecule. At the edges of these regions glycosaminoglycan rings occur with water attached to them. As a result of irradiation and heating under the electron microscope, water is removed and the rings flatten. The significance of those rings, containing up to 98% of water, is that the molecules are prevented from sticking together. However, this theory has yet to be confirmed. It is based, however, on the observation that the sticking together of the molecules will eventually occur under conditions when the interaction of glycosaminoglycan and collagen is excluded during precipitation. Observing a cross section of the collagen fibril, one can usually see its elements at a distance from one another. These distances disappear when hyaluronidase action takes place, enzyme that attacks glucosaminoglycan.

The supramolecular structure, which can be observed under an optical microscope and has previously been defined as a structure of the fifth order, was discussed when the topic of skin function as connective tissue was discussed (1.2).

Semiconductivity of proteins

Along a peptide chain, consisting of intermittent, resonating units

$$\begin{matrix} & O & \\ & \| & \\ - & C - N & - \\ & | & \\ & H & \end{matrix}$$

and saturated carbons C_α, electron delocalization may occur, e.g., due to the change of the medium pH. Hydrogen bonds may also participate in the electron delocalization and be instrumental in the effects of conjugation. A huge number of factors may participate in this conjugation.

General problems of electron delocalization in the protein molecule have been considered by Szent-Gyorgyi [59]. The system of peptide bonds may give, due to delocalization, energetic bands which are typical for semiconductors.

Eley and Spivey [60] have found that electric conductivity of dry proteins is temperature-dependent. Their specific conductance x is given by the formula:

$$x = x_o \exp - (\Delta E/2KT) \; \text{ohm}^{-1} \cdot \text{cm}^{-1}$$

where x_o is a constant, K is the Boltzman constant, T is the absolute temperature, and ΔE is the electron excitation energy necessary to transfer it to the conductance band. Specific conductances of some proteins are:

Hemoglobin   2.75 eV   Collagen   0.90 eV
Polyglycine   3.12 eV   Myosin   1.75 eV

It seems that the testing of collagen conductivity may be very useful for the future of leather making.

REFERENCES

1. Reich, G. Kollagen, Steinkopf, Leipzig 1967
2. Ramachandran, G. N., Reddi, A. H. Biochemistry of Collagen, Plenum Press, N.Y. 1976
   Bailey, A. J., Etherington, D. J. in Protein Metabolism 19[b]/I Vol of Comprehensive Biochemistry. Florkin M., (ed) Elsevier, Amsterdam 1980
3. Parry, D. A. D., Creamer, L. K. see ref. 7 ch. 1
4. Ramachandran, G. N. ed. Treatise on Collagen, Acad. Press, N.Y. 1967
5. Mikhailov, A. N. Collagen of Hides and Skins, Moscow 1971 in Russian
6. Villeé, C. A. Biology, 5 ed., Saunders, Philadelphia 1967
7. Burgeson, R. E., Hollister, D. W. Biochem. Biophys. Res. Commun., 87, 1124 (1979)
7a. Nimni, M. E. in Dynamics of Connective Tissue Molecules, (ed. P. M. C. Burleigh, A. R. Poole), North Holland, Amsterdam 1975
8. Sage, H., Bornstein, P. Biochemistry 18, 3815 (1979)
9. Epstein, E. H. J. Biol. Chem., 249, 3225 (1974)
10. Epstein, E. H., Munderloh, N. H. J. Biol. Chem., 253, 1336 (1977)
11. Schneir, M., Miller, E. J. Biochim. Biophys. Acta 446, 240 (1976)
12. Deyl, Z., Macek, K., Adam, M. Biochem. Biophys. Res. Commun., 89, 627 (1979)
13. Chung, E., Keele, E. M., Miller, E. J. Biochemistry 13, 3459 (1974)
14. Trelstadt, R. L., Kang, A. H., Toole, B. P., Gross, J. J. Biol. Chem., 247, 6469 (1972)
15. Wetlaufer, D. B. Nature 190, 1113 (1961)
16. Veis, A. The Macromolecular Chemistry of Gelatin, Acad. Press N.Y. 1964
17. Fruton, J. S., Simonds, S. General Biochemistry, Wiley N.Y. 1958
18. Karlson, P. Outlines of Biochemistry 3rd Polish ed., Warsaw 1971
19. Ogle, J. D., Arlinghaus, R. B., Logan, M. A. Arch. Biochem. Biophys., 94, 85 (1961)
20. Kefalides, K. A. Int. Rev. Conn. Tissue Res., 6, 63 (1973)
21. Risteli, J., Tryggvason, K., Kivirikko, K. J. Eur. J. Biochem., 73, 485 (1977)
22. Rozycka, D. Prace IPS 11, 9 (1967) in Polish
23. Heidemann, E. Leder 30, 27 (1979) and J. Soc. Leath. Tr. Chem., 64, 57 (1980)
24. Langerwerf, J. S. A. Leder 24, 113 (1973)
25. Pikkarainen, J., Kulonen, E. Nature 223, 839 (1969)
26. Lebedev, D. A. Usp. Sovr. Biol., 88, 36 (1979) in Russian
27. Suard, M., Berthier, G., Pullman, B. Biochim. Biophys. Acta 254, 52 (1961)
28. Novak, A. in Infrared and Raman Spectroscopy, T. M. Teofanides ed. Reidel, Dordrecht 1979
29. Scheraga, H. A. Ann. N.Y. Acad. Sci., 125, 253 (1965)
30. Siegel, R. C. Proc. Nat. Acad. Sci. USA 71, 4826 (1974)

31. Piez, K. A. in "Biochemistry of Collagen" see. ref. 2 this ch.
32. Hulmes, D. S. J., Miller, A., Parry, D. A. D., Piez, K. A., Woodhead—Galloway, J. J. Mol. Biol., *79*, 137 (1973)
33. Traub, W. FEBS Letters *92*, 114 (1978)
34. Naranyan, A. S., Siegel, R. C., Martin, J. R. Biochem. Biophys. Res. Commun., *46*, 475 (1972)
35. Housley, T. J., Tanzer, M., Henson, E., Gallop, P.M. Biochem. Biophys. Res. Commun., *67*, 824 (1975)
36. Fietzek, P. P., Kühn, K. Mol. Cell. Biochem., *8*, 141 (1975)
37. Fujimoto, D., Moriguchi, T. J. Biochem., Tokyo *83*, 863 (1978)
38. Mitchell, T. W., Rigby, B. J. Biochim. Biophys. Acta *393*, 531 (1975)
39. Light, N. D., Bailey, A. J. FEBS Letters *97*, 183 (1979)
40. Fraser, R. D. B., McRae, T. P., Suzuki, E. J. Mol. Biol., *129*, 463 (1979)
41. Ramachandran, G. N., Kartha, G. Nature *176*, 915 (1955)
42. Hofmann, H., Fietzek, P. P., Kühn, K. J. Mol. Biol., *125*, 137 (1978)
43. Menashi, S., Finch, A., Gardner, P. J., Ledward, D. A. Biochim, Biophys. Acta *444*, 623 (1976)
44. Privalov, P. L., Tiktopulo, E. Y., Tishchenko, V. M. J. Mol. Biol., *127*, 203 (1979)
45. Tiktopulo, I. E., Privalov, P. L., Andreeva, A. P., Alexandrov, V. Ja. Mol. Biol., *12*, 619 (1979) in Russian
46. Torchia, D. A., Vander Hart, D. L. J. Mol. Biol., *104*, 315 (1976)
47. Heidemann, E., Deselnicu, M. Leder *23*, 17 (1972)
48. Heidemann, E. Leder *24*, 127 (1973)
49. Hulmes, D. S. J., Miller, A. Nature *282*, 878 (1979)
50. Woodhead-Galloway, J. Acta Crystallogr. B. *33*, 1212 (1977)
51. Gross, J., Highberger, J. H., Schmitt, F. O. Proc. Natl. Acad. Sci. USA *40*, 769 (1954)
52. Rubin, A. L., Drake, M. P., Davison, P. F., Pfahl, D., Speakman, P. T., Schmitt, E. O. Biochemistry *4*, 181 (1965)
53. Williams, B. R., Galman, R. A., Popke, D. C., Piez, K. A. J. Biol. Chem., *253*, 6578 (1978)
54. Borasky, R. J. Am. Leath. Chem. Assoc., *62*, 768 (1967)
55. McLachlan, A. D. Biopolymers *16*, 1271 (1977)
56. Hulmes, D. S. J., Miller, A., Parry, D. A. D., Woodhead-Galloway, J. Biochem. Biophys. Res. Commun., *77*, 574 (1977)
57. Bruns, R. R. J. Cell. Biol., *68*, 521 (1976)
58. Zimmermann, B. K., Pikkarainen, J., Fietzek, P. P., Kühn, K. Eur. J. Biochem., *16*, 217 (1970)
59. Szent-Gyorgyi, A. Nature *148*, 157 (1941)
60. Eley, D. D., Spivey, D. J. Trans. Faraday Soc., *58*, 411 (1962)

Very recently Heidemann (Leder, 32, 142 (1981)) has made an effort to localize crosslinking bonds, formed as a result of chrome tannage. It is a theoretical calculation in which the electrostatic interaction was considered in clusters containing at least five amino acids containing active groups in side chains and located along the collagen molecule. Such clusters are charged negatively or positively. According to the calculation 39 clusters in a chain do have a negative charge (Asp and Glu containing groups). Thirty-seven of these groups might participate in chrome tanning.

3.

OTHER SKIN PROTEINS

The fibrous proteins of skin (scleroproteins) besides collagen include: elastin and keratin; albumines and globulines, which are nonfibrillar, are represented in the skin by glycoproteids, phosphoproteids, chromoproteids and melanins. Collagen is dominant in the skin. Nevertheless, we have to know the composition, structure and functions of other skin proteins to facilitate the understanding of the behavior of skin as a whole to proceed with the first steps of leather making.

The content of proteins in the skin of mammals varies from 30 to 35%, and depends upon skin origin and its properties. Collagen accounts for about 94–95% of the skin proteins, elastin 1%, keratin 1–2% and the remaining part is non-fibrous proteins. The earlier classification of proteins into structured and non-structured ones is somewhat outdated, as the molecule of every protein is currently considered to have a definite, unchangeable shape (this has nothing in common with backbone-proteins, or scleroproteins). Among scleroproteins, keratin is the next best known after collagen; elastin is much less known, whereas the existence of reticulin has become apparent.

Table 3.1 gives data about the elementary composition of hide proteins with some of their typical reactions. These data can only serve as guidelines, since they are reproducible only under specific conditions.

3.1 Keratin — chemical composition, structure and functions

Cytological studies on keratin fibers have recently been extensively reviewed by Orwin [1]. This author dealt especially with sheep wool, but his survey has

Table 3.1

Elementary composition (sulfur and nitrogen) of various hide proteins and their typical reactions

Protein	Total nitrogen %	Total sulfur %	Molish reaction	Hopkins-Cole reaction
Collagen	17.76	0.19	weakly positive	negative
Elastin	16.84	0.31	negative	negative
Keratin	16.47	3.86	negative	positive
Albumin	15.25	1.88	weakly positive	weakly positive
Globulin	15.61	1.08	negative	positive
Mucoid	11.93	3.90	strongly positive	weakly positive
Myosin	16.58	1.17	negative	positive

encompassed much more. Keratin is a protein of the epidermis and can be found in hair, fingernails, hoofs, horns, scales of reptiles, and bird feathers. It has been widely experimented, as it is the basic component of wool and because it has specific structural features. In leather production keratin is, in principle, removed from the hide during the liming and bating processes due to differences in the behavior of collagen and keratin with respect to the chemicals used. It must be pointed out, however, that keratin usually contains certain amounts of nonkeratines, and secondly, that there are many kinds of proteins referred to as keratin. Their separation into single components is more difficult than the separation of collagens, and we are still far from solving this problem. Dealing with keratin, we must remember that its origin, the way it is collected, and the kind of animal species affect its properties; so, we should not be surprised if differences in the properties and behavior occur.

To define keratin exactly seems rather impossible. The term has a historical origin because of certain common properties of keratins, their mechanical and chemical behavior with regard to the environmental effects, e.g., insolubility in water, dilute acids and alkalis. It is, however, a biologically composite material, containing stable keratinous proteins and a certain amount of non-keratinous ones, including a very small portion of soluble protein and nonproteinous matter.

Keratins consist of a mixture of proteins, each having a different composition. The main group consists of insoluble proteins whose insolubility is due to the cystein content (about 2% sulfur). The proportion of cysteine crosslinks is arbitrarily set at 3 moles per 100.

Keratins such as hair, epithelium, wool, horn, nails and feathers, contain up to 20% of non-keratins, only slightly crosslinked and easily degraded. Those proteins are not distributed in keratin at random; the non-keratinous proteins are embedded in the form of a network in keratinous ones yielding a biological 'composite' structure. The non-keratinous proteins form chemical and physical weak spots in this composition. For instance, when a fiber is stressed in water at 60°C to the breaking point, longitudinal fractures occur at the interface between the keratinous and non-keratinous proteins. However, other examples of the significance of weak spots are known.

In section 1.1. epidermis morphokinetics was discussed. An identic process occurs in a growing hair. In Fig. 3.1 the hair follicle is shown schematically [2] in various stages of development. Hair development has three phases: (1) the phase of active growth (anagen), (2) medium phase (catagen), and (3) rest phase (telogen). The follicle of a growing hair reaches deeper into the corium, where it widens into a hair bulb, consisting of live, mitoting cells, in which the hair substance is then formed. After the first growth phase the follicle becomes reorganized, and does not increase any more in the rest phase, but rather holds the hair strongly in the skin. Elements of connective tissue, surrounding the hair follicle itself, consist of a layer of the matrix and two parallel layers of collagen and elastine fibers. Then comes a less organized layer. This may be understood,

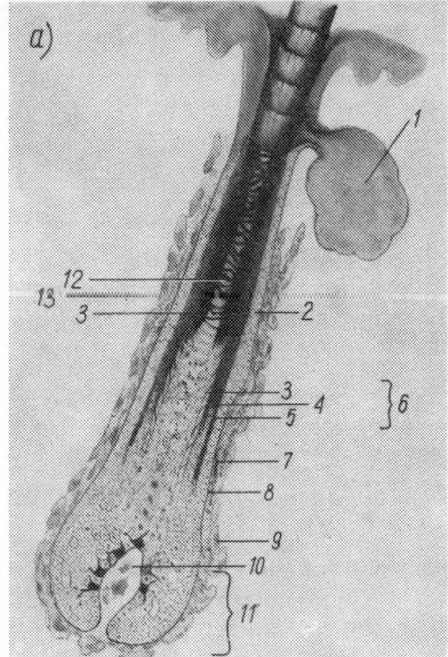
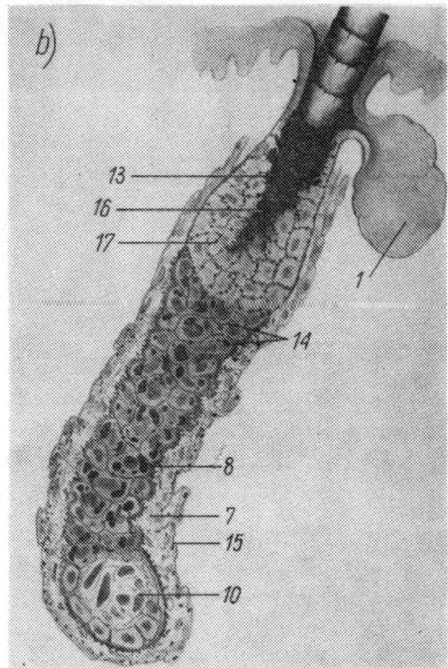
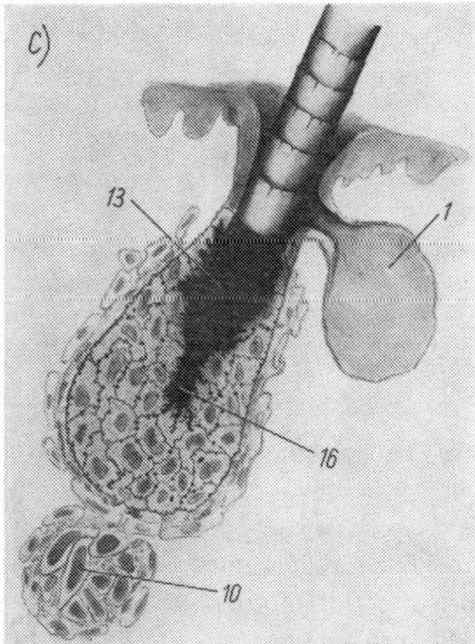

Fig. 3.1. Histology of hair bulbs: (a) in the phase of active growth, (b) in the intermediate phase, (c) in the rest phase. 1 - sweat gland, 2 - external lining layer, 3 - scale, 4 - Huxley's layer, 5 - Henle's layer, 6 - internal lining layer, 7 - collagen, 8 - basal layer, 9 - fibroblast, 10 - papillum of corium, 11 - hair bulb, 12 - core, 13 - autophagic vacuole, 15 - macrophage, 16 - club, 17 - germ.

if we realize that the quickly growing hair needs more nutritive substances than the slowly developing and gradually dying epidermis (section 1.1.) can supply. The transition of the hair base from the developmental phase to the rest phase seems to be made by the system's own enzymatic apparatus. The deeper part of the hair follicle becomes resorbed. In the source quoted [2] an exact histological description of the process is given. According to the cited authors, collagen fibers become destroyed by the system's own collagenase, becoming active in the appropriate moment. The pictures shown will explain the depths of hair positioning at various ages of animals and in various biological periods, e.g., in the hair coat change cycle. The lower part of the hair is clublike and contains a germinative substance, in which division of the cells occur. Keratinization occurs inside the follicle, the cortex cells of the hair extend immediately over this bulb zone. X-rays will show the α-keratin which is still accessible to the action of trypsine, acids and alkalis. Stabilization occurs in $\frac{1}{3}$ of the hair bulb length. It is also in this region that thiol groups disappear as they are transformed into disulfide bridges.

The flat overlapping cuticle cells are held to one another and to the cortical cells underneath by a cementing substance consisting of non-keratinous proteins (δ layer).

The amino acid composition of keratins depends on their histological and genetic origin as well as on the way they have been purified. The contents of glycine and proline in keratin are lower than in collagen and hydroxylysin does not occur. However, cystine and tyrosine are found there in considerable amounts. A specific feature of cysteine is the easiness with which it forms a disulfide bond-cystine, connecting two peptide chains.

$$2 \quad \begin{array}{c} H_2-C-SH \\ | \\ H-C-NH_2 \\ | \\ COOH \end{array} \quad \underset{\longleftarrow}{\overset{-H_2}{\longrightarrow}} \quad \begin{array}{c} H_2-C-S-S-CH_2 \\ | \qquad\qquad | \\ H-C-NH_2 \quad H-C-NH_2 \\ | \qquad\qquad\quad | \\ COOH \qquad\quad COOH \end{array}$$

The above reaction is a very important one for keratin as it limits its resistance to proteolytical enzymes. Keratins containing more cystine are more resistant to the action of those enzymes than those parts containing less cystine. This reaction is the first step of keratin extraction from biological materials. Transition of the disulfide group into two thiol groups is equivalent to a transition of the stable compound into the unstable one.

In order to test the reaction product, the thiol group is usually alkylated to increase its stability. This is done by oxydation or reduction. The fractions obtained in both methods are surprisingly alike. As a result of reduction one obtains kerateines, as a result of oxidation—keratoses. Each of these two fractions is separated into two groups: one containing more sulfur than natural keratin

(γ-keratose), and another containing less sulfur (α-keratose). The third, insoluble fraction of keratose is β-keratose. Further purification and separation of the preparations yields peptides which can be used, e.g., for the testing of the amino acid sequences. Chemical modifications of keratins, and especially those of sheep wool, were thoroughly experimented [3]. The epidermis keratin, containing less cystine, is called soft keratin, and hair keratin, containing more cystine is called hard keratin. In soft keratin chiefly thiol groups occur, whereas in hard keratin the latter are oxidized to disulfide bridges. Disulfide bonds hydrolyze easily, increase the resistance of keratin molecules, and may occur in one chain, making its structure more rigid (a) or forming bonds between the chains (b)

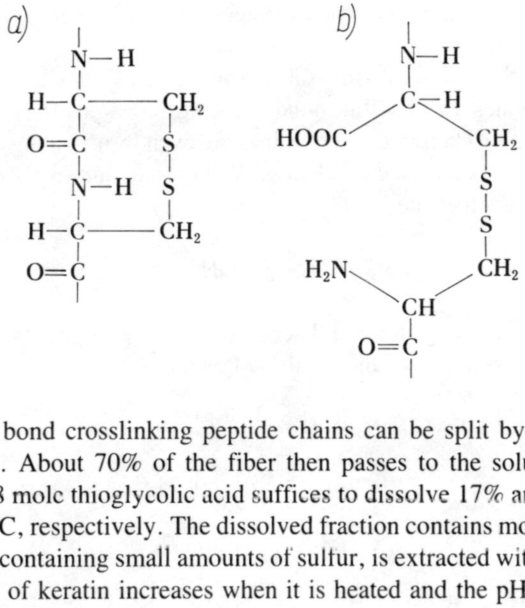

a)

$$
\begin{array}{ll}
 & \text{N}-\text{H} \\
\text{H}-\text{C} & \text{CH}_2 \\
\text{O}=\text{C} & \text{S} \\
 & \text{N}-\text{H}\quad\text{S} \\
\text{H}-\text{C} & \text{CH}_2 \\
\text{O}=\text{C} &
\end{array}
$$

b)

$$
\begin{array}{ll}
 & \text{N}-\text{H} \\
 & \text{C}-\text{H} \\
\text{HOOC} & \text{CH}_2 \\
 & \text{S} \\
 & \text{S} \\
\text{H}_2\text{N} & \text{CH}_2 \\
 & \text{CH} \\
 & \text{O}=\text{C}
\end{array}
$$

The disulfide bond crosslinking peptide chains can be split by oxidation with peracetic acid. About 70% of the fiber then passes to the solution. An ionic strength of 0.8 mole thioglycolic acid suffices to dissolve 17% and 26% of wool at 0°C and 40°C, respectively. The dissolved fraction contains most of the sulfur. This fraction, containing small amounts of sulfur, is extracted with 8-molar urea. The solubility of keratin increases when it is heated and the pH of the medium is increased.

The mean molecular weight of the soluble fraction is about 70,000. It is difficult, however, to estimate the molecular weight of insoluble keratin, as the problem of separation of the native molecule is not yet resolved.

The behavior of the disulfide bridge and of the thiol group, when treated with water vapor or LiBr in aqueous solution, is significant. Then an exchange reaction takes place:

$$-S_a^{\ominus} \; + \; -S_b S_c- \; \rightleftharpoons \; -S_b^{\ominus} \; + \; -S_a S_c-$$

This reaction results in the relaxation of stresses in the molecule, the breaking of the stretched bonds and their exchange. Keratin becomes an elastomer in which the role of the crosslinking system is taken over by covalent bonds. Then shrinkage of the whole fiber occurs, known as supercontraction. The removal of the factor causing this contraction may (but need not) cause the fibers to return to their previous form (reversible or irreversible denaturation). The degree of conversion depends on the amount of $-S^-$ groups present in the medium, i.e., it decreases at lower pH, when the majority of $-S^-$ groups is converted to $-SH$ groups. At high pH values the disulfide bond undergoes hydrolysis:

$$-S-S- \;+\; OH^{\ominus} \longrightarrow -S^{\ominus} \;+\; -S-OH$$

and the ionized thiol groups additionally formed catalyze the exchange. The presence of $-OH^-$ groups favors supercontraction due to hydrolysis of the disulfide bond and the exchange reaction described above.

Under the action of alkalizing substances such as Na_2CO_3, NaOH, Na_2S, cyanides and borates, the disulfide bond is transformed into the lanthionine bond. At pH values above 9 lanthionine crosslinks are even formed at low temperatures. This can lead to an increased brittleness of the fibers and to a lowered abrasion resistance. The lanthionine bond:

$$R_1 - CH_2 - S - CH_2 - R_2$$

can be formed according to several schemes.
(1) Hydrolytical splitting of the disulfide bond:

$$R-CH_2-S-S-CH_2-R \;+\; OH^{\ominus} \longrightarrow$$

$$\longrightarrow R-CH_2-SOH \;+\; R-CH_2-S^{\ominus}$$

(2) By α-elimination:

$$R-CH_2-S-S-CH_2-R \;+\; OH^{\ominus} \longrightarrow R-CH{=}S \;+\; R-CH_2-S^{\ominus} \;+\; H_2O$$

(3) by β-elimination:

$$R-S-S-CH_2-\overset{\displaystyle R_2}{\overset{|}{C}}H-R_1 \;+\; OH^{\ominus} \longrightarrow$$

$$\longrightarrow R-S-S^{\ominus} \;+\; CH_2{=}\underset{\underset{\displaystyle R_2}{|}}{C}H-R_1 \;+\; H_2O$$

These reaction schemes are, however, mostly based on experiments performed on models. (The intermediate products of keratin transformation have not yet been isolated.) Crewther et al. [3] have used a different nomenclature from that which has been generally accepted. They numbered the carbon atoms starting from the disulfide bond:

$$
\begin{array}{ccc}
| & & | \\
CO & & CO \\
| & & | \\
CH-\underset{\alpha}{CH_2}-S-S-\underset{\alpha}{CH_2}-CH \\
| & & | \\
NH & & NH \\
| & & | \\
\end{array}
$$

According to Zahn and Golsch [4] the reaction with the cyanide ion proceeds according to the following scheme:

$$R-S=S-R \;+\; CN^{\ominus} \longrightarrow R-S-CN \;+\; {}^{\ominus}S-R$$

$$^{\ominus}S-R \;+\; R-S-CN \longrightarrow R-S-R \;+\; CN-S^{\ominus}$$

A review of chemical reactivity of keratin has recently been given by Asquith [5].

When hair and wool are wet they reveal a considerable amount of elasticity, and the fraction of the plastic component is negligible on stretching. In view of the analogy of bonds and behavior, keratin is sometimes called a 'vulcanized' protein much like rubber has been termed as being vulcanized.

Keratin can occur in several forms which are usually called α-keratin, β-keratin and super-contracted (cross β-structured keratin). The differences between α- and β-keratins are revealed by the differences in the frequency of stretching vibrations of C-O and N-H bonds. Because of this frequency, the IR spectra of β-keratin and collagen are alike. Based on these experiments it has been found that in the β-structure the peptide bonds are almost perpendicular to the fiber axis. The fibers of α-keratin when heated with water vapor to 100°C stretch by about 100% and give an X-ray picture of β-keratin. This transformation of the α-form into the β-form is specific of keratin and myosin-muscle protein. The process can be reversible when appropriate parameters are applied. The fourth (amorphous) form of keratin is the denatured one. This keratin can be obtained under the action of Na_2S or NaHS, substances which break the disulfide bridges. Urea and $CaCl_2$ also produce denaturation of keratin and the breaking of hydrogen bonds. After such an operation keratin gives an X-ray picture very similar to those of other proteins indicating the loss of the crystalline structure.

The described behavior is typical of the KMEF protein group (Keratin-Myosin-Epidermin-Fibrinogen) of similar structure.

The β-structure was already known in 1928, and it has been shown that the peptide chains in it are almost completely stretched along the fiber axis. In Fig. 3.2 the β-structure of keratin is presented according to Ramachandran [6] together with the dimensions of its elements and crosslinking scheme; that structure is called a pleated sheet. In natural products it occurs in silk fibroin, which is a protein excreted by the silkworm, and in the keratin of feathers. The distances between the 'sheets' are about 10 Å. Feather keratin consists of filaments having helical symmetry with a unit length of 24 Å and unit twist of $-90°$ [7]. The pitch length of the basis helix is thus 96 Å. The β-sheet in the structure unit comprises four chains with eight residues in each chain. The β-sheet is twisted in the opposite direction to the basic helix of the microfibril. There are still many details of the β-structure that remain unknown to us. Astbury proposed an abbreviated structure for α-keratin (Figs. 3.2c, d and e). Further experiments have shown, however, that it has the Pauling-Corey's α-structure [8]. The α-helix has 3.6 amino acid residue per twist (18 residues in five twists). 1.5 Å of fiber length is assigned per one oncoming amino acid residue. If we assume 3.3 Å per amino acid residue to be in the β-structure, then the $\alpha \rightarrow \beta$ transition would give an extension of 120%. Every peptide bond in the α-helix participates in the hydrogen bond. The diffractogram of α-keratin is in close accordance with that which has been theoretically predicted. The normal keratin hair has such a structure. Its degree of crystallinity is fairly low; the proportion of crystalline regions being 10-15%, depending on the keratin origin.

If keratin fiber is stretched in hot steam for about 2 minutes and then released but still kept in steam, it will shrink to about ⅔ of its initial length. In this way it transforms into the supercontracted structure, losing all features of its crystallinity (Fig. 3.2f).

The elastic properties of keratin can be explained by its chemical composition, and number and kind of bonds. This is confirmed by X-ray investigations. Stress-strain curves for human hair keratin and neck fascia elastin are shown in Fig. 3.3 where we see that hair stretching shows some hysteresis, whereas elastin stretching is fully reversible.

Comparison of the properties of keratin and collagen has shown that:
(1) X-ray diffractograms of denatured feather keratin are similar to those of collagen and elastin (Table 3.2).
(2) Collagen and keratin IR-spectra are very similar in regards to both the band position and the dichroism. Numeric data are summarized in Table 3.3. In both cases the shapes of the bands are identical. The above observations find confirmation in Ref. [9] where overtone region ($4400 - 5200$ cm^{-1}) was examined. All three bands occurring in this region were found to be fully identical for both proteins.

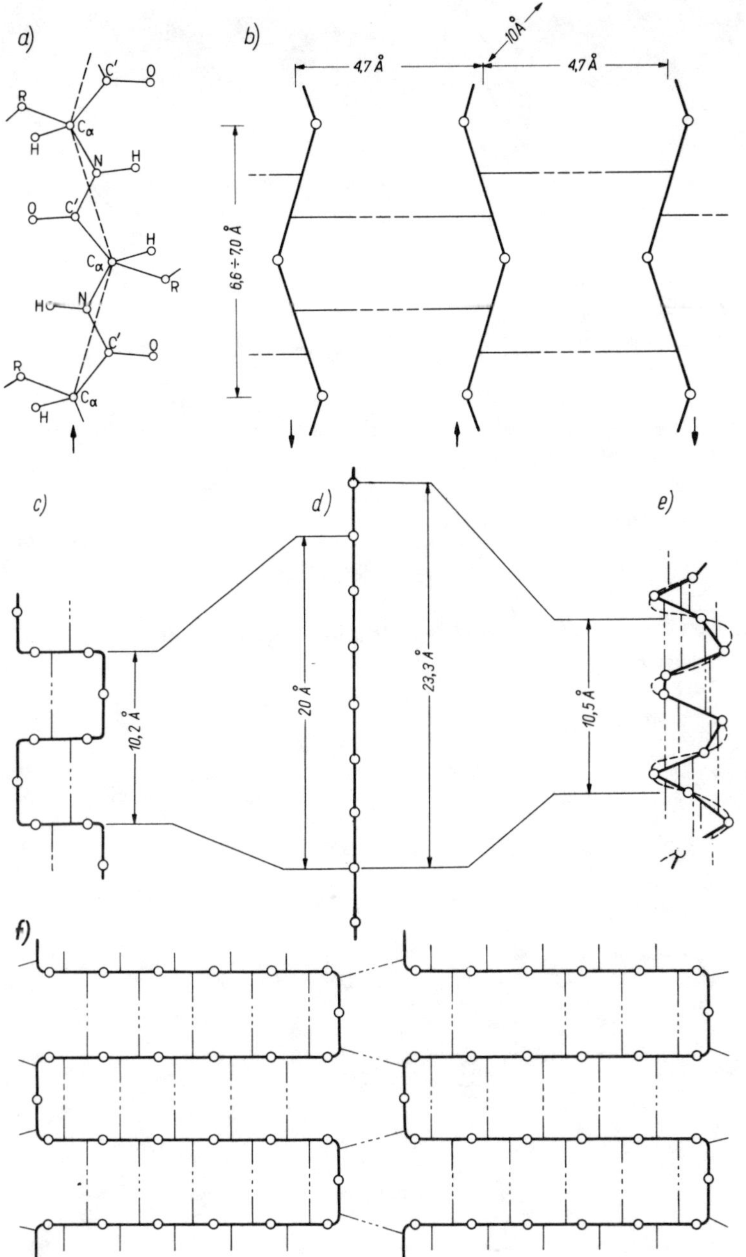

Fig. 3.2. Keratin conformation: (a) stretched β-structure, (b) chain crosslinked β-structure, (c) α-keratin structure (according to primary proposal of Astbury), (d) stretched chain of α-structure, (e) α-helix according to Pauling-Corey, (f) supercontracted structure.

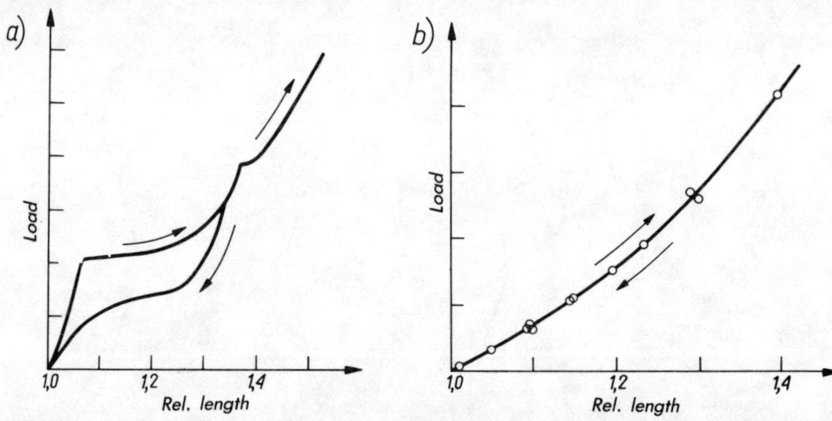

Fig. 3.3. Stress-strain plot: (a) for human hair (the hysteresis loop of deformation visible), (b) for elastin of ligamentum nuchae (neck tendon), where the behavior is typical for an elastic body.

Table 3.2
Reflexes in X-ray diffractograms of feather keratin, of degraded collagen and of elastin in Å

Feather keratin		*Elastin*		*Collagen*
degraded with urea and NaHS	*degraded with HNO₃*	*after thermal shrinkage*		*degraded with CaCl₂*
8.9	8.9	10	9	9
4.65 (s)	4.5 (d)	4.4 (d)	4.4 (d)	4.4 (d)
2.2 (v.w.)	2.2 (d)	2.2 (w.,d.)		

s—strong   w—weak   v.w.—very weak   d—diffused

Table 3.3

Dichroitic ratio of collagen, fibroin and feather keratin

Protein	*Vibrations of the bonds*			
	N-H stretching 3300 cm⁻¹	*C = O stretching 1650 cm⁻¹*	*Amide II 1535 cm⁻¹*	*N-H deforming, out of plane*
Feather keratin	0.95 ± 0.05	0.82 ± 0.05	1.25 ± 0.05	1.0
Tendon collagen (tail)	0.70 ± 0.05	—	1.20 ± 0.05	1.00 ± 0.05
Fibroin	0.54 ± 0.05	0.72 ± 0.05	1.42 ± 0.05	0.69 ± 0.05

(3) Ambady has shown [10] that diffractograms of samples on which sodium sulfate or carbonate has been crystallized, point to the identical orientations of crystals of these salts when built into the structures of both proteins.

(4) The optical birefrigence and rotatory power of both solutions are very similar. This suggests a considerable similarity of the collagen and keratin structures. The absolute values of both quantities are strongly dependent on the conditions of measurement, but are almost the same under identical conditions. The directions of rotation are opposite in the α-helix and in the collagen helix. The collagen helix is laevo-rotatory, and·its structure is much more rigid than that of keratin. This is because every third amino acid in collagen is glycine and every fifth is an imino acid which participates in the chain with two atoms (cf. p. 39).

Testing under an electron microscope has shown that hair keratin consists of ordered fibrils. The interfibrillar space is filled with the matrix-amorphous, probably non-keratinous substance. The structure of keratin fibers has been proposed on the basis of their picture under an electron microscope as well as on their wide and low angle X-ray diffractograms (Fig. 3.4). The IR keratin spectrum was analyzed by Benditt [11].

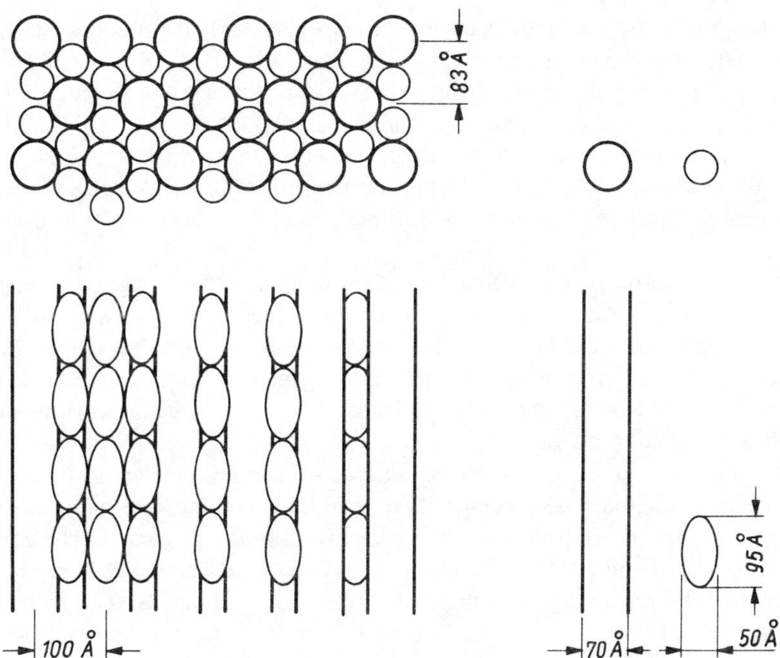

Fig. 3.4. Structure and dimensions of keratin fibers. The fibrile bundles are of various diameter. They have the structure of a multistrain rope fixed by disulfide bonds.

3.2. Elastin — chemical composition, structure and functions

Elastin is a fibrous protein occurring alongside collagen and contributing a considerable degree to the elasticity of connective tissue. Elastin fibers are formed in tissue of mesenchymal origin.

The paths of elastin biosynthesis were obscure until recently. Developing a method of tissue cells cultivation *in vitro* could help to demonstrate that elastin is synthesized in myocytes in aortal walls and in fibroblasts (cells synthesizing collagen, although in fibroblasts elastin is synthesized by a different mechanism than collagen). However, like collagen, elastin is formed in a cell as a precursor tropoelastin, containing no desmosines (see below) and much more lysin than mature elastin. Its molecular weight ranges from 72,000 to 74,000. Another elastin precursor is proelastin [12]. Its molecular weight is 130,000–140,000. Proelastin contains more dicarboxylic and hydroxyaminoacids as well as histidine, methionine and cystine. Those two elastin precursors were found to be immunochemically identical, so their transition in one another by enzymatic splitting of additional peptide may be presumed. However, this theory, which suggests a similar biosynthesis mechanism for both elastin and collagen, still requires confirmation.

Elastin is formed from tropoelastin when crosslinking bonds are formed. This is an extracellular process. According to Gallop and Paz [13], 38 out of 47 lysine residues occurring per 1000 amino residues in mature elastin are engaged in crosslinking bonds of allysine type. This process is catalyzed by lysil oxidase (probably the same which catalyzes collagen crosslinking).

Little is known about the transport of substances having high molecular weight through cell walls, like proteins or their precursors. This also relates to tropoelastin.

In tissue elastin is accompanied by other proteins, which may be separated by autoclaving or alkaline hydrolysis. They are glycoproteids containing 64–67% hexoses and some hexosamines. Their relation to elastin itself has not yet been made clear. Skin contains 1–2% of elastin; ligamenta (fasciae) contain much more of it; and *ligamentum nuchae* (located at the back of the neck) is the most common source of elastin.

Elastin is the subject of interest because of its specific chemical and physical resistance. Its amino acid composition changes somewhat depending on the origin. A portion of 10^5 g of protein contains 28 moles of dicarboxylic acids, 12 moles of diaminoacids, 33 moles of hydroxyaminoacids and 7 moles of tyrosine. The remaining 1120 moles are nonpolar amino acids. The hydroxyproline level is low, but constant, and the Hypro/Pro ratio which in collagen is about 1.0 equals 0.1 in elastin and may be considered to be the measure of purity of the preparation. Elastin as well as collagen contains only small amounts of thioaminoacids. According to elementary analysis, the sulfur content is 0.17%;

the glycine content is as high as in collagen; the content of aspartic and glutamic acid is smaller than in collagen. Elastin does not contain tyrosine, but it does give an absorption UV band at 275 nm which suggests the presence of aromatic rings. Some properties of elastin that are compared with those of collagen are shown in Table 3.4.

Two compounds of amino acid type are characteristic for elastin: desmosine (a) and isodesmosine (b) having a pyridine ring and four side chains. They have

a) and b) are chemical structure diagrams of desmosine and isodesmosine (pyridinium ring with four side chains terminating in CH(COOH)(NH₂) groups).

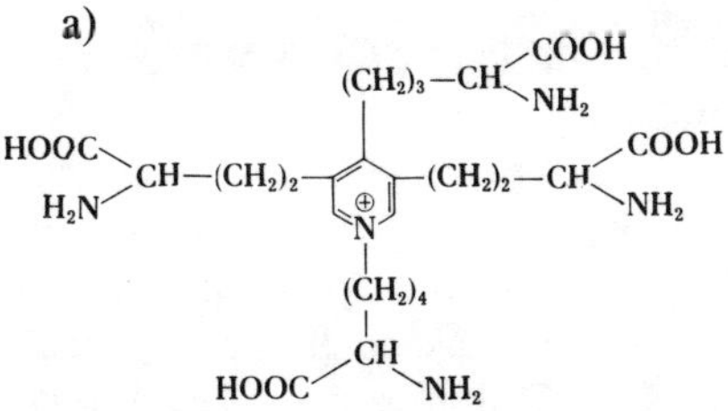

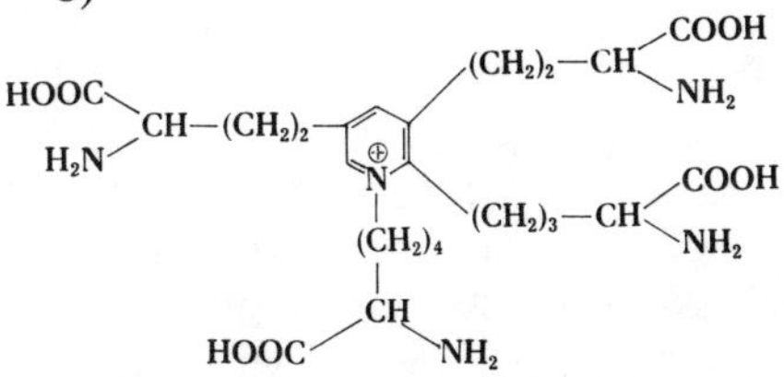

been discovered by Thomas et al. [14]. Most probably those compounds are decisive for elastin crosslinking, because they may form four interchain links of peptide character. In tendon elastin there is a reverse relationship between the lysine and desmosines contents: the amount of desmosine increases with tissue age, whereas the amount of lysine decreases; their total amount, however, remains at the level of 10–11 residues per 1000 amino acids residues and lysine is taken as ¼ of desmosine. The specific enzyme which attacks elastin is elastase, occurring in the alimentary tract of carnivora. Plant peptidases such as bromelain, papain and ficin attack elastin as well.

Elastin fibers subjected to the Romhanyi's aniline test are visible as an ordered ribbon of corrugated fibers resembling polyamide filament. In 1935, Leplat, staining tendons by indigocarmin, showed the histological importance of elastin in animal tissues: it formed rings, known for a long time as Henle's rings, making

Table 3.4

Characteristics of fibrous proteins of connective tissue

| Kind of fiber | microscope | Morphologic characteristics | | X-ray diffraction |
		polarisation microscope	electron microscope	
Collagen	thick	highly anisotropic	cross-striation; pattern with 640 Å period	reflexes
Elastin	branched, thin	slightly birefrigent	filament structure	shapeless (ring)

Fig. 3.5. Henle's rings formed of elastin on collagen fiber, swollen by acid. Staining with indigocarmin.

the changes in shape and thickness of tendons difficult (Fig. 3.5). A similar phenomenon has been observed in skin collagen fibrils.

To obtain pure elastin, so that any influence of the impurities on the characteristics of the biopolymer could be excluded, is as yet a difficult task because it is impossible to dissolve and precipitate it without it decomposing. Elastin dissolves in 88% cold formic acid, 0.25 molar hot oxalic acid, in sulfuric acid and in alcohol saturated with gaseous hydrochloride. It dissolves as well under a prolonged action of a 40% urea solution. We do not have, however, convicing criteria that would allow us to evaluate the degree of renaturation during precipitation. This is why some data concerning elastin can seem questionable.

From the above description we see that elastin is not removed from the fiber network in the liming and deliming processes. The opinions regarding its behavior during bating are divided; it probably depends on the enzymes used. However, in leather one can find destructs and even intact elastin fibers, especially around hair follicles, which has been confirmed by the microscopic pictures of Urban (Fig. 3.6).

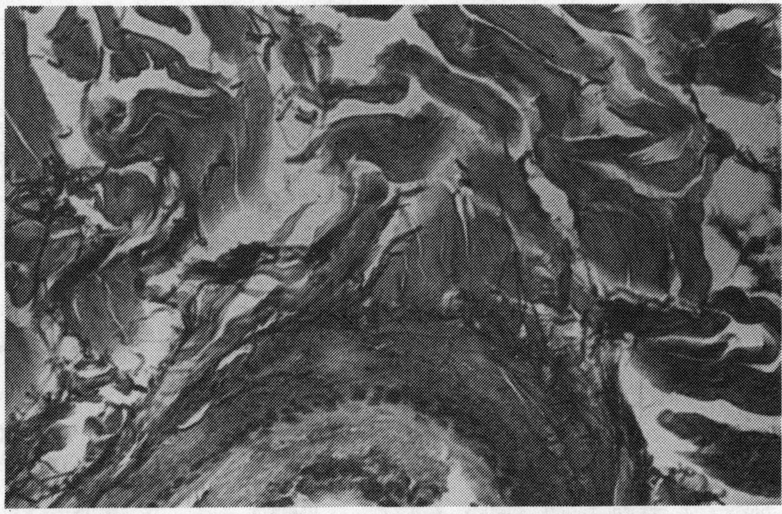

Fig. 3.6. Elastin fibers around hair pocket, visible as black lines. Magnification 250×, orceine − staining. Courtesy Dr. M. Urban.

The dissolving of elastin with oxalic acid yields to two fractions: α-elastin (molecular weight of 60,000–84,000) and β-elastin (molecular weight about 6000). The amino acid composition of these two fractions is almost the same that points to the homogeneity of elastin.

Elastin is thermally stable which is probably due to its low content of polar side chains. The main factor of this stability is β-elastin, which together with collagen is a basic component of arterial walls. The decrease of elasticity of those walls is a leading symptom of sclerosis. Elastin fibers can be extended as much as twice their length; their recovery proceeds without hysteresis (see Fig. 3.3). Because of its elasticity, elastin counteracts the creeping of tissue. Its Young's modulus is about 3×10^6 dyne/cm^2 (0.3 MPa), and the stretch resistance is about 1×10^7 dyne/cm^2 (1 MPa), which is typical of elastomers. Elastin sometimes accumulates at definite sites of the connective tissue, forming a kind of corset on the animal.

Elastin must be regarded as an energy storing material. It restores more than 90% of the input energy. The elasticity of elastin is closely related to the presence of water: when dry or in organic solvents elastin fibers shown no elastic properties. This fact is difficult to explain because of the high content of apolar, hydrophobic amino acid residues. An explanation is given by Weiss-Fogh and Andersen [15] who considered elastin as a 'liquid drop polymer' consisting of globular particles connected through crosslinking bonds (Fig. 3.7). These authors suggest that the hydrophilic residues protrude from the non-stretched elastin

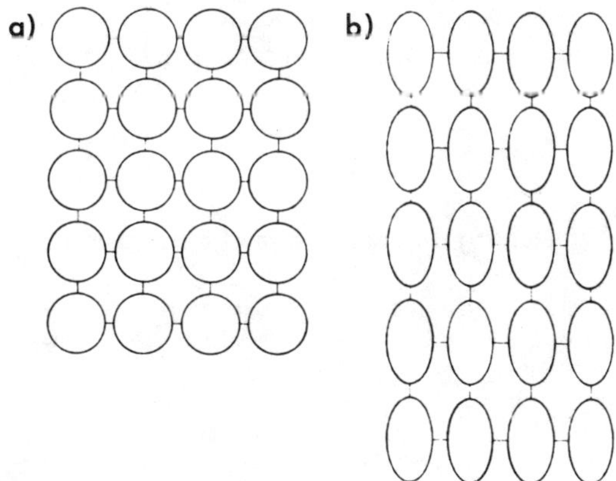

Fig. 3.7. Elastin structure as proposed by Weiss-Fogh and Andersen. The interior of the drop is hydrophobic, the hydrophilic groups being on the surface. Covalent bonds occur between the units. Particles are at rest, (a) particles are stretched by about 35% (b) [35].

particle and participate in its wetting, on stretching hydrophobic residues, and come into contact with the surrounding water which makes the structure, as a whole, water-repellent. After the stress is released, the particle returns to its initial shape due to the action of the hydrophobic residues unshielded when stretched. Weiss-Fogh and Andersen have tested, using a precise microcalorimeter, the heat and work of elastin stretching in various solvents (the results of some measurement are shown in Table 3.5). The data summarized show that the deformation of elastin causes reversible chemical changes, whose energy is several times greater than the energy required for stretching. This is the essential difference between elastin and vulcanized caoutchouck [16].

The traditional elastomer is an unordered network of kinetically free chains bound by stable crosslinking bonds. In the case of vulcanized *Hevea brasiliensis* caoutchouck, these are disulfide bonds, and in the case of, e.g., resilin (insect protein of elastomer character) di- and trityrosyl bridges occur. In the latter elasticity is caused by the entropy of configuration, and it decreases on stretching, the changes of internal energy of the system being insignificant. Elastin is a completely different type of elastomer. Insignificant swelling of elastin in water is temperature dependent. Swelling of elastin has a constant value below 50°C as it does between 50 and 70°C (though then that value is different). From measuring the mechanical properties of elastin in that temperature range it was concluded that about 85% of the isometric power was due to the configuration enthropy, and the remaining part was due to the change of the internal energy of the system. These considerations have been based on the assumption that no chemical reactions occur in the system.

Investigations carried out recently by Gotte [17] and based on electron microscope observations and exact analysis of the X-ray diffraction pattern have shown certain features of the α-helix structure in elastin. The helices are said to be double-coiled, each of them of approximately 15 Å in diameter. The distance between the double helices is about 50-60 Å. Their number in the

Table 3.5

Heat balance of stretching and shrinkage of elastin at room temperature

Sample	Stretching work mcal(J)	Stretching heat mcal(J)	Heat of recovery mcal(J)	Heat of recovery to heat of stretching ratio
1	1.33 (5.3)	−5.53 (23.1)	+5.60 (23.5)	1.01
2	1.02 (4.0)	−5.60 (21.5)	+6.10 (25.5)	1.09
		−5.71 (23.9)	+5.32 (22.3)	0.93
		−6.13 (25.7)	+6.00 (25.1)	0.98
3	1.08 (4.4)	−5.73 (24.0)	+5.50 (23.1)	0.96
		−5.92 (24.8)	+5.41 (22.7)	0.91

'microfilament' has not yet been determined. According to the quoted author some differences in the diameter, observable as bright and dark striations, may be due to the crosslinking bonds. He did not observe greater aggregates of a drop type. According to his observation this may be the result of irregular distribution of crosslinking bonds. Parry and Craig [7] could not obtain, however, any evidence of the existence of filaments in either the transverse or longitudinal section. Thus the problem of the elastin structure has not yet been made clear. Both possible structures are shown in Fig. 3.8.

Considerable difficulties are encountered when following the data of elastin in leather processing since it is uncertain whether or not the lack of band pattern (crossstriation) in the electron-microscope picture is equivalent to the fibers being noncollagenous (see, e.g., collagen fibers before and after liming, as photographed by Urban, Fig. 3.9a and b).

Numerous cases are known when histological staining of the fiber network, believed in microscopic technique to be typical of elastin, is in fact staining collagen. This may be explained as follows: Elastin staining dyestuffs are phenols which may react with collagen as vegetable tannins at sites where hydrogen

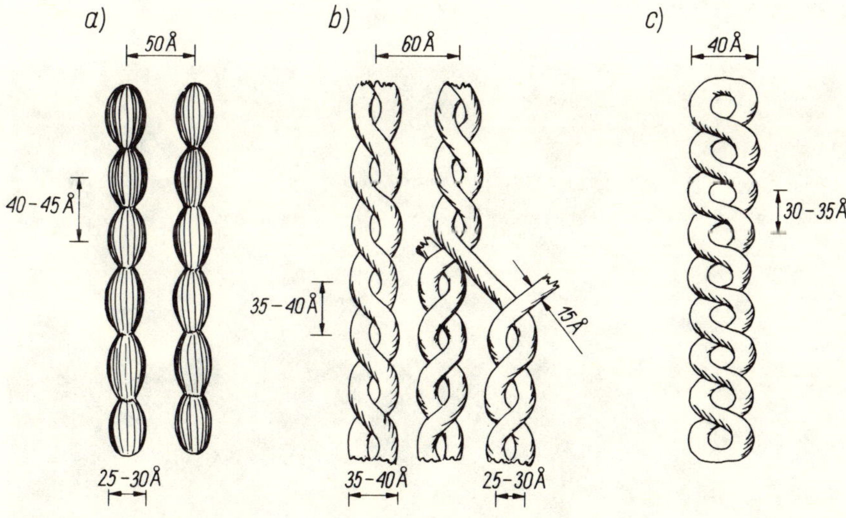

Fig. 3.8. Diagram illustrating possible arrangements of the filaments forming elastin fibers when seen in the electron microscope under stretched or relaxed conditions and negative stained: (a) stretched filaments, (b) relaxed filaments, (c) possible appearance of filaments in a highly relaxed state.

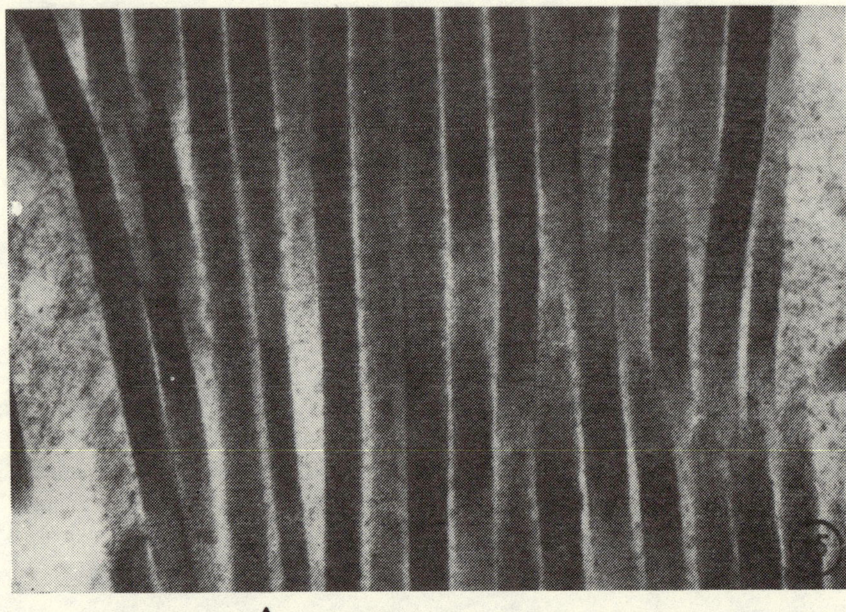

Fig. 3.9. Collagen fibers (a) after soaking, (b) after liming. Both stained in the same way. Courtesy Dr. M. Urban

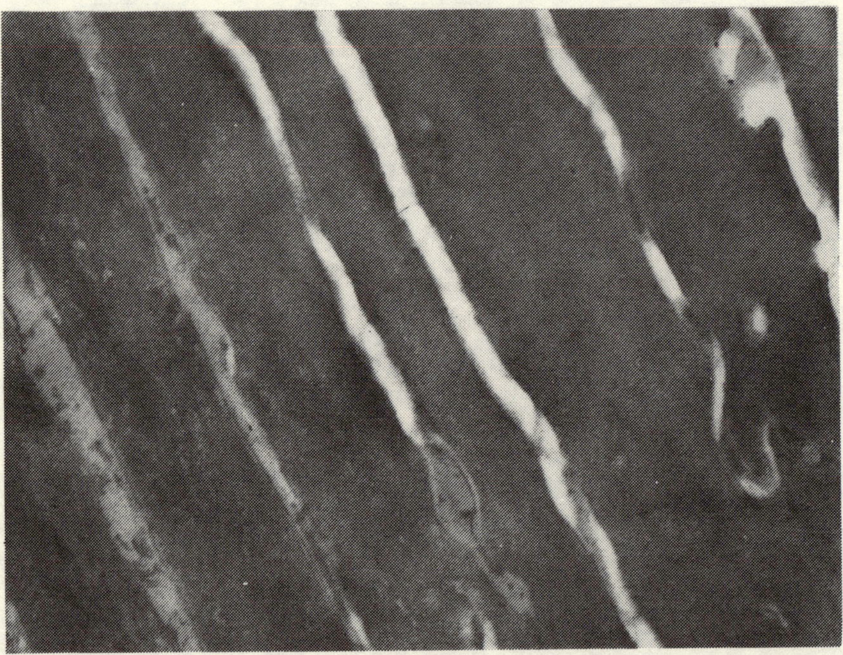

bonds may be formed. In young collagen one can suppose that the following sites bind dyestuffs in histological preparations:
(1) sites where hydrogen bonds are formed,
(2) unshielded ester groups of crosslinking bonds,
(3) functional groups in interfibrillar spaces.

Elastase also attacks collagen to some extent, so its action is not a specific way of identification, nor does elastin show typical X-ray reflections like collagen when shrunken or under the influence of lyotropic salts.

The known differences in chemical composition of these two proteins may be utilized for their distinguishing, e.g., phosphomolybdenic acid reacts with basic groups of the protein side chains: collagen contains about 13% of basic amino acids and elastin about 1%.

Elastin and collagen reveal considerable but different amounts of fluorescence. Heavy metals quench fluorescence, so it has been shown that this difference between elastin and collagen fibers may be used as distinguishing reaction. By using combinations of filters in a fluorescence microscope, one can distinguish with certainty collagen from elastin after their reaction to phosphomolybdenic acid. This interesting method has not yet been extensively tested.

3.3. Reticulin. Composition, structure, and functions

Reticulin, discovered some 100 years ago, was until recently a disputed point in the description of connective tissue protein components. The main difficulty consisted in its separation and purification, as the purity criteria of an unknown substance may be misleading, moreso as there was never a clear differentiation between reticulin and collagen except that reticulin was considered liable in staining black with silver nitrate. Some data, which were once the basis of differentiation between reticulin and collagen, are given in Table 3.6. From that table one can see how insignificant the differences are, except for the carbohydrate content. According to recent opinions (T. Stirtz, E. Heidemann, private communications) young forms of collagen IV (see above) containing large amounts of proteoglycans, under usual preparative conditions not separable from

Table 3.6

Comparison of some physicochemical data of connective tissue proteins

Fiber	Heating to boiling			Carbohydrate content	Specific enzyme
	in water	in acetic acid	in alkalies		
Elastin	resistant			0.2-0.4	elastase
Collagen	gelatinization			0.6-1.0	collagenase
Reticulin	gelatinization			4.0	?

collagen, were stained black with silver salt due to their sulfur content.

Reticulin has never been separated from skin. It was isolated from liver and kidneys, but these preparations are now believed to be a form of collagen [7].

There is one other point of importance that has not yet been made clear, i.e., the relatively high content of fat in the substance separated as reticulin. As much as 6% fat can accompany this protein fraction. This may contribute to the staining of these fibers by silver salts due to the reduction of fats at the double bonds. However, cross-striation period of the fiber and other feature points to the supposed identity of collagen and the hypothetic reticulin.

3.4. Remaining skin proteins

Among the fibrous protein occurring in skin are small amounts of myosin. It is the main component of hair muscles, responsible for the bristling of hair and fur.

The myosin molecule in the muscle is equivalent to collagen in skin; it is responsible for its shrinking and stretching. Filaments (fibrils) of myosin are aggregates of myosine monomers: one filament consists of 100–200 meres. The shape of the filaments corresponds to their function. The molecular weight of myosin is about 500,000, the monomere length is 1650 Å, and the filament length is about 15,000 Å.

Proteolytic enzymes (trypsin, chymotrypsin, subtilisin) split the molecule into two fractions of different molecular weights. The myosin molecule is linear and one of its ends, consisting of some 50 amino acids, is of a globular shape and ends probably with an ATP molecule. The structure of myosin is fairly well-known, though this protein is not of great importance for tannery processes, as it is sensitive to enzyme action and can easily decompose.

Apart from fibrous proteins some globulins, albumins and mucoids, generally determined as spheroproteids, also occur in skin. Albumins are water soluble proteins; globulins are water insoluble, but this is an arbitrary distinction. These two groups of proteins can be separated and advantage can be taken of their solubility in ammonium sulfate solution: albumins precipitate from the saturated solution, while globulins precipitate from the half-saturated one. Albumins and globulins are components of organic fluids-lymph and blood. Because of this they are of great importance to the nutrition of connective tissue. They are biologically active; their halflife time is much shorter than that of 'structural' proteins. They are decomposed by proteolytic enzymes and action of heat and sensitive to changes in pH and ionic strength. These proteins are discussed in detail in biochemistry handbooks as they are of great importance for living organisms. Their importance, however, disappears with death of the organism, and in leather making their removal is only important as they are liable to putrefaction. Mucoids are globular proteins which contain sugar in its molecule.

Some proteins, though they occur in very small amounts, are important because they act as enzymes (see section 9.2).

REFERENCES

1. Orvin, D. F. G. in "Fibrous Proteins" see ref. 7. ch. 1
2. see ref. 3. ch. 1
3. Crewther, W. G., Fraser, R. D. B., Lennox, F. G., Lindley, H. Adv. Prot. Chem., *20*, 191 (1965)
4. Zahn, H., Golsch, E. Hoppe-Seylers Z. Physiol. Chemie *330*, 38 (1962)
5. Asquith, R. S. in "Fibrous Proteins" see ref. 7. ch. 1.
6. Ramachandran, G. N. ed. Collagen. Interscience, N.Y. 1962
7. Parry, D. A. D. in "Fibrous Proteins" see ref. 7. ch. 1
8. Pauling, L., Corey, R. B. Proc. Natl. Acad. Sci. USA *37*, 246 (1951)
9. Fraser, R. D. B., McRae, T. P. J. chem. Phys., *29*, 1024 (1958)
10. Ambady, G. K. Curr. Sci. India *28*, 239 (1959)
11. Benditt, G. Biopolymers *4*, 529 (1966)
12. Foster, J. A., Mecham, R., Iberman, M., Farris, B., Franzblau, C. Adv. Exp. Med. Biol., *79*, 351 (1977)
13. Gallop, P. M., Paz, M. A. Physiol. Rev., *55*, 418 (1975)
14. Thomas, J., Elsden, D. F., Partridge, S. M. Nature *200*, 651 (1963)
15. Weis-Fogh, T., Andersen, S. O. Nature *227*, 718 (1970)
16. Dorrington, K. L., McCrum, N. G. Biopolymers *16*, 1201 (1977)
17. Gotte, G. L. Adv. Exp. Med. Biol., *79*, 105 (1979)

4.

NON-PROTEINOUS SKIN COMPONENTS

The most important non-proteinous component of skin as in every organ of the body is water. It is the solvent in which all reactions proceed in the organism. The properties of water, as a medium of all biologic reactions, are described in chapter 5. The present chapter deals only with the components which are of importance for leather making or which can change the properties of the raw material and/or leather. This group consists of polysaccharides of a definite kind (glycosaminoglycans), fats and inorganic compounds.

In skin saccharides occur chiefly as polymolecular substances. The fat content changes significantly from animal to animal, and the content of inorganic compounds depends on the living state of the animal. Among compounds, influencing life functions are hormones and vitamins, which affect the skin in negligible amounts. In the epidermis glycogen is another factor, whose amount is small and varies with age and various conditions. In fetus epidermis it occurs with fair abundance. It accumulates in the neighborhood of wounds; in corium it occurs rarely and irregularly. Glycogen is the source of energy in keratin or more exactly, keratohyalin synthesis.

4.1. Glycosaminoglycans

Glycosaminoglycans are typical polyelectrolytes of cellular and extracellular organic fluids. They control the viscosity of those fluids, act as buffers in tissue, participate in transport of ions and influence the water economy of the organism due to their hygroscopicity. Soaking and liming of skin are probably controlled by function of glycosaminoglycans. These substances occupy a special position among biological polyelectrolytes. They have a characteristic skeleton of molecules typical for carbohydrates, functional groups such as $-NH^+$, $-COO^-$ and $-SO_3^-$, and their specific distribution.

In glycosaminoglycans monosaccharide molecules occur bound by α- and β-glycoside bonds. This structure gives the molecule some stiffness. The only possibility of rotation of the molecule is around these bonds. Another factor determining conformation of the molecule is the distribution and charge of the functional groups, which affects the electrostatic repulsion of charges in the solution. In the compounds discussed, many hydrogen bonds, both intramolecular and intermolecular, occur. Glycosaminoglycans usually occur in extracellular spaces where they fulfill a structural function imparting plumpness and flexibility to animal tissue. Since all substances proceeding from cell to cell must pass through these spaces, glycosaminoglycans affect metabolic processes.

The term mucopolysaccharides was introduced previously for polysaccharides of animal origin which contain hexosamines, but was later replaced by the term glycosaminoglycans. In organisms these compounds usually form complexes with proteins (mucoids). It seems that the proteins are bonded to the saccharide part through sugar hydroxylic groups as well as to serine and tryptophane side chain hydroxyles. The group of glycosaminoglycans includes both acidic and neutral polysaccharides. In connective tissue we have: hyaluronic acid, chondriotin sulfate, chondroitin, dermatan and keratan sulfates and heparin.

All glycosaminoglycans of animal origin have in common the fundamental hyalobiuronic acid link composed of the D-glucuronic acid residue connected with 2-deoxy-D-glucose by a 1,3-β-glycoside bond. The residues of hyalobiuronic acid are connected in a chain by 1,4-β-glycoside bonds:

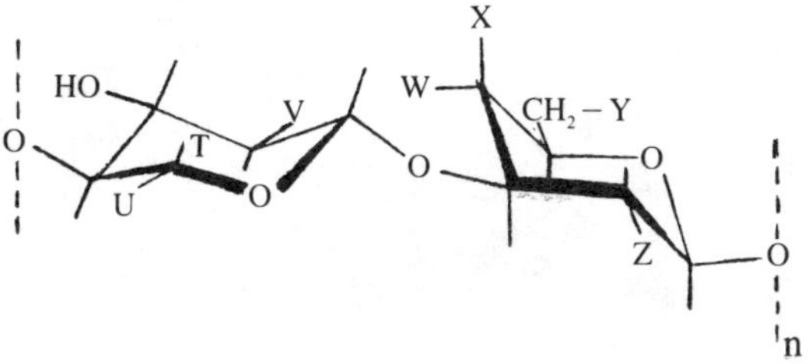

In this formula seven sites are substituted with functional groups. The substitutents are listed in Table 4.1, wherefrom the relationship between the particular glycosaminoglycans becomes obvious.

Hyaluronic acid is usually obtained from an umbilical cord or from the eye glass body; the procedure of its structure recognition has been described by Musil et al. [1]. Hyaluronic acid is a hydrophilic substance. The viscosity of its solutions was investigated very accurately, as it is considered to be one of the most important factors influencing the viscosity of organic fluids. The value of intrinsic viscosity $[\eta]$ is about 20-50 cm^3/g, depending on the origin and way of preparation of the substance. The same factors affect the molecular weight, which varies, according to different authors, from 5×10^3 to 8×10^6 [2].

The hyaluronic acid molecule is a long, non-branching polysaccharide chain with a considerable degree of hydration. The hydrodynamic volume of the molecule, i.e., the volume occupied by it in an aqueous solution, is almost twice as large as the real one. For leather producing operations, especially for soaking and liming, the most important is its interaction with water.

Chondroitin has a very similar structure to that of hyaluronic acid, except that galactosamine replaces glucosamine. It has not been explained as of yet, though it is generally accepted that chondroitin is a biologic precursor of chondroidine sulfates. Two isomers are known: 4-chondroitin sulfate (sulfate A) and 6-chon-

Table 4.1

Characteristic substituents of connective tissue glycosaminoglycans

Glycosaminoglycan	*T*	*U*	*V*	*W*	*X*	*Y*	*Z*
Chondroitin	H	COOH	OH	H	OH	OH	NHCOCH$_3$
Chondroitin sulfate A	H	COOH	OH	H	OSO$_3^-$	OH	NHCOCH$_3$
Chondroitin sulfate C	H	COOH	OH	H	OH	OSO$_3^-$	NHCOCH$_3$
Keratan sulfate	H	CH$_2$OSO$_3^-$	NHCOCH$_3$	H	OH	OSO$_3^-$	NHCOCH$_3$
Hyaluronic acid	H	COOH	OH	OH	H	OH	OH
Chondroitin sulfate B (Heparine, Dermatan sulfate)	CH$_2$OH	H	OH	H	OSO$_3^-$	OH	NHCOCH$_3$

droitin sulfate (sulfate C). The position of the sulfate group is the only factor distinguishing these two substances.

Chondroitin sulfate B has occurred to be identical with a compound known as β-heparin, now referred to as dermatan sulfate. Chondroitin and chondroitin sulfates have been isolated from animal cartilage tissues. Aqueous solutions of chondroitin sulfates have much lower viscosity than do the ones of hyaluronic acid. The intrinsic viscosity of these compounds is 1-2 cm^3/g, their molecular weight varying from 5×10^4 to 6×10^5. The dissociation constant pK is similar to those found for glucuronic acid.

Dermatan sulfate (according to the nomenclature of Comprehensive Biochemistry, Vol. 5, Elsevier, Amsterdam 1963), differs from other chondroitin sulfates in that it includes the L-iduronic acid unit instead of that of the D-glucuronic acid:

Thus we have (1.4)-O-α-L-idopyranosyluronyl-(1.3)-2-acetamid-2-deoxy-4-O-sulfo-β-D-galactopyranose. As it has been shown, certain amounts of D-glucuronic acid also occur in dermatan sulfate. D-glucuronic acid is an integral component of the molecule and occurs in certain fragments of it, alternately with L-iduronic acid. Dermatan sulfate dissolves in water giving solution of very low intrinsic viscosity $[\eta]$ = 0.5. Its molecular weight is about 2.2 × 10^4.

Keratan sulfate is found primarily in cattle cornea, in which it accounts for about half of the total glycosaminoglycans. That compound occurs in two forms: I and II. Keratan sulfate I has less sialic acid and methylpentose than the sulfate II (these components are considered to be accompanying components). Both keratans probably differ as regards to the place of connection to the protein: sulfate I is attached to asparagine and glutamine by a glycosylamine bond, while sulfate II is bonded to threonine and serine by O-glycoside bond. Sulfate II has a branched chain; the branches are saccharides or sialic acid. Keratan sulfates differ from other glycosaminoglycans because they do not contain uronic acid residue. They dissolve readily in water and are resistant to hyaluronidase attack.

Important from the technological point of view are some glycosaminoglycans of plant origin, which are applied as thickening agents in leather finishing. Pectins, derivatives of α-D-galactopyranosyluronic acid esterified in various extents, are biopolymers which occur frequently. Pectins are components of cell sap, cell walls and intercellular spaces in higher plants.

Pectic acid has a polygalacturonic chain in which at least half of the methoxy groups has been removed by action of enzymes, acids or alkalis. Depending on its origin, pectic acid can also contain saccharide units: arabinose, xylose, galactose or sorbose.

From sea algae, alginates have been isolated. These are polymers of anhydro-β-D-mannuronic acid with glycoside bonds in positions 1.4; other compounds of this kind have been isolated from algae, e.g., a compound with three basic units: arabinose-3-sulfate, galactose-6-sulfate and xylose, having bonds: 1.4 and 1.5 arabinose, 1.4 xylose, 1.3 and 1.6 galactose, whereas this quantitative ratio is 1:3:3. Alginates are complicated and versatile systems.

From the industrial point of view, gum arabic, containing L-arabinose, L-ramnose, D-galactose and D-glucuronic acid in a 3:1:3:1 ratio and tragacanth gum is important. Gum arabic also occurs in wattle bark but the ratio of sugar residues is different here and amounts to 6:1:5:1. The composition of tragacanth gum is similar, but it has not yet been accurately estimated.

From the physicochemical point of view, glycosaminoglycans are representatives of polyelectrolytes, i.e., polymers in which the ionizing functional groups are:

$$\left[\begin{array}{c} -CH_2-\,CH- \\ \quad\quad\; | \\ \quad\quad\; SO_3^{\ominus}\,H^{\oplus} \end{array} \right]_n$$

These may be of an acidic, alkaline or of amphoteric type. Depending on the groups occurring in the meres, we can speak of polyacids, polybases and polyampholytes. The charges of polyelectrolytes are pH-dependent. The ionic strength of the solution essentially influences the behavior of the polyelectrolyte in solution. The example of chondroitin sulfate has shown that the viscosity of the solution decreases with the increase of its ionic strength (Fig. 4.1).

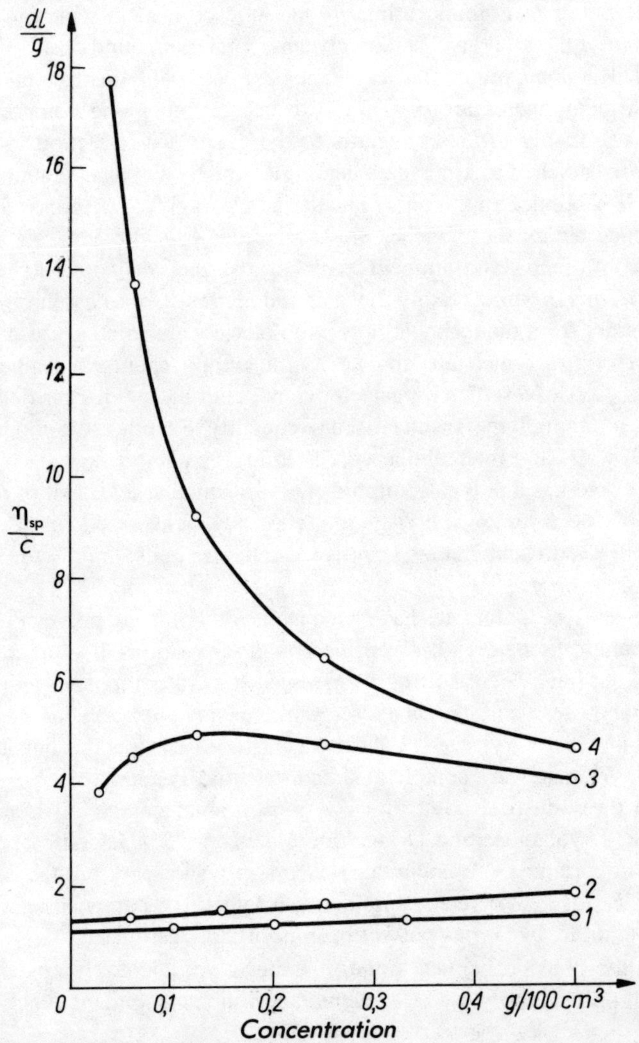

Fig. 4.1. Intrinsic viscosity of chondroitin sulfate solutions in dependence on proteoglycan (glycosaminoglycan) concentration and kind of salts present in solution: 1 - curve for solution in 0.15 m phosphoric buffer, pH = 7.0, in presence of 0.2 m NaCl, 2 - curve for solution in 0.022 m phosphate buffer at pH = 7.0, 3 - curve for solution in 0.001 m phosphate buffer, pH = 7.0, 4 - curve for solution in distilled water.

Polyelectrolytes of the glycosaminoglycan type and proteins differ in charge density on the molecule surface and in flexibility of the molecules. The increase of intrinsic viscosity $\left[\frac{\eta_{spec}}{c}\right]$ with decreasing salt concentration (ionic strength of the solution) is explained by the repellence of the charges of the same sign on the surface of the molecule, which makes it stiff. Repellence drastically decreases if ions with opposite charges occur in the medium.

The apparent degree of glycosaminoglycans ionization, as found from the analysis of titration curves, is expressed by the empiric equation:

$$pH = pK_\alpha + nlog\left(\frac{\alpha}{1 - \alpha}\right)$$

where α is the dissociation degree, and K_α the apparent constant. Spencer has given the relationship:

$$pK = npK^+$$

where pK^+ is a value that does not vary with ionic strength and concentration, whereas n and pK_α decrease with increasing concentration and ionic strength. From two different carboxylic groups we obtain two pK^+ values and $n_1 \neq n_2$. The pK values for various glycosaminoglycans have been given by Bettelheim [3].

According to former conception [4] the glycosaminoglycans were believed to be in close chemical connection with collagen in live organism. This conception was later discharged in spite of lack of the experimental proofs. However, recently introduced staining technique (with renium red) has shown that these two macromolecules are very difficult to separate, although the kind of bonds between them remains undefined.

Öbrink [5] has investigated the interaction of weakly crosslinked collagen (lathyric collagen) with glycosaminoglycans. Lathyric collagen has been chosen in order to provide collagen monomeres as the reaction substrate. Based on light dispersion, turbidimetric and by sedimentation measurement methods, the quoted author came to the conclusion that a condition of the reaction between collagen and glycosaminoglycans taking place was either due to the presence of a D-iduronic acid or the existence of a charge of density greater than one negative unit per disaccharide. Keratan sulfate does not attach to collagen, whereas hyaluronic acid does in only a very low extent. The bondless interaction of collagen and hyaluronic acid (mutual steric exclusion) depends on the increase of activity of collagen molecules. Investigation of the formation of collagen fibrils in vitro, carried out by the quoted author by the turbidimetric method, has shown that glycosaminoglycans accelerate the process under conditions close to physiologic ones. This influence is very pronounced in the nucleation phase; if the glycosaminoglycans are added later, they will slow down the process. Only hyaluronic acid accelerates fibril formation in both phases. According to the quoted author glycosaminoglycans decrease the thermal stability of the fibrils.

The behavior of glycosaminoglycans as polyelectrolytes has to be considered from two points of view: first as the behavior of substances which accompany collagen in connective tissue, and second, as substances binding cations.

Polyelectrolytes can bind cations in three ways: through undissociated -COOH group, through the formation of ion pairs, localized in the region of one or several charges or through nonlocalized binding (retention) of the ion in the region of polyelectrolyte potential.

Mathews [6] investigated the binding of the $Co(NH_3)^{6+}$ complex by acid polysaccharides of connective tissue. He applied the formula:

$$\frac{1}{r} = \frac{1}{nK(A)} + \frac{1}{n}$$

or

$$\frac{r}{(A)} = Kn - Kr$$

where r is the number of moles of the cation bound by a sector of the polysaccharide, n is the amount of bonds in that molecule, A the concentration of the cation in solution. The association constant K was found by Mathews from the above equation. According to Mathews the cation binding ability depends on the spatial positions of the anions: in dermatan sulfate both acidic groups have an axial position (for definition see, e.g., Hallas Stereochemistry); in chondroitin-4-sulfate only the sulfate group is in the axial position; in chondroitine-6-sulfate no anionic groups are in the axial position. Ion pairs can hardly be formed in keratan sulfate, which has only one charge per 1 nm of chain length. In heparin 3.5 charge units per same chain sector gives it the greatest surface charge density and best binding of ion pairs. The Ca^{2+}, K^+, Na^+ and methylene blue behave alikely under physiologic conditions, except that the binding of methylene blue is very strong and the association constant is about 2 orders higher than those for K^+ or Na^+ ions when the ionic strength is extrapolated to zero.

Dunstone [7] determined the increasing affinity of the cation series to the anionic groups of glycosaminoglycans of connective tissue and obtained the following sequence:

$$K^+ < Na^+ < Mg^{2+} < Ca^{2+} < Sr^{2+} < Ba^{2+}$$

and according to the affinity of alginic acid:

$$Na^+ < Co^{2+} < Ca^{2+} < Ba^{2+} < Cu^{2+}$$

Experiments of cation binding by connective tissue do not give reproducible results; e.g., we know conditions where K^+ is bound stronger than Na^+ and vice versa. We do not know exactly the behavior of glycosaminoglycans in leather making as it has not yet been experimented.

4.2. Fats

Fats or lipids are a great group of cell and tissue components. The general feature of fats is their water insolubility, tendency to form colloidal suspensions in it, and solubility in common organic solvents. Fats are distinguished by their high content of apolar hydrocarbon chains including fatty acids and alcohols, waxes, sphingosine and sterides. The polar components include monosaccharides, nitrogen containing alcohols, phosphoric and sulfuric acid residues. The bonds occurring in fats are chiefly of the ester type. Fats occurring in various organs and tissues are difficult to separate since their composition and structure are similar. Generally, they are classified into acylglycerides (neutral fats, called formerly glycerides), waxes and fat-like complex substances containing lipids—lipoids. Fats occur in the animal body as esters, and phospholipides. A very small part—about 1%—occur as free fatty acids. Fats stored in tissue are very quickly exchanged: their halflife time is ten to twenty days. It is about ten times shorter than it is for collagen. Digestion of fats in the body occurs in several steps, with characteristic dehydrogenation of the chain between carbons α and β and formation of α, β-unsaturated fatty acid which in turn becomes oxidized at carbon β, β-ketoacid formed decomposes according to the equation:

$$R - CH_2 - CO - CH_2 - COOH + H_2O \rightarrow R - CH_2 - COOH + CH_3COOH$$

This process is commonly called auto-oxidation. According to the present views, it takes five steps, and is controlled by known and identified enzymes.

Other possibilities of fat biodegradation are also known, e.g., *via* methyloxydation or ω-oxidation. Fats become rancid under the influence of oxygen (air) and ionizing radiation. That is a process of oxidation of unsaturated acids to peroxides, and then to aldehydes or ketones. If their carbon chains consist of 6–12 atoms, these compounds will have an unpleasant taste and odor. The action of bacteria and of mold lipases leads to hydrolysis of fatty acid esters and to oxidation of free fatty acids to ketones. Aldehydes, formed as a result of rancidity, are derivatives of fatty acids and may polymerize and give fatty spews on finished leather.

As it was already mentioned, fats, as a rule, occur in organisms as complicated mixtures. They are storage substances in organisms and take part in the form of composed compounds in the functions of the nervous system. Being storage substances fats are accumulated in certain animals in the skin, e.g., in pigskin, whereas they occur abundantly in the cutaneous tissue.

In skin numerous endings of neurofibers occur. They consist of the phospholipid type compounds (sphingomyelins) containing sphingosine (dihydroxyaminoalcohol) instead of glycerin:

$$CH_3(CH_2)_{12}-CH=CH-CH(OH)-\underset{\underset{\underset{H}{(CH_3)_2\overset{\oplus}{N}-CH_2-CH_2-O}}{\underset{O^{\ominus}-P=O}{|}}}{\overset{\overset{R_1-\overset{\overset{O}{\|}}{C}-NH}{|}}{CH}}-CH_2O$$

where R_1 is a residue of fatty acid, cerebroside, containing the hexose residue (galactosphingolipids)

$$HOCH_2-\underset{\underset{O}{|}}{CH}-\underset{\underset{OH}{|}}{CH}-\underset{\underset{OH}{|}}{CH}-\underset{\underset{OH}{|}}{CH}-\underset{\underset{O}{|}}{CH}$$

The above compound is kerasine, if R is lignoceric acid, cerebron if R is cerebronic acid, neurone if R is neuronic acid.

Gangliosides contain neuraminic acid, which is their typical component

$$\begin{array}{c}
COOH \\
| \\
C-OH \\
| \\
CH_2 \\
| \\
H-C-OH \\
| \\
H_2N-C-H \\
| \\
O-C-H \\
| \\
H-C-OH \\
| \\
H-C-OH \\
| \\
CH_2OH
\end{array}$$

and its derivatives, sialic acids. Lipids, occurring in organisms of animal species, differ in their composition and rotatory power because of the content of optic active components. Various sialic acids have been isolated, e.g., from cattle and horses. Lipids, among them sialic acids, were found in skin. Common triglycerols, i.e., storage substances from cells, occur in them in the form of droplets with a refraction index different than that of the surrounding matter; they are well visible under a polarization microscope. All the lipids become stained histologically with dyestuffs of the Sudan group, so one can sometimes speak of sudanophile substances. (The mechanisms of metabolism and synthesis of triacylglycerols are discussed in manuals of biochemistry.)

In almost all samples of animal lipids cholesterol occurs; it is a compound belonging to the steroid group, derivative of perhydrocyclopentanophenantrene:

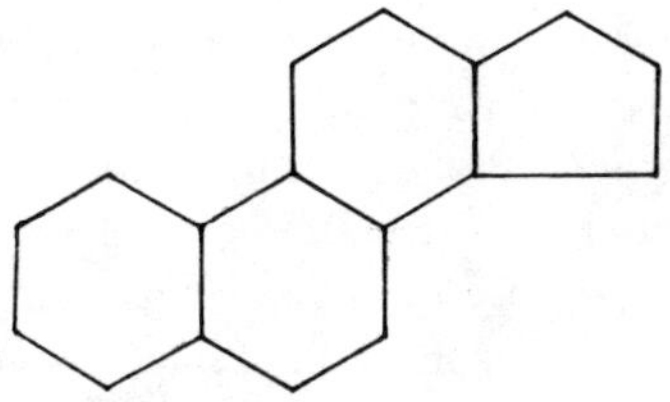

Cholesterol is an alcohol with a β-hydroxyl group in position 3, a double bond in position 5,6, and a side chain in position 17

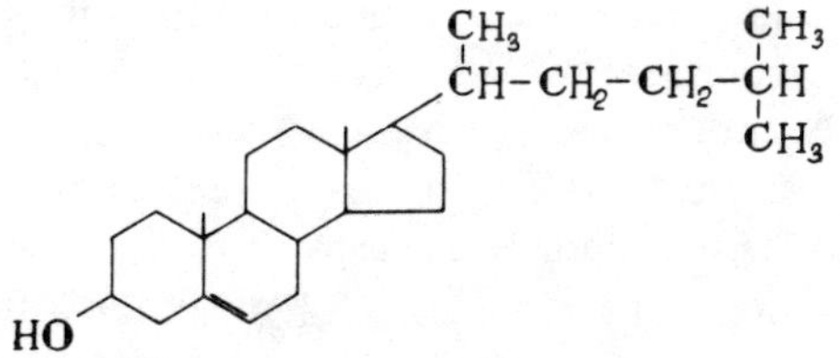

Cholesterol occurs in both blood and bile. It is to a great degree esterified, mainly to unsaturated fatty acids. This compound and its derivatives are of particular importance as hormones and detergents. When esterified to fatty acids, cholesterol forms a wax, less soluble in organic solvents than the triacylglycerides. Another wax, lanolin, occurs abundantly in sheep wool grease. Waxes are more resistant to the action of acids and alkalis than fats. Their saponification is more difficult than that of glycerides. The purpose of sebaceous glands of skin, excreting mainly waxes, is the coating of the epidermis and hair with a thin secretion layer which makes them elastic and water-repellent.

The approximate content of fats in skin varies from 0.5 to 4.0% in calfskins, cattle and horse hides; from 3.0 to 30% in sheepskins; from 4.0 to 40% in pigskins; and from 3.0 to 10% in goatskins as calculated in terms of dry skin. From this amount 75-80% are triacylglycerols; the balance are waxes and composed lipids.

From the point of view of tanning chemistry, fat in the skin is a component giving it flexibility, softness and stability. Natural fat is removed from skin in leather making processes; thus it is necessary to apply fat to it in the finishing processes. A significant amount of fat in the raw skins (pigskins) makes their processing difficult, because hydrophobic spaces are then formed, repelling water during soaking, and because insoluble calcium soaps are formed during liming. Raw skins containing much fat have to be degreased before processing.

Fats bound to other components, e.g., composed fats of lipoproteids, are not extractable to the same extent as when different organic solvents are used. Thus the determination of fats can yield various results, and the method itself is the subject of experiments and a constant point of interest of chemists-analysts trying to unify the results of various methods of fat extraction.

4.3. Inorganic components and their significance

Raw skins contain about 0.5% of mineral salts. In regards to cations, sodium occurs in greatest amounts, then comes potassium, calcium and iron (hemoglobin component); among anions hydrocarbonates, carbonates, chlorides, phosphates and sulfates are of greatest importance. It should be emphasized that today inorganic components cannot be considered apart from their bodily function, as they are of basic importance for the water economy and transportation in the live organism. The mechanism of enzymatic action, in which inorganic components play the basic role, will be discussed separately, though it is in some points identical to the transport mechanisms.

Ionic strength. According to Lewis and Randall the activity coefficients of dilute solutions of strong electrolytes are the same for all solutions of the same ionic strength. The ionic strength $\frac{i}{2}$ is equal to half of the concentration C of ions multiplied by square of their valency Z:

$$\frac{i}{2} = \frac{C_1 Z_1^2 + C_2 Z_2^2 + C_3 Z_3^2}{2}$$

The ionic strengths of, e.g., 0.1 molar Na_2HPO_4 and $MgSO_4$ solutions are 0.3 and 0.4, respectively. The Lewis and Randall rule is fulfilled in the case of ionic strength about 0.1. The ionic strength in biochemistry is an important and unique expression, and bringing solutions to the same value allows standardization of conditions.

Lyotropic Hofmeister series. Hofmeister found that cations and anions can be arranged according to their influence on protein solubility. The capability of certain salts to precipitate proteins when their concentration is high enough has found application in protein purification. At larger concentrations the logarithm of protein solubility is a linear function of ionic strength. The solubility of a protein in the solution of salt S is related to its water solubility S_0 and to the ionic strength $\frac{i}{2}$ of the salt by an empirical equation

$$\log S = \log S_0 - K \frac{i}{2}$$

where K is the salting out constant. In a saturated solution protein has a constant chemical potential due to which its activity remains constant. A change of solubility due to the change of ionic ratio produces a change of the activity coefficient, which is reversely proportional to stability. The Hofmeister series is a regularity observed in many cases, not only as related to protein solubility. Such series can also be set up with respect to other properties such as solution viscosity, electrophoretic mobility, enzymatic reactions and others. The anionic series is as follows:

$$\text{citrate} > \text{tartrate} > \text{sulfate} > \text{acetate} > Cl^- > NO_3^- > Br^- > I^- > CNS^-$$

and the cationic series, where the order is less definite, is:

$$Al^{3+} > H^+ > Ba^{2+} > Sr^{2+} > Ca^{2+} > K^+ > Na^+ > Li^+$$

No doubt the reason for such an order is the intensity of electrostatic field around the ions; small ions have more intensive fields than large ions of the same valency. The intensity of the field of small ions is the reason of greater hydration, which is an immediate reason of ordering. This rule is not that simple from the theoretical point, since, e.g., the strongly hydrated SO_4^{2-} ion has very pronounced precipitating properties, whereas in a concentrated aqueous solution, e.g., LiBr dissolves silk easily (fibroin).

The tanner should remember that the ability of particular ions to solubilize proteins is equal to their peptidizing ability in leather making. This rule is important in soaking, liming and bating, as a part of non-collagenous proteins becomes dissolved in a process which is parallel to softening and swelling, if the ionic strength and the kind of ions are appropriate. Peptidizing in this case is not equivalent to dissolving only: In this process a part of the weaker peptide bonds is split and the native proteins are thus converted into peptides with smaller molecules, which are easier soluble. This is due to the properties of the ions introduced.

The ion balance in the live organism and complex formation. The ion balance is an essential feature of living organisms. Disturbances in this balance may be the reason of a serious illness or even death. Biosynthesis processes are, as a rule, related to the ion action: the enzymatic system of the body is particularly sensitive to ion equilibrium. As of yet 15 inorganic cations are known to activate enzymes: Na^+, K^+, Rb^+, Cs^+, Mg^{2+}, Ca^{2+}, Zn^{2+}, Cr^{3+}, Cu^{2+}, Mn^{2+}, Fe^{2+}, Co^{2+}, Ni^{2+}, Al^{3+} and NH^+.

In general, an ion when placed in an electrostatic field of another ion with an opposite charge can be considered as complexly bound. When the ion radii are sufficiently small and the ionic charge high enough, ionic pairs are formed, bound electrostatically. There are two kinds of complexes: simple and chelated. In the simple complex the ion is attached to the site with a suitable charge, if the size of the ion is appropriate to fit that site; the ion maintains its character and the complex may dissociate in a dilute solution. Some simple complexes (metalorganic sulfides, i.e. derivatives of thiol -SH groups in proteins) have a high degree of stability. Such complexes may yield insoluble aggregates which dissociate with difficulty. If a substance combined with a metal ion contains two or more donor groups so that a ring (or rings) is formed, then the resulting structure is called chelatic or a metal chelate. The binding electron pair, formed between the acceptor (metal) and donor (complexing agent) may be chiefly ionic or chiefly covalent, depending on the metal and complexing agent.

The stability of ionic complexes is expressed by the association constant:

$$K = \frac{[MA]}{[M^+] \times [A^-]}$$

where $[M^+]$ is the metal ion concentration, $[A^-]$ is chelating factor concentration, and $[MA]$ is that of the complex. K is usually replaced by log K, as K has very large values. The association constants are usually determined by physical methods, e.g., often by determining the change of the pH of the solution due to complex formation, since the proton passes from the complex forming compound to the solution due to its replacing by the metal ion. The complexing reactions are often studied using electron spectroscopy (UV and Vis) as well as polarography and redox potential determination.

It is not always easy to tell whether simple or chelate complexes are formed. The most frequent criteria used for detecting the chelating reaction are: the change of color, change of properties of the metal ion, and formation of two optical isomers.

Amino acids form complexes with many metals. In Table 4.2 the log K values are given for complexes of several metals with glycine and alanine.

Table 4.2

The log K values for some ions in complexes with amino acids

Ion	Glycine	Alanine	Glycylglycine
Cu^{2+}	8.62	8.51	6.05
Ni^{2+}	6.18	5.96	4.49
Zn^{2+}	5.52	5.21	3.80
Pb^{2+}	5.47	5.00	3.23
Co^{2+}	5.23	4.82	3.49
Mn^{2+}	3.44	3.02	2.15
Mg^{2+}	3.44	1.96	1.06

REFERENCES

1. Musil, J., Adam, M., Houba, V. Macromolecular Components of Connective Tissue, Academia, Praha 1968 in Czech
2. Walton, A. G., Blackwell, J. Biopolymers, Acad. Press N.Y. 1973
3. Bettelheim, F. A. in Biological Polyelectrolytes, ed. A. Veis Dekker N.Y. 1970
4. see ref. 4 ch. 2
5. Öbrink, R. Eur. J. Biochem., *34*, 129 (1973)
6. Mathews, M. B. Arch. Biochem. Biophys., *104*, 394 (1964)
7. Dunstone, J. R. Biochem. J., *97*, 236 (1962)

5.

SKIN COMPONENTS AND WATER

Water is the most important inorganic component of the body. Its mass is usually more than half the animal body weight. It is necessary for the functioning of proteins and metabolism of the cells. Outside the cells it acts as a transportation medium, as solvent and as a substantial component of body fluids. Evaporation of water is the main way to remove heat from the organism. The organisms of land animals have rather scarce water resources, which could equilibrate its balance. Usually that balance is equilibrated by drinking or eating juicy food.

The breathing system plays an important role in the water economy in the organism, since water is always formed as one of the final oxidation products. Man excretes 300–400 g of water daily together with the air used for breathing. There are organisms known which are able to utilize the whole water formed in biological oxidation. Water consumption is controlled by thirst, secretion of sweat and urine. The skin of land animals is almost water-impermeable. That regulating mechanism has been looked into many times. The binding of water by substances being tissue components was and still is the subject of very comprehensive studies, since all the biological functions of the molecules of those substances are carried out in an aqueous solution. Leather production is carried out largely in water. Skin does not dry between the production operations. The discussion of the character and strength of bonds between water and skin components will follow after a short consideration of water structure.

5.1. Structure of water as substance

In solids the term structure refers to the spatial arrangement of molecules. Such a description is simple insofar as these arrangements are usually stable, i.e., they exist for a longer time than that required to take the measurement. Deviations from the positions calculated are small, as the time of measurement is by several orders shorter than that of correlation of molecular motions.

According to Eisenberg and Kautzmann [1] the discussion of water structure can be started by considering the structure of ice. There are several kinds of molecular motions around their average position. The crystal-forming molecules vibrate around these average positions; the system of these positions makes a network of a geometrical distant range ordering. The mean vibration frequency is equal to $\frac{1}{t_v}$, where t_v is the characteristic vibration time. The absorbtion band at 229 cm^{-1} in the far infrared is assigned to vibration motion in water (t_v), which corresponds to time 2×10^{-13} s. The rotational and translational displacements to which the molecules in ice are subject to are much less frequent and amount to $t_D = 10^{-5}$s. Hence $t_D \gg t_v$.

The notion of structure may have three different meanings when applied to ice crystal. Validity of that notion depends on the time during which the motion is observed. For example, snapshots of water molecules, made in times t_I, t_v and t_D, are shown in the pictures in Fig. 5.1. The molecules shown there are making the motions mentioned. In other words, the position of a molecule as a function of time changes periodically; the mean values of its positions give a regular crystal network.

In liquids, the structure is entirely different. The molecules oscillate around the temporary equilibrium positions. This can be shown spectroscopically. The oscillation frequencies in water are approximately the same as in ice. However, analysis of the results of experiments on self-diffusion, viscosity, dielectric relaxation and NMR relaxation methods shows that the mean equilibrium positions in water change very often, much more often than in ice, t_D at a temperature close to 0°C is about 10^{-11} s. The thermal motions of molecules in the liquid phase are also of two kinds: Quick vibrations around temporary equilibrium positions and translations of these positions which are much slower.

Going back to the three meanings of structure we can say that a hypothetical snapshot made in time t_I would show a fairly accurate picture of a water molecule, identical to that of the molecule in ice (Fig. 5.1a). A second snapshot made in time t_v would be approximately the same, but a little 'moved' (Fig. 5.1b.) The

Fig. 5.1. Structure of ice at various time intervals: (a) structure I, $t_I = 10^{-15}$ s, (b) structure V, $t_v = 10^{-13}$ s, (c) structure D, $t_D = 10^{-5}$. According to [1]

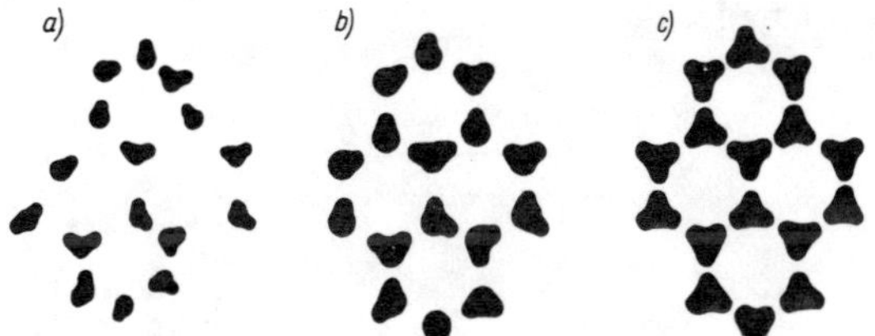

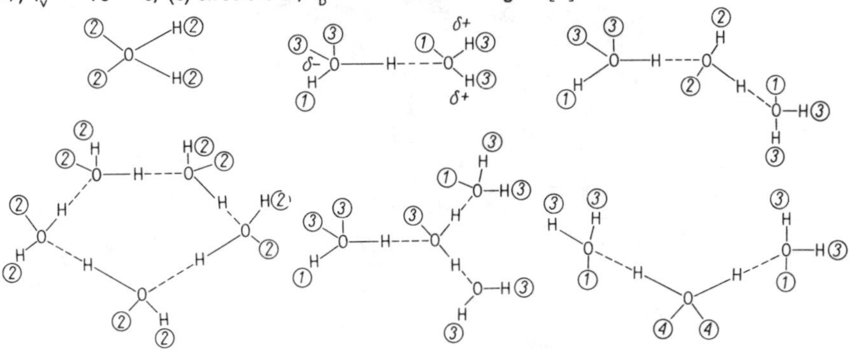

$① < ② < ③ < ④$

Fig. 5.2. Various degrees of water aggregation. The figures refer to the capability to form bonds at various points.

third picture, however, would be completely blurred, (Fig. 5.1c) since in time t_D the molecules would be strongly translated. No technical possibilities have been found to inform us about the structure of the water molecule in time t_I, however, data concerning the structure t_v can be obtained experimentally. One has to remember, however, that t_v changes very significantly and decreases as the temperature increases, like t_D. At very low temperatures, when t_D is of the order of days, water is in vitreous state.

The formation of vibrational structures depends on the forces acting between the molecules. In water like in ice hydrogen bonds -O . . . H-O occur between the molecules; the difference lies in the strength of these bonds and in the lattice energy, a definition resulting from the balance of water in the solid state, in terms of quantum mechanics.

Due to the action of hydrogen bonds water molecules are associated. The size and shape of associates are not defined yet; at least four opinions are known here. Most probable seems to be the theory of 'flickering clusters' shown in Fig. 5.2 which are supported by very accurate calculations of Nemethy and Scheraga [2] who presumed occurrence of 0-4 co-operating hydrogen bonds in one molecule (Fig. 5.3). In the quoted model there are neither dimers or trimers of water, but only clusters and single molecules. On the surface of the clusters molecules may occur bound by one, two or three hydrogen bonds. Inside the cluster there are four bonds per molecule.

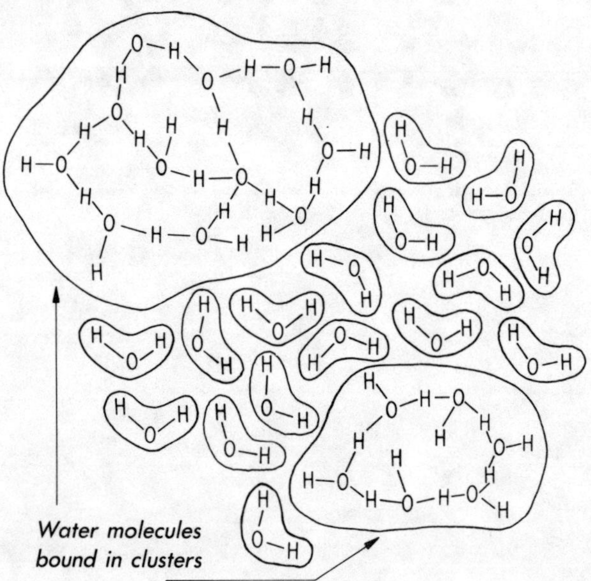

Fig. 5.3. Single water molecules and their groups, aggregate into clusters.

Considering the spectrum of water in the near UV as well as the data obtained by the method of X-ray diffraction it may be believed that '*long range*' interaction in water does not exist or at least there is no existing proof of such interaction. According to Symons [3] there are no particular kinds of '*free*' and '*bound*' water. He believes that water is a fairly homogenous system consisting of discrete '*oligomeres*'. He draws attention to the fact that not all hydrogen bonds in the water molecule system are equivalent, as they differ in energy levels. In Fig. 5.3 the possible kinds of those bonds are shown. This model is also based on the continuous regrouping of molecules and bonds. According to Privalov [4] long-range interactions do exist in water.

There are many handbooks and monographs dealing with water, among which the five-volume work edited by F. Franks deserves special attention [5].

5.2. Water bound to the macromolecule

Water bound to the macromolecule in the biological sense is in a state intermediate between solid and liquid from the point of energy. This view is supported by numerous experiments which can be carried out using various methods. In Fig. 5.4 those methods are shown as well as their applicability in dependence on the kind of information required. In view of their limited use, not all of these methods are discussed in the present work, as they are either expensive or unique

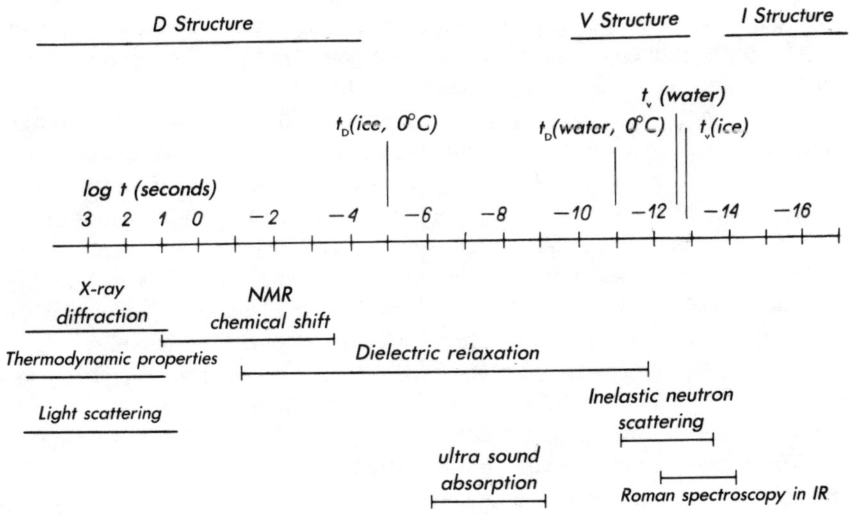

Fig. 5.4. Range of investigation of water molecule as a function of the measurement time range.

apparatus is involved, or due to great theoretical problems, requiring a specialized knowledge.

The term *water bound by a macromolecule* is not uniquely defined, and it can refer to [6]:

(1) Water remaining in the molecule after drying under definite conditions, or after the equilibrium is established in an atmosphere of known, constant humidity.

(2) Water, whose molecules do not have freedom of rotation, does not change its dielectric constant under the action of high frequency current.

(3) Water, with no solvent character.

(4) Water, translating with the large molecule in sedimentation, diffusion and other processes, i.e., increasing the hydrodynamic size of the molecule.

(5) Water, whose protons give a wider line in the NMR spectrum than those of bulk water.

(6) Water, that does not freeze at a constant, definite temperature.

(7) Water, that has a constant position in the macromolecule that can be defined, e.g., by X-ray analysis.

There are some similarities in the definitions contained in particular points as the definitions relate chiefly to the experimental methods, whereas the largest part of water in question participates in all the determinations. It is impossible to uniquely define what is understood as water bound to the macromolecule; one can only say, quite generally, that we have in mind water molecules whose one (or more) properties has/have changed due to interaction with the macromolecule. Of course the choice of experimental methods determines to some extent the result, as the property changes is subject to determination. What property is to be determined depends on the chosen method.

According to Varga-Manyi [7] it is correct to say that water bound in a certain way can serve as a solvent for certain materials to a certain extent only. For instance the bound water of $CuSO_4.5\ H_2O$ is not a solvent for any material, whereas water bound to the swelling organic substance can be a solvent for every organic substance. Hydration water bound to Na^+ and Cl^- ions is a different solvent for glucose and urea: if each ion binds 6 molecules of water in its first hydration shell, the quantity of urea solved in such a 'saturated' solution is the same as in bulk water until saturation, whereas for glucose it is not.

A biological system can be considered from the point of view of the effect of ions and nonionized molecules on the water structure in their surrounding, or, conversely, of the effect of water on the structure of molecules of biologic importance, e.g., biopolymers. There is still another point of view, namely the permeability of water and aqueous solution through the skin, which is of special importance in leather production, as well as in permeability of the skin of live animals.

5.3. Hydration of ions

The energy of ion hydration depends on the charge and kind of ion: for H^+ it is usually high, about 276 kcal/mole (1156 kJ/mole). Ion hydrates are hetero-dynamic systems, i.e., systems in which various intermolecular forces are acting. Among them there are long-distant interactions, dipole interactions and hydrogen bonds among water molecules. These short-time interactions seem to be very important; their strength decreases very quickly with distance. This is why the ion radius of the dissolved molecule is of importance, as the number of attached water molecules directly depends on it. The influence of ions on the water structure is greater than those of polarization, binding and pressing of molecules closest to the water molecules. A working hypothesis, proposed by Frank and Wen [8] distinguishes three zones of ion influence (Fig. 5.5).

(1) Internal zone, containing molecules undergoing polarization (A), immobi-lization and electrostriction (B) (electrostriction is a dielectric deformation of molecules in the external electric field proportional to the square field intensity).

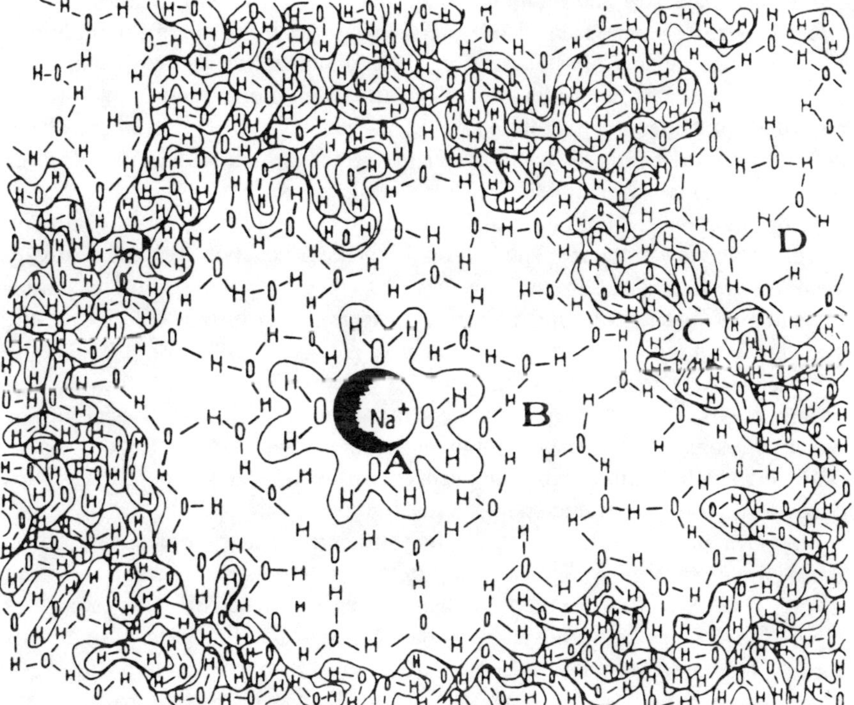

Fig. 5.5. Zones in which the water molecules are influenced in various degrees by the ion.

(2) Intermediate zone of disturbances C, where the lack of ordering of the molecules is greater than in liquid water.

(3) External nondisturbed zone D, with the structure of bulk water.

Destroying of structure around the ion in the intermediate zone is due to equilibration of two competitive factors influencing the ordering: the internal one (ion) and the external one (bulk water).

Interaction of ions and water molecules consists of water-cation and water-anion interactions. The absolute free energies and enthalpies of ion hydration are somewhat greater for anions. The configuration of cation, corresponding to the minimal energy, is such that the two O-H bonds are oriented to the outside of the ion whereas the centers of negative charges (at the oxygen atom) are towards the ion. Therefore this bond has a different character than the hydrogen bond. On the other hand, since the two free electron pairs of oxygen are bound to the cation, this molecule can form only two hydrogen bonds with other water molecules. In the case of anions the protons are oriented towards the ion, which makes the configuration of the closest water molecules more dependent on the size of the central ion as compared to the cation.

Ions of small radii and the multivalent ones (Li^+, Na^+, H_3O^+, Ca^{2+}, Al^{3+}, OH^-, F^-) increase the viscosity of water—they show a structure making ability. They produce, apart from polarization, the immobilization and electrostriction of water molecules as well as the decrease of entropy (due to 'additional ordering') in the second hydration layer. This '*second hydration layer*' has been defined as a shell formed by the water molecules, which without real loss of translation motions are still subject to ordering by the ion.

Large monovalent ions generally give a structure-breaking effect (enthalpy increase). In view of the repulsion of dipoles in the solvatation shell the relatively weak electrostatic field around those ions can produce polarization and immobilization of water molecules in the first layer, in the further layers only the structure breaking effects occur. In this way K^+, NH_4^+, Cl^-, Br^-, I^-, NO_3^-, IO_3^- and ClO_4^- ions increase the mobility of water. The groups forming hydrogen bonds, e.g., $-NH_2$ or $-OH$, may show an insignificant influence on the water structure, as when they build in into groups of water molecules they cause only small deformations, and their tendencies to form and disrupt the groups depend only on the parameters of the medium.

Nonpolar substances have a very strong structure-forming influence on water. A scheme of such an action according to the hypothesis of Scheraga is shown in Fig. 5.6. According to this hypothesis, when a cluster of water molecules (aggregate) comes into contact with the hydrocarbon, the water structure is formed and the water molecule coordination number is increased to 5 (Fig. 5.7). More details regarding this supposition as to the water structure around a hydrocarbon molecule can be found in the work by Luck [9].

Changes in the energy levels are due to transitions of the water molecule from

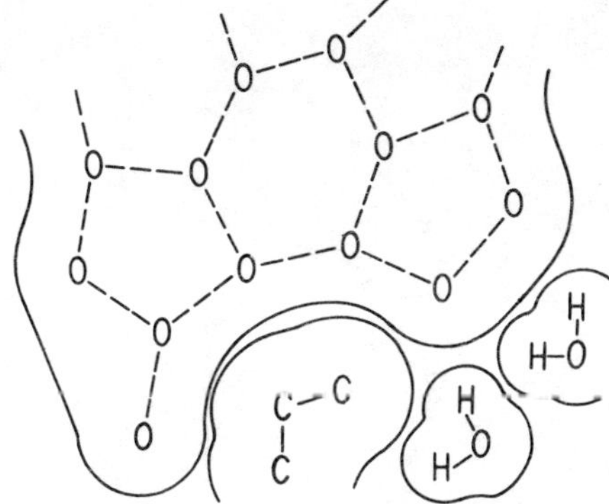

Fig. 5.6. Cluster of water molecules around the dissolved hydrocarbon molecules.

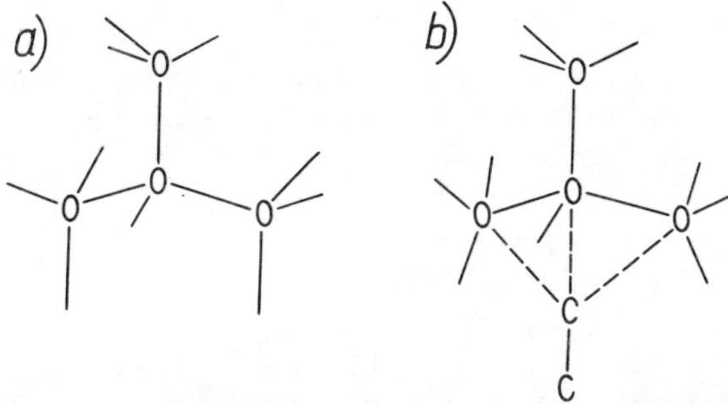

Fig. 5.7. Change of coordination number of water molecule: (a) normal molecule aggregate, (b) molecule aggregate in neighborhood of a nonpolar molecule. Dashed lines represent non-bonded coordination. According to [3]

the normal structure to the structure occurring in the vicinity of a nonpolar, dissolved substance (Fig. 5.8).

This is only observed in the first layer of water molecules. Taking into consideration that in water the majority of molecules has four hydrogen bonds—the energy level of the monomolecular layer surrounding a hydrophobic substance is lower—such layers will be formed spontaneously. This makes the structure forming influence of hydrocarbon residues obvious.

The increase of the coordination number of water molecules in the neighborhood of hydrocarbon decreases hydrocarbon-hydrocarbon interaction, whereas the water-water interaction remains unchanged. This occurs when the normal water structure defects (holes) are formed, while the density increases at other sites. In the case of full 'ordering' the hydrocarbons do not enter forcedly into it, since the energy of the hydrocarbon-water bond is smaller than that of the

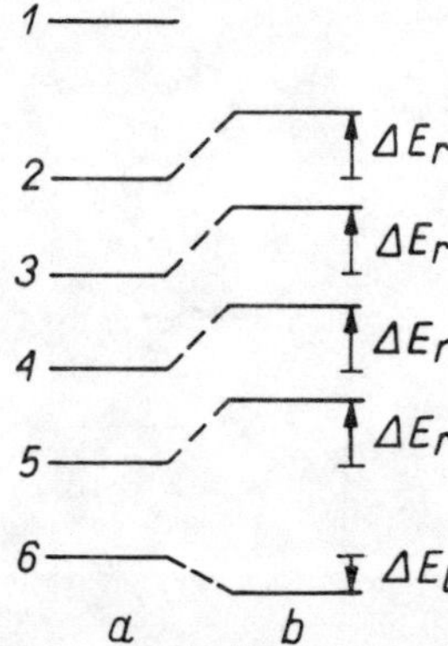

Fig. 5.8. Energy levels expressed in arbitrary units: (a) in molecules of bulk water, (b) in the first layer surrounding the dissolved substance; 1 - vapour, 2 - non-bonded molecules, 3 - molecules with one hydrogen bond, 4 - molecules with two, 5 - three, 6 - four hydrogen bonds.

water-water one. One must, however, keep in mind that hydrophobic interactions have an opposite (endothermal) effect to the formation of hydrogen bonds.

5.4. Collagen-water system

The collagen-water system is a part of the general problem approached already above, i.e., of interaction of biological macromolecules with water. In parts it was studied for many years, and very many opinions and theories were formed. However, before modern experimental techniques were introduced into research work and the macromolecule structure was recognized, hardly anything more than suggestions could be made. Perhaps the only exception were the calorimetric experiments of Kanagy [10] who was able to estimate the kinetics of water-to-collagen binding, expressing it in the form of exact figures as well as giving the relevant thermodynamic parameters. Now, due to the development of more sophisticated experimental techniques like, e.g., DTA, NMR in several modifications, dielectric measurements, X-ray analysis and obtaining accurate ideas about the collagen molecule structure, it became possible to have a much deeper insight into the problem. Its importance hardly needs explanation. Collagen both in live matter as well as in leather making processes is continuously involved in interactions with water. The last steps of leather making, including drying, are factors controlling the process limits, since it is known from experience that overdrying of leather can change its properties in an irreversible and undesirable way. In the last decade the introduction of man-made synthetic leather has shown the question of water economy of leather and the leather substitute during wearing from another point of view. As yet the efforts to imitate the peculiar way of water balance, which occurs in leather, remain without great success. Thus there are several important reasons in biology, medicine and industry, for attention being focused on the collagen-water system. It deserves notation that only few works have recently been devoted to the system taking account of the influence of tanning operations and agents, probably due to its extreme complication.

The first who was able to observe experimentally the behavior of water molecules, or more exactly water protons, and distinguish them from collagen protons, was Berendsen in 1962 [11]. He used a wide-line NMR spectrum technique, in which it was possible to observe water protons as rotating more quickly, i.e., giving much narrower peaks than slow rotating collagen protons, giving rise to very broad lines (see ch. 7). Berendsen's interpretation of the observation of changes in shape and number of narrow peaks and their position was that water bound to collagen forms a kind of chain, parallel to the collagen molecule chain. This interpretation was later questioned by Dehl and Hoeve [12] and discussed by many authors. Now this 'snake' model is considered to be true [13] again, especially because it is strongly supported by the Rich-Crich collagen molecule model II. The NMR wide-line spectrum of the collagen-water system has been elucidated and interpreted in almost every detail in numerous papers [e.g., 14, 15]. In those works, however, mostly tendon collagen was used, as it consists of macro-oriented fibers, which is necessary in orientation studies by the NMR method. The matter of strength of the collagen-water bond could be explained using the differential thermal analysis (DTA) method. This method, applied for the first time to investigate the considered system by Witnauer and Wisnewski [16] and developed later, permitted to differentiate between bonds of various kinds, as they had been calculated by Bull and Breeze [17]*. These authors have indicated the possible difference in the number and binding strength of water molecules, bound to peptide bonds or to the polar groups of side chains (-COOH, $-NH_2$, -OH) via hydrogen bonds.

The method of the testing of dielectric properties of collagen started by Hanss et al. [18], and developed by Grigera et al. [19], gave the possibility to evaluate the time of reorientation of water molecules bound to collagen. This may be considered as a step forward in the evaluation of the strength of the water-collagen bond.

Recent investigations of Bieńkiewicz et al. [14] carried out by the NMR method indicated the influence of $chromium^{3+}$ salt, used as tanning agent on the system.

The present view on the collagen-water system can be formulated as follows. There are two water molecules per tripeptide unit, firmly bound by hydrogen bonds to the helical part of the collagen molecule. Their residence time in their sites is about $0.1–1.0$ μs. This water accounts for more than 35% of collagen weight. The remaining part of water, in a not strictly limited amount, which is in weak interactions with a number of different sites, forms a multilayer with

---

DTA method is explained on p. 257.

liquid-like properties. These molecules, comparatively 'free,' participate in a rapid exchange between each other and have rotational correlation times of less than 10^{-10} s. This is not more than one order of magnitude longer than in bulk water. Such a system is in accordance with the Rich-Crich collagen model II and explains many observed phenomena [20].

The first 'kind' of water does not freeze at 0°C or, as it might be thought of for bound water, at temperatures from 10–20°C lower. This fact, now easy to understand, provoked some doubts and discussions. These water molecules had been already immobilized in 'cages' among peptide chains and involved in another pseudocrystalline lattice. It was possible to estimate the quantity of those molecules only through a very slow procedure, as every additional amount, crystallizing, could engage some bound molecules. Thus the figure of 35% was frequently questioned. The bond strength in the considered amount is different, and by some techniques is differentiated into several groups. For instance, Nomura et al. [21], using a very special technique of testing the rigidity of collagen membrane at low temperatures by dynamic mechanical spectroscopy and X-ray diffraction, have observed two peaks of rigidity at 150 K and at 200 K. Based on their experiments, they assumed four 'regions' of hydration of collagenous tissue: the first one below 7% water content, ascribed to structural water; the second between 7 and 25% is bound water; the third between 25 and 45%—a transition region, in which both bound and 'free' water were sorbed; and the fourth over 45%—with free water. The quoted authors defined as structural water the part, which was shown by the X-ray diffraction method as molecules located within the crystal, as bound water—having properties measurably different from those of bulk water, measured by the same technique. The point of view of the quoted authors may be still subject to discussion, as the 'free' water is not really free, which may be demonstrated, e.g., by the NMR technique, which was not used in their experiment.

A pulsed NMR (spin-echo) technique has been applied in the investigations on the collagen-water system as well. This technique is based on the use of radio frequency pulses (5–60 mHz) in the magnetic field. Due to the pulses the sample magnetization vector rotates over an angle of 90° and decays after the pulse is turned off in a way permitting to calculate the relaxation time. A detailed description of the principles underlying the method can be found, e.g., in the work by Poole and Farach [22]. By that method Fung et al. [23] have found the relaxation time T_1 for water bound to collagen in the range of 50–100 ms in dependence on the temperature and water amount. This relates, however, to the water content not exceeding 0.55 g per gram of collagen, which does not freeze until it reaches -90°C. The quoted authors were right to indicate that not all the water molecules, when in this amount, could be bound simultaneously to collagen though they could be described by a single distribution function.

If there is more water in the collagen surrounding, a more distant interaction takes place causing an increase of the average melting point to $-8--10$°C, which is in accordance with general observations. Aksenov [24] brought attention to the fact that the freezing of the aforementioned part of water results in an increase of the integral of the broad ('collagen proton') line. This happens only when the binding energy is close to 11–12 kcal/mole (46–50 KJ/mole) and indicates that the lifetime of such molecules is equal to 0.1-1 ms. The latter observation has been confirmed by the DTA method.

One question that remains open is what kind of bonds occur. In the opinion of the present author those bonds result from the general interaction between water molecules in the surrounding of the macromolecule and, on the other hand, from a certain influence of the hydrophobic regions of it, which press the water molecules more closely together.

Mrevishvili and Sharimanov [25], using low-temperature calorimetry and the NMR high resolution technique, concluded that 35% of water in the collagen structure did not freeze even at 4 K. Testing the heat capacity of hydrated collagen the quoted authors found a freezing peak at -8 - $-10°C$ when the water content exceeded 35%; when the water content exceeded 50% they observed a thermal effect at about 0°C. Thus they suggested first a 'hydration shell' consisting of 35% water, and a second one between 35 and 50%. The freezing temperature of -8–10°C for the water content exceeding 35% (but not limited to 50%) was estimated previously by Bieńkiewicz et al. [14]. The quoted authors payed attention to the pH of collagen investigated, and were able to demonstrate the pH-dependence of the freezing temperature and NMR-line shape, thereby connecting the water binding ability with ionization of side chain polar groups.

Whatever the way of water binding may be, the limit of 35% is repeated in recent publications, and because it is now justified by analysis of the Rich-Crich model, we can accept it as being definite. Certain differences may arise from the way the sample was prepared and the kind of material used; both factors vary greatly in various publications.

Bieńkiewicz [26] demonstrated that even a very long drying of samples at about 100°C is still insufficient to remove the total amount of water from skin collagen. The Fischer method of water determination allows us to detect 1–2% of water in collagen, dried to constant weight. Dubinska et al. [27] showed that this residual water may be removed, when the sample is dried to constant weight in a stream of nitrogen. Besides the quoted papers little attention was paid to the accurate estimation of water content, operating with collagen of a truly known hydration degree.

The formalism of calculation of binding water to collagen is based on the BET (Brunauer-Emmet-Teller) adsorption isotherm, according to which the amount of water X bound to the macromolecule at equilibrium water vapor pressure is:

$$X = \frac{V_\mu CA}{1 - A} \times \frac{1 - (n + 1) A^n + nA^{n+1}}{1 + (C - 1) A - CA^{n+1}}$$

where V is the gas volume filling the first monomolecular layer, A is the partial pressure of water vapor in the gas phase $\left(\dfrac{P_q}{P_s}\right)$, n is the maximal layer number, and C is the energy component $\left(\exp - \dfrac{E}{RT}\right)$, where E is the heat of adsorption of the first layer minus the heat of condensation. If we take $n = \infty$, the equation is thus simplified:

$$\frac{A}{X(1 - A)} = \frac{1}{V_\mu C} + \frac{(C - 1)\,A}{V_\mu C}$$

This equation fits experimental data for $n = 5$, and the relative humidity of 40% or less. For n to equal 6 it must agree with experimental data to the relative humidity of 45%; if $n > 6$ the experimental results are no longer comparable with calculation. Better accordance is obtained with D'Arcy-Watt adsorption equation [28]. However, their theory is based on the assumption that water dissolves in a solid, so that there is no real possibility to answer the questions of a binding mechanism. A further disadvantage of that theory is that four constants in the formula proposed have to be calculated for the given process by iteration.

Grigera and Berendsen [20] have performed their calculations of the collagen hydration model using the Guggenheim's approach, based on statistical mechanics. In this approach the assumption is made that the partition functions of a water molecule in the second and higher layers are equal $(= q)$ while the partition function in the first layer $q_1 = cq$. With N indicating the number of sites per given amount of adsorbate, the number n of adsorbed water molecules is:

$$n = \frac{N_c\,\dfrac{p}{p^*}}{\left[1 + (c - 1)\,\dfrac{p}{p^*}\right]\left[1 - \dfrac{p}{p^*}\right]}$$

N and n are expressed in g H_2O/100 g collagen, p^* is a parameter related to q

$$p^* = p_o q^{-1} \exp\,(\mu_L/RT)$$

where p_o is the saturation pressure and μ_L is the thermodynamic potential of the liquid. Thus p^* indicates how much the standard thermodynamic potential of a water molecule in the second or higher layers deviates from that of the liquid state; q is the partition function of the molecule.

The fraction of the total number of water molecules occupying each of the specific sites per three amino acids can be derived. In this case:

$$f = \frac{6}{N} \left[1 - \frac{p}{p^*} \right]$$

The values of f found for the Guggenheim and the BET models are not very different in the range of experimental water contents tested by NMR. The amount of 'bound' water in the first layer is close to 35%.

The formulas given here are in accordance with experimental data obtained for bound water. When the water content is higher, the calculations are no more fully valid.

From practice we know that the amount of water, equal or close to the bound water value, is still relatively low, but if collagen, skin, hide or leather contains water, it seems almost dry, or at the utmost slightly wet. Water, which makes skin swell, is far beyond the limit indicated. This is another kind of water being in weak interactions with collagen. As was just mentioned, some properties of this water will change; for instance its rotational correlation time is greater than that of bulk water. However, it is difficult to distinguish this water in the NMR spectrum, as usually the weighted average of behavior of all water protons is observed and the calculated values of reorientation times calculated are of the order of t = 0.1-1.0 μs. The results obtained by the measuring of dielectric properties in a wide frequency range (from 100 kHz to 23 GHz) by Grigera et al. [19], brought some proof of weak bonds existing between this water and collagen. Assuming that a fraction of water is bound to collagen with residence times of the order of 10^{-6} s, they concluded that the remainder exhibits rotational rates of the order of 10^{-10} s. This is the first clue of interaction of collagen and this water (which perhaps may be called 'swelling' water). The second one is given by lowering of its freezing temperature, which is about $-8°C$ according to the present author's data and between -10 and $-15°C$ according to Dehl [12]. Surely that decrease of the freezing points is again an average and depends on the amount of water (the present author observed it at a relative humidity of 100–120%) as well as on the investigation method. How great the amount of water weakly bound may be, has not been estimated yet, but one should realize, however, that it may be quite high. The long range interactions in water can be unexpectedly extended. The research performed by Hori may be quoted here [29]. He found that water between glass plates is easily supercooled to $-40°C$ in a 20 micron layer, due to the interaction resulting from a forced molecule orientation. The strength of hydrogen bonds between 'swelling' water and collagen is about 1–2 kcal/mole (4-8 kJ/mole), and may be determined by the spin echo NMR-technique [30].

There are no sharp limits between strongly and weakly bound water, nor between weakly bound and completely free water. This can be easily understood, if we consider the variability of bonding possibilities. The influence of additional components involved, like the presence of neutral salts (e.g., NaCl), has been less investigated; Grigera and Berendsen [19] were able to demonstrate that NaCl is not of significant influence; inhowfar these factors are absent often depends on the purification technique.

5.5. Glycosaminoglycans and water

When discussing the significance of water in connective tissue, one has to pay particular attention to glycosaminoglycans. The glycosaminoglycans-water systems are tested as frequently as protein-water systems. However, the evaluation of the tricomponent glycosaminoglycans-protein-water system has not been made yet because of the great number of independent variables it involves.

Glycosaminoglycans under biological conditions occur either as gels or as solutions of sol character. The difference is that the gel is, in principle, a solid. The phase transition between the concentrated solution and the gel may be defined as the point at which the solvent and solute become homogenous. This occurs when the macromolecule chains become a random network, or when microcrystallites are formed. In the polymer network many kinds of forces occur; ionic and hydrogen bonds as well as ion-dipole and dipole-dipole interactions.

Investigations on the sorption of water vapor on glycosaminoglycans have shown that the amount of water adsorbed depends on the amount of polar groups accessible in the polymer and on the physical state of the latter. Glycosaminoglycans can be ordered according to their sorption capability at low pressures of water vapour as follows: heparin > chondroitin-6-sulfate > dermatan sulfate > chondroitin-4-sulfate. The sorption capacity of these substances does not exceed 36%. The sorption isoterms suggest that monomolecular layers are formed like those in proteins: for per one disaccharide unit we have a maximum of 1.2 molecules in chondroitine-6-sulfate and 2.4 in heparine. Sorption processes are mostly equilibrium processes, as the sorption isoterms have a significant hysteresis. The width of the hysteresis loop is reversely proportional to the sorption capacity. It has not yet been estimated which functional groups of glycosaminoglycans bind water molecules. Investigation by IR-spectroscopy has shown that in this bond all polar groups are involved (OH^-, COO^- and SO_3^{2-}). Swelling of hyaluronic acid is greatest in water solutions: the swelling degree decreases with the increase of ionic strength of the NaCl solution. This may contribute to the swelling on soaking.

The gel can be obtained by concentrating the solution and entangling the polymer chains by cooling, solvent evaporation, addition of dehydrating substances or by combined application of these operations. These processes are of special importance for obtaining commercial concentrates of alginates, pectines and agar-agar. Irreversible gels are formed when the polymer chains are connected by covalent bonds; they are resistant to high temperature and addition of solvent. Reversible gels are formed when the polymers are bound by weak bonds. When heated these gels are converted to sols; they also dissolve when more solvent is added. Linear molecules gelate easier than the branched ones. Helical molecules show a tendency to form internal bonds, and straight molecules react with each other. For steric reasons branchings and side chains may hinder the formation of crosslinking bonds.

In reversible gels the interactions between macromolecule chains result from the formation of hydrogen bonds, when the distances between chains are small, and Coulomb forces result when those distances are greater. The occurrence of this phenomenon was shown by an example of agar gel by the NMR method based on the change in the signal width and on calculation of the surface under the lines. Such a broadening of the water proton line was not observed in hyaluronic acid when the gel concentration was increased. This broadening of the line is due to the change of the water structure by a hydrogen bond occurring between the water molecules and glycosaminoglycans polar groups. Hyaluronic acid gels show at pH 2.5 maximum values of the stability module (G') which is the measure of energy stored in the substance in the form of elasticity and loss module (G'') which is the measure of energy dispersion as a result of viscosity change. Solids swollen by the action of solvent belong to the simplest colloidal system. The ionotrope gels containing an additional small cation belong to two complex systems. It has been shown repeatedly that glycosaminoglycans bind in a stoichiometric reaction cationic dyestuff, the energy of that bond being of the order of 8 kcal/mole (33 kJ/mole) of dyestuff. This value was calculated from the shift of an absorption band towards greater wavelengths.

The transportation of a macromolecule through a compartment consisting of polysaccharide molecules, can proceed through or between the elements of the compartment (Fig. 5.9) [31]. Diffusion or hydrodynamic flow provides an ad-

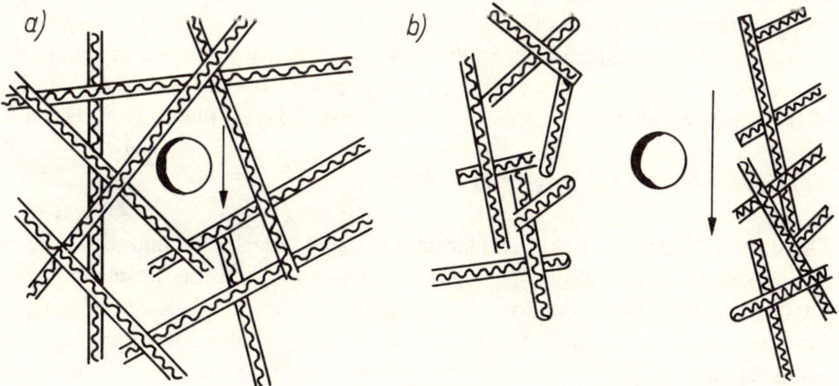

Fig. 5.9. Two ways of transport of a macromolecule through connective tissue: (a) through the glycosaminoglycan compartment, (b) between the compartment elements. According to [31]

equate driving force. Like in gel chromatography, the quickest are the biggest molecules, the slowest the small molecular components.

The macromolecules are transported quicker than the compounds having small molecules because the macromolecules do not penetrate the sorbent molecules. Polysaccharides may influence the transport velocity in either a steric or energetic way; in the latter case as polyelectrolytes. The theory of gel column chromatography is well known and its detailed explanation can be found in related manuals.

In chromatographic technique no glycosaminoglycan gels of animal origin have been used. It is, however, known that they largely influence the physicochemical properties of skin. It can be expected that the effect of glycosaminoglycans on the behavior of water in skin is greater than that of collagen, since the former are more spread out and thus more accessible. Hydration of polysaccharides, as it follows from the NMR spectrum of the substance in D_2O solution, is directly dependent on the amount of -OH groups occurring in a mere unit.

Aizawa et al. [32] found while investigating aqueous agarose gels by the NMR method at least three kinds of water bonds of various strength, temperature dependent, according to the line width analysis.

In many molecules of organic compounds the distances between oxygen atoms are of the order of 0.48 nm. That distance is almost identical to that between oxygen atoms in the ice lattice extrapolated to 25°C. Thus, assuming the existence of ice-structured clusters in water, one can expect a co-operative hydrogen bond between water and the organic molecules, e.g., biotine, 1,4-quinone, D-glucose or acylglycerides. These observations have led to the assumption that water functions as a 'structural cement,' e.g., in biological membranes.

Comparison of the transportation ability of ions in water and ice suggest that the mobility of alkali metal ions decreases and that of protons and hydroxyl groups increase as a result of the secondary action of the structure forming agent.

Water vapor pressure is, for spatial reasons, much lower over water in capillaries than it could be expected. One can presume that the degree of ordering of the water structure is higher in capillaries and in layers than it is in the bulk.

5.6. Water permeability of skin

The external surface of skin of land animals is permeable to liquids to a very limited extent. Quantitative investigations of skin permeability have shown that after the death of the animal, there is no substantial change in the permeability. Observation of the behavior of skin samples in the apparatus shown in Fig. 5.10, using the isotope tracer technique, made it possible to accurately determine the permeability. In this apparatus the internal surface of skin is in contact with a physiologic salt solution and the amount of the substances trapped in the tissue

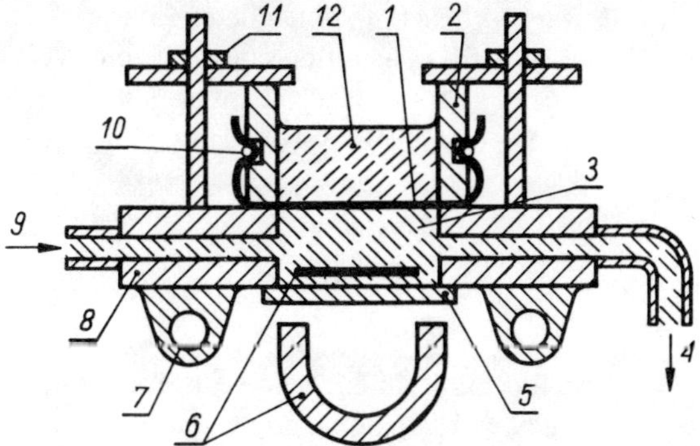

Fig. 5.10. Principle of the apparatus for testing skin permeability: 1 - skin, epidermis upwards, 2 - stainless steel ring, 3 - physiologic salt solution, 4 - outlet, 5 - glass plate, 6 - magnetic stirrer, 7 - thermostating water, 8 - brass block, 9 - inlet, 10 - clamp, 11 - gasket, 12 - solution to be tested.

is determined after the flow test is finished. Some edge effect is observed as the sample, as it may be seen in the figure, is bent at a straight angle. Of course, if the investigations have, e.g., a toxicological aim, it is always recommended to check the results on live animals. Penetration of water through skin is time-proportional, except for the initial period (Fig. 5.11). That period, denoted as

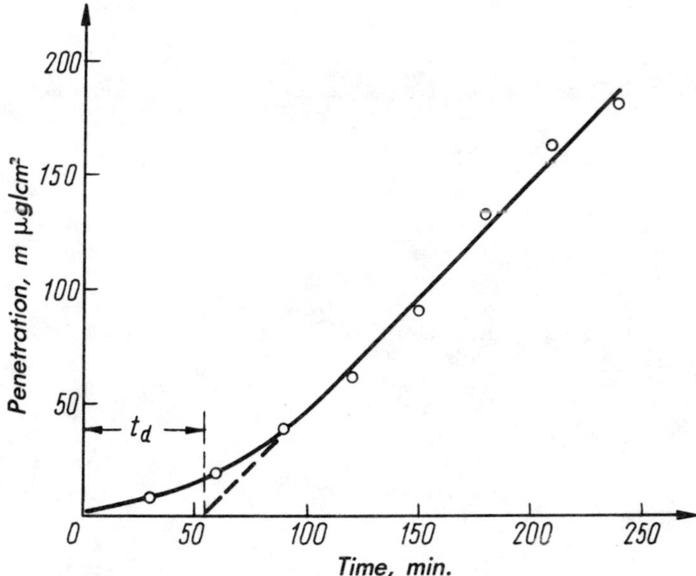

Fig. 5.11. Transport of Na$^+$ ions through human skin. Permeation of 0.9% NaCl solution, labelled with ^{24}Na was tested.

delay period t_d, is necessary for reaching dynamic equilibrium. The slope angle of the straight line is defined as the penetration rate r. Of course r is only constant if the penetrating substance does not produce changes in the skin (e.g., swelling under the action of acid). The permeability constant p is defined by Fick's first law:

$$p = \frac{r}{c}$$

and given in units of length per unit time; c is the concentration, most frequently expressed in $\mu M/min$ or in cm/s 10^{-8}. Some permeability data are given in Table 5.1. Not all the values are exact, as the results of various authors are in disagreement. The experimental technique plays an important role. It is seen from the table that human skin is less permeable to ions than skin of, e.g., rodents. This is confirmed by the electric conductance of skin. Even though this is opposite to the common opinion, one has, however, to take into consideration the skin

Table 5.1

Skin permeability to different substances

Penetrating substance	Skin	Permeability constant p, $\mu m/min$
Water	human by evaporation (I, E)	100-120
Water	human by penetration (I)	20
Water	human (E)	45
Water	rabbit (E)	147
Boric acid 10%	human (I)	0.6
AlCl (OH)$_2$ 20%	human (E)	0.02-0.12
HgCl$_2$ 1-10 mg/l	guinea pig (I)	15-45
Potassium chromate	guinea pig (I)	20-30
0.02-0.4 molar	rabbit (I)	52
KCl 1-2%	human (I)	1.1
Urea in H$_2$O	rabbit (E)	3
Ethanol	rabbit (E)	60
Ethanol	human (E)	4
Thioglycolic acid	rabbit (E)	110
Tricresyl phosphate (subst.)	rat (E)	0.004
Human serum protein	rabbit (I)	0.015
Paraffin	pig (E)	30

E—testing on skin cuts samples
I—testing on live organism

permeability of substances with small molecules and of macromolecules. This is a fundamental problem for soaking and other beamhouse operations. For these operations it is essential that Fick's law be obeyed when water penetrates into or through skin. The ability of ions and nonionic substances, e.g., surfactants, to penetrate skin is of great importance for the curing and soaking processes.

Molecules penetrate the skin in a passive way—by diffusion. Three arguments speak in favor of that:

(1) The specific permeability remains unchanged for a long time after the skin is removed from the animal.

(2) Diffusion obeys Fick's law, an exception is the sodium and potassium ions which are actively absorbed by the skin.

(3) It can be proved that stratum corneum of epidermis is resistant to the penetration of various compounds.

It has been shown that *stratum corneum* is fully responsible for the arresting of alkyl phosphoric compounds in skin. They penetrate easily, however, through skin from which *stratum corneum* has been removed.

Various factors affect skin permeability. There are differences between species, e.g., human skin is much less permeable than animal skins; it is particularly impermeable to ions, perhaps due to its nakedness and relative thickness of epidermis. In Fig. 5.12 the permeability of several kinds of animal skins are compared with human skin. The values shown in the figure are relative; they

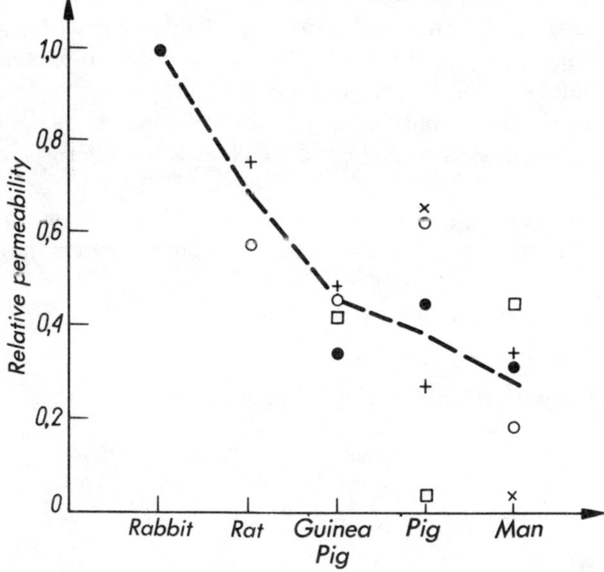

Fig. 5.12. Relative permeability of skin for various chemicals: ● - water, o - thioglycolic acid, x - physiologic salt solution, + - ethylene bromide, □ tricresyl phosphate. The permeability of rabbit skin was assumed as unity.

are referred to rabbit skin permeability assumed as a unity. The permeability of skin decreases with biological age; it also decreases to half its value as the temperature decrease by 10 K.

The above theory is contradictory to Laurent's opinion [31], who suggests that penetration through glycosaminoglycan gels proceeds according to the molecular sieve mechanism. Experimental proof is insufficient, however, to eliminate one of these opinions. One more model that could explain the behavior of skin as a factor controlling penetration of a substance into and from the organism is the model of biological membranes. Recently, a great effort has been devoted to the investigation of the pattern, both from the experimental and theoretical point of view. The membrane is considered to be the basic element in a system of transport functions of the organism. The Golgi apparatus, endoplasmatic reticulum, cell membrane and others are connected with the notion and operation of membranes.

Biological membranes consist of protein-lipid complexes where the protein to lipid ratio is specific for a given membrane. The kind of lipid substance depends on the origin of the membrane: in plants these are glycolipids, in animals—sphingolipids. Both kinds of lipids occur in bacteria. Membrane proteins have the molecular weight of 20,000-50,000. The general amino acid compositions of some membranes selected as examples are shown in Table 5.2. In the table the great amount of nonpolar amino acids should be noted. There are very small amounts of cystine and cysteine, so is the number of disulfide bridges. It is possible that the limited number of crosslinking bonds in the protein molecules is the cause of the pronounced flexibility of the membranes probably decisive for their operation in the course of active transport. It has not been decided yet whether the lipid-protein interaction consists in binding the polar groups of lipids with the protein, or whether hydrophobic bonds are formed between polar centers of amino acids of the protein and the nonpolar residues of the lipids. Two models of biological membranes have been based on those interactions.

Investigations of membranes being protective shells and organella of the cells are of particular importance because of their functions; they are, however, very

Table 5.2

Amino acid composition of membrane lipoproteids

Membrane	Acidic amino acids	Basic amino acids	Nonpolar A-acids	Cystin cystein
Hyalin	12.4	22.2	44.6	0.3
Red blood cell membrane	20.2	12.0	53.4	1.4
Mitochondria				
(Halobact. cutirubrum)	24.9	12.5	44.1	10.3
Cuticula	25.6	5.9	48.3	0.3

difficult from technical and preparative points of view. It seems that some information in this field can be obtained by applying resonance spectroscopy; information on hydration of the membranes can be obtained from IR-spectroscopic studies.

Water accounts for 30-50% of the membrane weight, however, little is known about its properties in this system. It was thought that water in membranes was organized in the same way as hydrated crystals or in ice. Recent works have shown, however, that the ice lattice model cannot be used without some reservations [33]. The total permeability of a membrane can also depend on water, since water, present in the lipid-protein system, participates in transport. According to Kavanau [34] the water molecules lying between the hydration layers, immobilized at the membrane, and water of fully liquid character, participate in protein transportation in the proximity of the membrane; those molecules travel subsequently in the immobile water layer (Fig. 5.13). In this way the protons, being hydronium ions in the layer adherent to the surface of the macromolecule, control water transmission through the membrane. An opinion exists [35] that in the biologic membranes hydrophilic pores are regulated by water.

The plasma membranes, forming an envelope around the cells and contributing to the maintenance of their shape have a thickness of about 6-10 nm. The proteins lipids ratio is usually 1:1, but it may vary. The main part of the membrane lipids

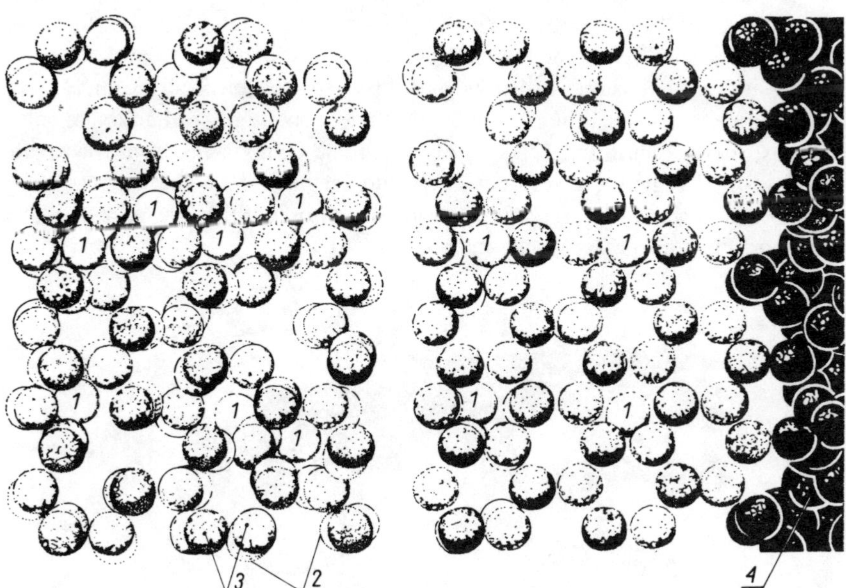

Fig. 5.13. Motion of free water molecules in the layer adherent to the macromolecule: 1 - free internodal water molecule, 2 - thermal vibration zone, 3 - water molecules in lattice, 4 - macromolecule surface.

are phosphoglycerides, glycolipids and cholesterol. The proteins occurring there are of two types: peripheral and integral. The peripheral glycoproteins are readily extractable by water or aqueous solutions. The integral ones are more tightly associated with other membrane components and can be solubilized by forcing detergents to disrupt the lipid bilayer.

A simple picture of a plasma membrane is given in Fig. 5.14. The lipid bilayer shown there is asymmetric in composition: the external layer contains saccharides localized on the external surface and phosphatidylcholine, whereas the internal layer is composed chiefly of phosphatidylserine and phosphatidylethanolamine. This composition probably results in a different fluidity of both monolyaers. A higher proportion of anionic substances on the internal side of the membrane may facilitate ionic interactions with peripheral membrane components and ones that are associated with membranes.

The globular peripheral proteins of the membrane are structurally asymmetric, having one end polar and one nonpolar, and are embedded in the hydrophobic interior. In the polar region the ionic amino acid residues and saccharide residues are clustered on contact with the aqueous phase.

Integral proteins traversing the membrane have polar ends and a hydrophobic center in between. The fluidity of the membrane is a function of the character of the acyl groups of fats of the membrane. The more unsaturated they are, the more fluid is the membrane. Saturated fatty acids and sterols harden the membrane and restrict lateral motion of the protein molecules. Such a simple model is modified by the conception that proteins are either immobilized or their mobility is hindered by the membrane—associated components at the inner membrane surface. It is suggested to add to this restricted mechanism three more: aggregation or association of membrane components through horizontal interactions, sequestration or exclusion of membrane components from (or to) the specific lipid region and restraint by peripheral membrane components at the membrane surface (see Fig. 5.14). All these possibilities are related to the transport phenomena across the membrane [36].

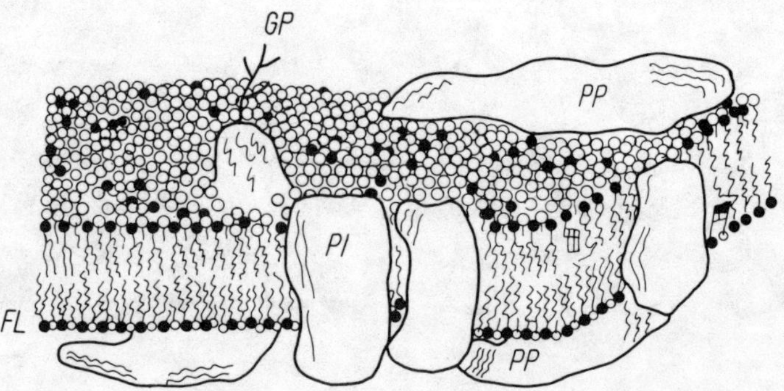

Fig. 5.14. A schematic view of the fluid mosaic membrane, with the FL phospholipid bilayer, PP the peripheral proteins, PI the integral proteins and GP glycoproteins.

REFERENCES

1. Eisenberg, D., Kauzmann, W. The Structure and Properties of Water, Oxford London 1969
2. Nemethy, G., Scheraga, H. O. J. chem. Phys., *36*, 3382 (1962)
3. Symons, M. S. R. Nature *230*, 91 (1972)
4. Privalov, P. L. Biofizika *13*, 163 (1968) in Russian
5. Franks, F. ed. Water—a Comprehensive Treatise, Plenum Press N.Y. 1975
6. Berendsen, H. J. C. in Theoretical and Experimental Biophysics vol. I, ed. A. Cole Dekker N.Y. 1967
7. Varga-Manyi, R. Arch. Biochim. Biophys. Acad. Sci. Hung., *14*, 87 (1979)
8. Frank, H. S., Wen, W. Y. Trans. Faraday Soc., *24*, 133 (1957)
9. Luck, W. A. P. ed. Structure of Water and Aqueous Solutions, Verl. Chemie Weinheim 1974
10. Kanagy, J. R. J. Res. Natl. Bur. Stand., *38*, 119 (1947)
11. Berendsen, H. J. C. Thesis Univ. of Groningen 1962 and J. chem. Phys., *36*, 3297 (1962)
12. Dehl, R. E., Hoeve, C. A. J. chem. Phys., *50*, 3245 (1969)
13. Hoeve, C. A., Kakivaya, S. R. J. Phys. Chem., *80*, 745 (1976)
14. Bienkiewicz, K. J., Berendsen, H. J. C., Andree, P. J. Rocz. Chemii *51*, 149 (1977)
15. Lindner, P., Forslind, E. Chemica Scripta *3*, 57 (1973)
16. Witnauer, L. P., Wisnewski, A. J. Am. Leath. Chem. Assoc., *59*, 598 (1964)
17. Bull, H. B., Breeze, K. Arch. Biochem. Biophys., *128*, 497 (1968)
18. Hanss, M., Herbage, D., Compte, P. J. Chim. Phys., *65*, 176 (1968)
19. Grigera, J. R., Vericat, F., Hallenga, K., Berendsen, H. J. C. Biopolymers *18*, 35 (1979)
20. Grigera, J. R., Berendsen, H. J. C. Biopolymers *18*, 47 (1979)
21. Nomura, S., Hiltner, A., Lando, J. B., Baer, E. Biopolymers *16*, 231 (1977)
22. Poole, Ch. P., Farach, H. A. Relaxation in Magnetic Resonance, Acad. Press N.Y. 1971
23. Fung, B. M., Witschel, J., McAmis, L. L. Biopolymers *13*, 1767 (1974)
24. Aksenov, S. I. Biofizika *22*, 923 (1977) in Russian
25. Mrevishvili, G. M., Sharimanov, Ju. G. Biofizika *23*, 242 (1978) in Russian
26. Bienkiewicz, K. J. Leder *22*, 222 (1971)
27. Dubinska, W. A., Nikolaeva, S. S., Mikhailov, A. N. Kozh. Obuv. Prom., *17*, 10, 39 (1975) in Russian
28. D'Arcy, R. L., Watt, J. C. Trans. Faraday Soc., *66*, 1236 (1970)
29. Hori, T. Taion Kagaku A *15*, 33 (1956) in Japanese
30. Westover, C. J., Dresden, M. H. Biochim. Biophys. Acta *365*, 389 (1974)
31. Laurent, T. C. Fed. Proc., *25*, 1128 (1966)
32. Aizawa, M., Mizuguchi, J., Hayashi, S., Suzuki, T., Mitomo, N., Toyama, H. Bull. Chem. Soc. Japan *45*, 3031 (1972)
33. Khanagov, A. A. Biopolymers *10*, 789 (1971)
34. Kavanau, J. L. Water and Solute-Water Interactions, Holden-Day San Francisco 1964
35. Askochevska, N. A., Petrinov, N. S. Usp. Sovr. Biol., *73*, 288 (1972) in Russian
36. Nicholson, G. A. Biochim. Biophys. Acta *457*, 63 (1976)

6.

PROBLEMS OF TRANSFER OF SUBSTANCES IN THE ORGANISM

Organic fluids are water solutions and suspensions of compounds of molecular weight varying within very broad limits. Their circulation in the body is partly due to hydrodynamic motion, mostly in capillar vessels; in part, it is evoked by the physicochemical action of dissolved substances. The understanding of the mechanism of motion of organic fluids requires remembering the basic laws of diffusion, osmose and transportation. These problems are particularly important in the case of skin, which contains no great vessels in which the blood could be mechanically pumped; circulation of organic fluids in the skin takes place via capillar and osmotic motion. Those are the phenomena connected with life of the organism as a whole. However, they do not stop suddenly at the moment of death, but rather they slow down gradually like other life phenomena; the slower and simpler the process, the less dependence on the 'central control system.'

Diffusion, a process discussed in detail in biochemistry handbooks, consists of the increase in entropy due to the decrease of the ordering of the molecules. This means that molecules of the solvent and solute mix spontaneously. The diffusion rate dn, i.e., transport per time unit depends on the difference of concentrations and diffusion constant D

$$\frac{dn}{dt} = - Dq \quad \frac{dc}{dx}$$

where q is the cross section through which diffusion occurs and dc is the concentration gradient.

Diffusion may also proceed through a membrane, if it is totally permeable. Its rate is, in this case, limited by the membrane thickness and pore diameter. Introducing a 'membrane constant' M_c depending on the membrane thickness d and active pore surface q, the diffusion rate can be expressed by the equation:

$$\frac{dn}{dt} = - M_c c$$

where c is the concentration difference. The constant M_c defines whether diffusion is easier or more difficult. Thus the notions of facilitated and hindered diffusion are derived.

When the membrane is semipermeable, dissolution of the substance by diffusion is impossible. However, in the part of the container where the dissolved substance is present, the pressure will increase. This phenomenon is known as osmosis, and the osmotic pressure π is given by the equation:

$$\pi = cRT$$

where c is the concentration, R the gas constant, T the absolute temperature. A measure of the osmotic work ΔG is the load on the semipermeable piston which equilibrates the osmotic pressure. That work is given by the equation

$$\Delta G = RT \ln \frac{c_2}{c_1}$$

where c_1, and c_2 are the values of concentration before and after dilution. Osmosis can either produce an increase or decrease of concentration. The latter case is simpler and spontaneous. If in the intercellular spaces the concentration of salt is greater than in the cells, water is then extracted due to osmosis from the cells until the concentration becomes equilibrated. The first case refers to oncotic pressure. Such a pressure can be observed in the case of membranes permeable to compounds with small-molecules but unpermeable to macromolecules. In such cases salts diffuse without hindrance while colloidal molecules behave according to osmosis rules. Oncotic pressure is low as the number of macromolecules is small; the organism compresses liquids to higher pressures. The osmotic and oncotic pressures cooperate when water is transferred from capillary arteria to the intercellular spaces, where the motion of water is the result of arterial pressure, and when it returns to the vein sectors of low hydrostatic pressure.

Proteins and other macromolecules of the organism are polyelectrolytes. This requires their electrical neutrality, if the system is to remain undisturbed; thus in the osmotic cell there must be as many ions of opposite charge as there are ions of macromolecules. Thus, if the internal space of the cell contains Na^+ cations and protein anions, and NaCl is added to the extracellular space, then Cl^- ions will penetrate inside only if Na^+ ions will as well, despite the produced concentration difference. This phenomenon, known as Donnan's equilibrium, can be formulated according to Karlson [1] as follows:

$$\frac{[Na^+]_e}{[Na^+]_i} = \frac{[Cl^-]_i}{[Cl^-]_e}$$

or

$$[Na^+]_e \times [Cl^-]_e = [Na^+]_i \times [Cl^-]_i$$

where e and i are the concentrations outside and inside the osmotic cell, respectively.

The macromolecule 'pushes' out of the cell the ions of the same sign; thus the difference of the pH inside the cell and in extracellular space is the greater, the greater is the polyelectrolyte concentration and the smaller is the concentration of electrolytes with small molecules (Fig. 6.1).

If the permeable pores of the osmotic cell are so small that the trapped molecules cannot change places (Fig. 6.2), we then have a 'single file' mechanism of transport. Transition of A-type molecules causes the transition of B-type molecules. The membrane enveloping the osmotic cell must be strong enough to resist the pressure difference. Measurement of oncotic (colloid) pressure in the salt solutions gives values the higher, the greater is the charge of the ions of the salt. The Donnan effect decays at the isoelectric point. The solution in the space between colloid particles has an opposite charge to that of the macromolecule. Diffusion results in the dilution of the solution. Osmosis can also produce increased concentration, but that would require very high pressures, which do not occur in the organism.

Migration of a molecule which proceeds in disagreement with the diffusion or osmosis laws is called active transport. We have passive transport when the direction of migration of a small molecule is in accordance with the diffusion or osmosis laws. Active transport is an oriented phenomenon which requires input of chemical energy.

Among the many functions of biological membranes, ranging from mechanical support to cells, through location of many enzymes, prevention of solutes passing from one compartment to another to the reception and transport of chemical,

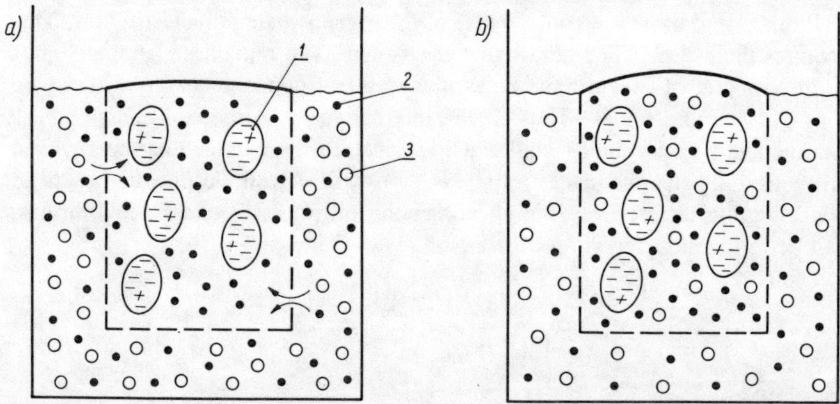

Fig. 6.1. The Donnan equilibrium (carrier transport): (a) NaCl is added to the external space, (b) Donnan equilibrium ie established, Na$^+$ ions accumulate in internal space, Cl$^-$ ions accumulate in the external space. The osmotic pressure is higher inside the cell than in the outside space, 1 - protein, 2 - Na$^+$ ion, 3 - Cl$^-$ ion.

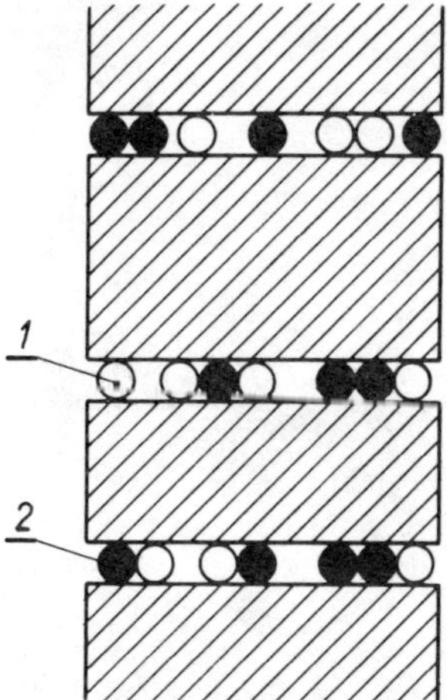

Fig. 6.2. Single file mechanism.

electrical and optical signals, the primary function is the transport of molecules and ions from one solution to another. One aspect of this function is to maintain a constant intracellular medium, especially with respect to cations. Another one is to transfer nutrients into and products (wastes or those with some mission) out of the cells.

The transport of nonelectrolytes and some ions across the membrane (Fig. 6.3) may be characterized as simple diffusion, though interactions with the membrane components, hydrophilic (moving through polar protein domains) or hydrophobic (moving through the lipid domains), take place. The high activation energy of these processes provides evidence for this, as well as obeying Fick's law.

The majority of substances, however, penetrate into cells by a process involving specific membrane proteins and show a similarity much like that of enzyme reactions. The process tends asymptotically to saturation at high substrate concentrations. It proceeds either with or without a source of energy.

According to the definition, active transport is directly coupled to an energy yielding chemical reaction (it does not obey Fick's law). By extension of the definition it may be coupled to a flow of another solute which has built up a difference in chemical (or electrochemical) potential by a truly active transport. The flow-to-flow coupling can be visualized in the form of 'carrier protein' binding both the driving (usually ionic) and the driven (usually nonionic) species in a ternary complex; the primary coupling is frequently concealed under the term *energized membrane.* This energization may occur through ATP or polyphosphate splitting, or substrate oxidation and possibly through interaction with the energy of electromagnetic radiation. The way energy coming from ATP

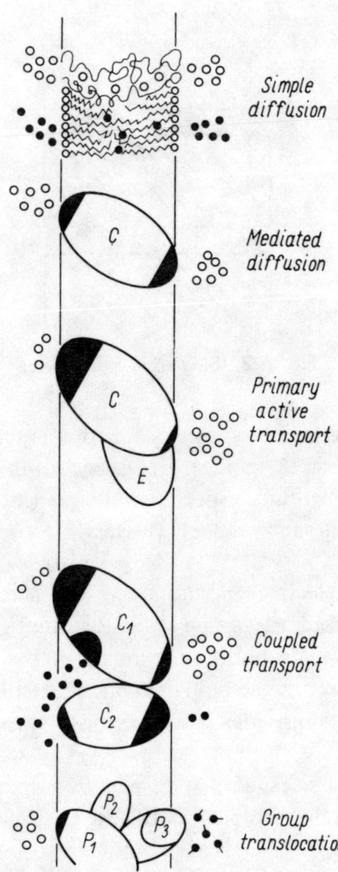

Fig. 6.3. Different types of membrane transport where C is the 'carrier,' E the source of energy and P the group translocation protein.

hydrolysis is used mechanically to move the molecule of the ion-binding unit is still unclear.

The term carrier has been used to describe an entity in the membrane which brings about the translocation of the substrate, unless it is simple diffusion. Saturable transport may possibly occur through the mediation of a single protein, changing its conformation and thus exposing its binding site to one or other membrane face (Fig. 6.4). If a change of affinity occurs simultaneously, active transport can ensue.

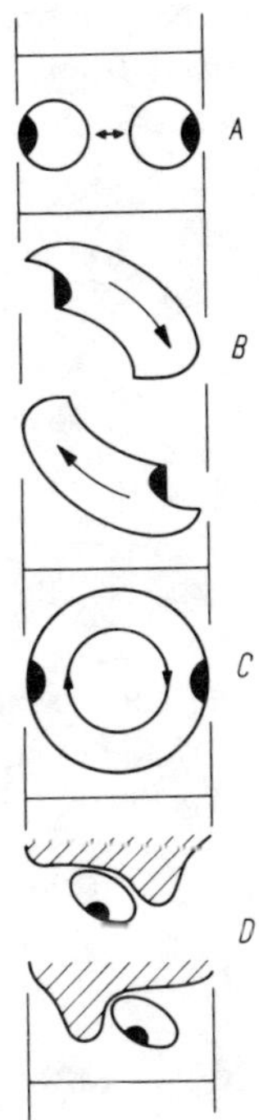

Fig. 6.4. Different modes of carrier movement in the membrane. A - translational diffusion, B - conformational flip-flopping, C - rotations, D - gate mechanism where the carrier itself does not move. Mechanisms B and D are more likely than A and C.

On the other hand, binding proteins exist that constitute only a part of the transport system. How this binding protein transmits the substrate to the carrier is unknown. It should be noted that more membrane components (Fig. 6.5) can account for the movement of the carrier across the membrane [2].

Many data have been gathered to show that a living cell accumulates ions transported in the direction of the growing concentration gradient and that this accumulation is connected to the cell metabolism. Intoxication of the cell, or lack of metabolism substrate, is a reason for loss of capability to accumulate ions [3]. After the death of the organism the cell walls allow penetration of ions both ways, so Donnan's equilibrium is established.

Apart from the ion transport mechanism problem information about energetic systems of the cell can be obtained from the following argumentation: The diffusion driving force is the electrochemical potential gradient. For 'counter-current' flow it is given by:

$$\mu_i = \frac{c_i d\bar{\mu}_i}{fdx}$$

and for flow 'with' the current:

$$\mu_o = \frac{c_o d\bar{\mu}_o}{fdx}$$

where c_i and c_o, $\bar{\mu}_i$ and $\bar{\mu}_o$ are the concentration of ions and electrochemical potentials on both sides of the barrier. Dividing one equation by the other we get:

$$\frac{\mu_i}{\mu_o} = \frac{c_i d\bar{\mu}_i}{c_o d\bar{\mu}_o}$$

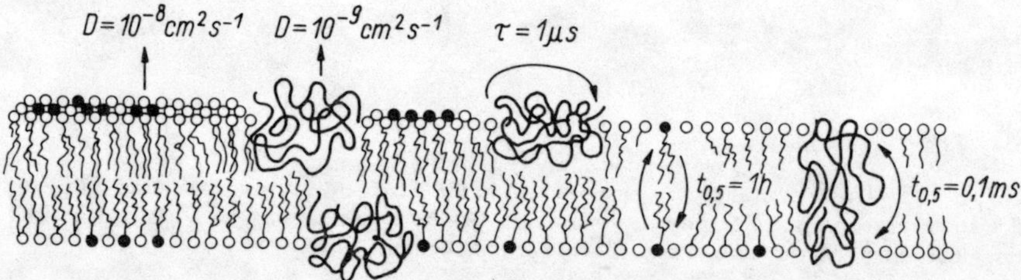

Fig. 6.5. Basic types of movement of membrane components. Lipids as well as proteins diffuse laterally with the diffusion coefficients shown. Some proteins (rhodopsin, bacteriorhodopsin) rotate in the monolayer or possibly entire bilayer about an axis perpendicular to the membrane with relaxation times ranging from factions of μs to ms. Phospholipids flip back and forth very slowly, with half-times of hours. Carrier proteins must perform this function much more rapidly, the value of 0.1 ms for the half-time having been computed for glucose transport in baker's yeast. According to [2]

Integrating the result and substituting the values necessary to calculate the potential we obtain the expression:

$$\ln \frac{\mu_i}{\mu_o} = \ln \frac{a_o}{a_i} + \frac{ZF}{RT} (\phi_o - \phi_i) + \frac{f'}{RT} \int_o^x \frac{1}{f''} \left(\frac{d\mu_{H_2O}}{dx} \right) dx$$

where a_o and a_i, ϕ_o and ϕ_i are the ion activities and potentials on both sides of the membrane, respectively, μ_{H_2O} is the chemical potential of water, f' is the friction between one mole of solute and one mole of water, f'' is the friction to which water is exposed when passing through the membrane, Z is the ion valency and F the unbound energy of the system. All the components of the equation can be easily determined except for $d\mu_{H_2O}$. The flow rate is unity when clean water does not pass through the membrane [4].

In reality the problem is much more complicated, as the situation that only one ion is transported does not exist. Each of the transported ions influences other ions present in the medium and, in turn, is influenced by them. An example how far the accumulation of ions inside the cell can go is the comparison of concentration of the potassium ion in the cell sap of sea alga *Valonia macrophysa*, amounting to 500 mval/l, with that in the surrounding sea water which amounts to 12 mval/l. Bacteria of Coli-type uptake lactose from the surrounding even if the concentration of this sugar in the cell is 2.10^3 times greater than in the surrounding [3]. However, in general, one can say that the potassium ion is the substantial one inside the cell, and the sodium ion in the extracellular systems.

REFERENCES

1. see ref. 18 ch. 2
2. Kotyk, A. Acta Biochim. Biophys. Acad. Sci. Hung., *12*, 135 (1977)
3. Bull, H. An Introduction to Physical Biochemistry II ed., Davis Philadelphia 1971
4. White, A., Handler, P., Smith, E. L., Hill, R. L., Lehman, I. R. Principles of Biochemistry, 6-th ed., McGraw-Hill N.Y. 1978

7.

SPECTROSCOPIC AND DIFFRACTOMETRIC METHODS OF INVESTIGATION OF MACROMOLECULES

In the investigation of macromolecule and changes in it among others the methods of absorption, emission and diffraction of electromagnetic radiation are used. In order to describe the information which may be obtained by these methods, some introductory remarks are necessary.

An atom or a molecule does absorb only radiation that has the energy equal to the energy of transition from one (quantified) state of energy E_i to the other, of higher energy level, say E_j. The energy of the absorbed photon, being $h\nu_{ij}$, should be equal to the difference of the energy level between two allowed excited states:

$$\Delta E = E_j - E_i$$
$$\Delta E = h\nu_{ij}$$

E where h is Planck's constant (6.62×10^{-27} erg s or 6.62×10^{-34} J s) and νn_{ij} is the frequency of radiation absorbed. This equation may be written in another way

$$\frac{1}{\lambda} = \bar{\nu} = \frac{\Delta E}{hc}$$

and because $\bar{\nu} = \dfrac{c}{\lambda}$. thus $\Delta E = \dfrac{hc}{\lambda}$

where λ is wavelength, in cm, $\bar{\nu}$—wavenumber, in cm^{-1}, and c—light speed (3×10^{10} cm $\cdot$ s^{-1}). Radiation energy is thus proportional to the wavenumber. In the spectroscopy it is often expressed in cm^{-1}. Energy units, as applied to various branches of spectroscopy, are shown in Table 7.1. The frequency ν of electromagnetic radiation is related to the wavelength λ by

$$c = \nu\lambda$$

where c is the velocity of propagation. In a transparent medium the velocity is reduced, compared to the velocity in vacuum; this reduction is given by refractive index. As radiation enters a medium of a higher refractive index, the wavelength is reduced and the frequency remains constant.

Table 7.1

Conversion of various energy units used in spectroscopy

Unit	$J/mole$	$kcal/mole$	eV	cm^{-1}
J/mole	1	2.39×10^{-4}	1.04×10^{-5}	8.34×10^{-2}
kcal/mole	4.19×10^{3}	1	4.34×10^{-2}	350
eV	9.65×10^{4}	23.06	1	80.67
cm^{-1}	12.0	2.86×10^{-3}	1.24×10^{-4}	1

Sometimes several quantum states of a molecule have the same energy levels. This is defined as degeneracy. For instance, p orbitals of the excited hydrogen atom are threefold degenerate and d orbitals are fivefold degenerate. The application of an electrostatic or magnetic field may remove the degeneracy, causing the orbitals to have different energies. Thus application of these fields provide important information about the structure. However, magnetospectroscopy is still, due to instrumental difficulties, not yet widespread.

Usually, the investigations on the structure of macromolecules are carried out in five regions of the spectrum of electromagnetic waves. The first of it covers the X-ray region, the second—UV and visible light, the third—IR region, the fourth—microwave region, and the fifth one—the radiofrequency region. The intermediate regions are less frequently, if at all, used mostly because of technical problems and interpretation difficulties. Various types of apparatuses are used in the investigation—depending on the radiation region.

The energy of X-rays is sufficient to knock out the electrons lying on internal orbitals from an atom (section 7.1).

The energy of 'vacuum' ultraviolet is sufficient to excite the valence electrons of internal shells, e.g., the electrons of σ bonds. This region gives large experimental difficulties and is relatively less known. There are also serious difficulties of interpretation, especially in the region below 150 nm. Waves of the region near ultraviolet and of visible light do excite the electrons from outer shells, which are more weakly bound to the nucleus; e.g., those taking part in the conjugated bonds (delocalized electrons), those of the lone pairs in the oxygen atom, or those displaced inside a molecule (charge transfer). Spectroscopy of this region provides information about the course of numerous reactions, e.g., building of the ligands into the molecules and complexes of chromium, around bond energy, or the presence and condition of ring molecules in bigger complexes (section 7.2).

The problems of electron transition to the excited states and of their coming back to the ground state are extended and completed by investigations in the field of secondary emission—i.e., luminescence, generated by absorption of UV or Vis-radiation absorption (section 7.3).

Analysis of the absorption of circular polarized electromagnetic radiation of the UV-wavelength, allowed to pass the solutions to be tested, gives information concerning the conformation of a protein molecule. We call this kind of investigation Optical Rotatory Dispersion (ORD) and Circular Dichroism (CD) (section 7.4).

Infrared radiation energy is not great enough to excite electrons; however, it brings the molecules to the higher oscillation and vibration states. These oscillations or vibrations result in periodical changes of the bond lengths, or angles between the bonds, or in a rotation of the molecules around the axis perpendicular to the bond. Absorption spectrum in IR may exactly define a small molecule. Information about protein molecule are more limited because of overlapping and addition of various vibrations, giving only resultants. Vibrations of the atoms that constitute a macromolecule are different due to crosslinking and to chain conformation. The spectrum is less rich in typical bonds than that of a small molecule. By using this method one may provide important information concerning characteristics of a macromolecule, its functional groups and crosslinks (section 7.5).

The relatively simple infrared technique may be complemented by the use of the Raman scattering method. Infrared radiation absorption is due to periodical variations of the molecule, whereas Raman scattering effect takes into account the induced moment. Thus, some frequencies can be observed both in Raman or infrared, others (alternatively) in Raman or infrared only, or can be observed by neither techniques.

The Raman scattering effect, very useful in investigation of organic molecules, has been developed very rapidly after introduction of the laser as a radiation source. Now it is fairly widely applied in macromolecular research. (section 7.6)

Microwaves supply the amount of energy sufficient for rotations of the molecules around their rotational axes. The energies of rotations allowed are small and they do not differ much but depend on angular speed and on the moment of inertia in relation to their axes. The use of microwaves in the investigation of macromolecules has not been reported yet.

The ranges from ultraviolet to the microwaves have their origin in three types of energy excitations and are described by the following general formula:

$$E = E_{electr} + E_{osc} + E_{rot}$$

which is only approximately correct, i.e., correct if we presume the unchanging distances between the atoms. This presumption is justified only for the equilibrium conditions or when the deviations are minute; for bigger vibration amplitudes it is not valid.

The macromolecule structure investigations by the use of radiofrequencies have been rapidly developing in recent years due to the advancement in electronics and technical possibility to obtain strong and highly homogeneous (10^{-6} %) and stable magnetic fields.

In the method of electron paramagnetic resonance (EPR), radiation causes the electrons transition to the higher spin states. In organic chemistry the EPR spectra inform us at first about the radical reactions in which the unpaired electrons occur. This method, seldom used to biopolymer testing, has recently grown in importance because of the growing interest in the changes in protein due to ionizing radiation. Also in some other studies the influence of a metal ion, when introduced to a biopolymer, has been examined. Energies of still longer waves correspond to the energy difference between the spins of nuclei and are the subject of nuclear magnetic resonance (NMR) method.

This method has several branches. One of them, the wide line (WL) technique, is preferably used in investigations on solid state. The free induction decay (FID) technique gives dynamic characteristic of some nuclei engaged in the bond. It is the most modern and theoretically most difficult one. The information obtained generally should undergo some mathematical treatment, preferably by the use of Fourier's analysis. The combining of signal detection system with computer makes it possible.

The NMR and EPR spectroscopy methods are discussed in section 7.7.

The NMR technique was formerly used almost exclusively to the investigation of protons. Now, due to the modernized apparatuses and techniques it became possible to investigate other nuclei with an odd nuclear spin quantum number. Almost every one of the elements contained in a protein molecule has isotopes with an odd nuclear spin number (^{13}C, ^{15}N, ^{17}O). The use of the NMR method to the investigation of the properties of these nuclei is now mainly a problem of apparatus ability, i.e., the detector sensitivity and homogeneity of the magnetic field.

The problems just outlined will now be discussed in more detail. However, the problems of apparatus and measurement techniques are left out here. They may easily be found in several excellent monographs.

7.1. X-ray diffraction

An atom is in an excited state when an inner shell electron is completely removed from the atomic structure. The atom is then left with a hole in its previously dense inner structure. The characteristic X-ray spectral line is emitted when an electron from a more outer shell falls into the hole. Since the change in the total energy state of the atom is held accountable for the emission of a photon, it may be preferable to say that the hole has traveled upward to a shell, rather than saying that an electron from that shell has fallen into the hole. This may occur in several steps before the hole reaches the outer surface. Eventually, an electron from some external source will be picked up in the final vacancy and the atom will then return to its ground state. In this process some typical spectral lines are emitted with definite frequencies characteristic of each discrete step.

To obtain a spectrum, the stream of electrons, coming from a heated metallic filament and accelerated in high vacuum by a potential of 10-1000 kV, is suddenly braked when it comes in contact with a metal plate (anticathode); with 1000-100 kV the X-rays of a wavelength below 0.1 Å are obtained. These rays are used in therapy and in metallurgical radiography. With 100-30 kV we have the rays of a wavelength of 0.1–0.4 Å, used in radiography of human body. In the range of 30-6 kV the wavelength of the radiation produced is 0.4–2 Å; the X-rays of this range are used in studies on solid structure. An orbital electron, say from shell K in an atom of metal, may be ejected by some accelerated, charged particles. An electron from another shell, e.g., L, will fall into this hole, emitting an energy quantum hν. The energy quantum of X-rays is great as compared to the energy quantum of visible light, because its frequency is great and wavelength very small. The X-ray wavelength depends on the position of the electron to be activated in the metal atom. Radiation, corresponding to the jumps of orbiting electrons to the K, L, M shells, is marked with the same letters. Radiation of shortest wavelength is emitted when the electrons jump from the L to K shell. It also has the most penetrating power. The energies of various shells of individual metals are different. The processes observed are defined in terms of a notation specifying the metal of the cathode, and the shell. For instance the symbol $CuK\alpha_1$ says that the source of radiation is copper, and the electron is emitted from the K shell; α means that an electron from the next (L) shell falls into the hole, and the index 1 shows which electron of L shell takes part in the process.

In order to obtain monochromatic radiation, metal foil filters or crystals as monochromators are used. These filters stop the part of X-rays having a lower energy level (longer waves) and at the same time they decrease the beam intensity. Lowered intensity of the beam does not imply a reduced energy of the photons; it only means that they are diminished in number. Photons of the beam are incident on the metal plate, and the lattice units of the metal have a length comparable to the wavelength of the photons. The diffraction of X-rays results from this collision (Fig. 7.1).

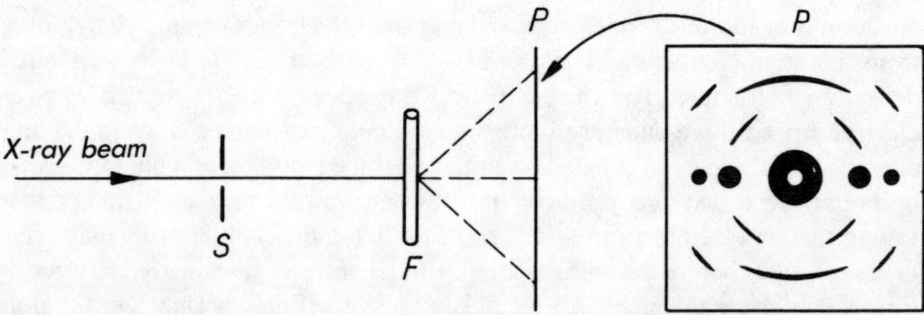

Fig. 7.1. Simplified system of X-ray diffraction investigation. S - collimation slit, F - fiber tested, P - film or counter.

This is a phenomenon of the same kind, like the diffraction of visible light made to pass through a diffraction grating.

Diffraction of X-rays may be shown as a reflection on the crystal lattice planes. To the reflected ray the coefficients are ascribed, which give the orientation of the imaginary reflecting plane of the crystal. To estimate the orientation of the reflection planes, some such planes are shown in Fig. 7.2.

The Miller indices denote the situation of the planes by the inverse values of identity period fraction, given by the planes on crystal axes, and expressed as whole numbers. The planes belonging to the same set have the same indices. The distances between planes are marked as $d_{h,k,l}$, where h, k, and l are the Miller indices of given planes set. In Fig. 7.3 a beam of parallel rays is falling upon a crystal at an angle α. In each of the crystal planes part of the beam is reflected. The difference in the paths traveled by the rays reflected, e.g., from planes C and D is 2a, if it is equal to the wavelength λ or to its multiple. In such a case the reflections are mutually reinforcing themselves, and we observe an intensity increase at the reflection angle. From Fig. 7.3 it is evident that $\sin \alpha = \dfrac{a}{d}$, and because $n\lambda = 2a$, according to Bragg's law the condition of maximal reflection is

$$n\lambda = 2d \sin\alpha$$

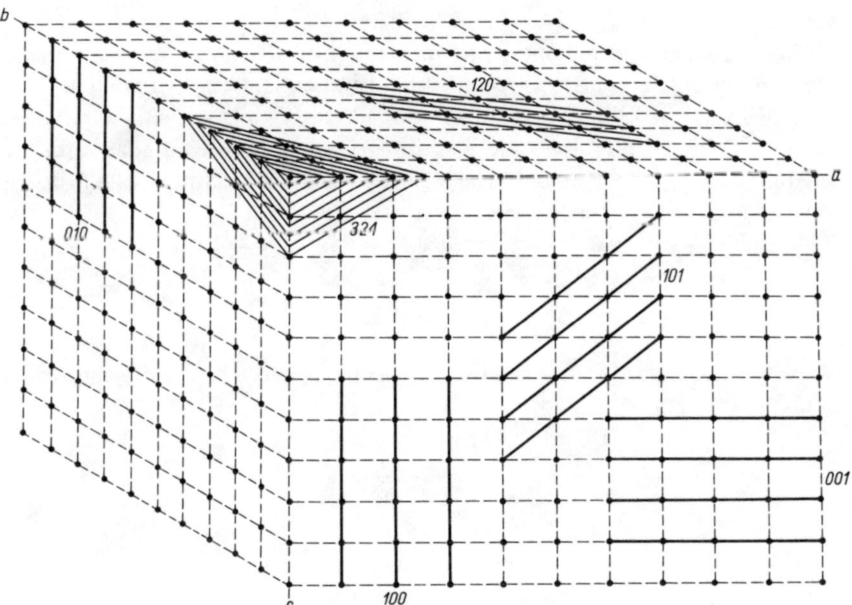

Fig. 7.2. Crystal lattice and imaginary planes, with appropriate Miller indices.

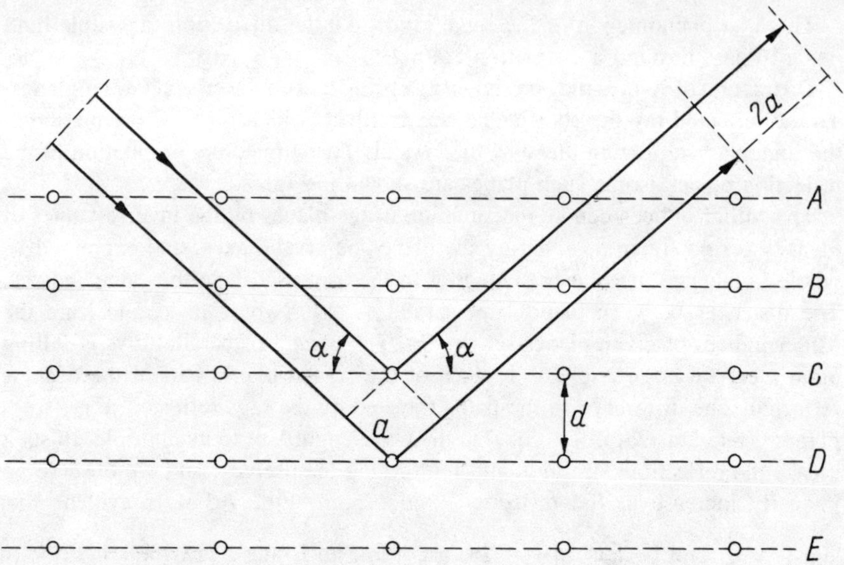

Fig. 7.3. Diffraction of the rays in the crystal planes. d - distance between grating knots, α - reflection angle of the ray in the crystal plane.

where d is the distance between reflecting planes and n is a multiple of the wavelength.

When investigating polymers, the diffraction of a beam of monochromatic rays, falling on a specimen is recorded; registration is done on a photographic film, placed on a cylindrical surface around the sample or with a radiation counter. The registered lines are a trace of cross section of a cone with the generating line of the cylinder. The reflection angle is measured as a distance between symmetrical bands along the central line.

$$\alpha = \frac{45°}{\pi R}\, f$$

where R is the radius of the chamber and f is the distance between symmetrical bands. The interplane distance is

$$d = \frac{n\lambda}{2\,\sin\alpha}$$

If α is increasing, the diffraction maxima will appear which are called diffraction maxima of the 1, 2, . . . nth order, according to the n value.

In the modern diffractometers the photographic plate or film is superseded by scintillation counters or Geiger counters. This makes it possible to measure the intensity in a more rapid and exact way, and to feed the results straight to the computer.

Qualitative evaluation of the X-ray diffraction pattern allows one to draw the following conclusions:

(1) Diffuse, concentric rings (Debye rings) suggest the presence of recurring periods, approximate to the intermolecular distances, and not indicating an orientation of the sample molecules. Such a pattern is given by solutions and unoriented fibers.

(2) Sharp, well-defined rings suggest the lack of orientation, but indicate regularly repeating periods. Such a pattern is observed in randomly arranged crystals and in crystalline powders.

(3) Arches, appearing instead of concentric rings, are the typical pattern of fibers. The length of the arches gives information about their orientation degree, and sharp, well-defined arches are the sign of constancy of the repeating period. The ordered structure and the plane distances, perpendicular and parallel to the axes, can be distinguished.

(4) Sharp and regular punctual reflexes are typical for crystals, i.e., for tridimensional ordered structures.

The position where the diffraction beam falls, depends on the distance between the reflection planes. The more diffracted is the beam, the closer are the planes to each other. As it also follows from Bragg's equation, the closer the diffracted beam is to the undiffracted one, the more spaced are the reflection planes in the crystal. The outer part of X-ray pattern features small elements of the structure, and the internal part features the larger elements. To understand this is equivalent to understanding the idea of the X-ray diffraction technique.

The recorded intensity, or intensity of the beam at a given point of the pattern, is proportional to the square root of the wave amplitude and reflects the electron density at a point of the reflection plane. Diffraction of X-rays at a very low ($< 2°$) angle may be observed, if in the sample tested the identity period is present (50-1000 Å), characterized by different electron density. The measuring technique is, in such cases, more sophisticated, however, as it is in the case of collagen, very useful information concerning its great identity periods may be obtained. There exists a correlation between this 'great' diffraction period and the cross striation period. However, the interpretation of the low-angle diffraction pattern is still unsatisfactory. One may investigate the influence of tanning agents or water on the great diffraction period.

Electrons, oscillating around their equilibrium positions, become, in turn, the source of new electromagnetic waves, having the same wavelength as the incident beam.

The intensity of the beam scattered on the electron is described by the formula:

$$I = I_o \frac{e^4}{m^2 c^4 r^2} \times \frac{1 + \cos^2\alpha}{2}$$

where I is the intensity of incident beam, e the electron charge, m the electron mass, c light velocity, r the distance between diffracting electron and the nucleus of the atom tested. Thus it is clear that only electrons may diffract the X-rays, for the nuclei are too heavy. The intensity of the beams diffracted on the atom is a sum of the intensity of all beams diffracted on electrons. A measure of the yield of diffraction is the 'atomic factor'

$$f = \frac{\text{amplitude of the wave dispersed by an atom}}{\text{amplitude of the wave dispersed by an electron}}$$

and the greater the intensity of the diffracted beam is, the heavier the atom is.

The so-called 'Compton diffraction,' caused by a change of wavelength due to the electron collision, gives *background* which is stronger when the atom is the lighter. The strongest are the backgrounds in the diffraction patterns of organic compounds which contain mostly light atoms. The wave amplitude being resultant of diffraction on all the unit cells of a crystal or molecule, is a vector sum of amplitudes of waves diffracted on single atoms. This summarized pattern is examined to establish the spatial structure of a molecule. Every place in the pattern is ascribed to the set of planes of the crystal unit cell.

X-ray diffraction is proportional to the electron density in the sample. Electron diffraction in the space investigated is usually calculated by the use of Fourier series. Two-dimensional cross sections used as 'maps' of the planes are calculated, finally coming to the spatial electron density and to structure visualization. For some proteins and peptides such as subtilisin and insulin, such spatial maps have been elaborated. Neither the experimental technique nor the calculation technique limit the structure studies by X-ray diffraction method. A crucial problem here is the number and the sharpness of reflexes, which may be obtained from the substance given. This, in turn, depends on the degree of structure ordering.

Investigation of the fibrous biopolymers may give diffraction effects, but here the number of reflexes is incomparable with the number of the reflexes obtained from monocrystals. Analysis of the diffraction patterns of fibrous biopolymers may give some values characterizing the helices, like, e.g., the length of one turn, or the number of amino acids in it. When comparing three diffraction patterns of collagen, keratin and elastin, it can be clearly seen that different amount of information can be derived from each particular pattern. As it is shown by other methods, the degree of ordering is different in all these biopolymers. Because of cylindrical symmetry of the helix, the points in the diffraction pattern are usually defined in terms of cylindrical coordinates.

In order to define the spatial structure of a molecule, the optical transformation method can be used. For this purpose a model of the molecule is constructed with the help of spheres of different diameters. The positions and size of the spheres simulate the positions and dimensions of individual atoms. A model that is placed in the way of a visible light beam is used to obtain a bi-dimensional structure model and is compared to appropriate structure patterns, obtained by the X-ray method. For collagen, such an experimental proof is provided by Ramachandran et al. [1].

The collagen and gelatin diffraction patterns obtained by the wide-angle technique, shows hardly any difference; just the shape of some of the reflexes is a little different: the equatorial spots, which are due to the interplane distances (11-12 Å) are halfmoon-like in the case of collagen, and are circles in the case of gelatin. However, this difference should not be used for purposes of identification.

The diffractograms of collagen of various origins do not differ among themselves. A wide-angle collagen pattern, as described by Millionova and Andreeva [2] is shown in Fig. 7.4, and the result obtained by these authors is quoted in Table 7.2.

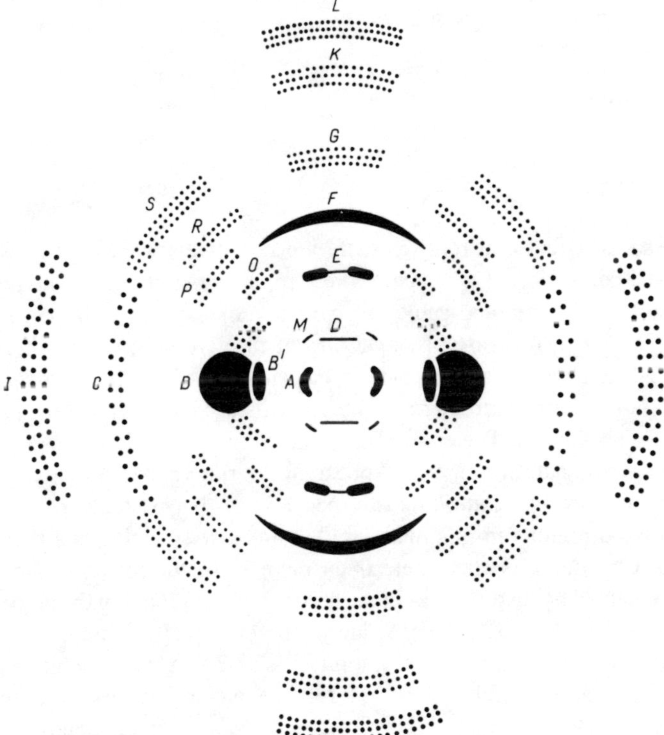

Fig. 7.4. Wide-angle diffractogram of oriented collagen (schematic). The reflexes are indicated by letters.

Table 7.2

Diffractogram of the rat tail tendon collagen

Reflex symbol	Distance d Å	Reflex intensity
A	11 ± 2	very strong
B	5.6 ± 0.1	medium
B′	0.12	medium
C	2.1 ± 0.2	weak
I	1.6 ± 0.1	very weak
D	9.3 ± 1.1	medium (linear)
E	3.9 ± 0.2	strong (sometimes 2 maxima)
F	2.9 ± 0.1	very strong
J	2.0 ± 0.1	weak
K	1.65 ± 0.05	weak
L	1.42 ± 0.05	weak
M	7.6 ± 0.5	medium
N	4.9 ± 0.3	weak (strong background)
O	3.5 ± 0.2	weak
P	2.9 ± 0.9	weak
R	2.5 ± 0.8	weak
S	0.24	weak

In the set of equatorial reflexes there are three weak reflexes corresponding to the distance of 38 Å. Existence of these reflexes may be justified, presuming the existence of 'five-strand-cable' of five macromolecules; at the same time the distance 38 Å is a measure of staggering of the five strands, as already mentioned.* A collagen diffraction pattern has sharply delineated elements clearly distinct against the background and a broad, diffuse ring, typical of liquids and amorphous substances. The ordered elements of the molecule are responsible for the first group of pattern elements. Appearance of rings and stripes in the collagen diffraction picture has contributed to the view of the existence of the ordered and unordered regions in this material. The intensity of the equatorial reflexes decreases very quickly when there is an increase of the reflection angle. This may be explained by disturbances in the ordered regions, having the perpendicular direction to the fiber axis. This is due to the chemical inhomogeneity of side chains. It is usually believed that the reflexes at 2.9 Å correspond to the helix height for one amino acid residue; the reflexes at 4.6 and 11 Å are due to the interchain distances: 4.6 Å where no side chains are present, and 11 Å in the direction where the side chains stand off from the peptide chain. The low-angle

*This argumentation, however, was contradicted recently (see Ch. 2 p. 59)

reflexes are meridional. In dry cattle tendon collagen they correspond to a distance of 640 Å, and in a wet sample, to 680 Å. Between 10 and 20 stripes of various intensity are usually observed, corresponding to the distances smaller than mentioned above. Water and tanning agents affect the positions and intensity of the stripes.

Hosemann and Nemetschek [3] have investigated the correlation between the cross-striation of collagen fibrils from rat tail tendon using phosphotungstic acid and glutaraldehyde and the low- and medium-angle diffraction patterns of these samples. The results of their work are of interest for tanning chemistry. Two reaction types between collagen and phosphotungstic acid found by the authors are:

(1) A reaction independent of steric relations which may take place over a wide pH range (2.2–6.0). In this reaction the easily accessible parts of the peptide chains structure participate. Perhaps this is a bundle of protofibrils of a diameter of 38.5 Å because this reflex disappears as a result of the reaction with phosphotungstic acid. Besides the reflex of 38.5 Å, some other equatorial reflexes (e.g. 12.6; 13.5; 17.7; 25 Å) become weaker or disappear.

(2) Intramolecular bond through the unstressed fibers. This reaction influences the reflex at 18.7 Å (in wet samples) or at 17.3 Å (in dry samples). Authors observed shifting of this reflex.

Using low-angle X-ray diffraction they have shown the reaction between collagen and phosphotungstic acid to lead to some stable compounds. Phosphotungstic acid bound in this way, may be removed only after a prolonged action of complexing agents, like phosphate buffer or polyvinylpyrrolidon. The phosphotungstic acid, however, bound in the other way, may be easily removed. According to the authors, both these reaction types do elucidate properties of cross-striation elements, having different degrees of stability. They believe that one may have better visible striation, if the reaction of the first type occurs, and this is because a good picture of cross-striated collagen may also be produced at pH = 5.5. Numerous diffractograms have been demonstrated in the paper quoted, showing a dependence between the diffraction pattern and the binding of phosphotungstic acids, glutaraldehyde and formaldehyde.

The diffraction pattern, having the shape of a set of parallel stripes over one another, is typical of one-direction order with regular parallel knots. The reflexes from collagen fibers have a disc shape. Thus the shape of the molecule was suggested as a cylinder of various identity periods. In this model the length of the reflexes is proportional to the cylinder diameter, i.e., to the dimensions of the reflecting planes. In Fig. 7.5 we may see the fibrils that form a smooth cylinder and corresponding X-ray low angle pictures.

In Fig. 7.5 there is a cylinder with visible constrictions, such as those observable with the electron microscope on prepared fibers. The fibrils, consisting of protofibrils and the corresponding fan effect, are shown in Fig. 7.5 c. In this effect the stripes, more distant from the non-diffracted band, are elongated. In

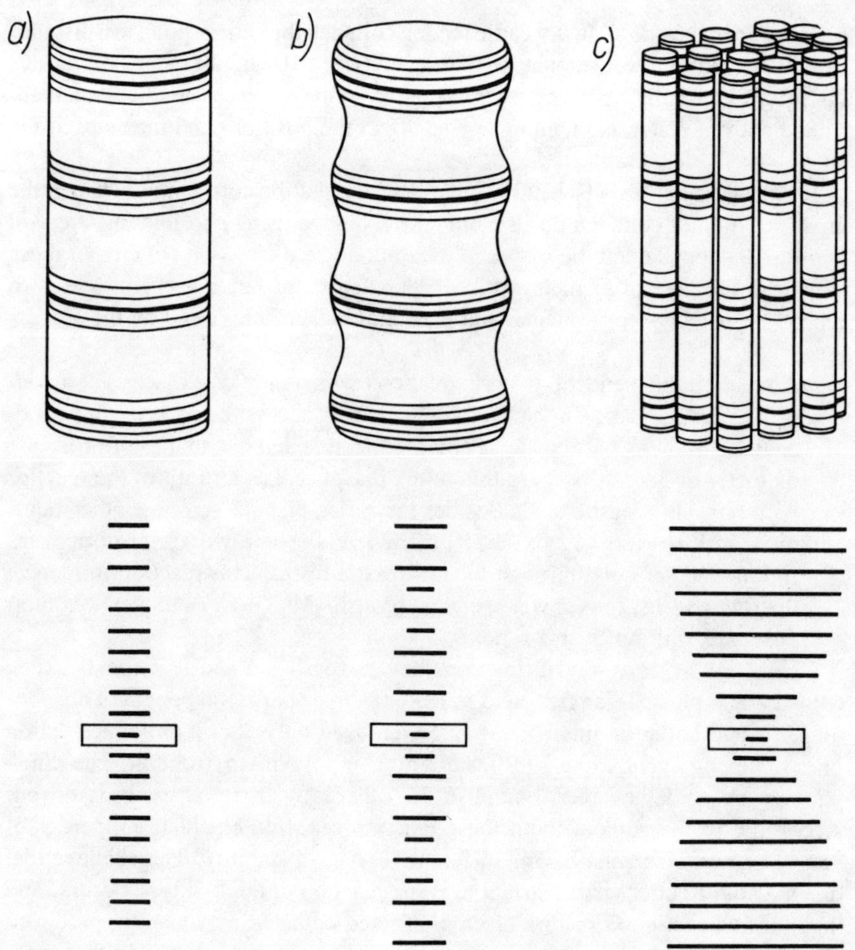

Fig. 7.5. Cylindrical models of the fibrils with the corresponding schemes of their low-angle diffraction pattern (a) smooth cylinder, (b) cylinder with constrictions, (c) fibril consisting of protofibrils.

dry samples the diffraction stripes give this effect. This speaks for splitting of the fibers into thinner elements due to drying. The distance of the reflex from the centrum of the pattern is dependent on the reflex order.

Accepting for the angles from Bragg's equation:

$$n = 2\,d\,\sin\delta$$

$$(\text{at } \delta < 1;\ 2\sin\delta = \sin 2\delta \cong \tan 2\delta = \frac{Y_n}{r}$$

at λ_{CuK_α} = 1.54 Å and r = 231 mm, we have checked the distances d, which should correspond to the cross-striation in electron microscope [4]. For bovine Achilles tendon collagen were obtained the following d values:

Collagen, untanned, bated	715 Å
Collagen, formaldehyde tanned	635 Å
Collagen, chrome tanned	614 Å

Thus we concluded that the length of the periods by crosslinking of collagen may be approximately believed as informing about the crosslinking bond energy and about partition of the charges in the molecule.

As far as it is known to the author, no other experiments have been made on tendon collagen treated according to the common tanning practice, i.e., including liming, bating, etc. Thus the rather unusual distances observed may be justified.

Recently, the influence of chromium tannage on the electron microscope picture of collagen fibers has been studied by Andreeva et al. [5]. They observed small knots of tanning agents at magnification of 35,000 × on chrome-tanned, previously pickled, fibers and a thin layer on the fibers (veil), however, not changing their picture. Collagen fibers, treated with chloroparaffin emulsions before tannage, give a more diffuse picture; their cross-striation periods remained unchanged. As a whole, no essential changes have been found in cross-striation patterns neither due to pretanning treatment, nor after tanning.

Numerous authors of recently published papers made use of the X-ray diffraction in collagen investigation. The wide-angle technique, as well as the low- and medium-angle ones, has been applied. Two directions predominate in the experiments mentioned.

The first one is an effort to establish the crystal unit cell and molecular packing. Several lattice types were proposed: tricline, tetragonal and hexagonal [5, 6, 7]. Wide-angle diffraction reflexes are analyzed in these studies. The use of very sensitive detection tools such as the Optronic scanning microdensitometer allows to detect and exactly localize the reflexes (by indicating the reflection maxima with great accuracy). However, it should be emphasized that an essential difference exists between crystal lattice diffractogram as it may be obtained by investigation of monocrystals and crystal lattice constants found from collagen X-ray reflexes. It must not be overlooked that simple protofibrils build up the crystal lattice only as a statistical average due to their unhomogeneities.

The other direction is to explain exactly the low-angle diffraction pattern and to combine it with the electron microscope observed cross-striation patterns. In a very detailed study Hulmes et al. [8] have investigated 41 meridional low-angle reflexes and tried to find out if they had a relation (as might be expected), to the amino acid sequence of the collagen chain. From computing the reflexes in comparison with the volume occupied by particular amino acid residues and the number of electrons in every residue, the authors concluded a great sensitivity of the reflexes to the conformation of the non-helical terminal peptides (telopeptides) and a lack of center symmetry of the collagen molecule. However, a relation between the low-angle reflexes and the amino acid sequence could not be proved, as the differences were too significant. Correlation between the low-angle reflexes and the quarter-staggering distance has been found and some differences between tendon collagen and skin collagen are indicated [9]. Similar results were found in other papers [7, 8].

7.2. Ultraviolet and Visible absorption spectroscopy

Ultraviolet and Visible (UV-Vis.) absorption spectroscopy provides some useful information about hide and skin proteins, the changes they undergo, as well as on tanning agents and their properties. In order to simplify the problems of spectroscopy in all these applications it is most frequently considered as a one-electron excitation, i.e., the excitation in which just one electron has been elevated to a higher-energy level. This is, of course, only approximately correct. In the organic compounds the absorption in the regions mentioned is usually connected with some types of atoms or their groups called chromophores. Typical examples of chromophores are: double bonds and their conjugated systems, systems and groups containing atoms with lone electron pairs, such as oxygen in the carbonyl group or nitrogen in the amino group. The behavior of chromophores is usually investigated by the absorption of nonpolarized light, as it is done in the absorption spectroscopy and in luminescence analysis, or with the aid of absorption of circular polarized light, which is the subject of optical rotatory dispersion (ORD) and circular dichroism (CD).

In the calculations done by the LCAO method (linear combination of atomic orbitals), the atomic orbitals are assumed to partly overlap in the molecule, which gives a new molecular orbital. Using the quantum mechanical methods it is possible to show that there are always two possibilities, viz.: the electron density between nuclei may increase, the electrons are pulled into the space there, and then we speak about the bonding orbital, or alternatively the electrons are pushed away from this space. In this case we call the formed orbital antibonding; it is usually marked by an asterisk. This may be explained in terms of classic laws of electrostatics, simply assuming that these two protons are repelling each other according to Coulomb's law. Negative charge of the electron accommodated between them attracts both protons. Because of this the repelling action is weaker. The system is stable if the negative charge is great enough. Energy of the electron depends on the orbital where the electron dwells—a fact that is impossible to explain in terms of classical physics. In Fig. 7.6 it can be seen how the bonding and nonbonding orbitals are formed from atomic orbitals. The definite positions of the orbital in space are forced by interaction with the other atom which, together with the first one, becomes a molecule despite that both orbitals have the same second (azimuthal) number. It is customary to call the orbitals that lie along the axis connecting the atoms—σ orbitals, and the per-

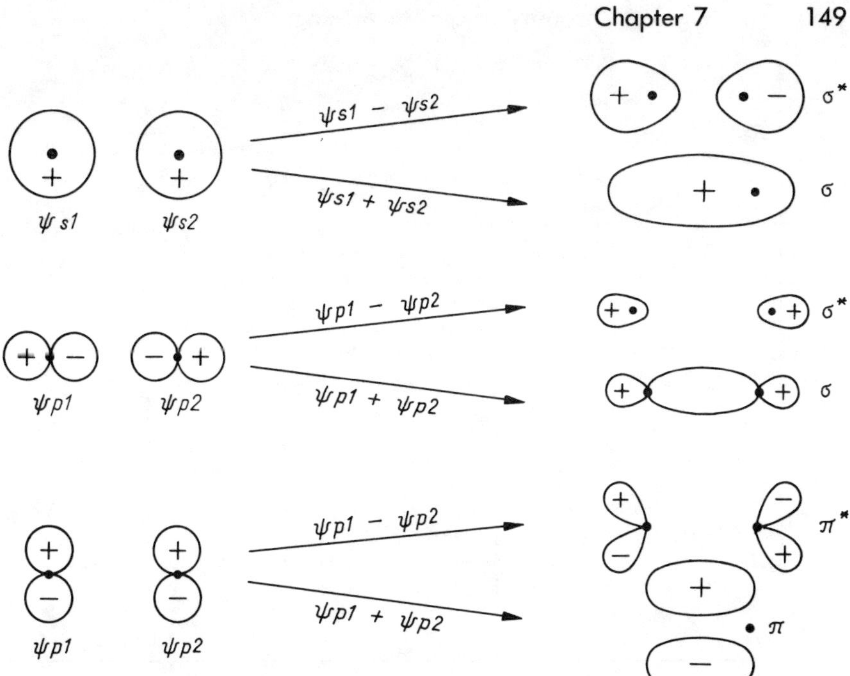

Fig. 7.6. Formation of molecular orbitals from atomic orbitals; ψ denotes wave function.

pendicular to the axis—π-orbitals. The molecular orbitals of the σ and σ^* type are derivatives of the atomic orbitals S, the orbitals π and π^*—from atomic orbitals P.

The molecules containing an odd number of carbon atoms must have an odd number of molecular orbitals. If the orbitals are 'paired', then there will be one orbital left. This orbital lies at the 'center of gravity' of the energy levels, i.e., at the nonbonding orbital energy level. This central orbital, common to all odd alternates is described as the nonbonding molecular orbital n.

The shape of these orbitals is the same as those of the atomic ones. The energy levels in the bonding orbitals are lower than in the nonbonding. Generally, the energy levels of the molecular orbitals depend on the kind of molecules, but for certain types of molecules the following sequence is valid

$$\sigma < \pi < n < \pi^* < \sigma^*$$

All electrons are situated at the lowest possible energy level. According to the Pauli's Exclusion Principle one energy level in one atom may have just two electrons with opposite spins. Thus the electrons are successively occupying the orbitals of the lowest energy levels. This is known as the ground state of the

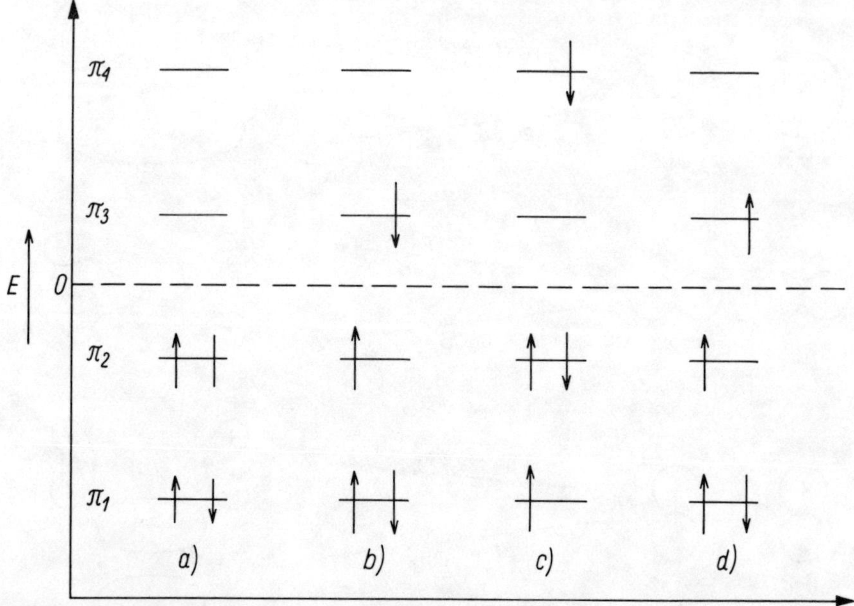

Fig. 7.7. Orbital diagram showing some of the electronic configurations of a simple molecule (1, 3-butadiene). According to [10] with permission

molecule. Absorption of light of suitable wavelength is a reason for transition of one electron from the σ, π or n level of the σ^* or π^* level. (Fig. 7.7) This means that the molecule turns into an excited state.

If all the electrons in the molecule are paired or the sum of spin numbers is zero, then we refer to it as the singlet state. This state is typical of the unexcited molecules. The excited molecules are in a singlet state when both unpaired electrons on the orbitals have the opposite value of the spin number. If the spin numbers of both excited electrons is equal ($+\frac{1}{2}$ and $+\frac{1}{2}$), then their sum is equal to one and it is called the triplet state. In space, the electrons having opposite spins may be close to one another; the electrons having the same spins have to occupy possibly the most distant positions. Electrostatic repulsion of the latter is weaker compared with the singlet state and, consequently, the energy of the triplet state is lower than that of the singlet state.

The absorption intensity of a given electron transition depends on its probability and on the size of the absorbing molecule. The molar absorption coefficient, or absorptivity, is given by equation:

$$\varepsilon = \frac{\log \dfrac{I}{I_o}}{cl} = kPa$$

where c is the concentration, I the thickness of the absorbing layer, k is a constant of the order of 10^{20}, P the probability of transition, and a the molecule cross section. Maximal possible molar absorption coefficient for a small organic molecule is of an order of 10^5 at $P = 1$. An actual measure of absorption is the integral of the area under the absorption line. The chance for various electronic transitions and thus for several absorption bands to appear in the spectrum seems to be high, when in actuality the number of observed bands is rather low, which is due to the selection rules. A simple explanation of the selection rules, allowed and forbidden transitions, may be found, e.g., in the work by Griffiths [10]. In short, a transition is allowed if the dipole moment component of the excitation has a positive value for both orbitals. In this case, the transition dipole moment contributions of the opposed points in space having the same signs and magnitude will reinforce each other. Integration over all space will then give a non-zero value for the transition dipole moment M and thus this electron excitation is an allowed process and the absorption band will be intense. When both the dipole moment components have opposite signs, the M value will be zero, because they are equal in their absolute value, thus cancelling out one another. Such a transition is forbidden. For the transition to be allowed one orbital must be symmetrical and the other, antisymmetrical.

The $n \rightarrow \pi^*$ transition may be allowed or forbidden. The analogous reasoning leads to the conclusion that allowed transitions of this kind occur when the overlap integrals of both engaged orbitals are of a non-zero value. The $n \rightarrow \pi^*$ transitions between the carbonyl oxygen atom orbital and the carbon orbitals are generally forbidden, thus these absorption bands are usually very weak. This also applies to the peptide bond.

The transitions between vicinal states are allowed. This is due to the symmetry of wave functions of the electron state. The wavefunction of the excited state belongs to the same symmetry type, like the wave function of the ground state. The transitions between states of various excitation degree are forbidden (e.g. singlet—triplet). These transitions may be observed in the spectrum of bands with a very low intensity: e.g., the intensities of absorption of formaldehyde, belonging to the allowed transition $\pi \rightarrow \pi^*$ and to partially forbidden $n \rightarrow \pi^*$ and $n \rightarrow \sigma^*$ transitions are of the ratio 1250:50:1. Some forbidden transitions may be observed due to the intra- and intermolecular interactions: e.g., the singlet$\rightarrow$triplet transition in the presence of paramagnetic substances. Electronic transitions which may be observed in the range discussed are quoted in Table 7.3.

An energy diagram of possible electronic transitions of the peptide group is given in Fig. 7.8. As we may see therein, the present technique of UV spectroscopy the $\pi \rightarrow \pi^*$ and $n \rightarrow \pi^*$ transitions are of importance. The principal significance of the last transition is due to the optical activity of proteins.

A transition to the orbitals of very high energy levels, e.g., a transition to the orbitals of the higher first quantum number, gives bands in the vacuum UV

Table 7.3

Kinds of electron transitions

Transition	*Spectrum range*
From the bonding orbital in the ground state to the orbital of higher energy level	
$\sigma \rightarrow \sigma^*$	vacuum UV
$\pi \rightarrow \pi^*$	UV
(K, A or E-bands)	
From nonbonding orbital to the orbital of higher energy level	
$n \rightarrow \pi^*$ (R-band)	UV, far or near,
$n \rightarrow \sigma^*$	visible light

range. Technical difficulty of testing of this range lies in absorption of solvents and optical materials in this range, i.e., below about 170 nm.

The proteins absorb electromagnetic radiation in the near and far UV range. Only few metal-protein complexes, like hemoglobin, absorb visible light at certain wavelengths. In the near UV the proteins have an absorption maximum, which is due to the aromatic amino acids: tyrosine, phenylalanine and tryptophane, and another one close to 190 nm which is due to peptide bonds. Between 210 and 250 nm no absorption maximum may be observed, but there is an absorption in this region, caused by various factors. The helix structure does influence the absorption there, as one may conclude from the Cotton effect (cf. below) observed in this region.

Typical absorption bands of the amino acids containing aromatic rings are also present in the region of 210-220 nm, but they are masked by the peptide bond bands. Most probably both 'aromatic' bands are due to the $\pi \rightarrow \pi^*$ transition in the conjugated ring bonds. Due to ionization of the tyrosine phenolic groups by increased pH of the solution, a bathochromic effect can be observed.

Briefly, we may define here the most commonly used vocabulary related to the spectra characteristics:

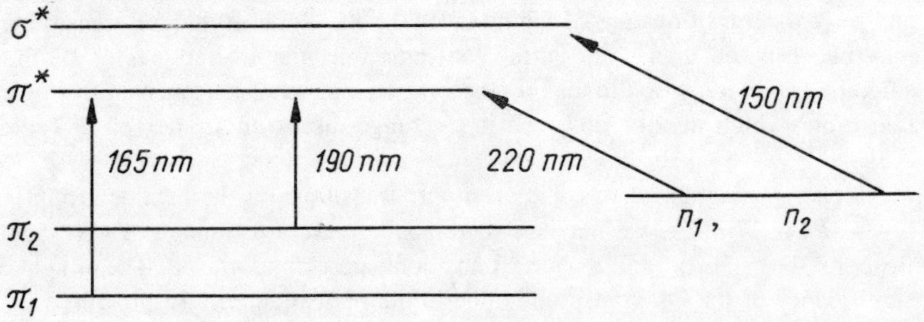

Fig. 7.8. Possible electron transitions in the peptide group.

Bathochromic shift (red shift)—the displacement of an absorption band towards a longer wavelength.

Hypsochromic shift (blue shift)—the displacement of an absorption band towards a shorter wavelength.

Hyperchromic effect—an increase in the intensity of an absorption bond.

Hypochromic effect—a decrease in the intensity of an absorption bond.

Solvatochromism—the change in position and/or intensity of an absorption band due to a change in the polarity of the solvent.

Halochromism—the color change due to a change in the pH of the solution.

Half-bandwidth. The width of an absorption band at one half-height of the peak.

Also important is the position of the aromatic amino acid in the protein molecule influencing the strength of its absorption bands. This influence may be depicted as follows (Fig. 7.9).

Tyrosine A does not affect the strength of the absorption band, tyrosine B, lying in the 'bay', affects it in a limited way, but tyrosine 'C' has a maximal effect on the band strength. Position of the residue may also influence the band position: the shift may be by 1–5 nm.

The strongest absorbing aromatic amino acid is tryptophane, the weakest —phenylalanine. The absorption band of the latter lying at 260 nm has a fine structure, i.e., the peak consists of several narrow peaks corresponding to the vibrational motions of the molecule. Extinction of the bands in the region of

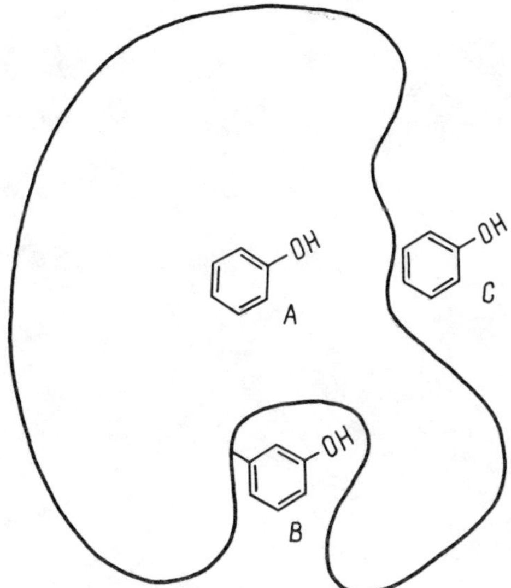

Fig. 7.9. The effect of the position of the tyrosine residue in the protein molecule on its absorption spectrum. A - small, B - moderate, C - distinct.

260–280 nm is—between 10^2 and 5×10^3. Thus the bands are weak. In native collagen which contains very little aromatic amino acids, these bands are absent. They may be observed, however, in the spectrum of the rat tail tendon collagen [11], if the samples before solubilization are treated with pepsin or pronase (which deprives it of telepeptides) and then the differential spectrum is taken. The extinction value is shown in Fig. 7.10 given as calculated per 1 mole of amino acid per liter.

Noteworthy is the change of absorptivity due to the heating from 23 to 40°C in the range of 190 and 227 nm. The increase of absorption at $\lambda = 227$ nm at higher temperatures is shown in the insert of Fig. 7.10. The breaking of the straight line at 38°C is due to denaturation which occurs at this temperature. The comparison of the absorption spectrum of thermally denatured and native collagen leads to this conclusion. This confirms (Fig. 7.11) the concentration of the aromatic amino acid in the nonhelical part of the collagen molecule, i.e., in telopeptides, because they may be removed by the action of a nonspecific proteolytic enzyme.

The peptide bond absorption bands observed at 190 nm originate from two $\pi \rightarrow \pi^*$ transitions from carbonyl group in the region near 198 and 189 nm. These bands overlap and come from both directions in relation to the helix: parallel and perpendicular. The resulting band is strong, its molar absorptivity is at 205 nm (where it is often measured) of an order of 3×10^3, and at 190 nm (where

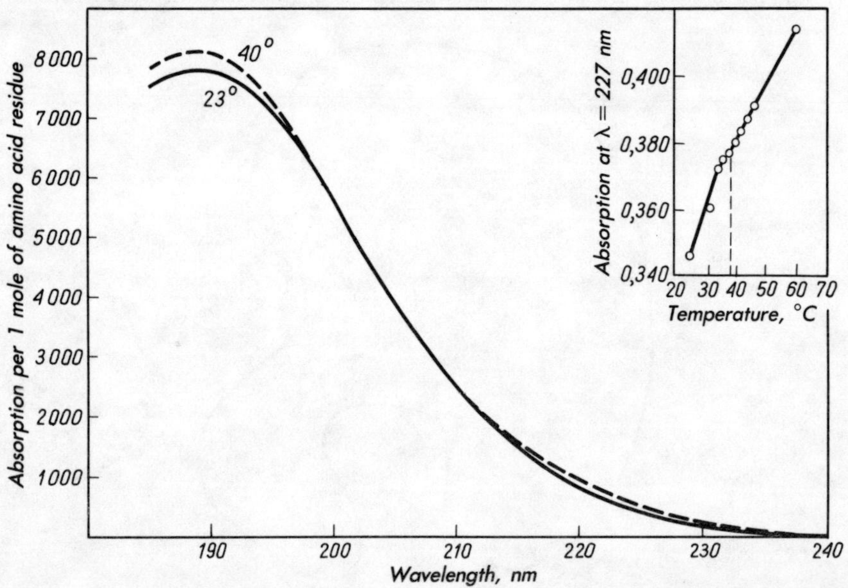

Fig. 7.10. Spectrum of collagen solution in the phosphate buffer (pH: 3.8) at temp 23 and 40°C, in the range of 185-240 nm. According to [11]

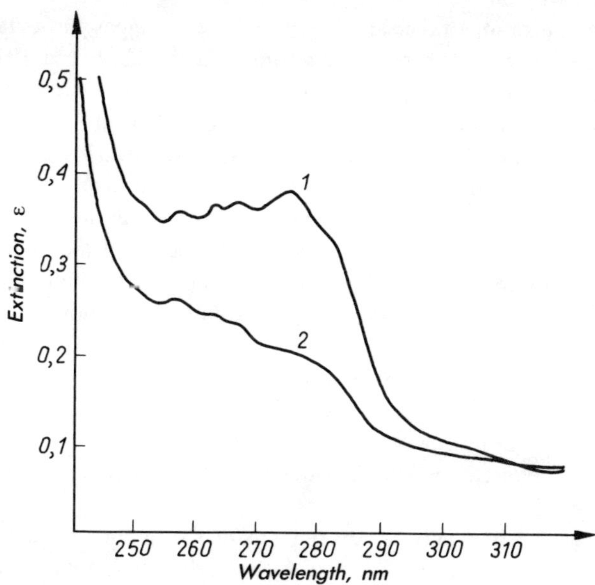

Fig. 7.11. The range of 240-310 nm of collagen spectrum 1 - native, 2 - treated with pronase to remove telopeptides. Concentration in this case is 10^5 times that in Fig. 7.10. According to [11]

the measuring may offer some technical difficulties) of an order of 8×10^3. For the UV absorption spectrum the polarizability of the medium is of crucial importance: the decrease in polarizability causes a blue shift of the chromophore absorption band. This rule, however, has some exceptions, because a reverse behavior of protein chromophores could be expected due to denaturation and the coming of the chromophore to the molecule surface: however, an interaction of two chromophores, with two induced dipoles of a stable orientation, always makes one band stronger at the expense of the other. The helix conformation formed causes a change in the protein spectrum, i.e., a decrease of the 190 nm band and the appearance of an inflection point at approximately 205 nm. This rule is widely used in the DNA conformation study. The changes observed in collagen spectra during denaturation in the regions of 190 and 223 nm may be used in evaluation of the denaturation degree of the sample, and by the same to the evaluation of the helix and coil parts. Exact analysis of the spectrum (Figs. 7.10 and 7.11) leads to the conclusion that this method may not be considered as a quantitative one, because the differences between spectra 1 and 2 are too little. Based on these differences, Wood [12] found maximal difference in the wavelength at 227 nm ($\Delta\varepsilon$ of an order of 0.1). Much better information may be obtained in this relation by other methods, first of all, from measurements of optical activity.

During gelatinization of the sample the spectrum changes particularly at about 191, 203 and 223 nm. The physical meaning of the change at 191 nm is not clear, because, according to Katz et al. [13], this band is in 30% due to the absorption of side chains. However, the change at 223 nm occurs at the point where optical rotatory dispersion changes its sign (see below section 7.4). From this it may be concluded that when the conformation change took place, the backbone chromophores (peptide groups) will change their hydration properties, which is revealed in a change in the number or energy of the hydrogen bonds.

The increase of absorption at 190 and 223 nm due to the helix-coil transition is a generally observed effect and may be explained as follows: the transitions $\pi \rightarrow \pi^*$ and $n \rightarrow \pi^*$ occur in planes perpendicular to each other, thus the degree of their participation in the helix is lower than that in the coil. In this latter case the participation possibility increases. This phenomenon can be better explained in terms of quantum mechanics. Some information may be obtained as well if we take the shape of curves into consideration. This point is valid also for other spectral regions.

The absorption bands may be described using the Lorentz or Gauss functions, the former being sometimes called Cauchy's curve. The Lorentz (Cauchy) curve is given by the equation:

$$I_{\nu} = \frac{a}{(\nu - \nu_o)^2 + b^2}$$

where b is the halfwidth of the peak $\Delta\nu_{\frac{1}{2}} = 2b$, and a is the peak height $I_o = \frac{a}{b^2}$. The Gaussian curve is given by the formula:

$$I_{\nu} = a' \exp - \frac{(\nu - \nu_o)^2}{q^2}$$

where q is the halfwidth of the peak at the height $\frac{I_o}{e}$ and $A' = I_o$. The total intensity of the curves is given by their integrals, the integral of the Lorentzian curve

$$I_{\infty} = \frac{\pi a}{b}$$

and that of the Gaussian curve

$$I = a' q \sqrt{\pi}$$

For the curves of intermediate shapes a factor r of the line shape is applied

$$r = \frac{\pi}{2} \times \frac{\Delta\nu_{\frac{1}{2}} \cdot I_o}{I}$$

For the Lorentzian curve r $=$ 1, for the Gaussian curve r $=$ 1.47.

An absorption band is often the result of the overlapping of several absorption peaks, lying so close to each other that the resolving power of the apparatus is no longer satisfactory. The band must not be separated, however, if there are no good reasons for this. Such a reason may be, e.g., the checking and proving of the physical sense of separation. One has to keep in mind that, e.g., mathematically it is always possible to separate the Lorentzian curve into two, and the value obtained in this way will not deviate from the real one more than it may be thought as being due to the thickness of the line drawn.

There are several methods of numerical analysis of the spectral curves. The use of a computer in these calculations is almost unavoidable, because otherwise the calculation is very laborious and time-consuming. Once again it should be emphasized that such a separation of spectral lines should be justified on physical grounds.

In UV, Vis and IR spectroscopies the spectra are obtained by recording the light absorption as a function of wavelength λ or wavenumber ($\nu = \frac{1}{\lambda}$) of the incident radiation. The former manner is preferred to be used in UV and Vis region, whereas both are equally used in IR spectroscopy. Either system has some advantages.

The wavenumber scale is directly proportional to the frequency of the light, thus proportional to the photon energy, or the excitation energy of the absorption band. These spectra recorded show the shape and position of the absorption bands as a function of transition energy. For the physical chemist, recording in wavenumbers is more meaningful; in addition to energy comparison, the symmetry of bands may be evaluated at once. A linear wavelength scale gives some compression of the spectrum at the short wavelength and expansion at its long wavelength end. On the other hand, in the Vis region the expansion of spectra is of value, e.g., for better color evaluation and for structural studies.

7.3 Luminescence analysis

Luminescence effects have for a long time been known and observed in nature. Discovery of ultraviolet radiation has shown a fascinating appearance of glowing of teeth, nails, and skin during their illumination. In the last several decades this was explained. In organic chemistry luminescence is associated with several groups of bondings, e.g., to those occurring in aromatic rings; in inorganic—instead it is characteristic of some elements.

Luminescence itself is an emission of radiation, accompanying the transition of a molecule from the excited to the ground state. It occurs most frequently in the UV and Vis region, but it may be also observed in the IR spectra. The excited molecule, due to the absorption of light, after a short time is deactivated among others because of a.v. collisions with other molecules. This deactivation

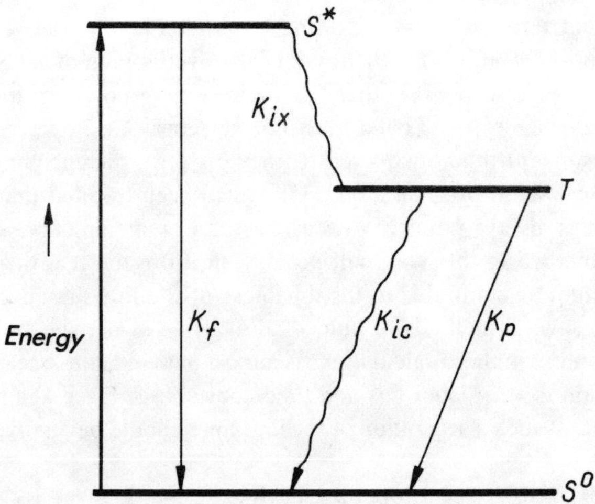

Fig. 7.12. Possible electronic transitions. Explained in the text.

is accompanied by energy radiation. If we denote the ground state of the molecule by S_o, its lowest single state by S*, then we have a picture as shown in Fig. 7.12.

The $S_o \rightarrow$ S* transition is accompanied by a $h\nu$ photon absorption. The S* $\rightarrow S_o$ transition is accompanied by photon emission, but of a lower energy, because a part of the energy becomes lost due to thermal motion. The emitted rays are thus of a greater wavelength.

The S* $\rightarrow S_o$ transition is referred to as fluorescence; transition from the lowest triplet state to the ground state is called phosphorescence. Phosphorescence is possible, when in the surroundings of the excited electron proper conditions exist, e.g., when it is deeply cooled with liquid nitrogen. K-values marked in the diagram are the constants of a given process. The mutual dependence of these constants is given in Table 7.4.

In the luminescence experiment the sample is irradiated and the wavelength

Table 7.4

Electron transitions in luminescence experiment

Process	Constant	Phenomenon
$S_0 + h\nu \rightarrow$ S*	1	absorption
S* $\rightarrow S_0 + h\nu_f$	K_f	fluorescence
S* $\rightarrow$ T*	K_{ix}	internal conversion
T* $\rightarrow S_0 + h\nu_p$	K_p	phosphorescence
T* $\rightarrow S_0$	K_{ic}	internal conversion, (heat emission only)

of the light should be chosen so as to achieve its maximal absorption. Thereby absorption is always greater than total emission.

Deactivation of the excited state starts as a rule at a time of about 10^{-8} s once irradiation has been stopped. Fluorescence decay in time is very similar to the capacitor discharging

$$A = A_o e^{-\lambda t}$$

where A is the charge, A_o the charge at the start, and λ is the process constant during the time t. The residence time of the excited state is $\frac{1}{\lambda}$ or the time during which $A = A_o e^{-1}$. Maximal peak of the fluorescence spectrum correlates with the lowest excited state. This state may become metastable, e.g., in liquid nitrogen as a result of changes in spins and in electron moments. Freezing of the milieu in liquid nitrogen may cause the transition K_{ix} and phosphorescence phenomenon. In the case of hide components such metastable states may take seconds or minutes.

The luminescence analysis in practical application consists of two parts: fluorescence analysis, which gives the feature of spectral emission of the sample in singlet state, and phosphorescence analysis, which also makes it possible to determine the duration and energy level of triplet states.

In the luminescence experiment the quenching effect may occur. Usually its reason is the removal or addition of some electrons to the observed milieu and, as result, the luminescence decreases. Among other reasons for quenching known are, e.g., transfer of the electron energy to the electrons of another substance present in the solution. Emission of electromagnetic radiation by proteins is weak (ca 7% of absorbed light), which is probably due to its intense quenching by the COO^- ions present.

Fluorescence and phosphorescence of globular proteins stems from the definite centers, i.e., from the amino acids containing aromatic rings. Emission of the protein is at room temperature dependent on the predominant amino acid in this protein. Maximal values of this emission (incident radiation $\lambda = 277$ nm) lies in the UV region. The emission due to tryptophane has its maximum by $\lambda_F = 353$ nm, that due to tyrosine by $\lambda_F = 303$ nm. Domination of tryptophane is very pronounced: its peak is always stronger even for higher tyrosine concentrations. Phenylalanine, whose fluorescence maximum lies also at ca 303 nm, gives no contribution to the luminescence spectrum of the protein, if it contains the remaining aromatic amino acids or at least one of them. This effect is not fully elucidated. It is believed that it is due to the resonance migration of energy between amino acids, and to the donor-acceptor transfer.

One may calculate the amount of donor-acceptor interactions with time, but the following data should be known:
(1) lifetime of the excited molecule;

(2) wavenumbers of the peaks: those corresponding to the donor absorption and to the acceptor fluorescence;
(3) overlap integral of the orbitals engaged.

From these data the critical distance can be calculated. For collagen, which hardly contains any tryptophan, 12 Å (1.2 nm) has been found as the critical distance between phenylalanine as donor and tyrosine as acceptor. This means, in other words, that because collagen shows a fluorescence the residues of these two amino acids lie closer in the protein (as an average distance between them). This may serve as a guidance in the evaluation of the chain conformation.

The proteins may be divided into two classes, depending on their tryptophan content: class A without tryptophan but containing tyrosine, and class B which contains both tryptophan and tyrosine. Collagen may be considered as belonging to the class A, because of the lack of tryptophan in it and because of the position of its maximal emission peak and of the lifetime of its triplet state. However, collagen has only two tyrosine residues in 1000 amino acid residues, and the reasons for its luminescence have not yet been elucidated. On the other hand, its phosphorescence is much higher than the fluorescence at low temperatures, which is inconsistent with the behavior of proteins belonging to group A, but it may have something in common with the trapping mechanism.

Trapping is an important mechanism which may be responsible for the phosphorescence effect. It may roughly be described as a succession of the following steps:
(1) excitation energy imparted to an electron ejects it from a luminescent center;
(2) the excited electron is trapped in the lattice defect (in crystals);
(3) additional energy, usually as infrared radiation (heat) must be imparted to the electron to release it from the trap;
(4) the electron returns to the center to be captured.
The mechanism is not investigated as yet in collagen.

At low temperatures at about 80 K the proteins have the fluorescence maxima at the same positions as in room temperature. The cold, glassy surrounding of the excited molecule allows to observe the transition from the singlet state to the metastable triplet state. Fluorescence and phosphorescence occur at the same time if tryptophan plays a dominant role in the emission. The phosphorescence maxima are positioned at $\lambda = 412$, 437 and 460 nm in the visible region. The structure of the triplet state spectrum shows a fine structure, which is not the case for the singlet state spectrum and therefore gives more information about the structure of the compound.

When investigating the rat skin the highest excitation found for irradiation was at 277 nm. The fluorescence band observed for this irradiation has a maximum at $\lambda_F = 305$ nm, and the corresponding phosphorescence band (of some higher intensity) at 395 nm.

This latter band lies at a wavelength not corresponding to any of the amino acids with an aromatic ring (e.g., for phenylalanine this wavelength is 460 nm). In the same paper, the authors [14] found that λ_P of collagen does not change according to the molecule crosslinking because it lies in the same region for α-chains and for crosslinked molecules. No investigations of this kind have been done for tanned collagen.

Phosphorescence of collagen is specific and corresponds to structures of the desmosine type. Luminescence spectrum of 1,2,4,5-tetramethyl benzene has an emission maxima at $\lambda_F = 305$ and collagen at $\lambda_P = 395$ nm the same maximum of exciting radiation, probably because of more complicated structure of chain, distorting emission due to additional internal collisions. The decay time in the α-collagen chain is somewhat lower than in the crosslinked molecule, probably due to the quenching effect in the latter. The results mentioned are shown in Table 7.5. In the same paper [15], the authors have tested the luminescence spectra of connective tissue strips containing tryptophan.

The $\lambda_P = 395$ nm band may also be ascribed to the carbonyl group excitation. However, this opinion is questionable. According to Grimes et al. [16] the phosphorescence of bovine tendon collagen in 0.45% NaCl solutions and also in liquid nitrogen (77 K) shows a maximum at $\lambda_P = 396$ nm and weaker ones at $\lambda = 435$ and 460 nm. The intensity of phosphorescence at room temperature is 10 times lower than at the nitrogen boiling point. Analyses of the decay curve led to the conclusion that two independent triplet states exist in collagen, one with a decay time of 2.3 ± 0.2 s and the other with 0.5 ± 0.1 s. It is known that collagen fluorescence increases with age at a rate of approximately 1% per life year. The reason for this is not known, although it is probably not genetic, because the increase may be observed when collagen is stored in vitro. Perhaps there may be a thermodynamic factor involved, because this increase is also

Table 7.5

Phosphorescence maxima and lifetime of excited states at 89 K for aromatic amino acids, rat skin collagen and its components

	Maximum		Lifetime of
	of excitation	*of emission*	*triplet state *
	λ, *nm*	λ, *nm*	*s*
Collagen	277	395	2.1
Molecules β12	277	395	1.8
" α1	277	395	2.2
" α2	277	395	2.4
Tyrosine	277	404	1.7
Tryptophan	292	436	4.8

Calculated from formula $P = P_o e^{-at}$, $a = \tau^{-1}$

observed when collagen is artificially aged by oxidation. When oxidized collagen is treated with collagenase and the product of splitting is separated by thin-layer electrophoresis, then fluorescence is observed in the tyrosine-containing fraction (tyrosine reaction: nitroso α-naphthol + nitric acid). Excitation and the fluorescence maxima of this reaction are at λ = 275 and 315 nm, respectively. The second fluorescence range of unknown origin lies at 345 and 440 nm.

Wolframm et al. [17] have investigated the collagen oxidation products using peracetic acid. They have observed various emission maxima in their fluorescence analysis, when it was excited with radiation of various wavelengths. When the exciting radiation of 250 nm is used, the collagen destructs give a weak band at 275 nm and a strong one at λ = 400 nm. Excitation by radiation at λ = 275 nm gives two fairly strong maxima at 303–304 and at 350 nm. In some fragments a third one at λ = 425 nm may be observed. According to the authors, part of the emission is due to the dityrosine formed. However, this compound has not been found in collagen. These data deviate from those previously mentioned, possibly on account of the oxidation method used which may give rise to new fluorescence centers.

As yet there is no answer to the question, whether or not the excited states may be found in tanned collagen, i.e., if the centers which may be excited have something in common with the centers engaged in the tanning. Numerous interpretations and doubts discussed above allow one to expect that the collagen investigation by the luminescence method may be useful for future research; at that time we will be able to recognize the reasons for these doubts.

7.4. Optical rotatory dispersion (ORD) and circular dichroism (CD) as applied to the helical structures studies

A universal method, applicable in the investigation of a large, complex-structure molecule does not seem likely to be easily developed for the time being. Absorption spectroscopy, luminescence analysis and ORD—each of these methods may provide useful information of its own kind. Rotatory power in relation to the wavelength of the light used is called ORD. The shape of rotatory power curves depends on the structure of the compound tested. We refer to this as a positive Cotton effect, if, with decreasing wavelength, the rotation increases to reach a maximum before changing its sign. For examples of this type of compound, we may consider the trans-isomers of noncomplex molecules. In the cis-isomers of the Cotton effect is negative. The point at which the curve intersects the axis, corresponding to the $n \rightarrow \pi^*$ transition and to the absorption maximum in UV, determines the point where the Cotton effect occurs. The substances that exhibit a Cotton effect transmit at various rates the circular polarized light counterclockwise and clockwise. They also absorb both components of the radiation to a different extent.

Suppose a wave of plane monochromatic light falls perpendicular to a piece of paper. At any point in the paper we draw the light vector, and it will go through a series of changes, returning to its original value in the period of the waves. We can represent any motion by giving the xy-coordinates of the end of the light vector; for monochromatic light these must be sine or cosine functions of the time. If ν is the frequency we therefore take

$$X = A \cos (2 \pi \nu t - \alpha); \quad Y = B \cos (2 \pi \nu t - \beta)$$

as the general description of the light vector. If $B = 0$ we have a vibration in which the vector always lies in the direction of X and ranges between $- A$ and $+ A$. We call this plane polarized in the direction x. If $A = 0$ we have a light plane polarized in the direction y. If $B = A$, and $\beta = 2 + \dfrac{\pi}{2}$, then $x^2 + y^2 = A^2$ and the vector of constant magnitude describes a circle: this is circularly polarized light. Either left- or right-handed rotation is possible.

The chromophores usually contain the systems of π-electrons in planar symmetry, so they hardly show any optical activity. In the biological systems the chromophores are usually activated to some extent by the asymmetry of their surroundings, e.g.,

O   O   CH$_3$

Inactive   **Active**

The theory worked out in 1962 allows us to estimate the optical activity of a polymer when the permanent and induced dipole moments of monomers and the positions of the charges are known. These are experimental data or they may be calculated by quantum mechanics methods. The knowledge of the polymer geometry, resulting from the measurement of optical activity is also necessary. The accordance between calculations and experiment results is usually checked by assuming the geometry of various polymer models.

Each of the chromophores of the molecule, taking all its bonds into consideration, contributes to the resultant rotation: their greatest activity may be observed in the wavelength regions around the absorption maxima. This property decreases with the wavelength.

The measurement of ORD has some disadvantages, e.g., distortion of the results due to absorption of the background. The CD method is free of this disadvantage. It is used in the testing of chromophore asymmetry. Cotton in 1896, during the investigation of rotatory phenomena, found that optically active substances absorb the circularly left and right polarized light to a various degree. The measure of this property is the absorption difference $\Delta\varepsilon = \varepsilon_l - \varepsilon_r$. The curve $\Delta\varepsilon = f(\lambda)$ has a Gaussian (bell) shape in a simple case. $\Delta\varepsilon$ is of measurable size in an immediate vicinity of an absorption maximum of a chromophore. The CD and ORD are in fact two symptoms of the same phenomenon, i.e., interaction between polarized light and assymmetric molecule structure. As one may show, the CRD curve is a differential curve of CD (Fig. 7.13).

The dependence between rotation (α) and wavelength of polarized light is described by the equation

$$[\alpha]_\lambda = \frac{n^2 + 2}{3} \sum_i \frac{k_i}{\lambda^2 - \lambda_i^2}$$

where n is a refractive index, λ is a specific absorption band, and k_i is a coefficient, proportional to the 'rotatory power' of the electronic transition observed. This value does not correlate with extinction at λ_i. At a greater distance (expressed in terms of wavelength) of the extremum one may replace the above equation by those of Drude

$$[\alpha]_\lambda = \frac{A'}{\lambda^2 - \lambda_c^2}$$

where A' is a specific constant of rotation for the system tested, and λ_c is the wavelength at which the Cotton effect occurs. The dependence between rotatory

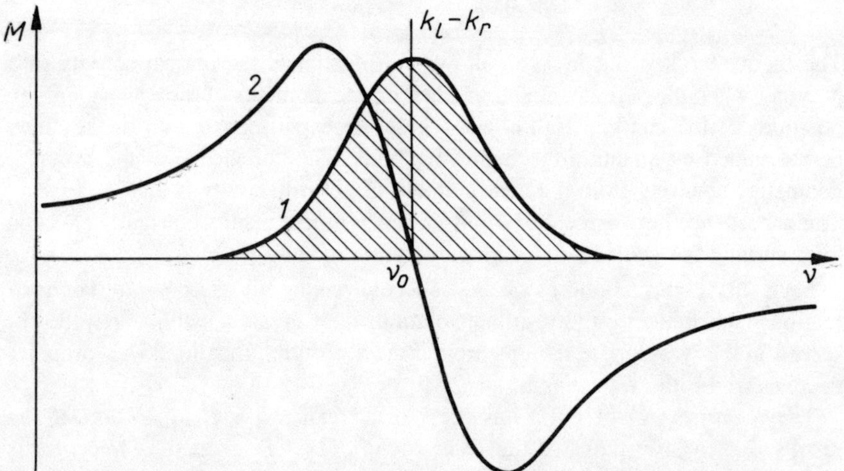

Fig. 7.13. ORD and CD of an isolated absorption band of Gaussian shape. Cotton effect positive ($k_1 > k_r$) M-molar absorption, v - oscillation frequency, v_o - oscillation frequency at the point of max. absorption, k - absorption coefficient of polarized light l - left handed, r - right handed.

power and wavelength may be expressed by Moffitt-Yang's equation.

$$[m']_\lambda = \frac{a_o\lambda_o^2}{\lambda^2 - \lambda_o^2} + \frac{b_o\lambda_o^4}{(\lambda^2 - \lambda_o^2)^2}$$

This equation contains three independent variables: a_o, b_o, and λ_o. The equation may be solved by plotting the values $[m']_\lambda$ $(\lambda^2 - \lambda_o^2)$ versus reciprocal of $(\lambda^2 - \lambda_o^2)$ in order to find the λ_o value, for which the function is a straight line. The b_o value, called 'peptide helix index,' is a tangent of the slope of the line, and a_o is found by substitution. By testing poly-γ-benzyl-L-glutamate (which is a polymer of α-helix structure), a straight line is obtained when $\lambda_o = 212$ nm, a_o value is for the dioxane solution equal to 135, and $b_o = -630$. This b_o value is specific for α-helix, and a_o is solvent-dependent. It is an oversimplification to presume that there may be only α-helices and coils in solution; however, further discussion did not give any better empirical or semi-empirical approximations than the Moffitt-Yang's equation.

On the basis of theoretical calculations it is possible to conclude that rotatory power due to the $n \rightarrow \pi^*$ and $\pi \rightarrow \pi^*$ transitions in α-helices and in 3_{10} helices is dependent on the chain distance in the peptides. This view has been experimentally confirmed for the $n \rightarrow \pi^*$ transition in α-helices.

The following optical properties are believed to be specific in the α-helix conformation:

(1) complex ORD-spectrum in the Vis and near UV region, obeying the Moffitt's equation with a b_o value close to -630;
(2) negative Cotton's effect with an inflexion point close to $\lambda = 233$ nm, due to the $n \rightarrow \pi^*$ transition of the peptide chromophore;
(3) at least one positive Cotton's effect, localized close to $\lambda = 192$ nm;
(4) hypochromism of the main transition of the peptide chromophore ($\pi \rightarrow \pi^*$ at $\lambda_{max} = 190$ nm);
(5) splitting of the 190-nm band into two components due to the resonance excitation between the residues present in the helix.

In a study on conformation of the chains of globular proteins containing fragments of right-handed helices, the rotatory power at every wavelength has been found as:

$$X = f_H X_H = f_\beta X_\beta + f_R X_R$$

where f_H, f_β, and f_R are the contributions of the helical (f_H) pleated sheet (f_β) and random (f_R) structures, and X are the values (which would be obtained) when the entire molecule would have a homogeneous structure. At the same time the dependence

$$f_H + f_\beta + f_R = 1; f \geqslant 0$$

is valid.

By solving the above equation (with a computer) for different wavelengths, concentrations, etc., by the method of least-squares, the values for the Moffitt-Yang's equation could be obtained. In the table 7.6 some results obtained are given and compared with those that are results from experiments on helical and β-structures of several proteins.

The values are in agreement and they show that the CD- and ORD- methods may well contribute to the conformation elucidation. According to the theory given the contribution of a helix is approximately linearly dependent of average rotatory power on one amino acid residue at a given wavelength: $[m]_\lambda = 233$, and on average ellipticity of the polarized light $[\Theta]_\lambda = 222$ nm. Calculations of the line shapes do agree with experimental data.

Another improved method of structure investigations in helical proteins proposed [18] is based on comparison of CD of solutions tested to the CD of polyamino acids in solutions of concentrated salts. The ORD method was used for the testing of keratin fractions, and also for fibroin. A review of the data concerning protein structure, as seen from the standpoint of optical activity, is given by Usmanov and Kochetov [19]. Investigations on optical activity and ORD of collagen and gelatin solutions, done between 1950–1960, have confirmed the strong levorotatory power of native collagen $[\alpha]_D = -280$ to $-400°$ dropping down upon its denaturation of collagen to gelatin to the value of 110-

Table 7.6

Comparison of experimental values of f_H and f_β (CD and ORD spectra) with values obtained by the X-ray method

Protein	Structure*	X-rays	CD	ORD1	ORD2
Myoglobin	H	0.77	0.77	0.77	0.77
	β		0.02	0.03	0.05
Papain	H	0.21	0.21	0.20	0.22
	β	0.05	0.10	0.14	0.19
Insulin	H	0.22-0.45	0.31	0.32	0.25
	β	0.12	0.18	0.06	0.07
Chymotrypsin	H	0.09	0.08	0.04	0.11
	β		0.10	0.12	0

*H—helix, β—pleated sheet

135°. This is one of the experimental proofs of the structural difference in the conformation of collagen and other proteins.

No significant differences are observed between collagen and poly-L-proline II helices. This is sometimes why the collagen helix is called the poly-L-proline II helix. It is not quite correct, however, using this expression*.

Poly-L-proline has a strong negative Cotton effect at $\lambda = 216$ nm, an inflection point at $\lambda = 203$ nm and a positive maximum at $\lambda = 194$ nm. The collagen curve is similar but its minimum is shifted to 205–210 nm. Denaturation by heating at 50°C during 30 min decreases the value of this minimum. From the curve (Fig. 7.14) a similarity, and at the same time, nonidentity of both helices may be seen. The negative Cotton effect shows the absence of α-helix in collagen and gelatin.

Recently Galatik and Blazej, however, have reported on the presence of α-helix certain elements there (5.3%), although it still seems to need more confirmation, as this is a rather unexpected effect although their calculation is based on the same principles as that done by Kühn et al. [20].

The CD of native and denatured collagen was tested as well: rat tail tendon,

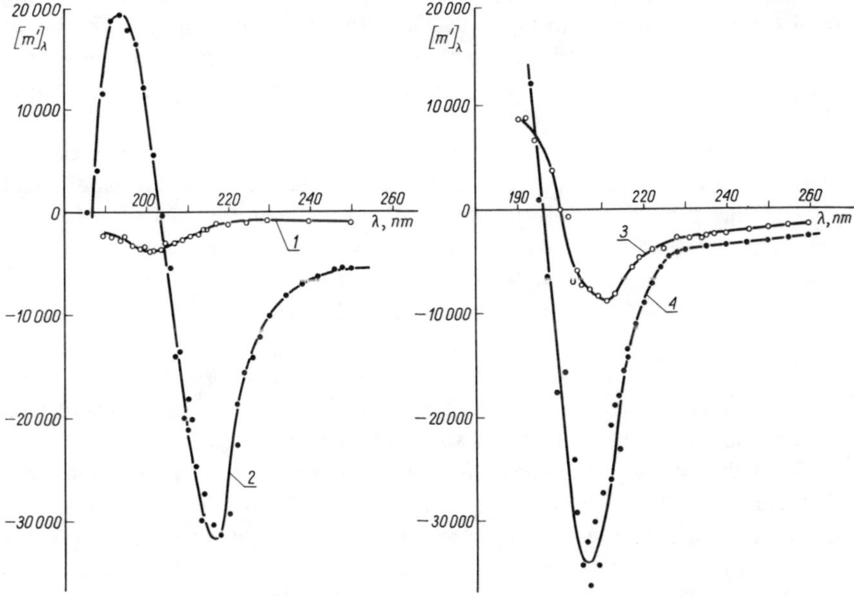

Fig. 7.14. ORD comparison of L-proline (1), poly-L-proline (2), gelatin (3) and collagen (4).

*Poly-L-proline may be obtained in two various forms, depending on the crystallization conditions (temperature, solvent, etc). These forms are called I and II. The results of comparison of ORD curves of poly-L-proline II and collagen are shown. A distinct similarity between these compounds may be seen, as well as some differences.

dissolved in 2% acetic acid has a strong negative peak at $\lambda = 195 \pm 5$ nm, and a weak positive peak at 221 ± 1. Denaturation by heating at 40°C results in the disappearance of the peak at $\lambda = 221$ and a decrease in the peak intensity at $\lambda = 195$ nm.

7.5. Infrared spectroscopy

Displacement of molecules in the space is called translational motion, or simply translation. A special case of this is Brown's motion. In a solid state instead of the Brown's motion a vibration of the molecules around their equilibrium positions occurs, such as in the crystal lattice. A measure of the translational motion is temperature; its energy is proportional to the absolute temperature T. An average translation energy of the molecule is kT, where k = Boltzman's constant $(1.38 \times 10^{-16}$ erg/K or 1.38×10^{-23} J/K). A progressive motion may be separated into three components, which may be viewed also as degrees of freedom. Another form of the molecule energy is the rotational energy, due to the rotation of the molecule around its axis. This energy may also be separated into three components; it is small however, and in practice it may be neglected. The third form of the energy in the molecule is the vibrational energy. The bonds in the molecule between valence electrons are not rigid. They behave like springs connecting the atomic cores (i.e., nuclei, together with their closed electron shells).

If the n atoms contributing to the molecule could be completely free, each of them would have three degrees of freedom. Translation of the molecule would give another three degrees of freedom; moreover, three degrees of freedom are due to the rotation in a nonlinear molecule, whereas two are in the linear molecule. A total number of the degrees of freedom is thus

$3 n - 6$ in the nonlinear molecules, and

$3 n - 5$ in linear ones.

The vibrational motions of the atom cores are at a first approximation similar to the oscillation of a harmonic oscillator (Fig. 7.15) where two masses m_1 and m_2 are connected to each other by a spring of a length r_e at the equilibrium state.

When the spring is stretched, the masses move in opposite directions to a distance $r > r_e$. Then a force F appears, acting in the direction opposite to that of the stretching, and the masses come back but, once running, they overcome the r_e distance and come closer to each other. And, so, in turn, the spring is compressed. Then the masses again move into the direction of the equilibrium positions r_e and over it. A repeated damping phase of this motion takes place. According to Hooke's law the force F is proportional to the stretching of an oscillator:

$$q = r - r_e$$

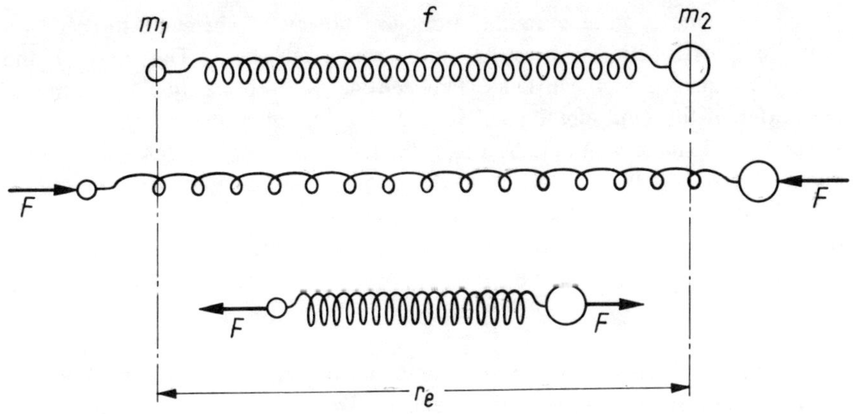

Fig. 7.15. Model of harmonic oscillator. The vibration phases of the models having masses m_1 and m_2 and connected by the spring of a force constant f.

in the equilibrium state $q = 0$; during vibration, however,

$$q = Q \cos 2\pi\nu t$$

where ν is the vibration frequency and Q is the vibration amplitude. In the harmonic oscillator, i.e., the one obeying Hooke's law

$$F = - fq$$

and the proportionality constant f, having dimension of force, is called the force constant. The force F is an opposite vector to stretching q, and hence a minus sign in the equation.

The potential energy in the differential form is expressed as follows:

$$dU = (-F)\, dq$$

and there the force constant $f = \dfrac{d^2U}{d\,q^2}$

The potential energy of vibration is

$$U = \frac{1}{2}\, fq^2$$

i.e. $U = 0$, if $q = 0$.

The molecule is an anharmonic oscillator, however, where the force F is a more complex function of q as this simple one given above. The energy in the molecule is quantified, the distances between energy levels are equal for harmonic and different for anharmonic oscillator, decreasing when the quantum number v increases. In an anharmonic oscillator the force constant is different at various energy levels. As it is visible from the above equation, the vibration energy curve is of a parabolic shape. In the anharmonic oscillator the curve shape is different; it may be approximated by Morse's function.

$$U = D (1 - e^{-\beta q})^2$$

where D is the dissociation energy of the molecule and β is a constant depicting the line shape (its deviation from parabola). The selection rule allows to pass over 1, 2, 3 or 4 levels, however, the probability of transitions there decreases. The transitions shown below are called tones or harmonics (Fig. 7.16).

The most probable, and therefore most intense, is the first tone (ground tone) then the first, second, etc., overtones. Intensity of the bands, corresponding to the overtones is decreasing with the overtone number. The position of the overtones is approximately given by the position of the primary tone in cm^{-1} and multiplier n. More exactly they may be found from frequency v_o and anharmonicity factor x from the formula

$$\Delta E = v_o [1 - 2x(v + 1)], \ cm^{-1}$$

It is possible to foresee the positions of absorption bands when the angles between bands and the force constants are known. It is also possible to estimate the force constant from the wavenumber of the oscillation observed.

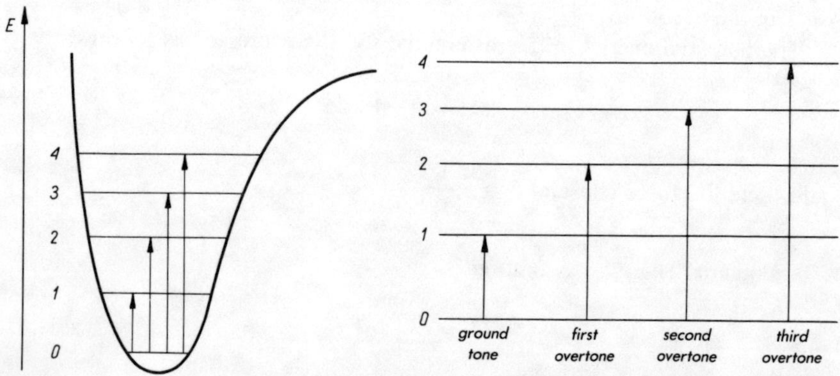

Fig. 7.16. Absorption transitions in anharmonic oscillator according to the selection rule; Δv = 1, 2, 3, 4.

A phenomenon of great importance in the IR spectroscopic analysis of macromolecular compounds is their dichroism. We may speak about dichroism of a substance, if the amount of polarized light absorbed by itself is direction-dependent. The measurement is usually carried out at two sample positions in the apparatus perpendicular to one another. The position of the bands remains unchanged, however their intensity ratio changes in the course of measurement. We refer to σ-dichroism, if the absorption of the light components polarized parallel to the direction of its propagation is greater than the absorption of the perpendicular polarized component. If there is a contrary case we speak about π-dichroism. The dichroic ratio δ is a quantitative formulation of this phenomenon and is a quotient of the extinction value for the parallel polarized ε_π, and for the perpendicular polarized ε_σ light

$$\delta = \frac{\varepsilon_\pi}{\varepsilon_\sigma}$$

The π dichroism occurs when $\dfrac{\varepsilon_\pi}{\varepsilon_\sigma} > 1$, whereas the σ dichroism – when $\dfrac{\varepsilon_\pi}{\varepsilon_\sigma} < 1$.

The dichroitic ratio provides information concerning the ordered, however, not fully crystalline states. In the fibrous proteins and in rubber it is a measure of the change in their orientations. In proteins the dichroism phenomenon is mainly related to the $=C=O$ stretching bonds and to the stretching and bending $=N-H$ bonds. Oscillation frequencies belonging to these groups are mainly dependent on the degree of their engagement in the hydrogen bridges. Differentiation of the protein structures by means of the infrared spectroscopy lies mainly in observation of dichroism of typical bands (Table 7.7).

The crystallinity of polymers is a very involved subject. When evaluating the

Table 7.7

Dichroism of typical IR bands and the protein structure

| Oscillation description | Structure | | Wave number in absence of hydrogen bonds cm^{-1} |
	H *(helix)*	*β* *(pleated sheet)*	
Stretching $C = O$	1650-1660	1630	1680-1700
Dichroism	π	σ	
Bending $= N - H$	1540-1150	1520-1525	below 1520
Dichroism	σ	π	
Stretching $= N - H$	3290-3300	3280-3300	ca. 3460
Dichroism	π	σ	

results of investigations, one may agree with Hummel's view: [21] "Crystallinity of polymers is a function of the testing method."

Identification of organic compounds of a small molecule is a task typically solved by the IR absorption spectroscopy. Application of this method to the investigation of proteins seemed to be limited because of a rather close similarity of the spectra of the majority of proteins and because of difficulties with testing substances which are soluble almost exclusively in water. These difficulties, however, may be overcome. The collagen IR spectrum has been described very thoroughly by Huc and Sanejouaud [22]. Because this spectrum is the most exact description of all the IR spectra of collagen available it has been characterized in Table 7.8.

The changes in position and in intensity of individual bands give the possibility to follow such processes, like denaturation, moisture absorption or deuteration. Using the MIR technique (multiple internal reflection) in which the surface is examined without sample dissolving, the hydrogen-deuterium exchange rate can be estimated. Also the quotient of intensity ratio for the $=$NH or $-$NH$_2$ and $=$ND or $-$ND$_2$ bands (amide I and II bands) in differently prepared samples can be studied. Such a work, e.g., [23] is an approach to the very difficult and important problem of study the protein deuteration. According to the authors, there is a possibility to investigate the rate of deuteration of the $-$NH$_2$ groups in collagen. Deuteration of other groups is too slow or negligible. Also in other papers the same problem was assessed.

Recently, Privalov et al. [24] used the IR method with radioisotope method to investigate the exchange rate of hydrogen in collagen, particularly that involved in hydrogen bonds. According to their conclusion, both the fast and slow· exchanging hydrogen atoms are involved in hydrogen bonds, and all of the detectable hydrogen bonds (1,7 per tripeptide unit) are connected with the peptide CO and NH groups.

7.6. Raman spectroscopy

The use of Raman spectroscopy to biological molecules has been developed only in the last decade. It has a great advantage over IR spectroscopy in that use of water, as solvent, always presented a problem. Raman spectrum of water is weak enough to be eliminated by subtraction methods. The laser beams used as a radiation source almost exclusively in modern apparatusses, produce a high-intensity, highly convergent and coherent monochromatic light.

Raman scattering in its simplest form supplements the infrared analysis. If there is a specific vibration frequency of the molecule, ω_k, which modifies the molecular electric moment, then absorption of an electromagnetic wave of this frequency can be observed in the IR region. The induced moment may also be set up. It can be shown that a molecule irradiated with an incident light of

Table 7.8

Acid-soluble collagen IR absorption bands

Frequency cm^{-1}	Properties of the bands	Assignment
3400	disappears on treatment with chloroacetic acids	bound water
3300	typical amine band	stretching $= N - H$
3060 2950 2900	equal intensity	asymmetrical stretching bonds of $= NH$ and $-CH_3$ groups asymmetrical stretching of $= CH_2$ groups
2870	uncharacteristic band	stretching, symmetrical of CH_3 groups
2850	very weak	stretching, symmetrical of $= CH_2$ groups
1710	increases in acids and on heating	COOH unionized group
1650	very strong	amide I-band and $=C=O$ stretching in helix structure amide I band in coil structure
1635	increases in acids and on heating	
1560	decreases in acids increases in alkalis	COO^-
1550	very strong, decreases in acids	amide II band of helical structure
1530	increases in acids and on heating	amide II band of coil structure
1445	uncharacteristic band	bonding oscillations of $= CH_2$ and $-CH_3$ groups
1407	increases in alkalis decreases in acid	$-COO^-$
1375	weak	bending, symmetrical oscillations of $-CH_3$ groups
1340	quite strong	amide band III CN $+$ NH of coil structure
1240	appears by acting of acids or on heating	amide band III of helix structure
1230	decreases by acting of acids or by heating	no data available
1080 1065		
940 920	very weak	bend mode of carboxyle $-OH$

definite wavelength emits waves of three frequencies.

ω_o—Rayleigh or elastic scattering,

$\omega_o - \omega_k$—Stokes Raman lines,

$\omega_o + \omega_k$—anti-Stokes Raman lines.

Infrared radiation is due to periodical variations of the electric moment of the molecule, whereas Raman scattering takes into account the variations of the induced moment (polarizability variations). Thus the frequencies observed in IR and in Raman scattering are not necessarily, although they may, be the same. When the polarizability changes, or its derivative is non-zero, the frequency is Raman-active. If the dipole moment of vibration in operation is unchanging, or if its derivative is equal to zero, then the molecule is IR-inactive (Table 7.9).

Looking at the table it can be visualized how the IR- and Raman spectra may complement each other. The IR- or Raman activity of vibration may be calculated and thus it is possible to predict whether or not a peak in spectrum will be observed.

This method allows one to evaluate the relative percentage of α-helix, β-structure and random coil in the protein molecule [25], similarly as it has been previously done for IR by Krimm [26].

Sometimes—and this is the case of collagen, it is necessary to resort to the superposition technique to separate the Raman lines from the peptide backbone and side chains. In this technique a mixture of amino acids contributing to the substance tested is used as a reference. By this technique Frushour and Koenig were able to obtain good quality collagen spectra [27] and carry out their assignments. The spectra are shown with IR collagen spectra in Fig. 7.17 and the

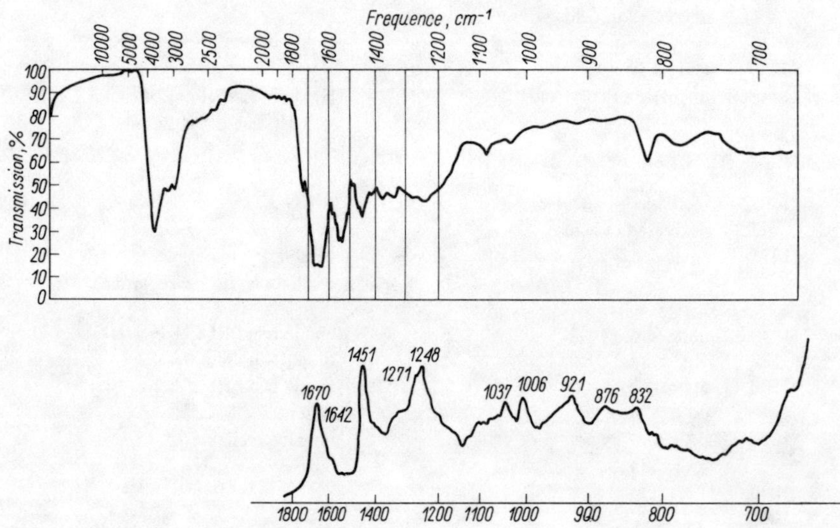

Fig. 7.17. - IR - (upper) and Raman (bottom) collagen spectra at the same wavelength.

assignments are reported in Table 7.10.

In the amide I and III band regions two lines appear in each region, and

Table 7.9

The Raman and IR activity of molecule vibration

Polarizability derivative	$\neq 0$	$= 0$	$= 0$	$\neq 0$
Raman activity	+	−	−	+
Dipole moment derivative	$= 0$	$\neq 0$	$= 0$	$\neq 0$
IR activity	−	+	−	+

Table 7.10

Raman lines in the bovine Achilles tendon collagen according to Frushour and Koenig [27]

Wavenumber, cm^{-1}	Assignment
1670 s	stretching $C=O$
1648 s sh	Amide I
1464 s sh	bending CH_3, CH_2
1451 s	″ ″ ″ , in-plane bend of carboxyl OH
1422 m	stretching, sym. COO^-
1392 m	
1343 m	out-of-plane deformation bends (wagging) CH_3, CH_2
1314 m	out-of-plane deformation bends (twisting) CH_3, CH_2
1271 s	Amide III
1248 s	Amide III
1211 w	Hypro, Tyr
1178 w	Tyr
1161 w	NH_3+
1128 w	stretching C-N
1101 w	stretching C-N
1087 w	stretching C-N
1067 w	stretching C-N, out-of-plane bend of carboxyl OH
1037 m	Pro
966 w	Phe
938 m	stretching C-C of residues
921 m	stretching C-C of protein backbone
918 w	stretching C-C of Pro ring
890 w	
876 w	stretching C-C of Hyp ring
856 m	stretching C-C of residues
821 w	stretching C-C of Pro ring
769 w	stretching C-C of backbone
622 w	Phe
396 w	Pro

s—strong, sh—shoulder, m—medium, w—weak

according to the superposition method, the 1271 and 1248 cm^{-1} lines are assigned to the amide III mode

Observing the relatively small changes in the amide III region during collagen → gelatine transition, the authors concluded that the 1248-cm^{-1} line in the collagen spectrum can be assigned to amide III mode of the collagen helix. However, there is no easy way to distinguish between helical and random conformations of collagen. The presence of two lines which do not disappear after denaturation is not explained and it is supposed to be due to local distribution of the proline-rich and proline-poor regions.

Comparing the Tables in which IR and Raman spectra of collagen are characterized in the region between 1670 and 920 cm^{-1}, only nine bands can be seen which are both IR and Raman active, with still some deviations in their positions (e.g., Amide band I 1670 and 1650 cm^{-1}, respectively). Thus more thorough study is necessary in order to say more about their interrelations. The relation between both spectra is still far from being conclusive which might rightly be expected, as this is common for organic compounds. Samples of the same origin and preparation should be compared. The problem of collagen deterioration by IR and laser irradiation should be solved in such experiments as well.

Spectra of keratin and elastin have also been analyzed and their assignments have been done by the same authors.

7.7. Radio frequency spectroscopy

The high degree of specificity of many biological reactions and processes depends on subtle differences in the structure and conformation of molecules. Nuclear magnetic resonance (NMR) spectroscopy is one of the few techniques available with the use of which detailed information can be obtained about biomolecular phenomena. With the NMR an individual nucleus in a molecule can be 'observed' by monitoring that nuclear line in an NMR spectrum. The various NMR parameters of that line: frequency, splitting, linewidth, and amplitude can be used to study the electronic and geometric structure of 'simple' molecules, macromolecules, molecular motions, interactions and rate processes. Quite often the molecular information obtained is of a qualitative value; however, in many cases NMR provides quantitative information not obtainable by other means.

The charge of an atomic nucleus may rotate around an axis creating a magnetic dipole along it (pair of poles). The spin number J (fourth quantum number) gives the angular momentum of the charge. This is a result of the sum of spins of protons and neutrons contained in the nucleus; if this sum is an even number, then J is a whole number. If both the proton and the neutron numbers are even, then $J = 0$. Such atoms have no spin. J has a half value, if the sum of protons and neutrons is an odd number. Nuclei such as ^1H, ^{19}F, ^{13}H have the value of $J = \frac{1}{2}$. The formula $2J + 1$ gives the number of orientations with which the nucleus may have in a homogeneous, external magnetic field. According to the quantum space principle, the nucleus may only take definite positions. So a proton may just have two orientations: parallel and antiparallel (i.e., lying in the opposite direction to the field). The parallel orientation has a lower energy level than the antiparallel one, so it is a more stable state.

When in the magnetic field of constant intensity two constant-energy levels of proton exist, and by adding to them certain energy quanta h, one may initiate the transition from a lower to a higher level. Energy is absorbed during this process. This may occur if the resonance requirement is fulfilled, i.e., when

$$\hbar\nu = g\beta H_o$$

which means that when the product of Planck's constant $\hbar = 1.054 \times 10^{-27}$ erg/s (1.054×10^{-31} J/kT) and the vibration frequency of exciting radiation is equal to the product of magnetic field intensity, constant g (dimensionless, for proton g $= 5.585$) and nucleus magneton ($\beta = 0.505 \times 10^{-23}$ erg/g), (0.505×10^{-27} J/kT) ($h = \frac{\hbar}{2\pi}$). The same formula is valid for electron paramagnetic resonance. (EPR), however, with different values for the symbols, viz.: g is for electron $= 2.002$, and electron magneton (Bohr's magneton) $= 0.927 \times 10^{-20}$ erg/g (0.927×10^{-24} J/kT). In the above formula the variables are: frequency and field intensity. One of these values, usually frequency, is made constant in experimental conditions. Then the field intensity only is changed. If the experimental parameters are chosen properly, a certain moment of absorption of oscillating radiation will be observed. To observe the proton absorption, frequency of 60 MH is usually used. The field intensity is then approximately 14 kGs. (1.4 T) It should be kept in mind that only a small part of nuclei (protons) usually becomes oriented in the magnetic field. Their number, however, is enough for the absorption to occur and make the spectrum observable, provided an adequate measurement technique is used. In the EPR-technique the standard frequency is of the range of 9500 MHz, which is equivalent to the wavelength of 3 cm. Magnetic field intensity H_o for free electron is here approximately 3400 Gs (0.34 T) the energy of radiation is thus about 1000 times higher than in the NMR method.

It would not be possible to observe radiofrequency radiation energy absorption when there would not be observed a transition of the nuclei from a higher to a lower energy level and the absorbed energy to the space lattice in time T_1 would be given back. This phenomenon is called spin-lattice relaxation*.

The reciprocal value of this time is equivalent to the line width: the shorter the time, the wider the line. This is valid for liquids. In solids T_1 is very long, and the absorption bands should be very narrow. In this case, however, there is another effect working, i.e., spin-spin relaxation, T_2. This is due to the energy transfer from one nucleus to another. Its spreading gives an essential broadening of the absorption band. Steady position of the nuclei tested (as compared with their free motion in liquids) is thus responsible for the shape of the curve shown in Fig. 7.18. This curve depicts the dependence of T_1 and T_2 values on correlation time τ, i.e., on the time of the revolving of the molecule over an angle of one radian or on the translation of the molecule to the distance, approximately equal to its dimensions. The shape of the absorption resonance line, being close to the Lorentzian line in liquids, or to the Gaussian line in solids if the resonating nuclei are randomly distributed, provides further information about the system under study.

The observation of solids was described the first time in 1948 by Pake: if there is two-proton system present (e.g., in water molecule), then the spectrum shape gives information about the distance between these nuclei and about orientation of their position as related to the applied magnetic field. Absorption of this

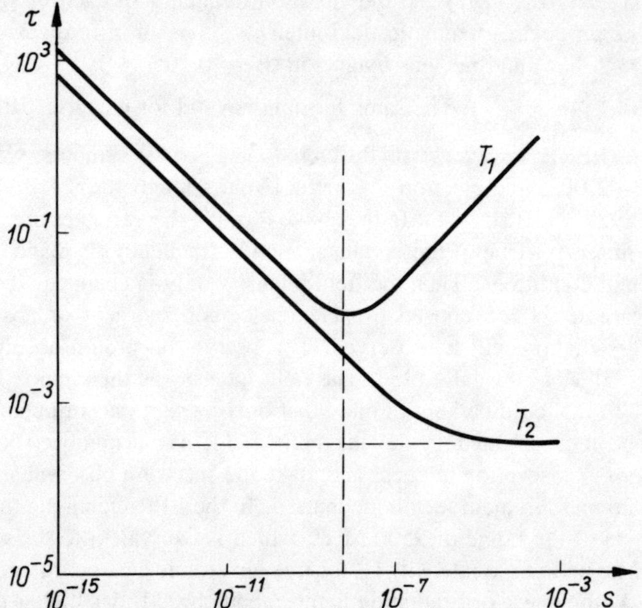

Fig. 7.18. Theoretical dependence of relaxation times T_1 and T_2 on correlation time τ.

*An ensamble of molecules, considered as a carrier of some thermal energy arising from their motion, is referred to as 'lattice.' This definition is valid for every state of matter, but it was initiated for crystalline state, where the oscillations are responsible for thermal energy.

system shows two symmetrical bands shifted from the resonance frequency by ΔH, and

$$\Delta H = \pm \ \frac{3}{2} \ \frac{\mu}{\vec{r}^{\ 3}} \ (3 \cos^2\theta - 1)$$

where μ is magnetic moment (for protons)$\mu = 14{,}1 \times 10^{-24}$ erg/Gs) (14,1 $\times 10^{-27}$ J/T), $\vec{r}$ is the distance between both protons, and θ—angle between $\vec{r}$ and magentic field. A similar, but more involved dependence shows the three-proton system. Thus from the line width of the substance tested one may conclude that the physical state of the molecule under observation from the line shape, can be inferred if the observed nuclei are situated randomly, or if there is a regular system, and how frequently they change their mutual positions.

Very often, two lines are also observed, as in the case of collagen-water system, which originates from the protons of the system in question: a broad line, coming from the protons of the peptide and a narrow line situated on top of the broad line, and for which the protons of bound water are responsible. The width of this line informs us as to what extent the properties of the bound water differs from those of bulk water. This is because the line width in the spectrum of ice is broader when compared to bulk water. The line width of bound water has the intermediate values, as the protons of hydration water are much less mobile than in free water. The width of the line from protons of bulk water is at room temperature in range of single milliGausses, that of bound water—of hundreds of milliGausses, and of ice protons—of Gausses. The changes in the water molecule, when approaching the protein molecule, are due to the marked decrease of the frequency of its translative movements (from 10^{-11} to about $10^{-5}\mathrm{s}^{-1}$). This frequency may be investigated by the self-diffusion, viscosity dielectric relaxation or NMR methods.

The NMR spectrum of wet collagen has, for the first time, been described by Berendsen. These investigations are described in more detail earlier in this book (sect. 5.4).

It is usually believed that the wide line in the proton NMR spectrum of solids is of a Gaussian (bell) shape. In analyzing this line, however, we found [28] that it differs from the Gaussian curve, and from the Lorentzian curve as well. We have further found [29] that in the oriented collagen from the bovine Achilles tendon the two-proton system may be observed. The interproton distance $\vec{r}$ in this system is close to the distance in water, compared with that is in the $-CH_2$ groups, and clearly differing from that in $-NH_2$ groups. The experimental data for dried, unoriented collagen show the interproton distance of 1.55 ± 0.02 Å; for water—1.58 Å; for $= CH_2$ groups—1.60 Å; for $-NH_2$ groups—1.76 Å. These results however still should not be accepted as finally proved, because they are not confirmed by another independent-method.

Another NMR-technique, i.e., high-resolution method, was used for investigation of the reaction between collagen destructs and metal salt [30]. First, a spectrum was measured in the gelatin solution in D_2O, then the gelatin solution after reaction with Co^{2+} ions. As the differential spectrum has shown, the complexes between ions and glutamic and aspartic acid have been formed. Accordingly, one may presume these residues to be responsible for the binding of cobalt ion to collagen.

As already mentioned, no important information has been obtained when collagen was investigated by the EPR technique. A few years ago, however, as Pepe [31] has shown, one may obtain the EPR signal from collagen after its irradiation with UV light ($\lambda = 365$ nm). This signal is observable during a few days after irradiation; then it disappears. Free radicals, arising in collagen after irradiation are responsible for it. This reaction, however, is not specific.

Torchia and VanderHart [32] have reported on anisotropic molecular motion of collagen fibrils, from their study of the ^{13}C-NMR line. This motion is said to be an average reorientation of the molecular collagen about the long axis of the helix. Reorientation may be presumed, as concerning the side chains through an angle of about 30°. It implies that the intermolecular forces do not depend strongly on the azimuthal angle and, hence, do not arise from a unique set of interactions between the side chain. For practice it might provide an explanation of the tanning process, if we presume, that the side chain may 'seek' complexing.

One may figure how difficult such an experiment from the fact that about 20,000 scans were needed to obtain an acceptable picture of the resonance shift of ^{13}C nuclei.

Recently Jelinski et al. [33], using collagen samples with built-in deuterated $(3,3,3\text{-}d_3)$ alanine and the quadrupolar spin-technique, were able to demonstrate a molecular motion in the fibrils. The molecule experiences reorientation around its long axis over an angular range of 30–40°, which is in accordance with the ^{13}C experiments. The T_2 is of an order of 110 µs, thus the frequency of the rotation is about 10^{-7}s.

REFERENCES

1. Ramachandran, G. N., Sasisekharan, V., Ramakrishnan, C. Biochim. Biophys. Acta *112*, 168 (1966)
2. Milionova, M. I., Andreeva, N. S. Biofizika 2, 294 (1957) in Russian
3. Hosemann, R., Nemetschek, Th. Kolloid Z., Z. für Polymere *25*, 53 (1973)
4. Bienkiewicz, K. J., Bukowska Strzyzewska, M. J. Polymer Sci. Part C *53*, 173 (1975)
5. Andreeva, O. A., Danish, L. V., Tverdokhlib, V. S., Shkaranda, I. T. Iz. Vys. Ucheb. Zaved., Leg. Prom., *23*, 4, 41 (1980) in Russian
6. see ref. 50 ch. 2
7. Fraser, R. D. B., McRae, T. P. J. Mol. Biol. *127*, 129 (1979)
8. Hulmes, D. S. J., Miller, A., White, S. W., Brodsky-Doyle, B. J. Mol. Biol., *110*, 643 (1977)

 9. Brodsky, B., Eikenberry, E. F., Cassidy, K. Biochim. Biophys. Acta *621*, 162 (1980)
10. Griffiths, J. Colour and Constitution of Organic Molecules Acad. Press London 1976
11. Doyle, R. J., Bello, J. Biochem. Biophys. Res. Commun., *31*, 869 (1968)
12. Wood, C. G. Biochem. Biophys. Res. Commun., *13*, 95 (1963)
13. Katz, E. P., Nishigai, M., Bonar, L. P. Coll. Curr., *4*, 11, 13 (1964)
14. Hoerman, K. C., Balekjian, A. G. Fed. Proc., *25*, 1016 (1966)
15. Hoerman, K. C., Balekjian, A. G., Boyne, P. J. Dental Res., *48*, 661 (1969)
16. Grimes, M. W., Gruber, D. R., Hang, A. Biochem. Biophys. Res. Commun., *37*, 853 (1969)
17. Wolframm, C. P., Iancu, C., Heidemann, E. Leder *23*, 233 (1972)
18. Rosenkranz, H., Schollan, W. Hoppe-Seylers Z. Physiol. Chemie *352*, 896 (1971)
19. Usmanov, R. A., Kochetov, G. A. Usp. Sovr. Biol., *6*, 75 (1973) in Russian
20. Galatik, A., Blazej, A. Kozarstvi *29*, 191 (1979) in Czech
21. Hummel, D., Scholl, F. Atlas der Kunststoffanalyse, Vol. I Hanser Munich 1968
22. Huc, A., Sanejouaud, J. Biochim. Biophys. Acta *154*, 408 (1968)
23. Heidemann, E., Magerkurth, B., Santhanam, P. Leder *21*, 29 (1970)
24. see ref. 45 ch. 2
25. Lippert, J. L., Tyminski, D., Desmoules, P. J. J. Am. Chem. Soc., *98*, 7075 (1976)
26. Krimm, S. J. Mol. Biol., *4*, 528 (1962)
27. Frushour, B. G., Koenig, J. L. Biopolymers *14*, 379 (1975)
28. Bienkiewicz, K. J., Florkowski, Z., Hennel, J. W., Lalowicz, Z. T. Leder *21*, 5 (1970)
29. Lalowicz, Z. T., Bienkiewicz, K. J., Korycinski, A., Salamon, B. Inst. of Nuclear Physics Report 819/PL, 110 (1973) in Polish
30. Rose, P. I. Science *171*, 573 (1971)
31. Pepe, I. M. Biophysik *7*, 115 (1971)
32. see ref. 46 ch. 2
33. Jelinsky, L. W., Sullivan, C. E., Torchia, D. A. Nature *284*, 531 (1980)

8.

THE PHYSICAL CHEMISTRY OF RAW HIDE AND OF CURING PROCESS

The changes in equilibrium state in the collagen-water system and the secondary processes, like, e.g., autolysis, bacterial growth and changes due to aging are related to the hide curing process. Among the essential changes the removing of water from the system and its saturation with salt or treatment with other curing substances may be named. The secondary processes start from the flaying moment; their intensity depends on the external conditions and they are insofar dangerous, as they may cause a quick and irreversible deterioration of hide.

8.1. Autolysis of hide/skin

The hide (skin) of a live animal contains 62-78% water. This amount is necessary for the life processes, to occur properly. Separation of the hide from the carcass does not change the water content of the tissue in a significant way. However, the death causes a dramatic change in the metabolic processes. The hide is cut off from the supply of oxygen and nutritional components. The way back is closed as well, i.e., the removing of the metabolites from the cell is stopped, which leads to the accumulation of toxic products. This in turn leads to the stopping or breaking of the part of the enzyme controlled processes due to the inactivation by poisoning (inhibition) of some coenzymes. The process of self-digesting or autolysis of the cells starts. In this process some intracellular enzymes from the 3.4. (peptide hydrolases), i.e., cathepsins group are essential. These enzymes may be found in lysosomes (i.e., grain-shape organella of the cell, which are like packages of enzymes wrapped in a membrane). Death causes a permeability change in this lysosomal membrane. Cathepsins, which in the living animal control the equilibrium state between cellular proteins and their metabolites, are coming out from lysosomes and decompose proteins to the peptides and aminoacids. Lysosomes also contain enzymes catalyzing decomposition of sugars, lipids and other compounds. Total digestion of the cell may be the end effect of autolysis.

Autolysis by itself does not cause changes in the flayed hide left at room temperature even for twenty-four hours. According to Tancous [1] in the cattle hide stored at room temperature during 9 months under aseptic conditions (under 150% of water, covered with toluene), the following phenomena may be observed:

(1) When equilibrium state is reached, i.e., after stopping the autolysis the hides

have lost 39% of nitrogen and 57% of water. Soluble protein destructs have been dissolved in the water. Water lost in the cells resulted from destruction of the membranes.

(2) 0.55% CO_2 on sample weight is liberated.

(3) 0.5 mole of organic acids have been found in the solution, among them mainly acetic and butyric acids, but some propionic, isovaleric, isobutyric and lactic acids were also present.

(4) Cathepsins have been found in the medium; among them amidase (EC 3.5.1.4 acylamide amidohydrolase) was identified, an enzyme splitting off ammonia from amides of monocarboxylic acids.

(5) According to histological examination, the cellular and intercellular fluid has been washed out, collagen and elastin fibers split; fat liberated from cells migrated to the other places.

(6) Leather, obtained from autolyzed hides was strong but spongy, stretchable but not flex-resistant enough.

These results point to a considerable deterioration of raw hide due to prolonged autolysis, without participation of bacterial processes.

As shown by the same author [2], the course of autolysis of salted cattle hides is dependent on the temperature and the amount of salt contained in the hide. The higher the temperature, the faster is the autolytic process. Its rate, however, decreases with the increase in salt concentration. As a measure of the autolysis rate the amount of liberated CO_2, soluble nitrogen or free fatty acids was accepted. There is a positive correlation among the factors indicated. In the same paper the influence of typical inhibitors on the enzymatic reaction was tested as well. The most dramatic inhibition of the autolysis may be achieved by sodium penta- and trichlorophenolate and alkylbenzyldimethylammonium chlorides. Table 8.1 presents the results of investigations on various preserving agents (biocides), as well as correlations among the factors in question. Histological evaluation was done according to O'Flaherty, whereas a possible maximum 15 points have been given, and 5 points for fully deteriorated condition. The quickest deteriorating components in hide are the proteins of fluids-blood and lympha. The red blood cells, as the entities highly specialized in oxygen transport, have their enzymatic system developed for this purpose only. Their activity is inhibited by stopped circulation, so they are easily attacked by external enzymes, because of lack of their own protective enzymes.

From the data given in Table 8.1 one may see that many of the very common preservatives, like boric acid or sodium carbonate, do not inhibit raw hide autolysis at all.

The yellow 'salt' spots on the hide are very often believed to be a product of bacterial activity. In fact they arise from the autolysis processes, i.e., due to the effect on alkaline phosphatases in the presence of calcium sulfate [3].

Table 8.1.

Efficiency of some biocides in hide curing [12]

Biocide	Soluble nitrogen	CO_2	Weight loss	Histological evaluation
	g per 500g of hide			
EDTA	1.00	2.32	39.0	10
Boric acid	4.92	3.52	85.0	6
Sodium trichlorophenolate	0.53	0.50	25.0	11
Alkylbenzyldimethylammonium chlorides	0.65	1.46	29.5	12
Sodium carbonate	1.38	2.73	51.5	6
Acetylosalicylic acid	1.13	2.42	57.0	6
Sodium pentachlorophenolate	0.56	0.40	24.0	14
Sodium fluorosilicate	1.38	1.68	65.5	8
Sodium hypochlorite	1.01	1.31	37.0	10
Sodium borate	2.09	2.77	56.0	6
Sodium sulfite	3.19	2.86	65.0	6
Sodium bicarbonate	2.44	1.98	63.0	7
Sodium azide	0.93	0.56	37.0	9
Sodium cyanate	1.16	0.80	53.0	10
Control	3.56	3.70	88.5	6

Correlation factor (r_{xy}): 0.8545, 0.8428, 0.7411, 0.9268, 0.8063, 0.7741

8.2 Bacteria growth on the flayed hide

The secondary process accompanying autolysis is the action of putrefactive bacteria, for which the autolysis products offer an excellent medium. These bacteria 'supplement' the deteriorative action of the 'own enzymes' of the cell which are unable to digest collagen. The question may arise here, why the 'own tissue' collagenase does not attack it. This enzyme is very easily inhibited, however, e.g., by blood serum. In the hide separated from the organism these factors do not work which may produce a natural shield against bacteria and other micro-organisms. Most important of these factors is a permanent supplying of oxygen in the living body by hemoglobin. Due to this the redox potential of the cell is maintained on the same level.

The anaerobic putrefactive bacteria may much easier attack the hide, once this oxygen supply stops. Various kinds of bacteria may live on the skin of the animal or inside it due to wounds, scratches, insect bites, etc., or they may travel through mucosae of the body openings and their hair pockets. Barriers existing in the living organism do limit the action, growth and propagation of bacteria. Breaking of these barriers results in penetration of the bacteria into the organism; these are the reasons for infectious diseases. The increase of bacterial activity however belongs to the physiology of death. Deterioration of tissues follows, i.e., of cells and of extracellular outgrowths. Conservation (curing) of the skin, flayed off a slaughtered animal should prevent it against deterioration and, first of all, to prevent bacterial growth. Basic condition for the process in question is a change in the water content in the skin. This change is accompanied by changes in other biophysiocochemical factors, e.g., ionic strength and redox potential. An additional change is the use of bacteriostatics (disinfectants) and bactericides.

Bacterial growth in the medium is proceeding in a specific way. The stages of this growth, shown in Fig. 8.1, are as follows:

(1) Latent stage—the number of cells is constant and even may be decreasing. In this stage the bacteria population is adapting itself to the medium offered in intended or accidental way. This is the longer, the lower is the nutritional value of the medium. Adaptation of the enzymatic apparatus of the bacteria is necessary. In this stage the metabolic activity of the bacteria is moderate, but its resistance to the influence of the harmful components of the medium is considerable.

(2) The stage of 'physiological youth' or of positive growth acceleration rate. Metabolism of the cell is of growing intensity, and the gain in their weight becomes increasingly higher: initially, due to the increased weight of the cell, and in the final part of the stage—due to increasing population. This stage is rather of short duration and the cells are sensitive to changes in the medium.

(3) Logarithmic growth stage, in which the increase of the biomass is constant; the cells are somewhat smaller than in previous stage, but their size is constant,

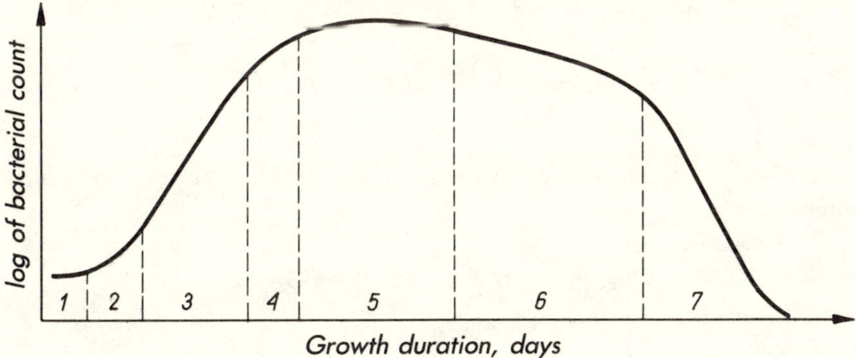

Fig. 8.1. Bacteria growth phases: 1—latent phase, 2—positive acceleration phase, 3—logarithmic growth phase, 4—negative acceleration phase, 5—equilibrium phase, 6—increased dying rate phase, 7—logarithmic dying rate phase.

as is their division frequency, and thus their average age. It is the main stage of the culture growth. Its duration depends on the factors of the medium and on the specific properties of the observed bacteria type. It may be shortened by insufficiency of nutrients, accumulation of toxic metabolites, substances being useless or changing pH or rH and other factors.

(4) Stage of deceleration of growth rate, in which the rate of divisions is decreasing with time. The processes reversing to the stage 2 processes may be observed to occur here.

(5) The equilibrium stage, in which, as in phase 1, the number of cells remains about unchanged. The number of dividing cells is approximately equal to the number of those dying. This stage may be of long duration, and a certain degeneration of the cells, changes in their size and shape may be observed here.

(6) Increase of dying rate stage, where the number of dying cells increases, and the number of the dividing ones decreases.

(7) The phase of 'logarithmic dying rate' takes place, if the dying process is very rapid. In this phase a full self-sterilization may be observed, i.e., the number of living cells may drop to zero, e.g., due to intoxication by metabolites or to the increase of the temperature of the environment.

The stages 1-5 may be described in mathematical formula in several ways. The basic equation is the following:

$$N = N_o \, e^{-a\Delta t}$$

where N is the number of bacteria at a given moment, N_o the initial bacteria number and Δt the growth time, a is the reciprocal of the average of individual age $1/g$ or the partition frequency, usually being approximately equal to $2/h$. According to Srinivastava and Avashti [4] a better approximation for stages 1-3 is given by the formula

$$\log \frac{a}{a_1} = L \left[(t - t_o)(1 - e^{\frac{-t}{t_o}}) \right]$$

where L is a constant, a_1 is the value of a at time t_o; for the stages 4 and 5 they proposed

$$\log \frac{a}{a_1} = L(t - t_o) + L(t - t_o)^2 \, \ldots$$

The bacterial processes are interesting for us principally as negative phenomena. Efforts should be made not to allow them to come to stage 2, or as soon as possible to reach stage 5 or still better stage 6. In this effort a 'rule of minimum,' formulated by Liebig, may be of use. This rule says that 'the production ability of the biological medium is depending on the substance which it contains in minimal amount and it cannot be increased by raising the content of any other component.' Use is often made of this rule in bioengineering processes, controlling, e.g., the yeast production process by regulation of the oxygen supply, or when titrating bacteriologically certain aminoacids, or depriving the medium of the ions of metals activating the enzymes (water demineralization), etc. The life processes are going until one of the medium components becomes exhausted. This is valid for nondisturbed media.

For bacterial growth certain humidity of the medium is required. It is, however, various and changing; usually it is 30-35%, for the molds it may be 12-15%. The temperature of optimal growth is for various strains different too. For thermophilic bacteria it lies between 45 and 60°C, its maximum is 70-80°C. In hot springs (98°C) some bacteria strains may be found in vegetative forms. Enzymes of thermophilic bacteria are more denaturation-resistant than the enzymes from other groups of bacteria [5]. The thermophiles mostly may form spores. The majority of parasites and saprofites* belong to the mesophilic group**.

Optimal growth temperature of psychrophilic bacteria, living in the seas, lakes or in the soil, lies below 20°C. Minimal temperature of possible growth is usually by ca. 5°C higher or about equal to the freezing temperature of the medium. Under laboratory conditions most psychrophilic bacteria behave like mesophilic ones.

In too high or too low temperatures the bacteria die or form spores. Spores are inactive, surprisingly resistant to the hostile conditions of the medium. Having formed the spore, the rest of cell gradually disappears. Spores are much more resistant to heat, drying, mild disinfectants and other harmful agents than is the original cell. They serve to maintain the organism over a period of time when the environment is hostile. When more suitable conditions return the spore germinates and again a cell develops from it like the one that originally formed the spore. Spores are usually formed by the rodshape cells only and not by all kinds of cells. The spore is formed within a cell and for this reason is sometimes referred to as an endospore. Spores may remain alive for a very long time. The spores of *Bacillus anthracis*, e.g., may stay in soil, in the walls of a building,

---

*Among the organisms, unable to synthesize food from inorganic substances, i.e., heterotrophic, one may differentiate parasites and saprophites. The saprofites are taking their nutrition in a dissolved form directly through the cell wall. Yeast, molds and majority of the bacteria are in this group. Parasites are living inside or on the surface of other organisms called host. The parasite bacteria usually require temperature, close to that of the host temperature, and the nutrition medium composed of sugars, amino acids and vitamins.

**Optimal temperature for the growth of mesophiles is 20-40°C. The majority of disease-causing bacteria require temperatures at very narrow range, e.g., *Spirochaeta pallida*, whose optimum temperature is 37°C, will be killed by temperatures of 40-41°C. Most pathogenic bacteria are unable to form spores (except for *Bacillus anthracis* and some others).

etc., for 25 years or more. Fortunately this bacteria, dangerous for people or animal, may in practice be hardly met with now due to the veterinary control of slaughterhouses in the developed countries. The majority of bacteria find their optimal living conditions at neutral, or slightly alkaline pH, the majority of molds—at acidic ones (about 5).

The mechanism of spore formation has not been as yet fully elucidated. As far as we know, however, they contain the enzymes resistant to high temperatures, not only inside spores, but even after germination. Germination process itself is controlled by thermostable enzyme L-alanine racemase. It transforms L-alanine into D-alanine, which inhibits endospore germination and in this way limits the number of vegetative cells. Other enzymes are also found in endospores, e.g., catalase and ribosidase, which catalyzes the adenosine → adenine + ribose reaction. These enzymes are present as complexes with peptides, Ca^{2+} ion and a dipicolinic acid polymer. The endospores contain 6-12% of this compound. Its volume decreases with the lowering of the thermal stability of the spore.

The influence of dipicolinic acid on thermal stability and on survival rate of the endospores has been known for a long time.

Among the numerous representatives of the bacterial flora on the animal skin (over 80 kinds are described, but probably there are many more) the most dangerous are those that deteriorate proteins. The degree of attack of the putrefactive bacteria on the skin is most frequently evaluated by estimation of the amount of water-soluble nitrogen compounds present. Among the proteolytic bacteria the most common are: *Proteus vulgaris*, *Bac. putrefaciens*, *Bac. subtilis* and *Bac. mesenthericus*.

Among the organisms growing on hides and particularly on tanned stocks (wet blues) some yeast strains have been identified. Together with molds and mildews they prefer acid medium of a pH 3-5. They may cause the same kind of damage on the finished material—stains and spots. Recently Kallenberger [6] has isolated from wet blues three yeasts genera: *Rhodotorula*, *Cladosporium* and *Torulopsis*. They are responding to fungicide action in about the same way as the molds do. This is in accordance with present author's observation. Beside yeasts, certain *Pseudomonas* strains are accompanying molds on wet blue leathers [7]. The presence of yeasts in vegetable tannin liquors, rich in sugars was known much earlier. The first place of bacteria attack is the flesh side of flayed hide, i.e., the side of the opened tissue. The grain side is protected by keratin 'armour' which, at least in the first deterioration stage makes the bacteria growth impossible, because it is dry and covered with a hair coat. Flesh side contains just a small amount of bacteria. Therefore it is justified to stack the hides 'flesh to flesh.' Among proteins contained in the raw hide, collagen is most resistant to the action of proteolytic enzymes. The reasons for being so are discussed in the chapter: 'Enzymology of proteolytic process.'

Perhaps one more question still should be posed. Bacteria die rapidly if thin colonies of them are exposed to direct sunlight. Thus sunlight is regarded as one of the most powerful destroyers of pathogenic germs. The part of sunlight that has the lethal effect is the ultraviolet—with a wavelength of ca. 260 nm. Most nonspore-forming bacteria or those that have not gone into the spore stage are killed after a few minutes exposure to the sunlight.

8.3. Bacterial damage to the skins and suppression of the bacterial growth

The most important types of damage to the skins cured by salting are the colored spots: red, violet and so-called yellow salt spots. From the red spots sets of various bacteria, mycoides and sarcinoides may be isolated, among them *Torula*, *Actinomycetes*, *Micrococcus rubescens* and *Sarcina luteus*. In the group of bacteria giving red color to the medium, halophile strains are present as well, like e.g., *Halobacterium cutirubrum*, *Halobacterium innocens* etc. Violet spots caused by various bacteria strains may be frequently observed in pig rawskins. They give violet spots on the hide under anaerobic conditions. Isolation of them has been fairly difficult. Studies of the redox potential needed for their bacteria to grow have not been carried out. Tancous [5] has isolated from the red spots on the flesh side of hides, cured by brining, a dyestuff of approximate structural formula:

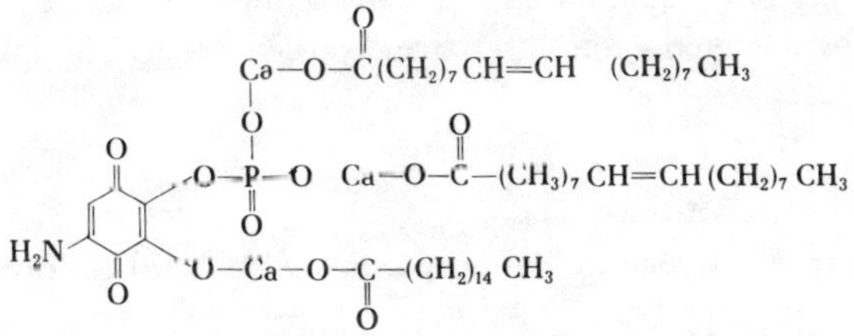

$$
\begin{array}{c}
\text{Ca} - \text{O} - \overset{\displaystyle \text{O}}{\overset{\|}{\text{C}}} (CH_2)_7\,CH{=}CH\quad (CH_2)_7\,CH_3 \\[2pt]
|\\
\text{O}\\
|\\
\cdots\text{O} - \overset{\|}{\underset{\|}{\text{P}}} - \text{O}\quad Ca - O - \overset{\displaystyle \text{O}}{\overset{\|}{\text{C}}} - (CH_3)_7\,CH{=}CH\,(CH_2)_7\,CH_3 \\[2pt]
\text{O} - Ca - O - \overset{\text{C}}{\underset{\displaystyle \text{O}}{\|}} - (CH_2)_{14}\,CH_3
\end{array}
$$

(benzoquinone ring: H_2N substituted p-benzoquinone bearing the phosphoric acid ester and $O-Ca$ groups)

This dye is therefore a calcium soap and phosphoric acid ester of dihydroxyamino-p-benzoquinone. The halophilic bacteria staining the medium, even if not dangerous for the hide as its destructors, may be considered as an indicator of the presence and activity of bacterial flora on the cured hide. These hides may be disinfected in different ways; however the best solution is to process them as soon as possible. The most frequent and the oldest preservant is sodium carbonate. Its addition gives a shift of the pH of the system to 9.5-10.5 so it makes more difficult the propagation and developing processes of the bacteria, having their optimum pH in a lower range. This preservant does not protect the hides against the appearing of violet and red spots, because the bacteria that give the spots are basophilic. However, it protects the hides against yellow salt spots. Better results may be achieved by the use of a mixture of sodium carbonate and naphthalene as antioxidant. Recently the use of trichlorobenzene instead of naphthalene has been recommended. A lot of various bactericides and bacteriostatics have been considered as preservants. In the acidic range oxalic acid has shown some advantages, sodium bisulphate and bisulphite have been used, as well as sodium fluoride and fluorosilicate and many organic compounds such as phenols, phenolates, cresols and chlorinated cresols. A review of the data concerning hides and skins preservation, may be found in Stather's handbook [8] and in current literature.

8.4. Changes in collagen due to aging

As it has been shown by Pietrzykowski [3] who investigated the changes in cured calf skins and cattle hides, collagen is aging in them during storage, quite similarly like in the living organism. As evaluation criteria he took shrinkage temperature, degree of swelling in 0.001 molar HCl and NaOH, and resistance to trypsin. The volume of observed changes is not always the same, but one may agree that the author is right in that the crosslinking of collagen is increasing during storage. Some of his results are shown in Table 8.2. As it may be seen there, the shrinkage temperature of the stored skins increases and their ability to swell in acids and alkalis and response to trypsin decreases as well.

Earlier, some other authors have already shown similar dependences in investigating the collagen aging in vivo and post-mortem. The influence of non-

Table 8.2.

The changes in physicochemical properties of calfskin collagen [3]

Skin	Storage time, months	T_S, °C	Swelling in 0.001N acid %	Swelling in 0.001N alkali %	Amount of protein, solubilized by trypsin %
Fresh	0	68.5	190	202	5.7
Dried	1	70.0	164	183	3.3
	3	71.0	161	180	3.1
	6	71.5	155	175	2.5
Salted	1	69.5	176	183	8.3
	3	70.0	175	181	8.3
	6	70.0	172	182	6.7

collagenous proteins on drying* and soaking of raw skins has also been investigated. The globular proteins make the soaking of the hides more difficult, because they, or their destructs, do close the ways by which water comes into the skin. High concentration of NaCl makes the growth of bacteria more difficult, except halophiles, which are growing better in the presence of halogenides.

When discussing the aging of collagen and the influence of this process on hides or skin properties, it should be kept in mind that various tissues and their components (among them collagen) are maturing at a different rate. Hence, the pigskin collagen should not be compared with the cattlehide collagen. Usually pigs are slaughtered after 6-9 months of life, cattle—after 3 years. A great deal of the organs of these animals are at the same maturity level (intestines, sexual organs, etc.). Collagen is maturing slowly and at about the same rate in all mammals, thus its degree of maturity, manifesting itself, e.g., in crosslinking is more age-dependent, than that of components having a shorter lifetime and greater turnover number.

This view has been demonstrated recently by Wada et al. [9] who compared the solubility of collagen from pigskins, calfskins and cattle hides. They were able to demonstrate that about the same solubility in $CaCl_2$ or urea (conc. 1 mole) characterizes cattlehide collagen as 3-year-old pig- and young, 6-month-old pigskin- and calfskin collagen.

8.5. Hides and skins curing by salting and dehydrating

The main problem in preservation of skins is to remove a significant part of the water, contained in them and the saturation of the remaining water with salt, usually NaCl. Also important, however of some minor significance, are: the use of bactericides and bacteriostatics in skin preservation, and the influence of salt or other substances added in the following part of the processing on the final products. The consensus now is, which can hardly be argued, that the best method of hide or skin processing is to process them right in the slaughterhouse to a certain stage, like, e.g., the chrome tanning (the wet blue leathers or the half dried crust leathers) or the pickling and storage of stock in the pickled state. In this way the typical curing process, for long time investigated and giving no absolutely safe solution, may disappear. Cooling of the rawskins to -25 or $-30°C$ and their storage at these temperatures may be used as well. There is no bacterial growth on the frozen hides, but they may easily be mechanically damaged (broken) when stored or handled. A short-time preservation has been proposed by immersing of the hides in solutions of bactericides, or by spraying of the hides with such solutions, e.g., by spraying with a concentrated solution

---

*Preservation of hides and skins by drying, applied mainly in the tropical areas, will be discussed below.

of sodium hypochlorite. Hides preserved in this way however are of lower quality and yield of leather. Bailey [10] proposed the use of acetates for short-time preservation for cattle hides, and Vermes [11] for pigskins. However, one has to be careful when applying these methods, because these may influence the leather quality.

Saturation of the system with salt

According to Cooper et al. [12] the fresh cattle hide contains 1.3% of NaCl (calculated on hide substance). Preservation by salting may be done in two different ways: spraying dry salt on the flesh side of the hide and 'flesh-to-flesh' (dry salting) stacking or immersing the hide in a saturated salt solution (brining). In both cases there is an osmotic penetration (transport) of the salt into the hide, because salt from the surface dissolves in the water present inside the hide, and water from the inside of the hide comes to the salt on the hide surface. Penetration of the salt from the outer (grain) side into the hide is much less intense, because penetration through epidermis is much slower. The osmotic process goes until equilibrium state is achieved. Classical investigations are those of Kritzinger and v. Zyl [13] on dry salting, and of Strandine et al. [14] on brining. Penetration of salt into the hide (which was separated into 10 layers after experiment and then tested) is shown in Figs. 8.2 and 8.3. The diffusion process in brining is similar.

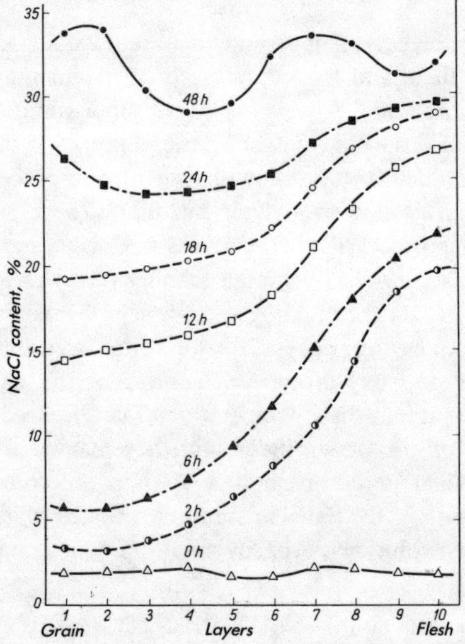

Fig. 8.2. Salting of cattle hides. Salt diffusion into the individual layers of the hide, calculated in terms of their water content in dependence on time. According to [13]

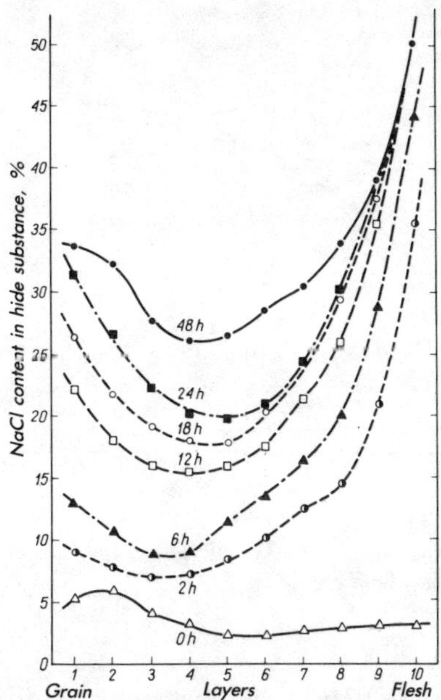

Fig. 8.3. Salting of cattle hides. Salt diffusion into the particular layers of the hide, calculated in terms of the hide substance. According to [13]

Salt penetration at room temperature takes about 48 hours, concentration however remains the lowest in the middle layers. The osmotic curves may also have different shape. Cooper et al. (l.c.) have calculated saturation with sodium chloride of the individual layers, based on the data of Kritzinger and v.Zyl concerning water and salt content (Figs. 8.4 and 8.5). The results are shown in

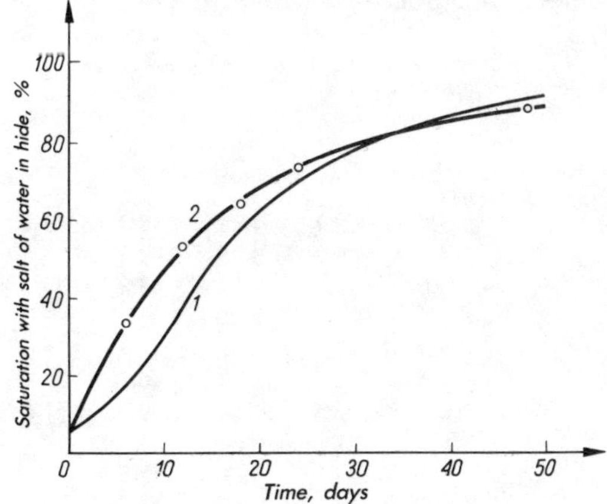

Fig. 8.4. Saturation of the water in hide with salt during dry salting in dependence on storage time: 1—saturation of the grain layer, 2—saturation of the whole hide.

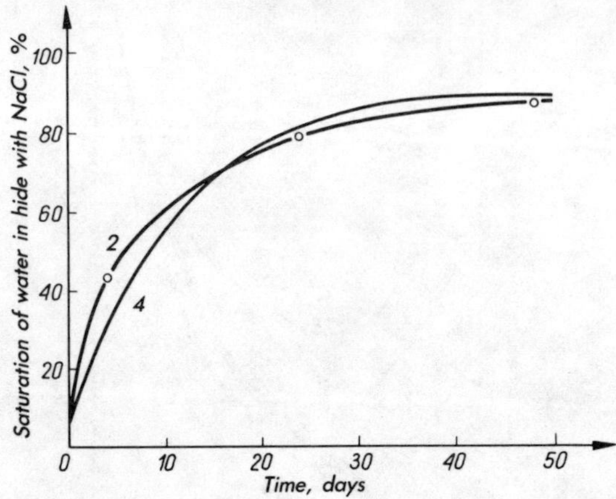

Fig. 8.5. Saturation of the water in hide with salt during brining: 1—saturation of grain layer, 2—saturation of the whole hide.

Table 8.3.

Principal condition for efficient preservation by salting on the factory scale is, besides the correct concentration, the salt quality. The difference between the use of rock salt, (big crystals) and salt obtained by evaporation of brine lies in the size of crystals; small crystals do readily dissolve in water, or in the solution on the hide surface, and the solution, now becoming brine, may too easily flow down from hide stack before proper amount of salt is adsorbed inside of the hide. If the crystals are too big, their transition into brine may be so slow

Table 8.3.
Saturation of water with NaCl in particular layers during dry salting

Layer *(starting from grain-side)*	*Salting time* *h*						
	0	*2*	*6*	*12*	*18*	*24*	*48*
1	5.4	9.7	16.1	40.3	53.1	73.0	93.5
2	5.6	9.2	15.9	42.2	54.0	68.9	94.2
3	5.6	10.8	17.5	43.0	55.2	67.2	84.5
4	5.9	13.6	20.0	44.1	56.6	68.0	80.6
5	4.7	17.5	26.1	46.1	57.7	68.8	81.9
6	4.7	22.5	32.4	50.2	61.8	70.3	89.4
7	5.7	29.5	42.0	58.3	68.1	75.7	93.0
8	5.7	40.0	49.8	64.2	74.3	78.8	91.6
9	5.2	51.0	56.6	71.1	78.0	80.7	86.1
10	5.1	54.4	60.8	73.8	80.5	82.0	88.4

that autolytic processes may be not inhibited by it. Hence the rule is to use coarser salt for preservation of hides, and finer salt—for skins. Composition of the salt is also of importance. A great amount of magnesium and calcium sulfates contained in the salt ($\sim$ 2%) promotes appearance of 'salt' spots. Probably this is due to the activation of alkaline phosphatases in autolytic processes. The salt spots do not arise when brine is used for preservation. This is perhaps due to precipitation of Ca^{2+} and Mg^{2+} compounds, as supposed by [3].

The results shown in Table 8.3 allow us to draw a conclusion that 48 hours of salting is enough to saturate all the water contained in the hide, except for the strongest bound part. In discussing the problems of the collagen-water system it was pointed out that the amount of bound water is not unmistakable estimated. Raw hide preserved by sodium chloride contains water saturated to 85-90%. In this solution most microorganisms do not find the conditions for their growth and die or change over into the spore forms. However there exists a group resistant to such concentrations: these are halophilic bacteria. One may inhibit their growth, using various preserving agents added to the salt. The purpose of this is to kill the halophile bacteria which may grow in concentrated salt solution.

Recently a method of curing skins by the use of dry salt and drumming was elaborated. An extensive study on this subject was given by Vivian and Rands [15]. According to these authors the salting time of 6 hours appeared sufficient for salt penetration. The observed weight loss was about 20-30% in process, after 3 days of staling-still 1-4% are lost. The yield of the final product is, however, only slightly smaller (0.13% less) than in the reference samples. No contraindications were reported. The use of drum salting was also discussed at the 5th Raw Hide Conference in Gottwaldovo, Czechoslovakia, in 1978 by J. Klinger.

The kinds of aerobic bacteria which may be found in brines have been investigated. It is obvious, however, that the bacterial flora of the hides is changing depending on the animal species, climate and the animal growing conditions. Thus the investigations done in this field may be valid in a specific habitat only and must not be extended to other climates, species or races of animal. The investigations mentioned [16] have been done on the brines with an addition of boric acid and naphthalene as preservants. Authors have divided the strains tested into true halophiles (growing on agar with 20% NaCl content), relative halophiles (growing on agar with 7% of NaCl) and halophobes (growing best on agar, containing 0.85% NaCl).

Aseptic storage of hides is unfeasible because there is no possibility of sterilization of such big masses of rawstock without thermal sterlization as yet. Thus the question of preservation of raw hides has to be considered from the point of minimalization of damage which may happen to the hides. As a measure of these kinds of damage one may consider the collagenolysis, or damage not influencing collagen itself but having adverse effect on the processing, on the appearance or on the wearing properties of the final products. There are only a few kinds of bacteria that produce collagenase. The tissue collagenase from mammals has been isolated and characterized for the first time about 1970. Before then its existence has been considered as doubtful. In the already mentioned paper of Woods there are named as collagenase-producers: four strains of Clostridium, three—*Bacteriodes*, two—*Pseudomonas* and *Staphylococcus aureus*. Also known are three strains of *Bacillus* and eight of *Achromobacter*. Among them *Pseudomonas, Bacillus* and *Achromobacter* are aerobic, the remaining are typically anaerobic. As Woods et al. have shown, investigating collagenolytic activity of bacteria grown on raw hides, it is great in the medium containing 0.85% NaCl (this concerns anaerobic and aerobic strains as well); in the medium containing 7% of NaCl the aerobes are growing slightly worse, and in the medium containing 10% of NaCl the aerobic are still growing, whereas anaerobic are not. The collagenolytic activity of bacteria is of special importance during soaking of cured hides. Decreasing salt concentration makes it possible for the anaerobic strains to develop their activity, because they may grow under these conditions. The losses because of this collagenolytic activity are not very big in a normal processing of hides :1-2% of collagen). The activity of these bacteria is in fact inhibited in the cured (salted) hide.

One more point is of importance: the tanner usually checks the nitrogen content in the soaking bath to measure the amount of proteins dissolved and/or decomposed. This method is unable to provide information about collagen dissolving or decomposition because it is unspecific. At the moment the only possibility of obtaining information about collagen content in solution is to determine the hydroxyproline content in it. This method, proposed by Langerwerf and then evaluated in several more experienced European laboratories, is slightly more difficult, its application however gives more advantages than troubles and costs [17, 18].

Salting problems are connected to the waste water question and to the salt price. Usually salt may be used only once or twice, then it becomes contaminated with bacteria and dirt. No useful method of salt recovery has been found. Among the efforts to replace salt in curing by other chemicals that are less contaminating to the waste water and used in smaller amounts, the curing method with formaldehyde was reported recently by Sharphouse and Kimweri [19]. Formaldehyde is a powerful crosslinking agent reacting with very many proteins, and thus widely used as a disinfectant, as it kills almost every microorganism.

Authors recommend the use of 0.25% formaldehyde as preservant. In this concentration it makes the leather slightly firmer than is usually obtained. This difficulty might be well overcome by post-tanning treatment. The formaldehyde curing process requires a careful pH control (adjusted by acetic acid) and exact dosage; an excessive amount of formaldehyde may cause difficulties in unhairing. The amount proposed increases the T_s from 64 to 68°C. Addition of some (7%) salt makes the hides mellower and with flatter grain. A T_s increase to 75°C is then observed.

Water removing

The problems of the hide-water-preservant system as a whole have not as yet been given a comprehensive treatment. The choice of the salting method, compounds inhibiting bacterial growth and the effect of these factors on the quality and yield of the final products is a subject of processing and process control. The aim of curing is to remove water from tissue to such an extent that no irreversible changes in the collagen properties should take place.

Removing of water from the spaces among the macromolecular compounds of raw hide is in fact an evaporation process from the surface of concave meniscuses (Fig. 8.6).

Decreasing tension of the liquid or its head growing shorter causes a decrease in the diameter of the capillary. Tension π inside of the capillary depends on its diameter D and on the surface tension σ of the liquid

$$\pi = \frac{4\,\sigma}{D}$$

When the liquid is removed from the pores, the porous body changes its shape. This process depends on the modulus of elasticity of the body. Collagen, which is the main factor of the resistance of the hide, is rather soft. Hydrated collagen in a fresh hide has a modulus of elasticity of ca. 2.5 kg/cm² (245 kPa). Tension in the capillary tube of a 1 micron diameter is 1.6 kg/cm² (157 kPa) in the one with a diameter of 0.1 micron it is 16.0 kg/cm² (1570 kPa). For simplicity of the calculation we assume the cylindrical form of the interfibrillar pores. In fact they have various shapes. Internal tension may reach 360 kg/cm² (35.3 MPa) during shrinkage. Due to this the capillary walls are approaching one another. In Fig. 8.7 the curve of internal tensions, arising in the hide due to drying is shown. These pressures are so powerful that some of the collagen fibrils, lying most unfavorably, may be torn away. At drying the pores in the raw hide may

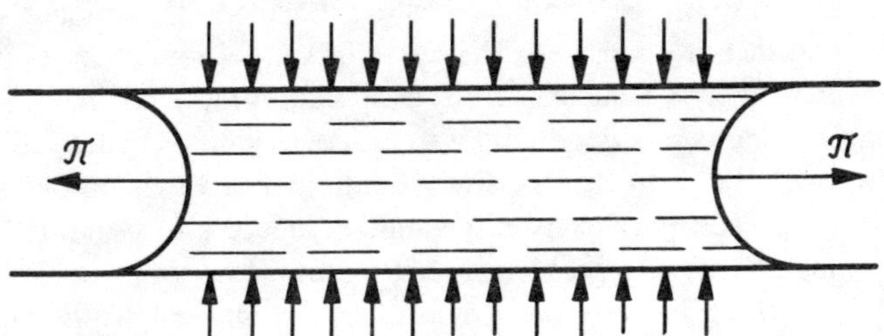

Fig. 8.6. Drying of a 'soft' capillary.

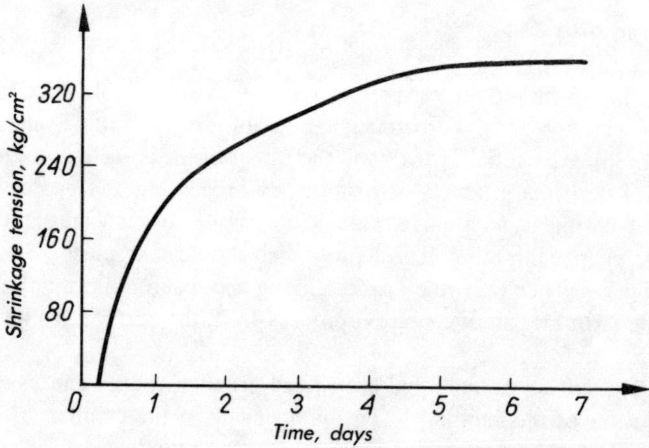

Fig. 8.7. Internal shrinkage tensions produced during drying of the bundle of collagen fibers.

be completely closed, and in this way the active centers, interacting with water, are coming closer to each other. Due to the fibers sticking together, the physical resistance ability of the hide markedly increases. The hide is becoming horny. It transmits certain amount of light due to joining together of the components of the same refractive index, which have been previously separated by spaces having different optical properties (water, air) or due to the decrease of the degree of dispersion.

Curing of skins by drying only is now applied exclusively to the fur skins as a primitive, uncontrolled way of preservation in a hot climate. Rate of such curing reaction is a function of skin surface, relative humidity of air used, its temperature and flow rate. Vapor pressure over the raw hide and the surface, occupied by water molecules on the tissue/air interface is depending on the motions of the molecules on the surface. Is the diffusion of water from inside of the tissue equal to the uptake of the molecules by air stream, the tension of vapor over the skin should not change. This is not an equilibrium condition, because the tissue structure changes when water is evaporating. The moment follows when the vaporizing rate becomes dependent on the amount of water coming to the surface in time unit. The curve of weight loss changes its shape at this moment. It means a change in the evaporation mechanism.

The drying mechanism is very similar to that of drying of hydrophilic fibers: it is changing with the degree of the water binding. The views on the water-to-skin binding have been previously discussed. It should also be borne in mind when discussing the drying problems, that at the same time more than one mechanism is working because the extent of hydration, and so the amount of water, bound by various mechanisms, is different in various layers of the skin and it depends, among other things on the diffusion rate of water to the surface. Accordingly, we do not encounter in practice the weight-time graphs consisting of two or three sections of straight line of various slopes. However, there is a dependence between the amount of water and of hexosamines in the connective tissue [20]. Very detailed inquiries at this point may be groundless because of marked individual differences of the hides examined. As we have had the opportunity to show, the characteristics of the dried sample may be variable in the sorption and desorption processes with time or temperature or both. According to these parameters, the process rate may change in its course, e.g., it may vary in the samples due to temperature, or time.

The authors of [21] have indicated the possibility of erroneous conclusion, which may be drawn because of the use of conventional methods of the sorption determination in research work. It happens frequently that the comparison of the samples investigated gives various results depending on the time of the examination. The characteristics of the material are only possible when full run of the curve which shows the changes of mass as time function is known. Using the logarithm of time we may usually have straight lines, or straight line sections (see Fig. 8.8). The process of conservation by drying, as already mentioned, is

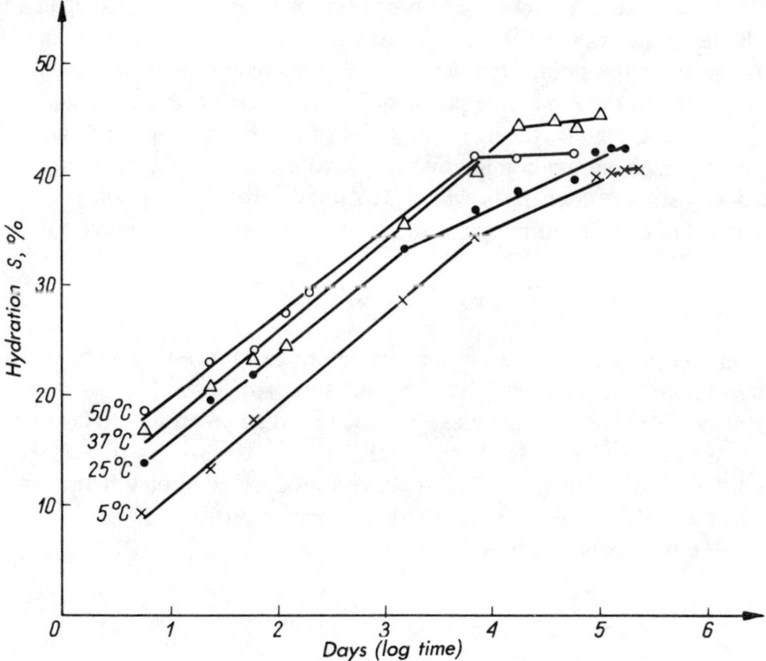

Fig. 8.8. Hydration of side leather samples at various temperatures as a function of the logarithm of time (in hours). According to [21]

usually done in an uncontrolled way, thus the parameters quoted in the above paper seem to approximate it better. Too quick and rapid drying causes blistering of the hide, due to the closing of the passes, by which water usually goes out. The bacteria, which might come to the inside of the tissue, remain there and cause its decomposition.

Dehydration of the hide by use of methyl alcohol, or ethyl alcohol followed by ethyl ether is described in detail by Reich [22]. Such a dehydration is a process that is different from drying. It gives the product which is very leather-like. This process is not used in industry as yet with the exceptions of U.S.A. and Czechoslovakia, where it has been introduced in one or two factories, however as a process of quick introducing of tanning agents into hide, followed by quick tanning by water addition.

An extensive study on the use of acetone for raw hide preservation was given by Maire [23]. It covered mainly the engineering aspects of the Ushakoff's process of dry tannage, solvent purification, solvent handling and economical calculations. Introducing sodium chloride into the raw hide, no matter how, we principally change the properties of the system. The changes occurring may be only approximately described. A detailed approach is still unfeasible. The changes in electric and thermal conductivity of leather, its stiffness, compactness, etc., are observed. The change in properties of salted hide, as compared to the fresh one, indicates the change in the way of water binding in the whole system. Looking from the point of water balance, the amount of water contained decreases, and in the remaining part changes its character. It forms shells around the Na^+ and Cl^- ions. Independent of the structure which we may take for the unbound water, an ion coming into it, increases the complexity of the system. To demonstrate the changes which will follow, we will introduce the partial molar volume V_o in infinite dilution. It is composed of two terms, vis.,

$$V_{o\,(MX)} = V_{o\,(M^+)} + V_{o\,(X^-)}$$

The influence of ions on water may be described as (see section 5.3) cf. [24]:
(1) structure breakers, which destroy the water structure;
(2) electrostrictive structure makers, which exert an ordering influence on water due to their strong electrostatic field;
(3) hydrophobic structure makers, which introduce additional hydrogen bonds to the water molecule close to their nonpolar surface.
Applying this model we have

$$V_{o\,(ion)} = V_{o(cryst)} + V_{o\,(electr)} + V_{o\,(unord)} + V_{o\,(bound)}$$

where $V_{o\,(cryst)}$ is the partial volume of crystal, $V_{o\,(electr)}$ the partial volume of water that undergoes electrostriction, $V_{o\,(unord)}$ partial volume of unordered water of disturbed structure, and $V_{o\ (bound)}$ a partial volume of water, bound to ion by long distance interaction (Fig. 5.6). Is the difference $V_{o\ (ion)} - V_{o\ (cryst)}$ negative, so the ion is a structure maker. For the structure breaker the value is positive. E.g., for the Na^+ ion the $V_{o\ (ion)}$ value $= 6.6\,cm^3$/mole, $V_{o\ (cryst)} = 2.1\,cm^3$/mole; for the Cl^- is $V_{o\ (ion)} = 23.2\ cm^3$/mole, $V_{o\ (cryst)} = 14.9\ cm^3$/mole. Thus the Na^+ is a structure maker, Cl^- ion—a structure breaker. As a whole the partial volume, engaged in the 'breaker' ion is greater, thus the general influence of the ions derived from the NaCl dissociation is structure-breaking.

This reasoning is right for infinitely diluted solutions which, in practice, never occur. Thus we must consider it as a rough approximation. The influence of a charged ion in an electrostrictive space prevails over a dipolar interaction between the water molecules, and the dipoles are completely oriented towards the central ion. The water molecules in this region are exchanging; when compared to the liquid water they are immobilized to a certain degree, and their volume is decreased. Has an ion a considerable region of electrostriction, i.e., great charge and small radius it belongs to the group 2. In the more distant solvation layers water is less immobilized and less closely packed. In the transitory space between electrostricted and liquid water the influence of the ion charge is small enough that only part of the water molecules be oriented, but the usual structure of liquid water is not maintained there. In this region there are less hydrogen bonds than usual. If an ion has a large hydrophobic surface, its influence on the surrounding water molecules is small. In this case an interaction between water molecules is greater, and such an ion belongs to the hydrophobic structure makers.

Freeze-drying

The freeze-drying is the most modern way of preserving skins; it is, however, at the moment used only for most precious fur skins. According to this method the skins are dried after freezing; evaporation occurs in high vacuum and the liquid state is omitted. The substantial difference between air drying, dehydration with acetone and freeze-drying is that during air drying the skin surface is the evaporation surface as well, whereas in freeze-drying the evaporation occurs on the fiber surface and among the fibers where the ice crystals are present. The fibrils, contributing to the fibers, are stuck together, while acetone comes inside of the fibers and makes their surface wet. Then the fibrils are separated and a great internal surface arises. Dehydration occurs by diffusion of water from inside the fibrils to the surface.

Another difference between freeze-drying and drying by removing of liquid phase is lack of any motion of liquid in the first case. In Fig. 8.9 one may see, how the freeze-drying process goes. The level of sublimation is inside the material and it lowers with time. In the drying curve the first section is missing, corresponding to this drying stage in which dehydration is a constant value due to the

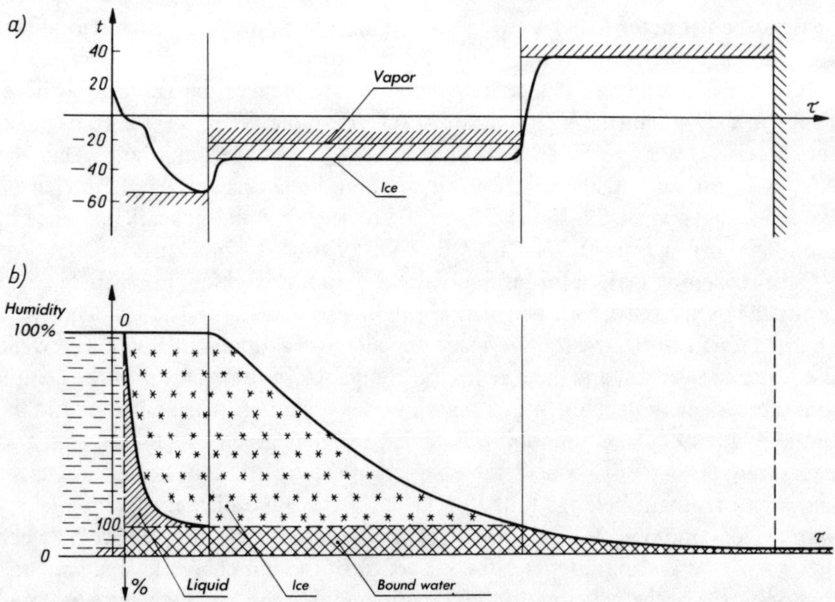

Fig. 8.9. Freeze-drying: (a) as a function of time, (b) as a function of humidity. Initial humidity assumed as 100%. The successive stages are: freezing, sublimation, desorption.

transferring of water onto the material surface by capillaries. A resistance of the pore system from the start of drying does influence the motion rate of vapor. Thus the freeze-drying rate (defined as the weight loss related to the surface and time unit $kg/m^2 \times h$) is from start of the process slower than during drying by liquid removing. This is a result of growing resistance of capillary spaces due to the travel of the drying surface, i.e., due to the change of the pathway length of vapor.

The freeze-drying process is more time-consuming than other drying methods. Its advantage is the quality of the product. Freezing is done quickly and small ice crystals are arising which do not destroy the cell structure. Goods dried in this way are very stable, and after moistening they take shape equal to that of a fresh one. Almost no changes in chemical and physical properties are observed. Freeze-drying is very common now in food industry (meat, fish, fruit, vegetables). Application of this to hides and skins has been investigated in detail by Heidemann and Riess [25]. Quick and deep freezing before drying, e.g., by immersing into the liquid air, prevents destroying of tissue and the fibers from sticking together. Drying is slow in its end stage and if no additional cooling is applied, the hide will be heated to over 0°C, because the evaporation heat withdrawn is not sufficient for maintaining a constant temperature. In this case the melting of ice crystals causes the collagen fibers to stick together. This obvious fact was the reason for the irreversible changes observed by Heidemann and Riess in collagen at about 25% moisture content in freeze-drying process.

REFERENCES

1. Tancous, J. J. J. Am. Leath. Chem. Assoc., 65, 176 (1970)
2. Tancous, J. J., Jayasimhulu, K. J. Am. Leath. Chem. Assoc., 68, 132 (1973)
3. Pietrzykowski, W. Thesis, Agricult. Academy Poznań 1973
4. Srinivastava, R. C., Avashti, P. K. Z. Naturforschung 26b, 804 (1971)
5. Tancous, J. J. J. Am. Leath. Chem. Assoc., 67, 344 (1972)
6. Kallenberger, W. G. J. Am. Leath. Chem. Assoc., 73, 6 (1978)
7. Bienkiewicz, K. J., Oberman, H., Malik, K., Sokotowska, B., J. Soc. Leath. Techn. Chem.—in print.
8. Stather, F. Gerbereichemie und Gerbereitechnologie 4 ed. Akademie—Verl. Berlin 1967
9. Wada, K., Shirai, K., Kawamura, A. J. Am. Leath. Chem. Assoc., 75, 90 (1980)
10. Bailey, D. G., Hopkins, W. J. XV IULTCS Congress Hamburg 1977
11. Vermes, E., Sipos, T. ibid.
12. Cooper, D. R., Galloway, A. C., Woods, D. R. J. Soc. Leath. Tr. Chem., 56, 127 (1972)
13. Kritzinger, C. C., van Zyl, J. H. M. J. Am. Leath. Chem. Assoc., 49, 91 (1954)
14. Strandine, E. J., De Beukelaer, F. L., Werner, G. J. Am. Leath. Chem. Assoc., 46, 19 (1951)
15. Vivian, G. W., Rands, M. B. J. Soc. Leath. Tr. Chem., 60, 149 (1976)
16. Woods, D. R., Atkinson, P., Cooper, D. R., Galloway, A. C. J. Am. Leath. Chem. Assoc., 65, 125 and 164 (1970)
17. see ref. 24 ch. 2
18. see ref. 23 ch. 2
19. Sharphouse, J. H., Kimweri, G. J. Soc. Leath. Tr. Chem., 62, 179 (1978)
20. Hvidberg, E. Acta Pharm. Tox., 16, 55 (1959)
21. Bienkiewicz, K. J., Pisalska, J. J. XIII JULCS Congress Vienna 1973
22. see ref. 1 ch. 2
23. Maire, M. S. J. Am. Leath. Chem. Assoc., 71, 488 (1976)
24. see ref. 8 ch. 5
25. Heidemann, E., Riess, W. Leder 14, 37 (1963)

9.
ENZYMOLOGY OF PROTEOLYTIC PROCESSES

In the previous chapter we have shown the necessity to discuss the enzymatic processes which the hide of an animal undergoes after slaughter. Among these are:

(1) autolytic processes, where mainly cathepsins and other enzymes of the hide itself are involved;

(2) deterioration processes, where the proteolytic activity of bacterial enzymes predominates;

(3) the processes intentionally carried out with participation of enzymic preparations, like soaking and unhairing;

(4) bating, which is a last enzyme attack on the proteins, remaining still in the collagen fiber 'wove,' after all the operations of the beamhouse have been done.

The knowledge how the enzymes are acting on connective tissue and particularly on collagen is necessary not only for their application in the beamhouse. Leather during its use is also attacked by bacterial and mold enzymes, especially in warm, moist climate. Partial hydrolysis of collagen, and also stability of bonds between collagen and tanning agents are studied with the use of enzymes. The influence of enzymes on tanned collagen, which is the main component of leather, belongs, however, to the other field, i.e., it is related to the problems of leather aging. The enzymic system of a living organism, consisting of a great many enzymes and enzyme systems of diverse activity is in its majority represented in the raw hide, like most other components of the organism. However, the mutual proportions among enzymes and other components are different in the hide than elsewhere in the organism. This is due to the function and properties of the hide. To depict in this chapter the action and properties of all the enzymatic system is neither possible, nor advisable. Hence the basic properties and way of action are shown just for major proteolytic enzymes, which are of importance in the above named processes. Before discussing particular enzymes, their basic properties as catalysts in biochemistry and their reaction kinetics are treated.

9.1. The properties of enzymes and the kinetics of enzymic reactions

Substances which are major components of living organism are unstable at the level of body temperature (or in the heterothermic animals, in the temperature of the surrounding), but they are not changing however, despite that they are not at equilibrium. They react with air oxygen only when appropriate amount of the energy (activation energy) is supplied, and are burned like in the breathing process. The mixture of the substances is the less reactive, the higher is the activation energy. Enzymes in the organism do trigger the chemical energy stored by acceleration of spontaneous reactions. The spontaneity of chemical reaction is always expressed by change of standard free energy ΔG. What an enzyme has to make is to increase the speed of spontaneous reactions to the proper degree. All enzymes are proteins.

An addition of a catalyst, i.e., an enzyme in a biochemical reaction, lowers the activation energy, thus speeds up the establishing of the equilibrium state K. Thanks to this the reaction may be carried out at the temperature of organism. It is also possible to perform a further consecutive enzymic reaction, that follows the first one and in the same medium, when the product of the first reaction is removed. Thus the equilibrium state in the first reaction will not be achieved, because one of its products is continuously removed and has to be produced anew. Thus the common action of the enzyme set in the cell or in the tissue leads often to the multireaction chain, known in the biochemistry as cycle: breathing cycle, transamination cycle, fermentation cycle, etc. A visualization of enzyme influence on the energy level decrease in a cycle of biochemical reactions is shown in the Fig. 9.1. The energy levels of catalyzed reactions are

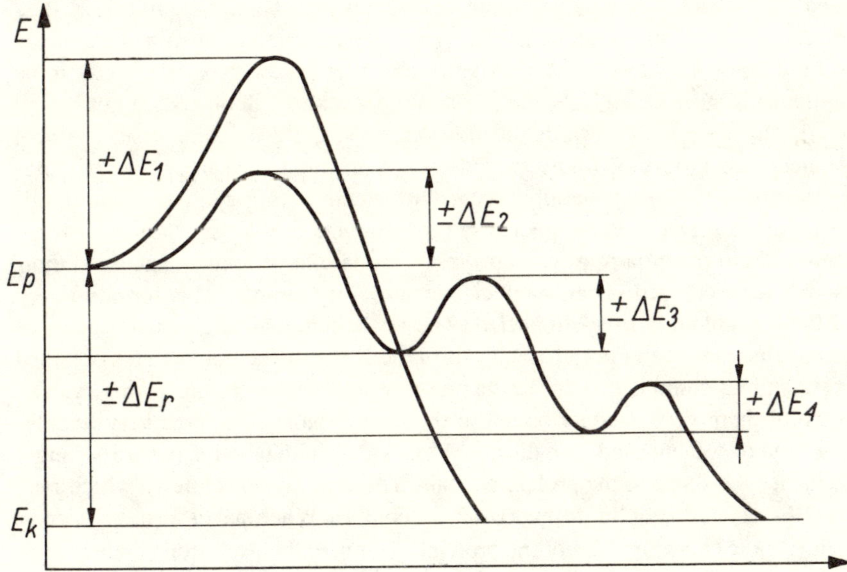

Fig. 9.1. Activation energy of a catalyzed and non-catalyzed reaction E_p—energy level of the substrate at start, E_r—decrease of free energy of the reaction, E_k—energy of the reaction product, ΔE_1 = activation energy of the noncatalyzed reaction, ΔE_2, ΔE_3, ΔE_4—activation energy of particular stages of catalyzed reaction.

gradually lowered, easing the reaching of the end point of the reaction chain. The organism is never in the state of chemical equilibrium, where its pool of free energy would be $\Delta G = 0$. Free energy is related to the equilibrium constant by the dependence:

$$\Delta G = - RT \ln K$$

Neither energy nor work may be obtained from the system, being in the state of chemical equilibrium. It would be a dead, closed system. The living system is in a steady state (state of dynamic equilibrium), i.e., it is open, in a stationary condition, where the reaction substrates are continuously added and the products are removed. A phenomenon of transport at the system borders is determined by the steady-state concentrations of individual substances. If an enzyme in a closed system does not influence the position of the equilibrium state, it may influence in the open state the steady concentration of one of the reaction partners: e.g., concentration of the substrate is decreasing, if the amount of enzyme is high and substrate supply is limited. There are forces in the system, acting against this effect, e.g., by inhibition of the production or activity of the enzyme in question. In the steady state the concentrations are different from the concentrations, thermodynamically defined for the equilibrium state. Instead of the notion of 'chemical equilibrium' also that of 'thermodynamic equilibrium' is used. The reactions in a living organism tend towards chemical equilibrium; they are the source of life processes. The specificity of action of an enzyme appears in binding of the substrate. For instance amino acid decarboxylase binds some aminoacids more strongly, some others more weakly, still others are not bound at all. This specificity towards substrates is very clearly visible in case of optical isomers (enantiomers). Many enzymes have been obtained in crystalline form· by the use of methods generally accepted in protein chemistry, like various kinds of chromatography, electrophoresis, centrifuging, etc. Several dozens of enzymes like trypsin, pepsin, chymotrypsin, elastase have been isolated and thoroughly characterized. Their number increases very rapidly. This formerly very difficult problem is now simplified. The crystallographical characterisation of an enzyme, based on Fourier's analysis of the X-ray diffraction patterns obtained is sometimes complicated due to the presence of water molecules in the crystals, some of them very strongly bound to the protein molecule, some very weakly. The kind of solvent used, and the conditions of crystallization may very strongly influence the shape of the protein crystals. There are proteins known which may have 12 crystallographic forms (β-lactoglobulin). What makes easier the crystallographic characteristics of the proteins is *a priori* the exclusion of some space groups, as they may not appear in a molecule consisting of L-amino acids exclusively.

Isolation of the enzyme in a crystalline form is the optimal case. It is not always possible and in such a case the classical methods of testing of enzyme properties have to be applied. Then the testing of reaction kinetics is of basic importance. A measure of the enzyme activity is the reaction rate. It is defined as the amount of substance, which has reacted per time unit. The unit recommended by the International Union of Biochemistry (IUB) is micromole/minute under optimal conditions. These units may also be applied to the non-purified enzyme preparations.

In the tanning industry the conventional units of enzymic activity are used. They are defined by Küntzel [1]. The enzyme preparation is the more purified the more enzymic activity units per mass unit it exhibits. It is an absolute value, dependent on another constant, specific for each enzyme, called turnover number. It is a number of substrate molecules, which will be transformed by one molecule of enzyme at optimal substrate concentration. To estimate this figure, one needs to know the enzyme molecular weight. If the concentration of only one substrate is changed, then the activity of an enzyme is given by the equation

$$v = \frac{VS}{K_m + S}$$

or the rate of enzymic reaction is proportional to the substrate concentration S, to the maximal rate V of the reaction in question and inversely proportional to the sum of Michaelis constant K_m and the substrate concentration. If the S values are very great, so is the rate of enzymic reaction v approximately equal to V.

If the concentration of the substrate is low, so is $v = \frac{v}{K_m}$; S, V and $\frac{v}{K_m}$ are the basic parameters of kinetics, changing independent of the concentration of other substrates, inhibitors, activators, parameters like pH, temperature and ionic strength. The K_m value is defined by concentration B, when the rate of enzymic reaction is equal to $v = \frac{V}{2}$. If the concentration of S is close to K_m then the substrate engages the main part of catalytic potential of the enzyme, its activity, however, may be controlled. If it is not so and the concentration of the substrate is higher than B, which corresponds to K_m, then the reaction rate is independent of concentration and the dynamics of the enzymic process may not be controlled. The dependence of reaction velocity of the substrate concentration is shown in Fig. 9.2. Point C in the graph is the value estimated as enzyme saturation.

Michaelis constant is not equivalent to the dissociation constant of the enzyme-substrate complex: it expresses the apparent equilibrium constant under dynamic equilibrium conditions. The dissociation constant $\frac{[E][A]}{[EA]}$ corresponds to the chemical equilibrium. The Michaelis constant is also a measure of affinity of

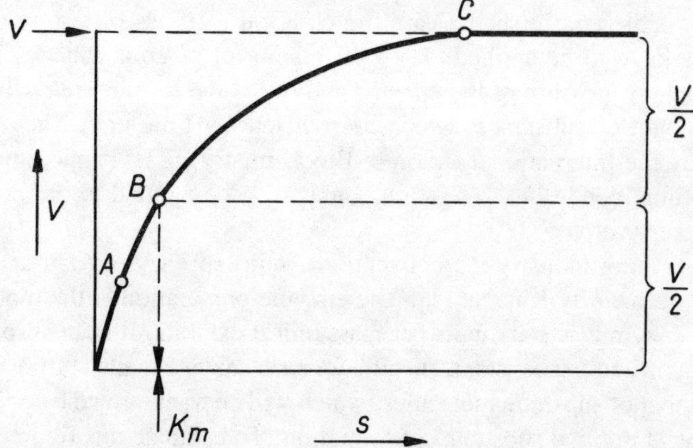

Fig. 9.2. The rate of enzymic reaction in dependence on substrate concentration S and on the Michaelis constant K_m; v—rate of enzymic conversion.

the substrates under conditions of dynamic equilibrium. It may be equal to, greater, or smaller than the dissociation constant, depending on the mechanism and on the velocity constants.

One can calculate the Michaelis constant and the maximal reaction velocity, e.g., using graphic estimation of the reciprocal values according to Lineweaver and Burke:

$$\frac{1}{v} = \frac{K_s}{V} \times \frac{1}{S} + \frac{1}{V}$$

Taking as variable the values of $\frac{1}{v}$ and $\frac{1}{S}$ we obtain this expression as an equation of the straight line. The values $\frac{1}{V}$ and $\frac{1}{K_m}$ are to be calculated from the graph (Fig. 9.3). The expression in this form is a basis for practical characteristics of enzymic reactions.

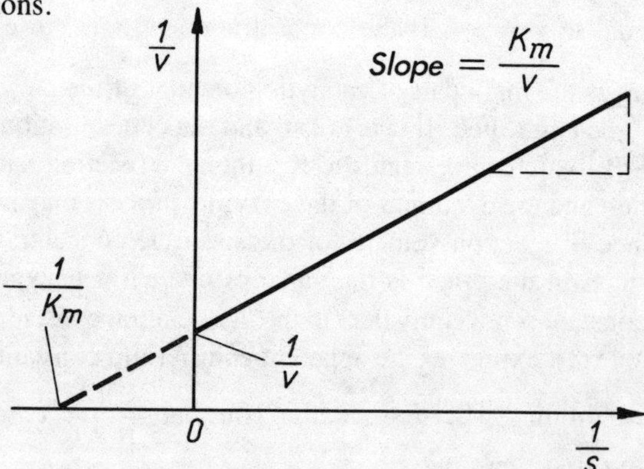

Fig. 9.3. Calculation of Michaelis constant and of maximal reaction rate from the Lineweaver-Burke graph.

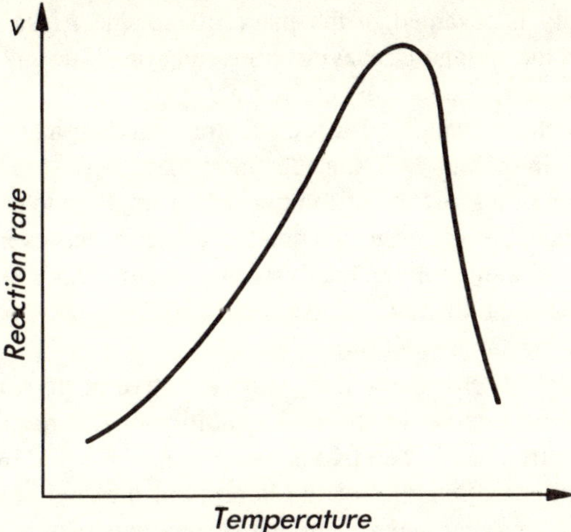

Fig. 9.4. The rate rule for enzymic reactions.

Enzymes have their maximal activity under optimal conditions. Of greatest importance among the factors controlling these conditions are: pH, presence of appropriate ions and ionic strength. Temperature does influence the catalyzed reactions and uncatalyzed as well, however the van't Hoff's rule is valid only up to a definite temperature. Enzymes are destroyed (denaturated) at the temperature of 40-50°C. Enzymic reactions show characteristic 'bell-shape' temperature dependence (Fig. 9.4).

Denaturation of the enzyme may be due to the temperature increase, or to other factors, like e.g., the change in ionic strength of the solution. Changes of the enzyme molecule shape are the result of translations of the active sites (Fig. 9.5).

The concept of active sites is related to the role of factors activating enzymes. The lock-and-key theory, formerly believed to be an explanation of the enzyme action mechanism, has later been replaced by Koshland's model [2], in which

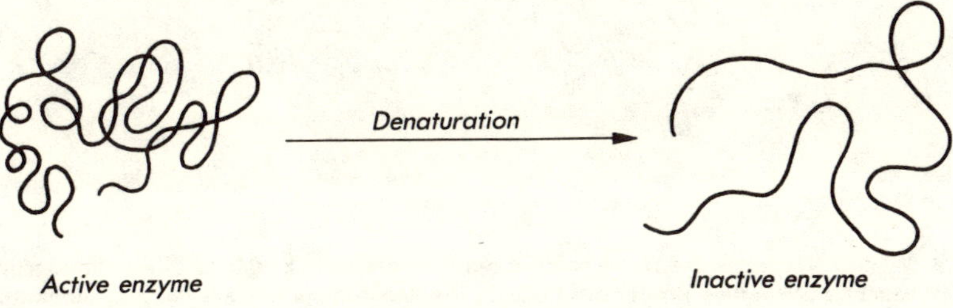

Fig. 9.5. Enzyme denaturation (scheme).

a certain plasticity is assumed in the place of enzyme attack (Fig. 9.6). The substrate induces the change in enzyme conformation. It means a change in the position of amino acid residues or of other groups in the enzyme in such a way that would make them available for the substrate binding and for catalysis.

As one can see in the Fig. 9.5, the substrates may cause some conformational changes. Addition of a substrate of too big or too small a molecule does induce incorrect arrangement of the groups. Then the substrate acts as inhibitor. In such a case a substrate binding without catalysis may occur. This model explain why in the peptide chain the distant, but conformationally close (loop) amino acids may take part in the same reaction.

A current theory of 'enzymatic trap' is a derivative of 'lock and key' theory, but maybe it is more close to the real conditions. The reacting substrate is according to this stretched, when two active groups are caged by enzyme active centers. Due to this a stretching of the bond (and thereby weakening) occurs, resulting in bond breaking and attachment of products of water dissociation to the ends. This theory may well explain such kinds of enzyme action, like, e.g., proteolysis.

An essential effect on the enzyme activity is that of the pH. There exists an optimal value of this parameter for every enzyme, being in connection with dissociation of its active group. The pH dependence may be more or less pronounced, which is depicted by the shape of the line of this dependence on the graph. The graph is similar to that of the temperature dependence line. Optimal pH for the majority of the enzymes is in the neutral range: the extremal values are typical of some enzymes from mammal intestines: pepsin has an optimum at pH 1.5-2.5; trypsin—in the range of pH 7.5-10.

Ionic strength of the solution, in which enzymatic reaction takes place, affects it in a way which is specific for the enzyme tested. Increase of ionic strength decreases the reaction rate or increases it, depending on individual enzyme character.

To the metabolism of living organism anabolic and catabolic reactions con-

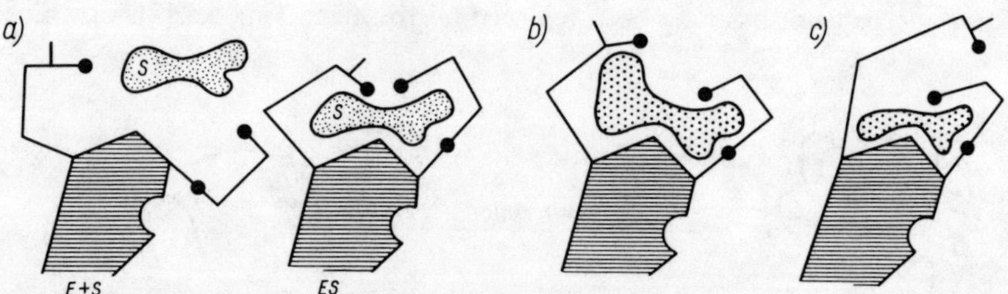

Fig. 9.6. Induced change of enzyme conformation according to Koshland: [2] (a) the functional groups reacting are in their proper position, (b) the substrate molecule is too big, the catalytic center cannot get into contact, (c) the substrate molecule is too small, the centers overlap.

tribute. In the anabolic reactions more complex substances are formed from less complex: they are incorporated into the cell and contribute to its growth. Catabolic reactions give simpler substances, resulting from the deteriorating processes. In anabolic processes energy accumulates, whereas in catabolic it is liberated. In living organisms the processes of both types are steadily occurring, with some kind of equilibrium between them. Activity of fundamental enzymes is increasing or decreasing due to the action of small molecules, being frequently the intermediate products of metabolism. Modificators which decrease the catalytic activity are considered to be negative, and increasing or stimulating activity—positive.

A condition for the enzyme action is in most cases the activity of certain ions. About 27% of the enzymes known contain a metal ion at their active site, or they require addition of an ion for activation. The majority of peptidases are activated by bivalent ions Mn^{2+}, Zn^{2+}, Co^{2+}.

The participation of the metallic ions in the enzymatic reactions may consist in the following:

(1) immediate participation in catalysis, e.g. in oxydation-reduction reactions, or in electron transport;

(2) formation of a complex with the substrate; this complex is the proper substrate for the enzyme;

(3) formation of metaloenzyme (ME) which gives with the substrate (S) the complex (EMS);

(4) changing of the equilibrium constant of the medium;

(5) changing of the peptide chain conformation.

Usually we consider as modificators the naturally occurring physiologically active molecules present in the medium. The nonphysiologically active molecules, acting as negative modificators, are called inhibitors. The inhibitors are classified according to their possibility to replace the substrate, or if they react with the enzyme at any other allosteric point. These may be the reaction products, substrates or other molecules which may replace substrates at the active sites of an enzyme. Ions of heavy metals rank also in this group, as they form covalent compounds with enzymes, as well as the compounds that cause denaturation. These drastic agents have an irreversible effect by inhibiting reactions and usually bringing all vital processes to a stop. Used in technology they are not, however, a subject of interest of chemical kinetics.

Stoppage of biosynthesis due to accumulation of the products of it is called repression. It is involved when the products control the genetic activity of the cell by themselves.

Inhibitors act in a competitive or incompetitive way. This may be found out by checking the dependence of inhibition on the concentration. Competitive inhibition is observed, when the inhibitor molecule competes with substrate for the active site of the enzyme. Usually it takes place when the molecule of the inhibitor or its fragment has a structure similar to the substrate molecule. Reaction may be inhibited completely, if the inhibitor concentration is high enough: increase of the substrate concentration suppresses the inhibitor. Competitive reaction obeys the law of mass action, so that this way of inhibition may be identified by analyzing the Lineweaver-Burke graph (Fig. 9.7).

The $\frac{1}{V}$ value is not changed here, the slope of the line however is dependent on the inhibitor concentration. The angle between the line and the x-axis increases with enzyme concentration. Deviation of the line is indicative of the incompetitive reaction. In a typical competitive inhibition reaction (in Fig. 9.7a) the apparent value K_m' of the substrate, denoted K_m, increases. The course of inhibition is given by the formula

$$ y = \frac{1}{K_m \left(1 \; \frac{(I)}{K_i} \right)} $$

whence the K_i is calculated by determination of K_m without inhibitor I. Noncompetitive inhibition is illustrated in Fig. 9.7b, where it is visible that the increase of substrate concentration does not decrease inhibition. This phenomenon may be observed in case of action of the enzymatic 'poisons' like, e.g., the salts of heavy metals. Mechanism of competitive and incompetitive inhibition is shown in Fig. 9.8.

Allosteric inhibition occurs when inhibitor I becomes bound to the molecule in another place than substrate A. Allosteric inhibition is shown in Fig. 9.9. It is visible there how the inhibitor, when bound to one place of the enzyme molecule, changes its conformation, which means it makes it allosteric*. A molecule of the substance responsible for allostery is named effector ligand or allosteric modificator. These ligands do attach themselves to the proteins with

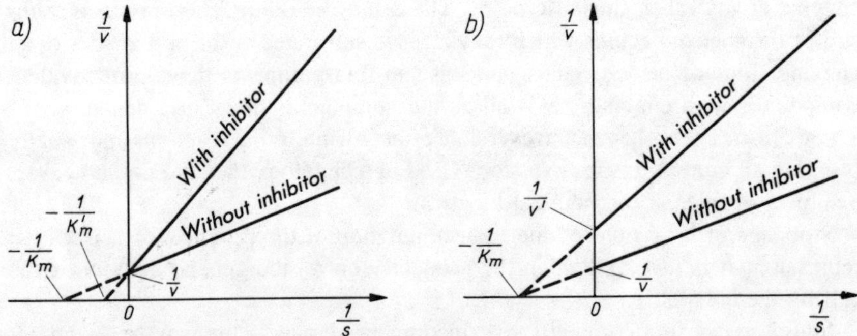

Fig. 9.7. Inhibition: (a) competitive, (b) incompetitive; S—substrate concentration, K_m—Michaelis constant, v—rate of enzymic reaction, V—maximal rate; top line—with inhibitor, bottom line—without inhibitor.

*Allostery (allos—other, stereos—solid body) is due to the changes in protein conformation and the resultant changes in its biochemical activity.

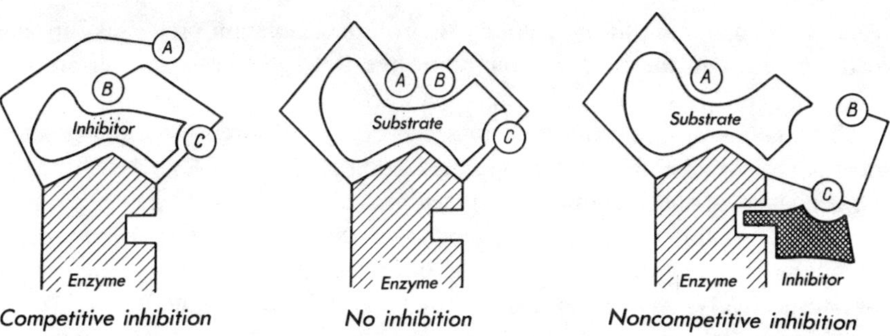

Fig. 9.8. Competitive and noncompetitive inhibition according to Koshland [2].

noncovalent bonds. Increase of the substrate concentration does not neutralize the reaction inhibition. In the Lineweaver-Burke graph one may see in this case an increased value of $\dfrac{1}{V}$ and of the slope of the line. If the reaction catalysis is slowed down or stopped at all, V of the reaction also decreases. Subunits of the enzyme may influence one another, and in this reaction allosteric inhibition may take place as well as the allosteric activation. Probably the sigmoidal shape of the curve of oxygen dissociation in hemoglobin is due to the allosteric inhibition. This mechanism of inhibition is working in very many reactions; it may also be caused by the reaction products particularly when a chain of enzymic reactions occurs.

Inhibition by the reaction products, i.e., feedback inhibition, may occur by various ways. This happens when at great concentration of products the opposite direction of reaction becomes more privileged, but maintaining of the products molecule at enzyme is also possible. It blocks the access of other molecules to the substrate. The substrate concentrations, differing from the optimal ones, may inhibit the reaction as well.

Inhibition may also be weaker, then it may be concluded from the concentration of inhibitor. This occurs when the steric accessibility of the inhibitor or its solubility is limited. Also another substrate may come into reaction, which is similar to the typical one. The product changed in this way may not react with further link of the enzymic reaction chain. Now we believe that the processes

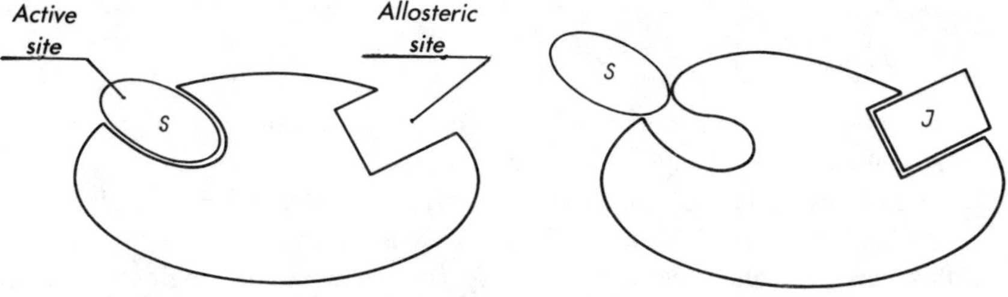

Fig. 9.9. Allosteric inhibition.

of enzyme adaptation to the medium (known as adaptation processes) are due to the changes in the system, controlling the RNA synthesis in the enzyme-producing cell. In mammals the hormones act as inductors.

The kinetics of enzymatic reactions can be systematized from the point of view of their course. One of the systematization proposals (Cleland) is based on the sequence of the incoming substrates and the products outgoing from the complexes. In this classification a course of the reactions is also considered (incoming-outgoing-incoming-outgoing or incoming-incoming-outgoing-outgoing, respectively) as well as participation of the modifiers and a number of possible parallel runs.

9.2. Classification and nomenclature of enzymes

International Union of Biochemistry has recommended in 1961 (the most recent version has appeared in 1978) to use a rational system of enzyme classification. According to this system the name of an enzyme consists of two parts: the first one is a substrate name, the second one having a suffix—ase—informs about the type of the catalyzed reaction. Additional information, if needed, is given in parentheses. Reactions and enzymes are divided into six classes each of which falls into subclasses. Every enzyme is determined by a systematic, four-member code number.

The first digit indicates class, the second, the subclass, the third one, the sub-subclass. This division is based on the catalyzed reaction type, type of donor, acceptor, etc. The fourth digit is the serial number of the enzyme in its sub-sub-class.

The enzymes already classified belong to the following classes:

(1) Oxidoreductases—enzymes catalyzing oxidoreduction between two substrates

$$S_{red} + S'_{ox} \rightleftharpoons S_{ox} + S'_{red}$$

This reaction is accompanied by hydrolysis of a high energy bond.

(2) Transferases—enzymes catalyzing the transfer of the groups other than proton, from one substrate to another

$$S-G + S' \rightleftharpoons S'-G + S$$

Enzymes catalyzing transfer of aldehyde and acyl groups or of groups containing sulfur or phosphorus are in this class.

(3) Hydrolases—enzymes catalyzing hydrolysis of ester, ether and glycosidic bonds, as well as C–C and C–X bonds (X is a halogen group element). In this class are also enzymes, splitting peptide bonds belonging all to the subclass 3.4. No systematic names are introduced into this last group, just

the traditional ones are maintained. This is due to the versatile activity of some enzymes in this group and to the differences in their activity, depending on their origin and way of isolation.

(4) Lyases—enzymes catalyzing the cleavage of substrate in the way other than by hydrolysis, with double bonds untouched. Enzymes from this class cleave the C–C, C–O, C–N, C–S and C–X bonds by elimination.

(5) Isomerases—enzymes catalyzing internal transformations of optical, geometric and spatial isomers, e.g., transformations of aldoses into ketoses.

(6) Ligases—enzymes catalyzing a linking of two molecules, combined with hydrolysis of pyrophosphate bond in ATP or other high-energy bonds. Ligases do catalyze reactions with the formation of C–O, C–S, C–N and C–C bonds.

The group specificity is manifested by the action of an enzyme on certain chemical groupings, e.g., pepsins and trypsins on peptide bonds and esterases on ester bonds. This specificity may be more detailed: carboxypeptidases and aminopeptidases split one amino acid residue from the carboxyl or amino end of the chain, respectively.

9.3. Enzymes involved in leather making

Proteolytic enzymes belonging to the sub-class 3.4 are called peptide hydrolases or generally peptidases and proteinases. The present classification in this group is different now than it was in EN (Enzyme Nomenclature) 1964 [3].

There are two sets of sub-sub-groups: peptidases (exopeptidases 3.4.11-3.4.17) and proteinases (3.4.21-24). The peptidases are divided according to their specificity into those hydrolysing single amino acids from the N-end of peptide chains (3.4.11), those hydrolysing single residues from the C-end (3.4.16-17) and those splitting off dipeptide units from N-ends (3.4.13) or the C-ends (3.4.15). The class 3.4.16 displays maximum activity in the acid range and is inhibited by the substitution of a serine residue by organic fluorophosphates (serine carboxypeptidases), the class 3.4.17 requires for activity divalent cation (metallocarboxypeptidases).

The proteinases (proteolytic enzymes, endopeptidases, peptidyl-peptide hydrolases) are divided into sub-sub-groups on the basis of the catalytic mechanism (active centers, effect of pH). The enzymes of sub-sub-group 3.4.21 (serine proteinases) have serine and histidine involved in the catalytic process. Those of 3.4.22 have a cysteine in the active center (SH proteinases). Those of 3.4.23 group have a pH optimum below 5 (asparticpeptidases) and those of sub-sub-class 3.4.24 are metalloproteinases, with a metal ion involved in the catalytic mechanism.

A number of proteolytic enzymes have been isolated, sometimes highly purified but without information about their catalytic mechanism. They are listed in sub-sub-group 3.4.99.

Basic information on important enzymes of 3.4 sub-class

Pepsin, isolated in several forms, has been identified as pepsin A (EC 3.4.23.1), B (EC 3.4.23.2) and C (EC 3.4.23.3). These enzymes are formed in stomach mucosa as preenzyme, called pepsinogen. (General name for preenzymes is zymogenes). Preenzymes are inactive; the active form is a result of removing of a polypeptide chain from the N terminal end. Thus the removed part is usually called 'activation peptide.' Pig pepsin, e.g., is formed by the removal of 42 from the 363 amino acids of the pepsinogen chain. It seems that we may regard the loss of a portion of the N-terminal end of the originally formed protein as a quite general phenomenon in the biosynthesis of enzymes or perhaps of more proteins (see, e.g., procollagen, proelastin).

Pepsins are active at low pH. Pepsin A predominates quantitatively. Pepsin B is characterised by its ability to cleave gelatin. Crystalline pepsins and pepsinogen have been isolated from stomach juice of cattle, birds and fishes. Several enzymes are also known to be active in a way which is very close to that of pepsin, like, e.g., renin (chymosin) from abomasum of calf stomach. From molds enzymes have been isolated of almost the same activity.

Amino acid composition of pepsins and pepsinogenes of various origin is known, as well as amino acid sequence: pepsinogen from pig stomach contains 3613 amino acids, whereas pepsin has 321. There are three disulfide bridges in the molecule and a large number of carboxylic groups in its side chains. Pepsin contains proline, aromatic and hydroxyamino acid residues in great quantity.

It seems that all pepsins and pepsinogens known are formed from one peptide chain crosslinked with 3 disulfide bridges. Pepsin becomes HCl—activated or autoactivated. The pig pepsinogen may be reversibly denatured by heating to 60°C (pH = 7) or by alkali treatment (pH = 11). Thermal denaturation is accompanied by an increase in the levorotatory power; the transition temperature is about 50°C. Pepsin is irreversibly denatured at pH about 6.0. Heating to 60°C, however, or treatment with 4 molar urea solution does not cause denaturation.

Pepsin B is considered to attack the Leu-Val, Ala-Leu, Tyr-Leu, Phe-Phe and Phe-Tyr bonds, whereas pepsin C—mainly the Tyr-Leu, Phe-Phe, Gly-Ala, His-Leu and Ala-Leu bonds. The point of pepsin attack depends considerably on structure, and on steric relations in the surrounding of the bond attacked. Mainly the sites are attacked, where hydrophobic groups occur in the neighbouring side chains. Aliphatic alcohols, acetonitrile, dimethyl formamide, phenolates and phenolic esters can be considered as competitive inhibitors of pepsin action. Asparagine carboxylic group is recognised as an active enzyme center.

The now proposed hypothesis of the way of pepsin attack [4] is the reaction of enzyme carboxylic group with the protonated amino group. In a reversible reaction then an intermediate is formed, which undergoes a four-center exchange reaction, the R-COOH group becomes expelled and imino enzyme is formed.

With water or with carboxylic acid this compound gives $R'NH_2$ (or transamination by a transfer of the imino group occurs) and the carboxylic group regenerates in the enzyme. A basic element of this mechanism is protonation of amino group, which precedes or is simultaneous with formation of a tetrahedral derivative. The range of pH where pepsin is active suggests carboxylic groups as a possible proton source. In this reaction two enzyme carboxyls participate, one of them as a donor, the other as a nucleophilic group (in a dissociated form):

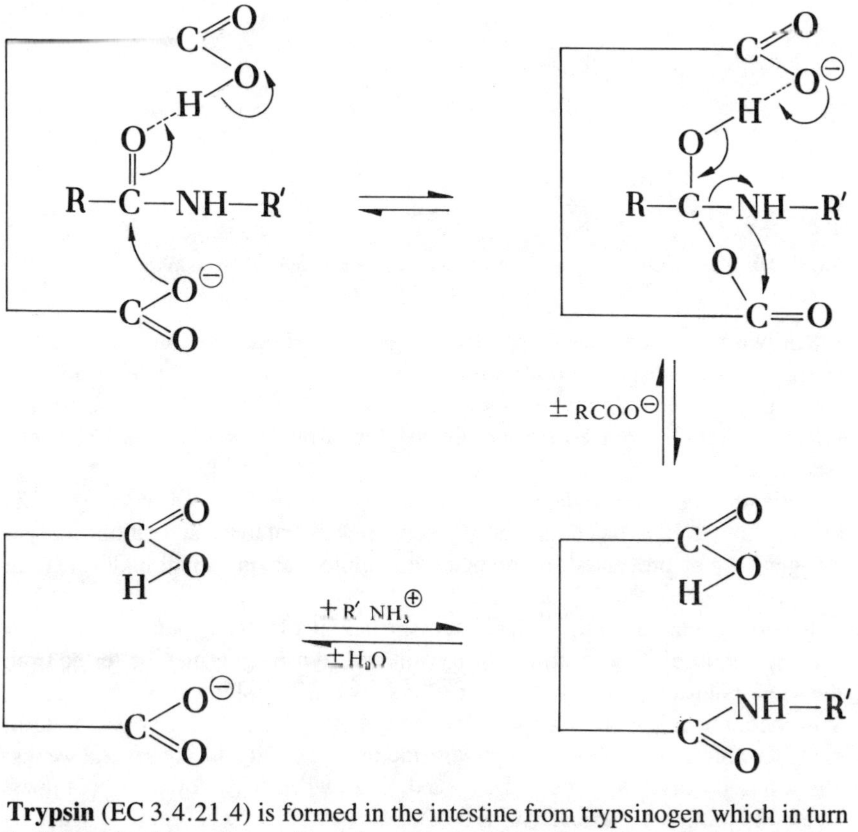

$\pm RCOO^{\ominus}$

$+ R'\,NH_3^{\oplus}$

$\pm H_2O$

Trypsin (EC 3.4.21.4) is formed in the intestine from trypsinogen which in turn is formed in the pancreas. Trypsin and trypsinogen have been isolated in crystalline form from various vertebrates and their amino acid composition and sequence have been compared. Trypsinogen becomes trypsin when a splitting of Val-(Asp)$_4$-Lys hexapeptide occurs. This reaction is controlled by small-intestine enteropeptidase EC 3.4.21.9 or trypsin itself and it probably involves deshielding of the catalytic center (Fig. 9.10).

The formed enzymes α, β and ψ trypsin do not differ much in their activity; α trypsin is derived from β trypsin by additional cleavage of a peptide bond.

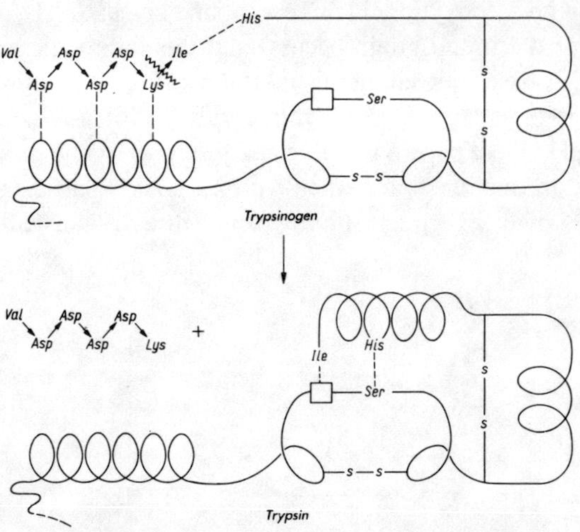

Fig. 9.10. Formation of trypsin from trypsinogen by splitting of hexapeptide.

When two bonds are cleaved ψ-trypsin is formed. Human trypsin contains 223 amino acids and four disulfide bridges: cattle trypsin has 229 amino acids and six disulfide bonds. Amino acid sequence of this enzyme has been investigated fully; also the difference between the enzymes from various animals has been recognised.

Activation of trypsinogen to trypsin requires the presence of Ca^{2+} ions. In its absence about 50% of compound is inactive. Autocatalytic activation of trypsinogen may be prevented, if competitive inhibitors having small molecule (like e.g., p-aminobenzamidin) are added.

Isoionic point of cattle and pig trypsin lies at pH 10.8; optimum of their activity lies at pH 7-9. Stability of trypsin and trypsinogen may be tested from the view point of structure or enzymatic activity. Trypsinogen is stable in the acid medium; in a neutral or alkaline environment it turns to trypsin. Trypsin is most stable at a pH about 3 when it maintains its activity during several weeks. However, it will slowly be self-digested, even when it is stored in liophilised state. In solutions, particularly in the alkaline ones, it loses rather quickly its activity: at pH = 10.7 after 2 hours a 0.01 solution loses 19% of its activity.

Trichloroacetic acid and urea if of a high concentration, cause a reversible denaturation of trypsin in alkaline medium. Urea at lower concentration (up to 6.5 molar) or 30% ethanol do not denaturate trypsin. Trypsin and trypsinogen form a variety of equilibria among the molecule conformations, in dependence on pH and temperature [5]. Activity of enzyme is in this case insignificantly changed. Serine and histidine form the activity center of trypsin. Serine hydroxyl group is acetylated by the substrate. Trypsin catalyses the substrate cleavage reaction in the following way:

$$E + S \underset{k_1}{\rightleftharpoons} ES \xrightarrow{k_2} ES' + P_1 \xrightarrow{k_3} E + P_2$$

where ES is an enzyme-substrate complex, ES'—a product—acylated enzyme-substrate intermediate, P_1—the remaining substrate group, and P_2—carboxylic acid. Acetylation (k_2) is a nucleophilic reaction. The intermediate product (acylated enzyme) is an ester in which carboxylic group of the substrate is bound to the enzyme hydroxyl. Deacetylation (k_3) is also a nucleophilic reaction, depending on the group and its characteristic $pK_a = 7$. It is a group belonging to histidine residue. Action mechanism of chymotrypsin and trypsin is the following:

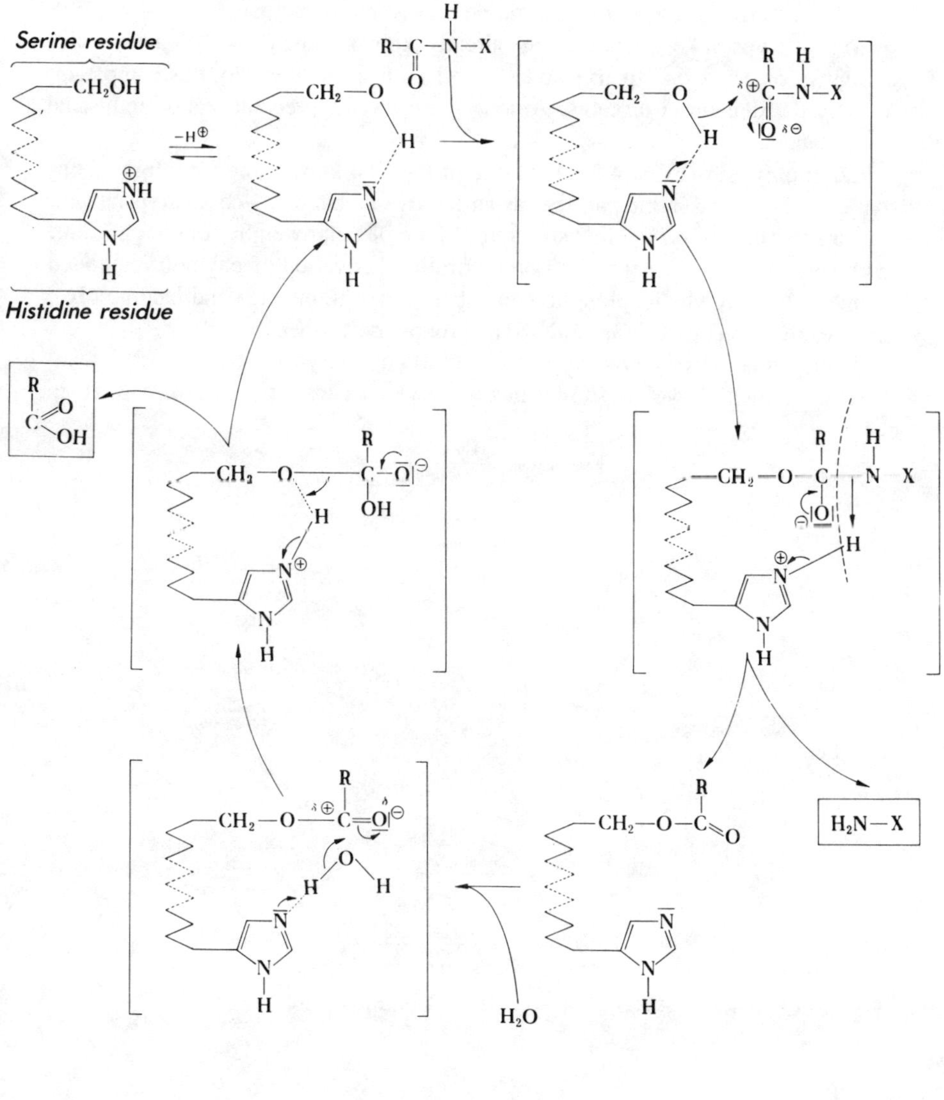

Specific is the inactivation of both enzymes by influence of diisopropyl-fluorophosphate, which is covalently bonded to serine (183) in a 1 : 1 stoichiometric ratio. Trypsin thus inactivated may be reactivated by hydroxylamine or other nucleophilic agents. Trypsin is also inactivated by denaturation, temperature increase, pH change, and chemical modification which is in fact a decrease of its disulfide bridges content. Trypsin action is inhibited as well by the small-molecule compounds, which may react with negatively charged asparagine residue, or by those forming strong bonds with hydrophobic sites in the enzyme. Among these are aromatic amines, ω-amino-carboxylic acids, guanidine derivatives and some others.

A typical point at which the enzyme acts is the ester and peptide bonds in which the carboxylic groups of arginine and lysine participate.

Many 'trypsinlike' enzymes are already known. Their way of action and composition are close to trypsin. As well as trypsin, they do attack most intensely the denatured proteins. Among them are the preparations of industrial importance.

Chymotrypsin (EC 3.4.21.1) is one of the best known and investigated enzymes. It is formed in the pancreas as an inactive preenzyme—chymotrypsinogen. It is a substance of tyrosinlike structure, of a molecular weight of about 25,000, with small deviations depending on its origin. Activation of chymotrypsinogen is due to trypsin which splits the bonds between arginine (15) and leucine (16); as a result, the end $-COO^-$ and $-NH^+$ groups are formed.

Transformation of chymotrypsinogen into chymotrypsin, based on the reactions described above, is shown in Fig. 9.11. π-Chymotrypsin formed as an

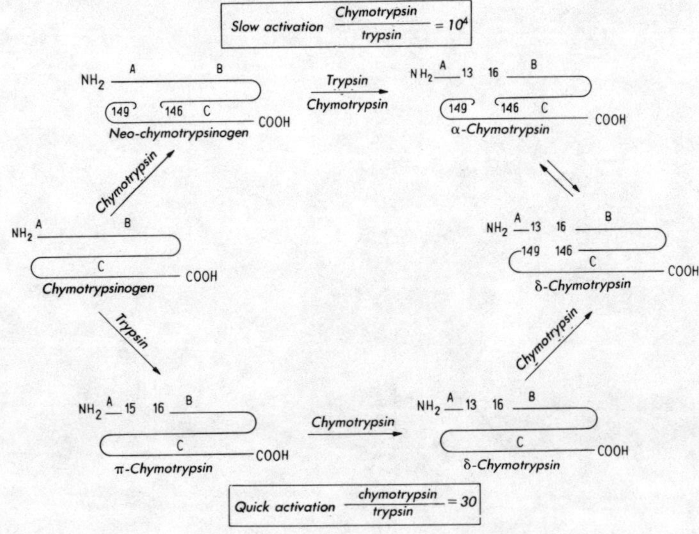

Fig. 9.11. Transformation of chymotrypsinogen into chymotrypsin.

intermediate is already fully enzymatically active. All the subsequent steps are the products of π-chymotrypsin self-degradation: γ-chymotrypsin is the main product of quick activation of chymotrypsinogen by large amounts of trypsin; α-chymotrypsin is formed when slow activation by small amounts of trypsin occurs. In the same reaction γ-trypsin is very slowly formed.

As it has been shown, a free NH^+ group in isoleucine (16) is necessary for maintaining enzyme in an active condition because this group forms an ion pair with the carboxylic group of the asparagine side chain (194); α-chymotrypsin consists of three chains connected by five disulfide bonds and numerous hydrogen bonds. A-chain contains 13 amino acid residues, B-chain—131 residues, and C-chain—97 residues. The amino acid sequences of these chains are determined, and physical properties of the molecule are well elucidated. The enzyme is most active at pH = 8; its activity rapidly decreases by changing of the pH value. One observes a significant jump in the CD values determined at λ = 230 nm, and of constant K_A which occurs at pH = 8-9. Three of the chymotrypsin amino acid residues do participate in the reactions catalysed by it, viz., serine (195), histidine (57) and asparagine (102). The active center in chymotrypsin, according to Blow et al. [6], is shown below:

about pH = 4

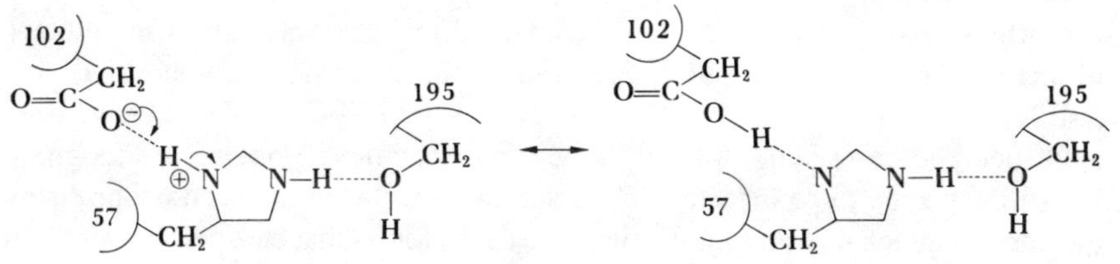

about pH = 8

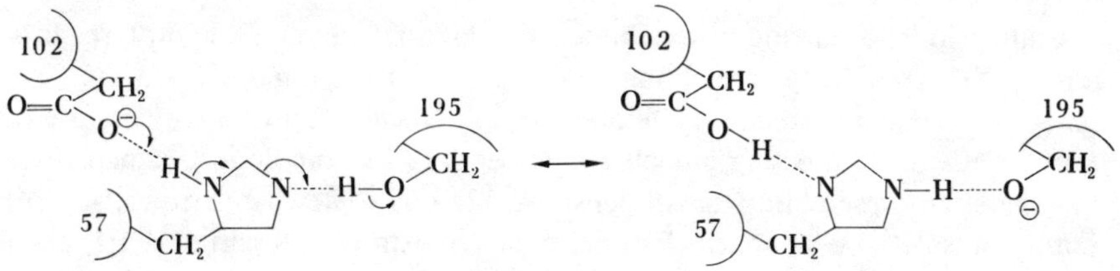

Indol and its derivatives, dioxane and phenyl iodoacetate, are known as inhibitors of chymotrypsin. Compounds, forming stable bonds with active centers, and some amino acid sequences in oligopeptides containing i.a. proline, have an inhibiting effect, just as other peptide hydrolases. They may make the access to the catalytic center sterically impossible.

Papain (EC 3.4.22.2) is a proteolytic enzyme isolated from the latex of a tropical plant *Caruca papaya* from South America. This enzyme is isolated in crystalline form, its composition, amino acid sequence and properties recognised. It is a compound of molecular weight 23 406, containing in its molecule 212 amino acid residues, among them 28 of glycine, 19 of tyrosine, 18 of valine, 6 of halfcystine, 1 of cysteine and 5 of tryptophane. Due to its high content of aromatic amino acids, papain has a specific fluorescence spectrum whose form is pH dependent. Papain is resistant to some denaturing agents: it is denaturating just very slowly when treated with 9 molar urea solution. In neutral pH range it resists heating and action of 50% ethanol or 30% dioxane. Papain is activated, when its blocked cysteine thiol group becomes released, and in this process cyanide, cysteine or glutamine residues are activators. Optimal pH of papain action is that of 5.5-6.0; papain attacks very many peptide bonds, which was formerly ascribed to the enzyme unspecificity. According to Schlechter and Berger [7] it depends on the possibility (or impossibility) of placing of a longer (25 Å) segment of the peptide chain in an enzyme slit, where the catalytic center is situated. Papain is active in the temperature range of 5-66°C in a pH range between 2.8 and 10.8; hence it is an enzyme of very broad activity range.

Because of its specific pH optimum papain closes the gap between pepsin and pancreatic enzymes, and due to this it is frequently used in protein investigations to estimate how the tanning agents are bound by collagen, etc. Papain preparations of high purity are remarkably stable.

Cathepsins-intracellular enzymes contained in lysosomes are of importance in anabolic and autolytic processes. Among these enzymes Cathepsin B (EC 3.4.22.1), D (EC 3.4.23.5), G (EC 3.4.21.20) and L (EC 3.4.22.15) have been identified and described, although some more enzymes belonging to this group are still expected to be detected. Their action on collagen is uncertain; probably they do not attack this protein. Burleigh et al. [8] found that cathepsin B degrades collagen at acidic pH. Cathepsins participate in the degradation of various components of connective tissue, like proteoglycans and other backbone components of connective tissue.

Cathepsin B, occurring in vertebrate animal tissues (thymus), hydrolyses proteins with a specificity resembling that of papain, its optimal pH is 3.7-4.0. In its active center it contains cysteine. Enzyme is stable up to a temperature of 80°C. Cathepsin D, acting probably together with its homologue, cathepsin E, has specificity resembling that of pepsin A, but it is somewhat narrower. Its pH optimum is 3.5-4.0. It attacks Leu-Ser bonds in extrahelical part of α 1(I) chain at 45°C [9]. It is less stable than Cathepsin B.

Together with cathepsins, carboxypeptidases A (EC 3.4.17.1) and B (EC 3.4.17.2) occur in almost every mammalian tissue. They are Zn-activated metallo carboxypeptidases. Their action is different than that of cathepsin in that they release C-terminal amino acids—carboxypeptidase A with the exception of arginine, lysine and proline: carboxypeptidase B-lysine and arginine, thus their action is complementary. Their pH optimum lies in the acidic pH range, and its active center contains cysteine. Degradation steps for glycosaminoglycans caused by the tissue enzymes are elaborated in detail. Nine enzymes are known to split specific bonds of these compounds; less thoroughly recognized is the sequence of action of proteolytic enzymes in the autolytic degradation. The interested reader is referred to current biochemical literature, as new information appears very frequently.

Elastase (EC 3.4.21.11) is an enzyme formed in pancreas of human beings, farm animals, birds and fishes; it is also produced in *Pseudomonas* species (bacteria). It cleaves preferentially peptide bonds involving the carbonyl groups of amino acids with unchanged non-aromatic side chains. It is formed from inactive proelastase; elastase is homologous with trypsin. Proelastase is activated by trypsin. Elastin hydrolysis occurs after a pronounced lag phase. Its amino acid sequence is known—elastase contains 240 amino acid residues and four disulfide bridges in one single chain; serine is its active center. Its activity optimum is at pH 8.8; aqueous solutions are relatively stable above pH 6. NaCl content in the medium influences elastase activity: 50-70 mmoles/dcm^3 decreases the activity by about 50%. Copper salts are strong inhibitors. Tris-buffer in concentration of ca 10 mmoles increases its activity by about 25%.

The enzyme molecule is small (M.wt. about 25,000) whereas that of elastin is high. Thus the mechanism of enzymatic action is probably a 'travel' along the molecule of protein and cleavage of bonds at appropriate positions. In new literature its synergetic action with trypsin and chymotrypsin is strongly emphasized [3]. Due to the similarity of action mechanism and of occurrence, elastase is probably used together with other pancreas enzymes in leather making, when pancreatic enzyme preparations are applied; so far as it is known to the author, no investigations in tanning chemistry have been done in which pancreatic enzyme preparations would be separated from elastase, and both fractions tested in bating or unhairing processes.

Collagenases (EC 3.4.24.3 and 7-9) are the enzymes cleaving the collagen molecule in its helical part. Splitting off the nonhelical part is not considered as collagenolytic activity. Roughly we can split this enzyme group in two parts: bacterial collagenases (Clostridium collagenase EC 3.4.24.3 and some others related to it), and animal collagenases participating in the living organism in the metabolic collagen turnover. For hides and skins the first group is very important and dangerous, as they may cause total deterioration of raw material. Discovered during the World War I in *Clostridium histolyticum*, collagenase was identified as the main factor causing gangrene and millions of deaths of human beings before the sulfonamide and antibiotics era. Still now the contamination of hides and skins with collagenase-producing bacteria leads to unavoidable, sometimes heavy losses in collagen substance.

The most dangerous and undesirable action of bacterial collagen hydrolase in the tanning industry is weakening and destroying of the barriers formed by connective tissue of the host. Such a function have the collagen hydrolases of *Clostridia* species and some others (see section 8.5). These bacteria often produce also elastases, polysaccharases and phospholipases and their attack is multilateral so the connective tissue may be totally deteriorated. Clostridial hydrolases as a rule do act outside of the cell. Nagai [10] has given very convincing results of his investigations on the reactions of collagen-hydrolase and synthetic peptides-hydrolase. Clostridial collagenase degrades the helical region of native collagen into small fragments. Its preferential cleavage site is -Gly in the Pro-X-Gly-Pro-X sequence. As its activator Zn^{++} is recognized. About 200 cleavage sites in collagen molecule are accessible to clostridial collagenase attack. This enzyme has molecular weight about 11,000. Its amino acid sequence is known, its activity optimum lies at pH 8-9. When pH exceeds 11 its activity ceases. Thus the recommended method of processing for *Clostridium* infected hides is to increase the liming pH up to this value [11]. The mechanism of *Clostridium* collagenase action is not exactly known. Probably the sites of attack are the 'pockets' in the molecule in which the glycine is situated. The Zn^{2+} ion 'drops' into this pocket and glycine-involving peptide bond comes then to form a complex with enzyme, and thereby to cleavage. Many kinds of bacteria have been described which produce this kind of enzyme; only small differences in their composition, molecular weight, and pH optimum have been reported.

Animal collagenases, present in almost every organ of the body, have an entirely different action and significance. Their concentration is usually very low. They cleave the collagen molecule not into small peptides, but at one site only leaving 75% of the molecule at one side (N-terminus) and 25% on the other (C-terminus) (The Gly-Ala-bond between 772 and 773 residues in α1 [I] chain is cleaved, Gly-Leu in α2). Due to their low concentration and unsatisfactory investigation methods they have been discovered some 10 years ago. The first enzyme of this type, the famous 'tadpole enzyme,' was detected in 1962, then the others followed. Until now but a few have been obtained in the amounts allowing their characterisation and more detailed description. Their significance in collagen metabolism is essential; they are however very easily inactivated—e.g., by blood serum components, cysteine, EDTA, by temperature decrease, etc. They are active at physiological pH and ionic strength. It is not elucidated yet, whether or not every kind of collagen requires a specific tissue collagenase. This does not seem likely, if we remember that damages caused by warble fly *(Hypoderma bovis)* or blowfly *(Lucilla sericata)* larvae are due to the same enzyme—and they may happen in several kinds of animals, having different collagens as native. Investigations on this group of enzymes, very interesting and important from the scientific point of view, are of minor importance for leather making.

After collagen molecule cleavage at one site the action of other proteolytic enzymes follows because at this place the molecule becomes unfolded, for the reasons not yet exactly elucidated.

Proteolytic enzymes from microorganisms

Usually in the tanning industries crude enzymatic preparations are used. They are extractions from tissues or from bacteria, molds, or other fungi cultures. As a source of industrial preparation the microorganisms are widely used, because they are quick growing, have no high demands concerning the medium, and besides the enzymes are obtained more easily from them than from organs of higher organisms. Pepsinlike-active peptidases, having pH optimum in the acidic pH range, are obtained from organisms belonging to the families: *Aspergillaceae*, *Penicillium*, *Mucoraceae*, *Saccharomycetaceae* and *Cryptococcaceae*. Generally typical of these enzymes is relatively low content of basic- and sulfur-containing amino acids. As a whole they contain 290-340 amino acids. Their amino acid sequences are mostly not investigated, isoelectric point of these enzymes is about pH = 3.0-3.8, molecular weight about 35,000. Temperature, at which half of the enzyme becomes inactivated is relatively high and is about 60-65°C. Peptidase from *Aspergillus niger* is more stable than other peptidases. As it seems, the bivalent cations Ca^{2+}, Co^{2+}, Mg^{2+}, Zn^{2+} increase the thermal stability of these enzymes. Investigation of their conformation has shown the possibility of concluding about their activity from their optical rotational power or from other physical constants, depending on the chain conformation. Bacterial peptidases to very small extent do attack the ester bonds; they are rather active in relation to peptide bonds. As investigation on synthetic peptides has shown, there are groupings of special susceptibility to the enzyme attack, e.g., Glu-Tyr. Acid peptidases are less susceptible to the action of typical proteolysis inhibitors. One may observe an expressed similarity of catalytic groups and active centers of animal pepsins and mold peptidases.

Another group of bacterial peptidases are the enzymes whose optimal pH region is alkaline. They are susceptible to organophosphorus compounds and to diisopropyl phosphofluorate. These enzymes have serine residues in their active center; they destroy ester bonds. Their activity is inhibited neither by chelating agents nor by compounds containing thiol group. Metal ions and reducing agents do not activate these peptidases. They may be found in various kinds of bacteria, molds and yeasts. These enzymes have molecular weight of 20-25,000, the isoelectric point at pH 9-10.5 and are trypsin- or chymotrypsin-like. Their proteolytical activity has a very broad range and is close to that of trypsin, chymotrypsin and elastase. In this group there is subtilpeptidase A (subtilysin), typical subject of enzymological investigation. It has been isolated from *Bac. subtilis* (EC 3.4.21.14).

In the group of bacterial peptidases there are also enzymes, susceptible to the chelating compounds, like EDTA. These compounds do contain a Zn atom in the molecule. Their molecular weight is 35,000-40,000. *Bacillus subtilis* produces them in significant amounts. They are not resistant to temperature, losing about 90% of their activity on heating for 15 min at 60°C. Ions of calcium group stabilize them, their optimal temperatures being 50-52°C. These enzymes act most intensely against the bonds in which leucine is involved from the N-end. Significant activity of amylases against proteins is due to their proteolytic impurities. As Steven et al. have shown [12], the crude α-amylase splits the teloptides from collagen molecule. An exact characteristic of these impurities is not known as yet.

Lipases

Lipases are enzymes belonging to the hydrolases (group 3.1), attacking the ester bonds or more exactly they belong to the carboxylic ester hydrolases (group 3.1.1). In leather making they hydrolyze the fatty acid esters. Although not used separately, they occur in enzyme mixtures, particularly in preparations of mammalian origin. Many lipases are separated, identified and their specificity recognized. They act on uncharged substrates and the main factors which affect their specificity are the distance between them (chain length) and the shape of the hydrophilic groups on either side of the ester link [3]. As examples for this enzyme group the following may be mentioned: Pancreatic lipase (EC 3.1.1.3) acting only at the interface between water and insoluble ester, which hydrolyses the outer ester links, lysophospholipase (EC 3.1.1.5) which splits the lysolecithin to glycerophosphocholine, and fatty acid anion, monoacylglycerol lipase (EC 3.1.1.23), which hydrolyses glycerol monoesters of long-chain fatty acids, both occurring in animal tissues, lipoprotein lipase (EC 3.1.1.34) which hydrolyses triacyl glycerols to diacyl glycerol and fatty acid anion, occurring in blood plasma and milk, and tannase (EC 3.1.1.20) which hydrolyses digallate to two gallate molecules; this enzyme occurs in molds.

There is an increasing tendency to investigate the significance of these enzymes and some other esterases for beamhouse operations as they influence the results particularly in the case of fat-rich skins (pig, sheep).

According to this tendency, the lipases after better recognition of their industrial use may be successfully used in solving the degreasing problem.

REFERENCES

1. Küntzel, A. Gerbereichemisches Taschenbuch, 6 ed., Steinkopff Dresden 1955
2. Koshland, D. E. Jr., Neat, K. E. Ann. Rev. Bioch. *37*, 359 (1968)
3. Dixon, M., Webb, E. C. Enzymes, 3rd ed., Longman London 1979
4. Knowles, J. R. Trans. Royal Soc. London *B 257*, 135 (1970)
5. Lazdunski, M., Delaage, M. Biochim. Biphys. Acta *140*, 417 (1967)
6. Blow, D. M., Birktoft, J. J., Hartley, B. S. Nature *221*, 337 (1969)
7. Schlechter, I., Berger, A. Biochem. Biophys. Res. Commun., *32*, 898 (1968)
8. Burleigh, M. C., Barrett, A. J., Lazarus, G. S. Biochem. J., *137*, 387 (1974)

9. Scott, P. G., Pearson, H. Eur. J. Biochem. *114*, 59 (1981)
10. Nagai, Y. J. Biochem. Japan *50*, 486 (1961)
11. Galas, E., Pilawski, S. Prz. Skorz. *27*, 6, 6 (1972) in Polish
12. Steven, F. S., Grant, M. E., Ayad, S., Weiss, J. B., Leibovich, S. J. Biochim. Biophys. Acta *214*, 564 (1970)

10.

COLLAGEN SWELLING IN WATER SOLUTIONS. MELTING AND SHRINKAGE TEMPERATURE

The soaking process opens the operations done in the beamhouse. The purpose of soaking is to bring the hide to the same condition in which it was immediately after separation from the carcass. Recovered softness makes it easier to introduce small-molecule substances into the hide. During soaking the mechanical impurities: scud, blood, salt and other preservatives used, are removed, a part of nonstructural proteins, and remnants of fat and meat. The hide becomes swollen in the process. Collagen, quantitatively, dominates in the raw hide, and thus it is identified to some extent with hide. One has to remember however that the raw hide is a tissue, which is undestroyed, or destroyed partly, containing mainly collagen but also other protein components, nucleic acids, fats, glycosaminoglycans and inorganic salts. This undamaged tissue forms a structure, as described in chapter 1.*

Mature crosslinked collagen is water insoluble but it swells, unless it has been kept at high temperature. Swelling of collagen, when already released from tissue and forming just a network of fibers, takes place in further steps of the hide processing as well. Addition of acids, alkalis, neutral salts, and other lyotropic agents results in changes of fiber length, thickness, and weight. Quantitative measuring of swelling is done, when electrostatic interactions, water binding, and crosslinking bonds are investigated. In the presence of a wetting solvent a system consisting of dry polymer molecules does attach the solvent and swells until it becomes a homogeneous solution. However, if there are disturbances in the polymer system, like, e.g., the crosslinking bonds, connecting the polymer chains, swelling becomes limited. Collagen is just such a system. Solvent which is able to make crosslinked polymer swell is in equilibrium with the surrounding solvent, when molecular free energy is equal on both sides. Extent of swelling is, in such a system, inversely dependent on the crosslinks number. In a fiber network the solvent may occupy the inter-or-intrafibrillar spaces, the general regularity, however, remains. Swelling of collagen is depending on two factors osmotic and lyotropic ones. Osmotic swelling (Donnan

---

*As it has been pointed out recently, the structures consisting of collagen and glycosaminoglycans remain through the tanning process probably intact. They change the properties of collagen, indicated by the knowledge of the macromolecule itself.
[A. N. Mikhailov Chemistry and Physics of Hide Collagen, Moscow 1980-in Russian]

swelling) occurs due to a high concentration of bound, nondiffusing ions located inside the structure. It takes place when pH of the solution is off the isoelectric point and the ionic strength of the solution is small. Greatest swelling effects may be observed at pH $= 2$ and 12. It is reversible by straining of the fibers, changed pH or increase of ionic strength of the solution by increase of the salt concentration. Lyotropic swelling which is due to the neutral salts at considerable ionic strength, decreases the cohesion of the fibers and is not completely reversible. Both swelling components may act together, and their contributions are condition-dependant. Swelling mechanism has not been fully elucidated because it is influenced by numerous factors. After tanning leather may also swell, which is very important for the wearing characteristics. This concerns first of all the swelling due to water action, which is discussed in the chapter 'Hide components and water.'

Swelling may be considered as the first step in the helix $\rightarrow$ coil transition [1]. The question, how far phase transition applies to biopolymers, may be answered as follows: in the 1st-order phase transitions the free energy of the system has continuous values, whereas the internal energy, density, volume and specific heat change have discrete values. This kind of change is combined with emission or uptake of heat. An example are state changes or allotropic transitions, which may be described by the Clausius-Clapeyron equation. In the 2nd-order transition the specific heat and coefficient of thermal expansion change discretely. In turn there is no change of volume and latent heat of transition. An example of the 2nd-order transition are: Curie point, ordering processes in alloys or phase transition helium I $\rightleftharpoons$ helium II at 2.2 K. Because a distinct discontinuity in collagen T_S occurs one has to take this phenomenon as a 1st order phase transition, with consequences coming therefrom. Among these consequences is, for instance, consideration of the biopolymers as crystalline or amorphous substances. This is not very exact in physicochemical sense, but it makes it possible to understand them. The proposed mathematical models are very complicated and not always in accordance with the results of the investigations.

Swelling may be tested by using optical birefringence, X-ray diffraction or by estimation of the changes in shape or elasticity or other properties of the substance. The problems of X-ray diffraction have been discussed in section 7.1.

A method of optical birefringence testing has been described by Reich [2]. The change in dimensions is to be estimated by use of catethometer (e.g., the change in the tendon length) or microscope with graduation of the subject.

The measurement of the elasticity makes it possible to characterize the change of crystalline fiber into an amorphous one.

In heating the hide, one observes the shrinkage of over 50% of the sample length. This is best observed if the sample is immersed in water. The temperature of this shrinkage (T_S) depends on the degree of crosslinking; it is lower for the raw hide, higher for leather. The non-swollen collagen is, as observed from the mechanical view point, a highly ordered polymer, which is synonymous with its crystallinity. It has high elasticity modulus and a limited liability to deformation. Dependence of the elongation on stress τ (stress/strain isotherm) is for the crystalline state expressed as

$$\tau = M(\alpha - 1)$$

where the modulus value M is of an order of 5×10^5 kg/cm^2 (5.10^4 MPa), and maximal elongation (at break) is of an order of 1.1. In the amorphous state the values for high elastic materials are applied: modulus value is about 5 kg/cm^2 (0.5 MPa) and α is about 3. The shrinkage transition is accompanied by essential change in physical properties.

No theory of swelling of crystalline polymers exists. The theory of swelling of amorphous polymer of Flory-Huggins is more adequate for gelatin-type gels. This theory, allowing thermodynamic characteristics of solutions to be given makes it possible to predict the behavior of polymer in the discussed system only to very limited extent, i.e., because it is valid only for one-component solvent. If, however, as is common practice in the tannery, beside collagen and water, other substances are present in the medium, e.g., salts or surfactants, the data may be based on the experiment only. It is difficult from theoretical standpoint to separate swelling and shrinkage, as usually the changes in the fiber length are connected with changes in the fiber volume, at elongation and at shrinkage as well. Both kinds of swelling are usually occurring jointly. Only their contributions to the whole process change.

10.1. Osmotic swelling

Osmotic swelling is due to the pH change, when the ionic strength is small and the temperature low. Changes of pH in the range between 4 and 8 do not affect markedly the length and diameter of fibrils. Outside these pH values almost tenfold increase of fiber volume may be observed. If the pH value drops below 2, then the volume decreases. The increase of ionic strength of the solution suppresses collagen swelling.

In Fig. 10.1 some properties of collagen fibers, crosslinked by formaldehyde in solution of various pH are shown. The changes in length are visible there, expressed as a quotient of fiber length at a given pH and the length at the isoelectric point, as well as changes in diameter, volume, and in birefringence. The changes in the acidic pH range are more sharply expressed—they occur in a much narrower pH range. They may be observed in the fibers, which are not additionally cross-linked in hide tissue and in gelatin, although their range may differ significantly. The described behavior of collagen fibers is due to the Donnan effect, coming from the increase of charge bound at the protein surface, as the pH value is drifting away from the isoelectric point. According to the

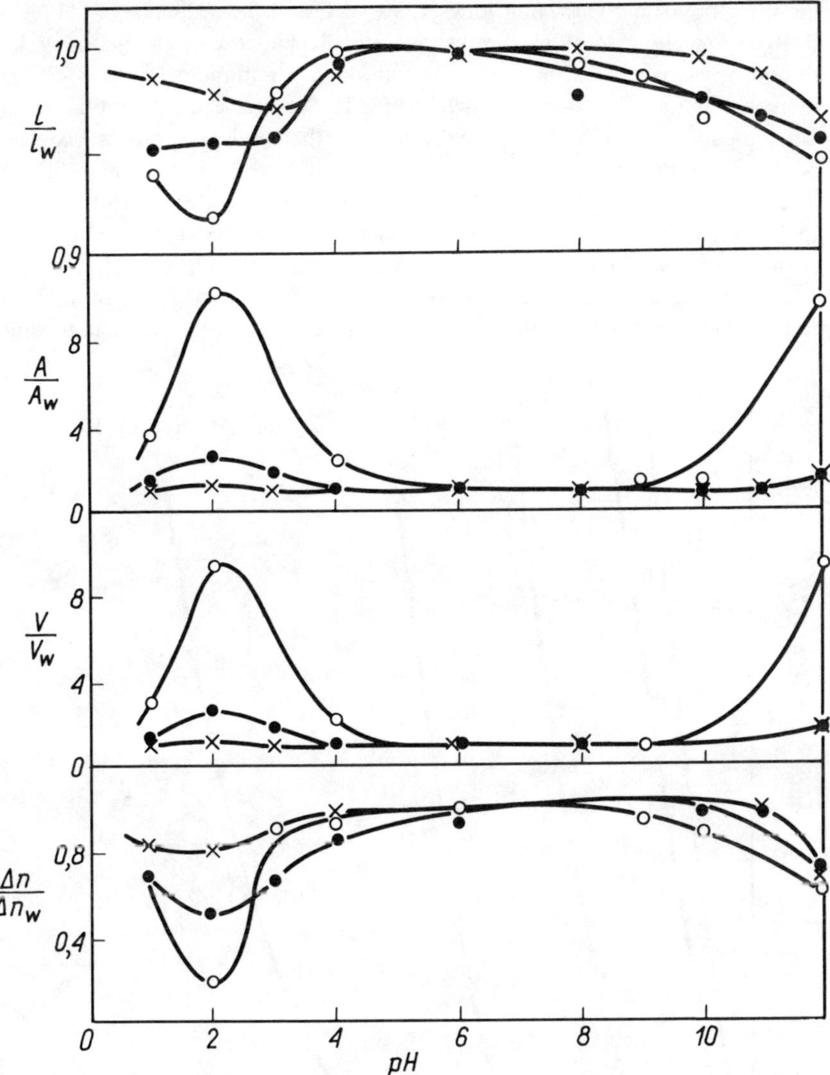

Fig. 10.1. Length, $\left(\frac{L}{L_w}\right)$, diameter, $\left(\frac{A}{A_w}\right)$ volume, $\left(\frac{V}{V_w}\right)$, and optical birefringence, $\left(\frac{\Delta n}{\Delta n_w}\right)$, of crosslinked collagen of rat tail tendon. Collagen was allowed to swell for 30 minutes at 25°C in salt solutions of different pH. o—no electrolyte added, ●—0.1 molar NaCl, x—0.5 molar NaCl.

criteria of Donnan's theory, occurrence of localized charges causes formation of excess ions, having opposite charges inside the gel, which in turn initiates action of osmotic forces. Donnan's effect does not elucidate satisfactorily the mechanism of attachment of solvent molecules to the biopolymer, although from the thermodynamic viewpoint it describes very well the influence of pH on the degree of swelling. As it has been shown, the dependence of elongation on load diminishes when pH decreases, passes through a minimum at pH = 2, then increases again. The curves, shown in Fig. 10.2 characterize the rat tail tendon collagen in an HCl solution. The values of the modulus as it is visible by comparing Figs. 10.1 and 10.2 do correlate with swelling of tendon collagen. The X-ray reflex at 13.5 Å is also shifted to 15 Å, when pH is shifted from 7 to 2. For collagen, crosslinked by formaldehyde this reflex is observable at 10.7 Å in dry sample, in a wet one at 14.3 Å (pH = 2). The position of reflex 2.86 Å does not change. As from this one may conclude, the extent of molecule ordering is unchanged, being a result of osmotic swelling in an acid medium.

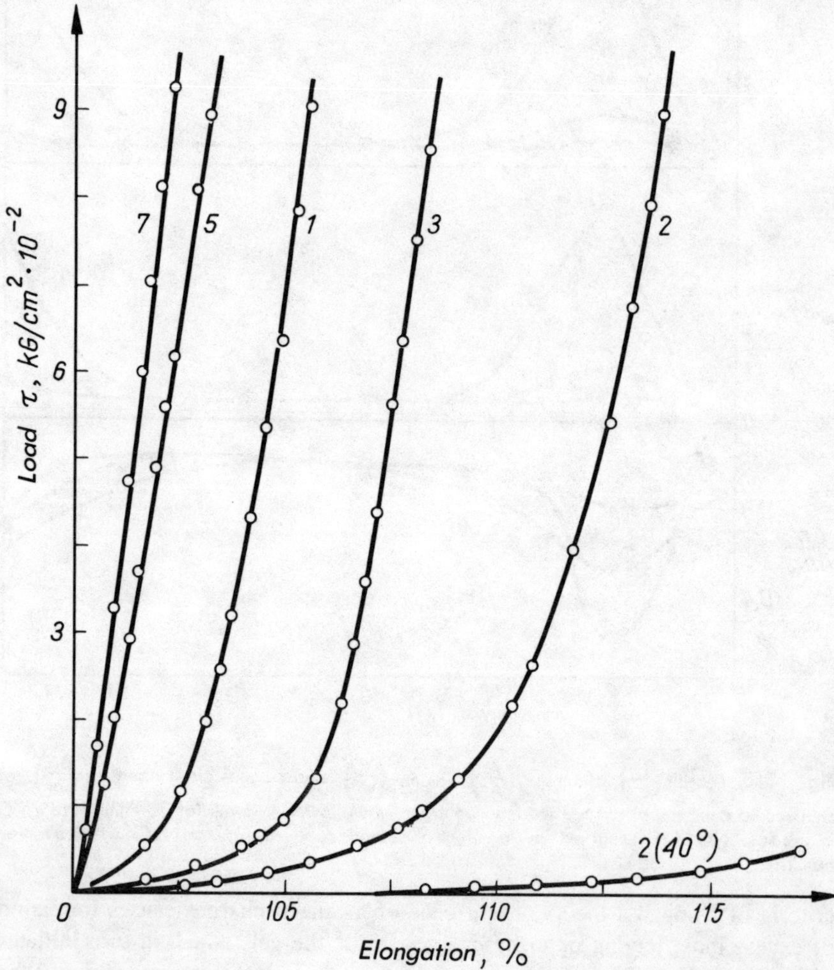

Fig. 10.2. Dependence of the resistance of tendon collagen on pH of the medium (stress/strain curves).

According to our present knowledge it seems reasonable to transfer the definition of polar and nonpolar chain segments to the changes in the side-chain polar groups. Investigations on the properties of membranes have shown the transition of collagen to a disordered (amorphous) state in 0.01 molar KCl solution at pH 3.5–2.0. Within pH limits 5–3.5 when swelling of collagen membrane increases, then filtration (permeability) index decreases. At pH = 3.5 one observes a structure disordering and due to this the said index increases. This difference is significant: decrease of filtration index when swelling increases speaks for de-

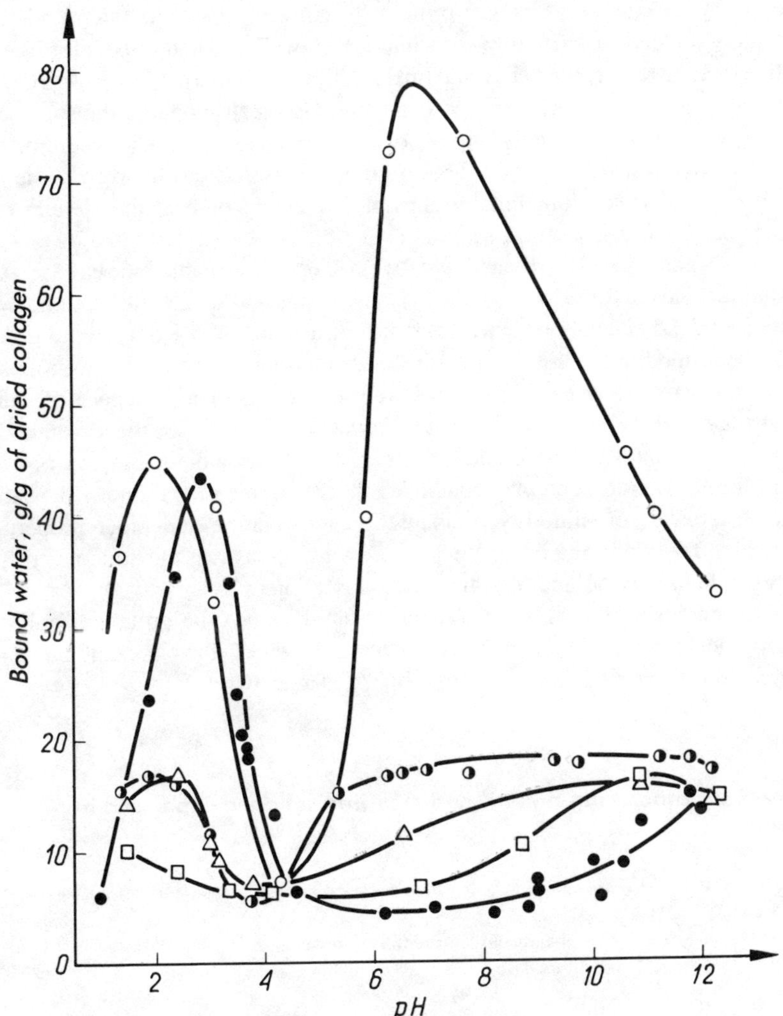

Fig. 10.3. Osmotic swelling of native and modified collagen according to Toyoda [3]. ●—native sample, ◑—succinated sample I, ○—succinated sample II, □—deaminated sample, △—acetylated sample.

crease of the pore size. The pore size among the unordered molecules is increasing when the swelling increases. This is very similar to the behavior of the sponge. It is observed during osmotic and lyotropic swelling, during leather processing and wearing. As investigations on osmotic swelling carried out by Toyoda et al. on native and modified collagen [3] have shown, collagen deamination deprives it of its swelling ability in an acid medium, whereas the ability to swell in alkaline medium remains (Fig. 10.3). Acetylation decreases the ability to swell in acid solution, when related to native collagen. Introduction of additional carboxylic groups by succination, changes principally collagen reactions with acids and alkalis. Collagen, esterified by succinic acid, swells in acidic medium like the native one, however, the pH of maximal swelling is somewhat lower. In a neutral medium, close to the pH corresponding to its isoelectric points, the succinated collagen swells very strongly, almost twice as much as in the acid medium. This modification is involving the ε-amino groups of lysine and hydroxylysine. Influence of individual functional groups of side chains on binding of water molecules has been discussed in section 5.4.

Toyoda and others (l.c.) found in native collagen from cattle hide ca 25.8/1000 lysine residues and 6.3/1000 hydroxylysine residues. After modification, there were found 3.5 and 0.4 respectively, which means that 87.5% of ε-amino groups have been modified. This speaks for basic importance of functional groups of side chains for water binding in the swelling process. The influence of amino groups alone does not determine this: acetylated collagen loses the same number of ε-amino groups of lysine and hydroxylysine. From amino acid analysis and from formol titration one may conclude that the degree of reaction is the same. The differences in solubility, denaturation temperature and other properties, tested by Toyoda in the quoted paper, do not demonstrate the structure differences, which may be due to collagen modification.

The conclusion seems to be obvious that the carboxylic groups of collagen determine to some extent its behavior towards water, their content however, increases only from 1.23 to 1.53 mmoles/g (Table 10.1).

Table 10.1

Some data concerning natural and modified collagen expressed in mmole/g

Collagen	ε-amino groups		Free carboxyl groups		Amides	Total carboxylic groups
	titration	amino acid analyse	titration	amino acid analyse		
Native	0.35	0.33	0.96	0.84	0.39	1.23
Succinated	0.05	0.04	1.02	1.07	0.44	1.53
Acetylated	0.07	0.04	0.90	0.77	0.46	1.23
Deaminated	0.09	0.03	0.60	0.67	0.56	1.23

10.2. Lyotropic swelling

Lyotropic swelling may be observed in the solutions of the salts, in which the forces, causing Donnan's phenomenon, are insignificant. One may observe it at every pH-value, if only the salt concentration is high enough (over 0.5 molar), or in solutions having lower salt concentration, and a pH close to neutral. The collagen fibers when osmotically swollen develop a glassy, translucent look; swollen lyotropically, they maintain their natural 'silky' look. The volume increase is much smaller during lyotropic swelling, length increase is smaller as well. Quantitative characteristics of the process are given in Fig. 10.4, where

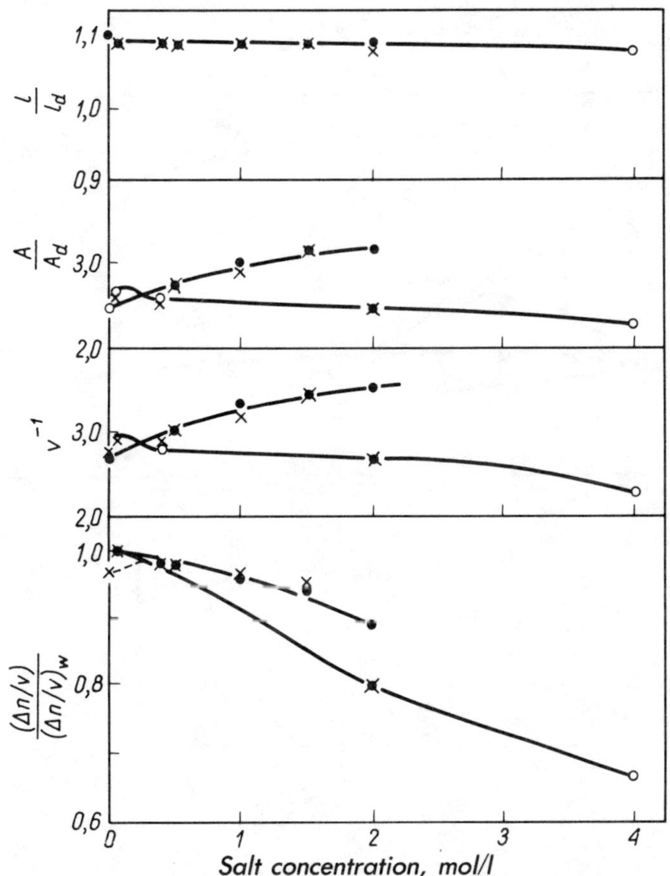

Fig. 10.4. Relative length, $\left(\dfrac{L}{L_w}\right)$, cross-section, $\left(\dfrac{A}{A_w}\right)$, volume, $\left(\dfrac{V}{V_w}\right)$, and optical birefringence $\left(\dfrac{\Delta nN/v}{\Delta nN/v_w}\right)$ of the crosslinked rat tail tendon collagen, as a function of salt concentration during swelling and deswelling. ●—KCSN, ○—KCl, circles to increasing concentrations and crosses to decreasing concentrations.

the changes in length, cross-section area, swelling degree and birefringence are shown; KCSN and KCl solutions are used, and crosslinked collagen, in order to avoid changes due to destruction. The dependence of lyotropic swelling of crosslinked tendon collagen on kind of salt, concentration and temperature is shown in the Fig. 10.5.

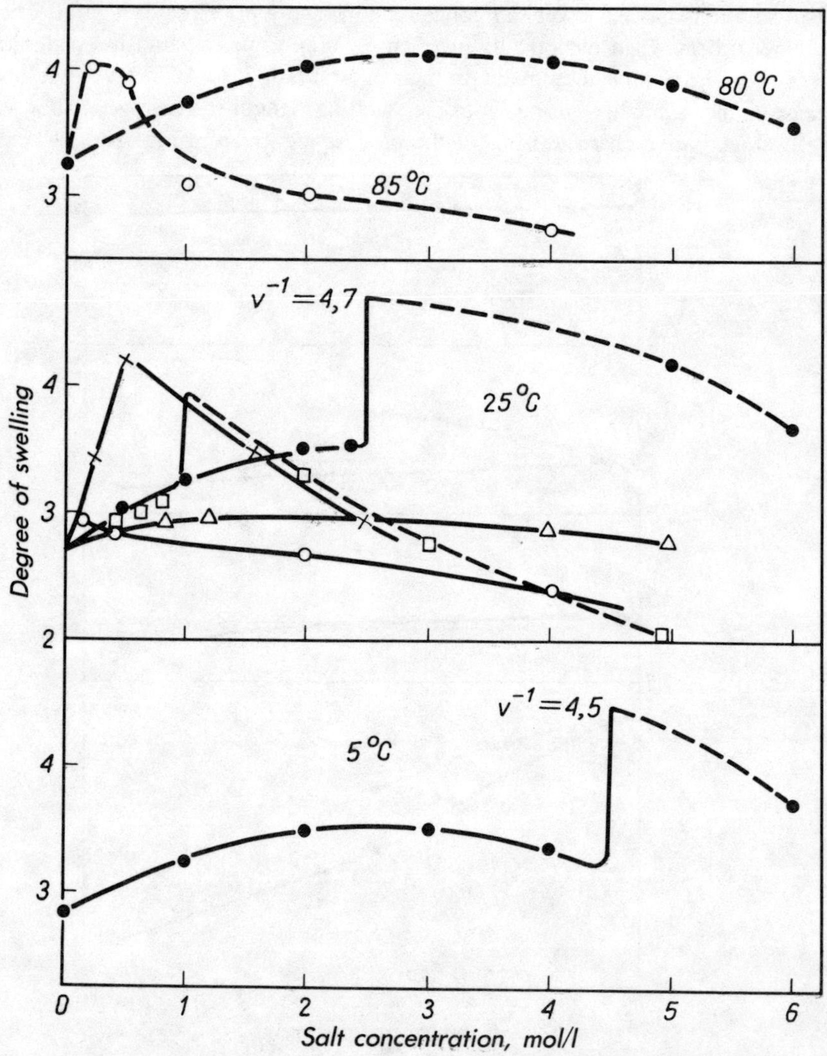

Fig. 10.5. Swelling of crosslinked tail tendon collagen as a function of salt concentration. Dotted lines—measurements for denatured collagen
○—KCl, △—NaCl, ●—KSCN, □—Ca(SCN)$_2$, X—AlCl$_3$.

The jumps observed at some points of swelling are connected with shrinkage or melting (phase transition). The broken line characterizes behavior of samples with unordered molecules. Increasing salt concentration causes at first swelling increase (salting in) and then decrease (salting out) of swelling. Gelatin behaves like collagen. Comparing the swelling effect of various salts they have been ordered in a lyotropic or Hofmeister's series. The most commonly used anions, ordered according to their swelling power, follow below:

$$F^- < Cl^- < CH_3COO^- < NO_3^- < Br^- < SCN^- < SO_4^{2-}$$

(sometimes the position of Br^- and NO_3^- is changed), and the cations are ordered as follows:

$$K^+ < Mg^{2+} < Na^+ < Cs^+ < Li^+ < Ca^{2+}$$

There is no theory as yet, explaining the positions of the ions in the lyotropic series. They are neither in exact agreement with solvatation properties of ions nor with their thermodynamic characteristics, probably because there are always two ions in solution, whereas the theoretical considerations are concerned with only one of them. The immobilization of one ion only among the others is changing with concentration in a way which is difficult to assess. There are different views on this point; it was believed that of dominant importance should be the degree of crystallinity of collagen. According to another theory it is dependent on the salt ability to keep maintained among the macromolecule fragments. Also a view exists according to which the strong swelling agents like KSCN decrease the degree of order, which may make the ordered regions react in a similar way, as the unordered do.

The elasticity measurements, done on the swollen tendon, show the decrease of modulus together with increase of the swelling degree. Testing of birefringence of fibers swollen in the solutions of KCl and KCNS, gives surprising results: one of these salts (KCl) makes deswelling easier, the second one—swelling. This puzzle is solved to some extent, if we look at the investigations of membrane properties. Behavior of membranes is shown in Fig. 10.6.

In the upper part the swelling of a membrane is characterized in its cross-section, where a discontinuity is observed, when the concentration reaches 3 molar KSCN. This point corresponds to the phase transition of melting, which is confirmed by optical testing. As one may believe, the solid and dotted lines show the respective properties of samples having ordered or unordered molecules. From this observation one may conclude that swelling does increase the homogeneity of the structure, maybe by closing of the openings, and during deswelling the structure becomes more heterogenous. As to the investigations on

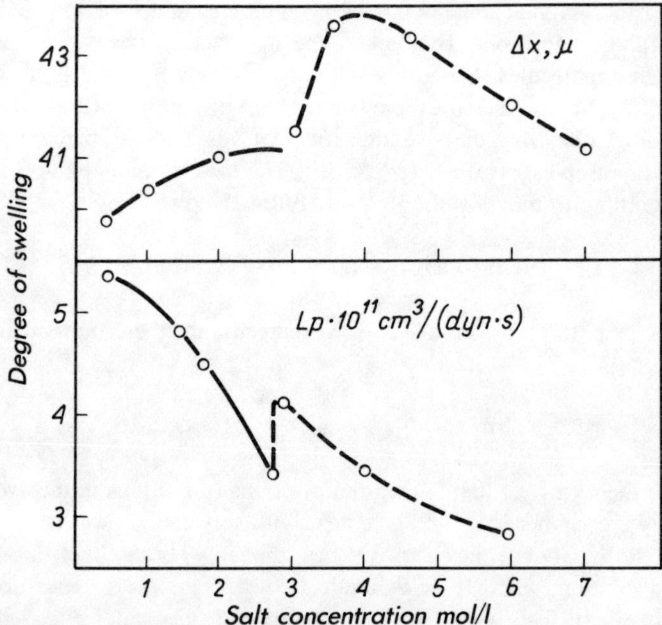

Fig. 10.6. Thickness (Δx) of the collagen membrane and filtration index (L_p) as a function of concentration.

absorption of various salts on gelatine, one may conclude (Fig. 10.7) the anions and cations may be ordered in a following way

$$SO_4^{2-} < CH_3COO^- < Cl^- < NO_2^- < Br^- < SCN^-$$

$$K^+ < Mg^{2+} < Na^+ < Cs^+ < Li^+ < Ca^{2+}$$

and the correspondence with the lyotropic series is obvious.

The salts may be divided into three groups according to their interaction with collagen. In the first groups there are salts that decrease T_S and cause collagen to swell, as $CaCl_2$ and KSCN do; in the second, there are salts that have little influence on T_S and swelling, as, e.g., NaCl; in the third, the substances that decrease T_S and give rise to swelling when applied at low concentrations, whereas at high concentrations they act in reverse way, e.g., sodium and ammonium sulfates.

There exists a dependence between adsorption of salt on gelatin and its reaction with collagen. The greater is the adsorption, the more it decreases T_S and swells collagen due to destroying of the hydrogen bonds. Increase in the thickness and plumpness is a measure of swelling; decrease of compressibility is a measure of plumpness of the hide. These properties in acidic solution have been tested

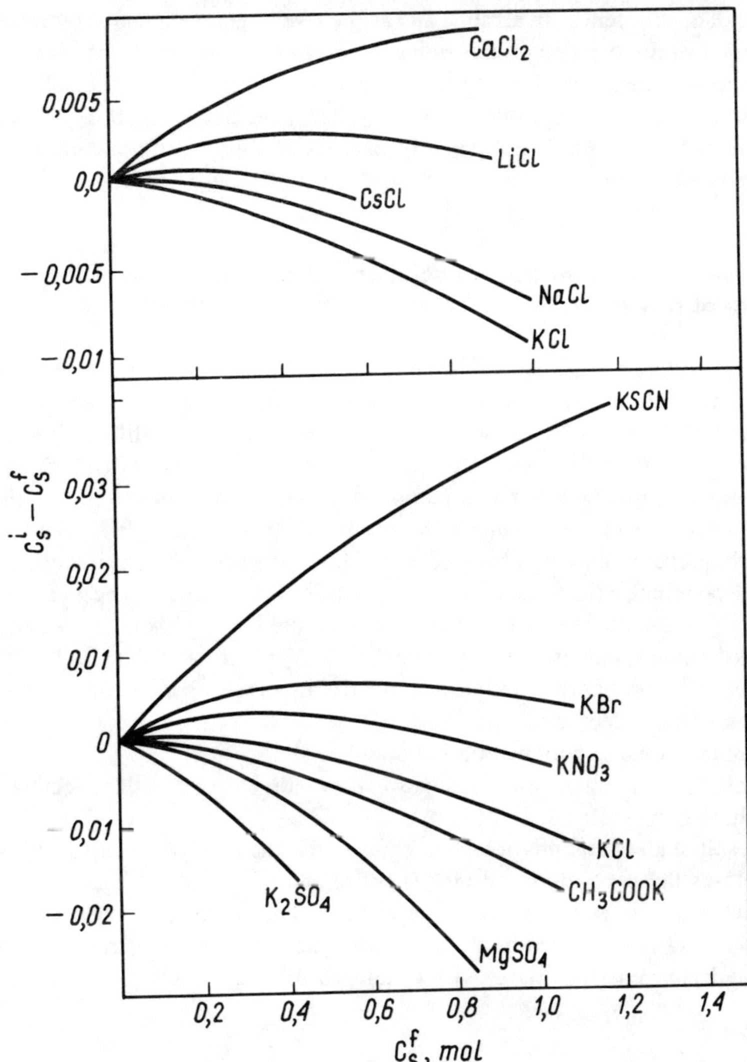

Fig. 10.7. Apparent absorption of salts in gelatin, as a function of concentration of the external solution in equilibrium with the gel. C_s^i = initial concentration, C_s^f—final concentration.

by Schubert [4]. He found that the swelling in acidic solutions is smaller than the plumpness increase, and the opposite is the case in alkaline medium; the dynamics of changes are greater in acidic medium. He has observed an obvious difference in reactions with various swelling-making acids: formic, acetic, lactic, and others. Hydrochloric and sulfuric acids have the strongest swelling power. Maximal swelling effects for the acids used in tannery occur in their 1.5%

solutions. The differences in alkaline and acidic swelling are explained by Schubert in terms of dissociation of non-collagenous protein compounds, by increase of the own charge and repelling of the surrounding charges. In an acidic solution mechanical resistance of the fiber network is highly involved, and if depickling process in NaCl solutions is not properly carried out it may result in damage or destroying of the network. The plumpness due to swelling in inorganic acids is an effect which has to be controlled.

10.3. Swelling in nonaqueous solutions and stabilization of the structure of raw hide

Swelling of collagen in nonaqueous solutions was often investigated. The purpose of these investigations has been, e.g., to find out the length of the carbon chains (in alcohols) which may fit between the collagen side chains, stiffening them and making the structure immobile. Gustavson [5] has considered the use of some specific compounds for this purpose (phenols, urea). According to the changes in X-ray reflections under the influence of alcohols a 'filling-in' of hydrocarbon chains may be observed as well; the distances between them are shown, depending on the axial distances of protofibrils (Fig. 10.8). The processes occurring give stabilization of space structure of the hide, without introducing new crosslinking bonds into it. This may be observed, if:

(1) surface tension of water in which leather is immersed is decreased by surfactants (this effect is minor);
(2) the hide is treated with neutral salt solutions;
(3) the hide is dehydrated with organic solvents (ethyl alcohol, ether, acetone);
(4) the hide is lyophilised.

The result of all operations or their combination is the removal of the substance which closes the pores during drying (matrix) and of considerable amounts of water, whereas the pore size does not change. The influence of dehydrating agents on hides is expressed by the use of two factors: volume decrease during drying and apparent specific gravity (Table 10.2). During water removal the

Table 10.2.

Effect of dehydrating agents on hide properties

Dehydrating agent	Volume decrease on drying, %	Apparent specific gravity, g/cm^3
Control	80	1.21
Alcohol	60	0.75
Alcohol-ether	29	0.35
4 N Na_2SO_4	31	0.64
4 N NaCl	64	1.22
4 N Na_2CO_3	44	—
4 N K_2CO_3	50	—

unstable bonds may be formed between the molecules of substances used and collagen. The characteristic property of a dehydrated hide is the loss of its properties after re-soaking in water. This process is easily reversible, unlike the tanning. If however, the dehydrated hide is impregnated, e.g., by silicones it remains in this state even after multiple wetting and drying.

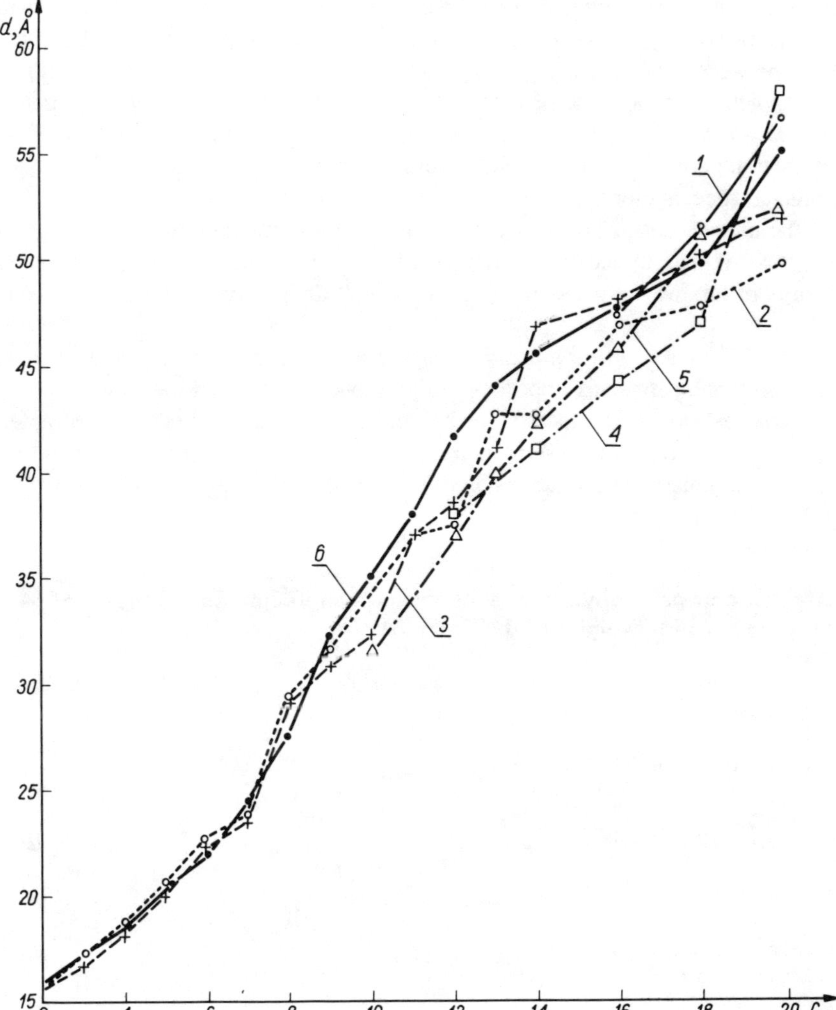

Fig. 10.8. Interaxial distances of protofibrils in the rat tail tendon collagen, crosslinked by aldehydes and swollen in n-alkanols as a function of the number of C-atoms in the chain. 1—native collagen, 2—collagen treated with formaldehyde (pH = 3.6), 3—treated with glyoxal (pH = 7), 4—treated with glutaraldehyde, 5—treated with crotonaldehyde (pH = 7), 6—treated with acrolein (pH = 3.25).

From the viewpoint of collagen structure, the separation of single fibers is considered to be dehydration. Soft fibers of the pelt are attached to one another and stick together, when water from interfibrillar spaces is evaporated. Hide, dehydrated by acetone extraction in a Soxhlet apparatus and then lyophilized, was investigated by Heidemann and Riess [6]. In observing collagen during lyophilization they found that during this process no concentric shrinkage occurs. Such a concentric shrinkage may be observed, during acetone treatment, and thus is the density of lyophilised hide greater than that of the acetone-dehydrated one, because the fibrils are sticking together to some extent. They have shown incompletness of acetone dehydration; the degree of its effectiveness depends on the water content in pure acetone and on the number of evacuations of the Soxhlet apparatus siphon, or on the number of exchanges of acetone layer over collagen (decantation).

The internal surface of the leather increases proportionally to the dehydration progress as determined by the BET method (Table 10.3). For obtaining good results in this way it is essential that very well dehydrated acetone (99.8%) be used.

The collagen swelling process discussed above lacks one consistent theory. At the moment it seems impossible to give one covering all the observed phenomena, as very many factors are influencing the macromolecule of complex structure. We may expect some more generalized explanation from thermodynamical and kinetic investigations on swelling and shrinkage processes. However,

Table 10.3.

Effect of calf pelt dehydration with acetone on its internal surface (250% acetone per hide weight) [Ref. 25. ch. 8]

Number of decantations and evacuations	Amount of water removed from leather after 24 h, %	Surface m^2/g
Decantations (exchange)		
1	56.71	—
2	23.26	—
3	10.64	0.44
4	6.94	13.39
5	4.26	23.55
Evacuations of siphone in Soxhlet apparatus		
1	1.943	28.81
2	0.859	34.60
3	0.216	35.18
4	0.036	32.21
5	0.003	35.26

Sample—100 g pelt; after drying—28.8 g of dry hide.

this may be only an averaging explanation, based every time on experiment done under given conditions on a definite material. Such experiments, supported by mathematical and physical considerations as well as on exact knowledge of collagen molecule structure and properties, have been started in the last decade and the results may not be generalized until sufficient volume of knowledge is accumulated.

10.4. Phase transitions in collagen and shrinkage temperature

It was Povarnin who first used the method of shrinkage temperature determination [7] as a criterion of effectiveness of the tanning process. Many authors have used T_s determinations when working on crosslinking of hide collagen and other proteins such as wool. Küntzel defined collagen shrinkage as a first-order phase transition. Recent investigations and thermodynamical considerations have confirmed his view, but with some limitations. The shrinkage temperature is decreased by water and hydrotropic substances. Water may be considered as diluent—and the hydrotropic substances either may attract water to the macromolecule, or disrupt the hydrogen bonds, thus decreasing the energy pool of collagen. Introduction of additional crosslinking bonds in tanning and in tissue aging increases the shrinkage temperature. There are numerous methods of the T_s determination. The phenomenon itself is dependent on structural properties of collagen tested, and on several more factors, like, e.g., solvent amount, salt content, and others. The important problem is how to determine the T_s of chrome-tanned leather when this temperature is over the water boiling point. It is not solved yet (cf. p. 263).

Physicochemical sense of phase transitions

Behavior of collagen in a protein-water-salt system according to pH, kind, and concentration of salt has been extensively tested. A scheme of phase transition, as suggested by Flory [8], is shown in Fig. 10.9.

Collagen may occur in three phases: ordered molecules, defined as crystalline and shown in the Figure as C, in solutions as random coils, RC, or as suspension of loose helices, H. This latter form is considered to be identical with tropocollagen solution, from which reconstituted collagen may be obtained. The reversibility problem for the transitions I and II has not as yet been elucidated conclusively. It has been shown, however, that when parameters are properly chosen (cooling, stirring, etc.) one may shift the position of equilibrium points. The visible changes are the following:

Transition I—precipitation or dissolving of the precipitate (reverse direction)
Transition II—change of solution viscosity
Transition III—precipitation or dissolving of the precipitate.

The curves for reduced viscosity, typical of transition II are shown in Fig.

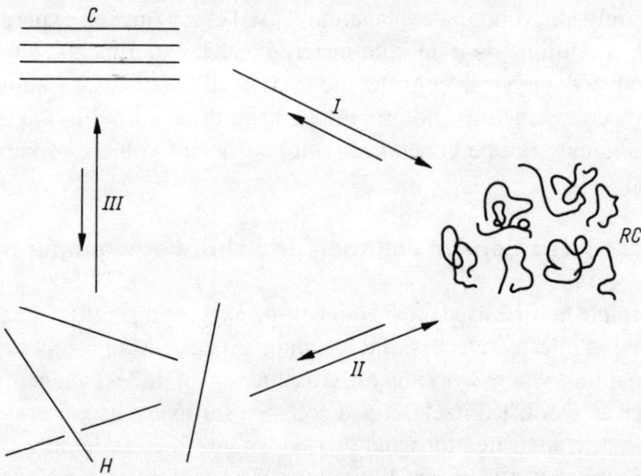

Fig. 10.9. Phase transitions of ordered collagen, denoted as crystalline (C), into the solution in the form of random coils (RC) or helices (H). I—phase transition C $\rightarrow$ RC at T_I, II—phase transition H $\rightarrow$ RC at T_{II}, III—phase transition H $\rightarrow$ C at T_{III}.

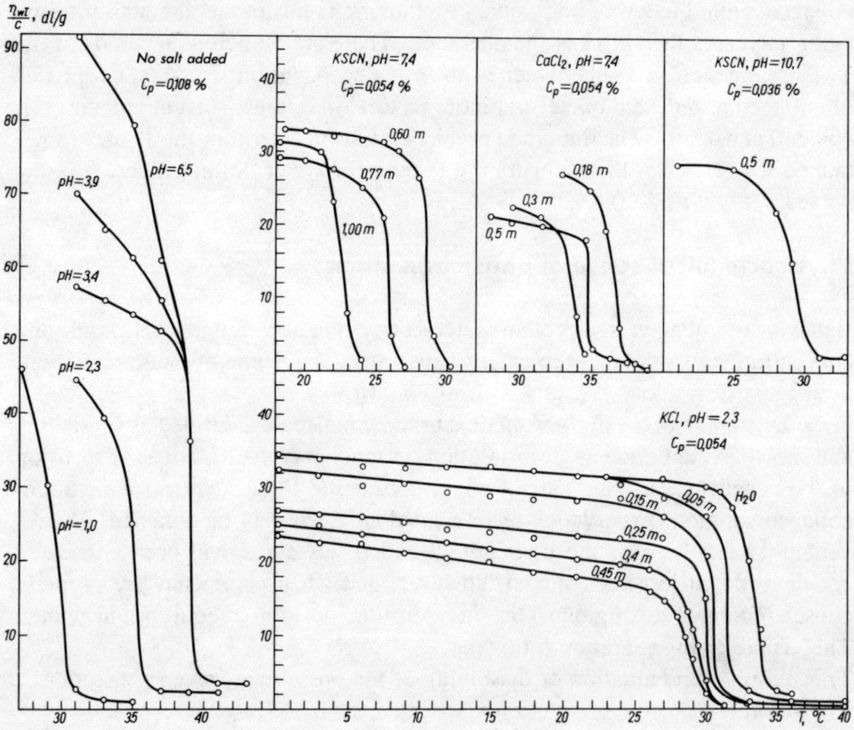

Fig. 10.10. The curves of reduced viscosity of the solutions of soluble collagen as a function of temperature, pH and salt content, C—concentration.

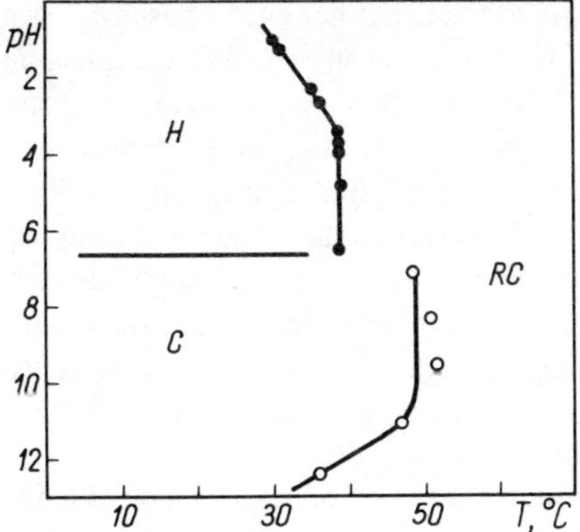

Fig. 10.11. Temperature of phase transitions for 0.1% collagen solutions in water as a function of pH of the medium.

10.10. An abrupt change is visible there, as it occurs at a constant and reproducible temperature under constant experiment parameters.

The influence of pH on transitions between individual phases is shown in the Fig. 10.11, and the effect of salt, in Fig. 10.12.

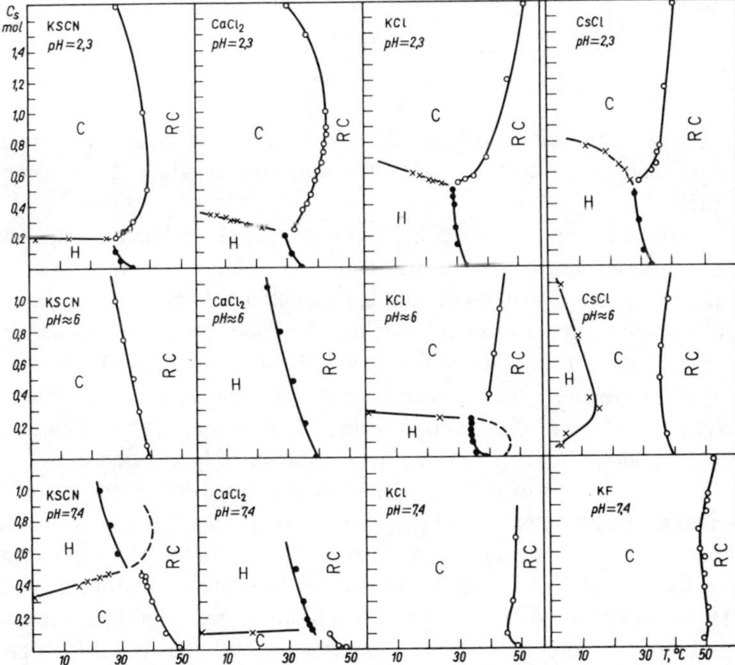

Fig. 10.12. Phase transitions of the tricomponent collagen-water-salt system as a function of molar concentration (C_s). The polymer concentration was about 0.05%.

In the last Figure one can see that not each phase occurs under the conditions tested (e.g., in a $CaCl_2$ solution at pH 6), and a triple point is missing, like, e.g., in a solution of CsCl at pH 6. These kinds of reactions may be called 'pseudophase transition' because of their incomplete reversibility. Also the extrapolation of the curves to the triple point is not exactly done, because of measuring difficulties in its surroundings. The most clear triple points have been observed at acidic pH values. An exact analysis of these Figures gives much information about collagen behavior in an aqueous medium. For the technologist a conclusion concerning absolute necessity of having reproducible parameters by T_S determination is of crucial importance. This transition for non-swollen collagen is synonymous with transition I.

The helix and random coil are two opposite states of a macromolecule. There are many possibilities for intermediate forms. Let us assume the helical state of maximal internal energy as a, and the random state of maximal rotational freedom and maximal conformational entropy as c. Now in state a every residue has the possibility to form two hydrogen bonds with two other mers in its proximity. In state c the links of a chain are totally free; in an intermediate state b—the residues (links) are bound to but one neighbor, thus having greater entropy than the bound mers in state a. The difference in energy level between states a and c $= \Delta F$ is expressed by equation

$$\Delta F = Z (\Delta H - T\Delta S)$$
$$a \rightarrow c$$

where Z is polymerisation degree, ΔH—enthalpy change and ΔS—entropy change per one mer. There is a certain temperature at which $\Delta F = 0$ and the helix 'melts.'

The composed character of the helix-coil transition may be explained by the fact that the large, fairly homogenous helical region is more stable, i.e., it has lower internal energy, than an ensemble of smaller helical fractures, having in general the same number of mers. This may be understood if we remember that in this latter case there are more links present, bound by only one hydrogen bond untouched. The tendency to form one helical row is opposed by another tendency to decrease the degree of ordered structure, to increase entropy. Minimum of free energy content of such a system is a resultant of both tendencies.

When a definite part of helical segment is converted into random coil, then the second tendency starts to increase. In other words, the increase of the amount of disturbed regions is a cooperative process. Thus in thermodynamic terms we consider the protein molecule in solution as a cooperative one-dimensional system. This is right for a single molecule of solubilized collagen. In hydrated fiber bundles however there are still other, intermolecular bonds and, although they are cooperative as well, a tridimensional system has to be considered in the case of collagen shrinkage. A change in the hide structure stability is related to the change in helical molecule structure or to randomized positions of fiber bundles in it. The change in structure may only follow certain strengthening of the bundles and chains. When the chains become strengthened, some motion of residues each against the other occurs and the moment when the heat energy supplied overwhelms the energy contained in interresidual bonds is the starting moment of shrinkage.

A new structure organization appears in the now randomly distributed chains. To define this order, single mers (residues) have to be considered as independent links. Starting with this argumentation, and applying Gaussian formula for thermodynamic probability of occurrence of possible chain conformations, Gorbatshev et al. [9] have elaborated a theoretical answer to the question of shrinkage temperature prediction and dependence. According to these authors, it depends on the number of crosslinking bonds, not splitting at the shrinkage start. The authors however assumed arbitrarily the number of crosslinking bonds in native collagen as 1 per 10 unit (links) and taking as an independent variable the ratio of non-disrupted to disrupted bonds, which is very difficult to check in practice. The shrinkage of collagen may thus be derived from the melting point of a substance suspended in a liquid. Because the substance may be either suspended as single polymer chains, or as bulk mass of fibrils, we have two evaluation criteria: denaturation temperature T_D, manifested by an abrupt change in viscosity and optical rotation of a colloidal collagen solution upon heating to approx. 40°C, and the shrinkage temperature T_S, manifested by change in solid sample length and/or stress. Both phenomena may be fitted into the usual polymer melting point relation

$$\frac{1}{T_m} - \frac{1}{T_m^0} = \frac{R}{\rho V \, \Delta H_t} \, (V - \chi V^2)$$

where ρ is the density of the amorphous polymer, ΔH_t is the heat of fusion per gram of crystalline polymer, V is the molar volume of the solvent, and χ is the parameter characterizing the interaction between solvent and collagen. The T_m^0 value represents the melting point of pure polymer. Under the conditions used by authors, i.e., melting point determination in ethylene glycol—it represents the polymer in a condition equivalent to its solvated state. This value has been obtained by extrapolation to the zero solvent content and equals to 145°C. χ was set at 0.30. The authors have supposed that the diluent is partly firmly solvated to certain polar groups and the remainder is loosely held. They emphasized the difference between melting point in glycol and in water—as the interaction with water is much stronger and thus the melting occurs at such a high temperature. This view explains that the differences between T_D and T_S are in fact none, and T_D may be considered merely as low concentration limit of the denaturation whilst T_S as denaturation in highly concentrated solution. The same conformational transformation occurs in both cases. Considering shrinkage of dry collagen—we have to keep in mind that the natural system consists of collagen and water. Removal of bound water means disturbing the system. Investigations done by Kawamura et al. [10] on tanned and untanned collagen have shown that there is a continuity in this last point, gradual dehydration of skin or leather leads to gradual T_S increase (Figs. 10.13a and b). This problem is entirely related to the thermodynamics of collagen hydration and now of particular importance for two reasons: the first one is theoretical, and concerns the question by which bonds and how strongly is water bound to collagen, and the second, of more practical significance, how the collagen-water system works in leather during wearing, particularly in shoe leather. The second topic is covered by technology and wearing hygiene and thus beyond the scope of this chapter. We may only refer the reader to the works of Muller-Limroth and his coworkers [11] who have carried out the most extensive works on this subject.

The first question was raised in the late forties by Kanagy [12] who investigated

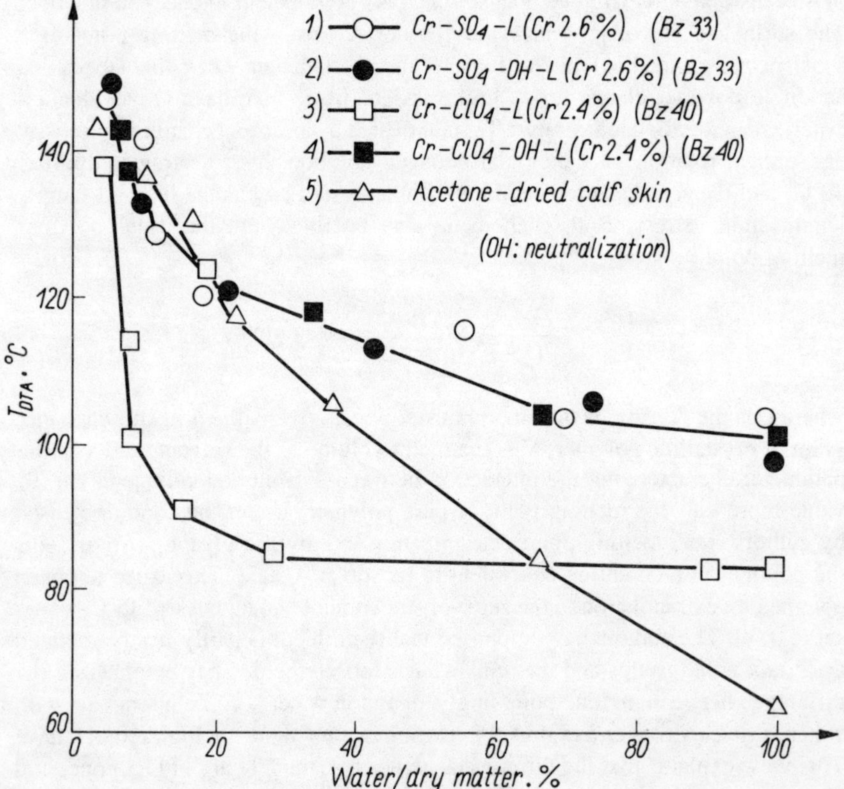

Fig. 10.13.a Relations between the T_{DTA} value and water content of leather. According to [10]. T_{DTA} means shrinkage or denaturation temperatures estimated by DTA method.

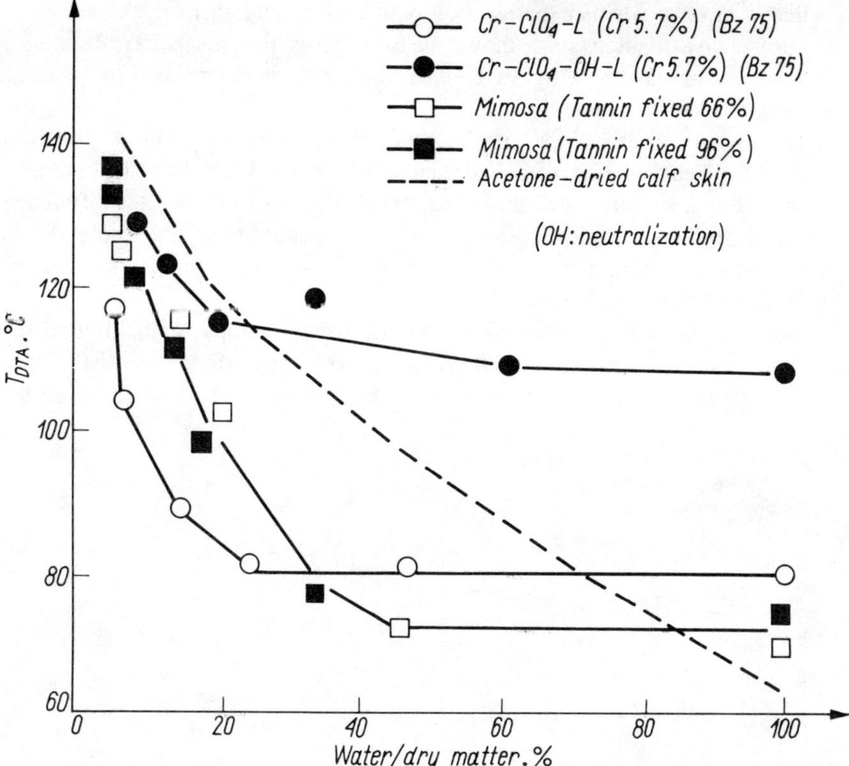

Fig. 10.13.b Relations between the T_{DTA} value and water content of leather. According to [10]

the heat of water to collagen binding, applying classical calorimetry and the BET isotherm. He was the first to propose the concept of water layers and indicated the reverse dependence of binding heat on the amount of water. More detailed studies have been recently done with application of thermal gravimetry (TG) differential thermal analysis (DTA), NMR and dielectric properties investigation methods.

DTA method, applied for the first time to the shrinkage temperature measurement by Witnauer and Wisnewski [13], gave the possibility of a more detailed analysis of leather collagen at this point. There exists a fairly good correlation within narrow limits of T_s, measured according to conventional method and DTA results. As a rule the T_s lies between onset and peak temperature. DTA method made it possible to demonstrate that the run of shrinkage may be different in leathers tanned by various tanning agents, rate of temperature increase, etc.

The advantage of the DTA method is that one may define the exact temperature and reaction heat (by integrating the usual differential curve) in one run. This indicates at once whether the reaction is of exo- or endothermic character.

Further investigations have shown the differences in heat capacity of collagen and the heat of water binding, two points which cannot be separated for practical purposes.

This is very clearly shown in the paper of Andronikashvili et al. [14] who studied the heat capacity of dehydrated and hydrated (35% of water) collagen over wide temperature range from 4 to 300 K. Fig. 10.14 shows this difference. The contribution of water molecules to the heat capacity of the whole system is not additive, thus a new state of substance with different thermodynamic properties is postulated.

The heat capacity of thermodynamic quantities (entropy, enthalpy and free energy) of native, dehydrated collagen is shown in Fig. 10.15, and the temperature dependence of the difference of heat capacities of hydrated, native and

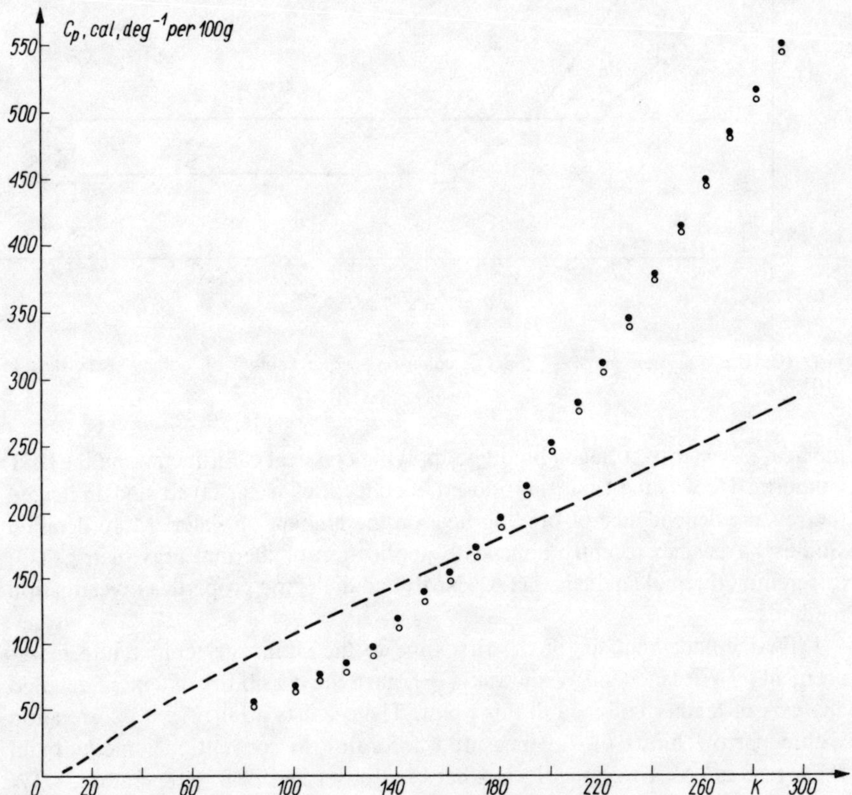

Fig. 10.14. Temperature dependence of heat capacity for hydrated native collagen (black circles) and for hydrated denatured collagen (open circles); the water content in specimens is 0.35 g H_2O/g protein. The dashed line is the heat capacity of dehydrated protein. According to [14]

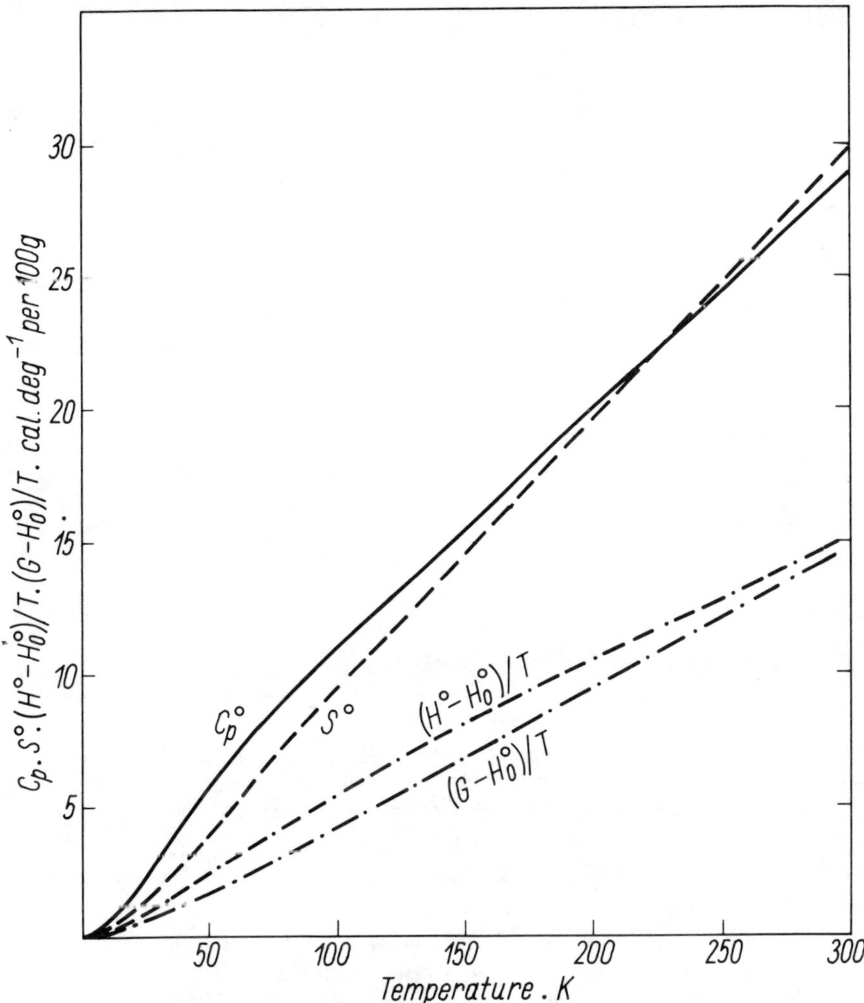

Fig. 10.15. Temperature dependence of thermodynamic quantites entropy $S°$, enthalpy $\frac{(H—H_o)}{T}$, the function of free energy— $\frac{(G - H_o)}{T}$ for native, dehydrated collagen. According to [14]

denaturated collagen in Fig. 10.16; this latter Figure permits to suppose that a strict correlation of water molecules along the macromolecular axis is highly disturbed in denatured collagen, where the mutual arrangement of water molecules, remaining bound to polypeptides chain after denaturation no longer exists.

This was recently confirmed by Sharimanov et al. [15] by calorimetric and NMR methods. Authors concluded that in the denaturation process besides intramolecular hydrogen bonds, the bonds joining together the water molecules bound to protein become split. Thus the cooperative process of rebuilding the

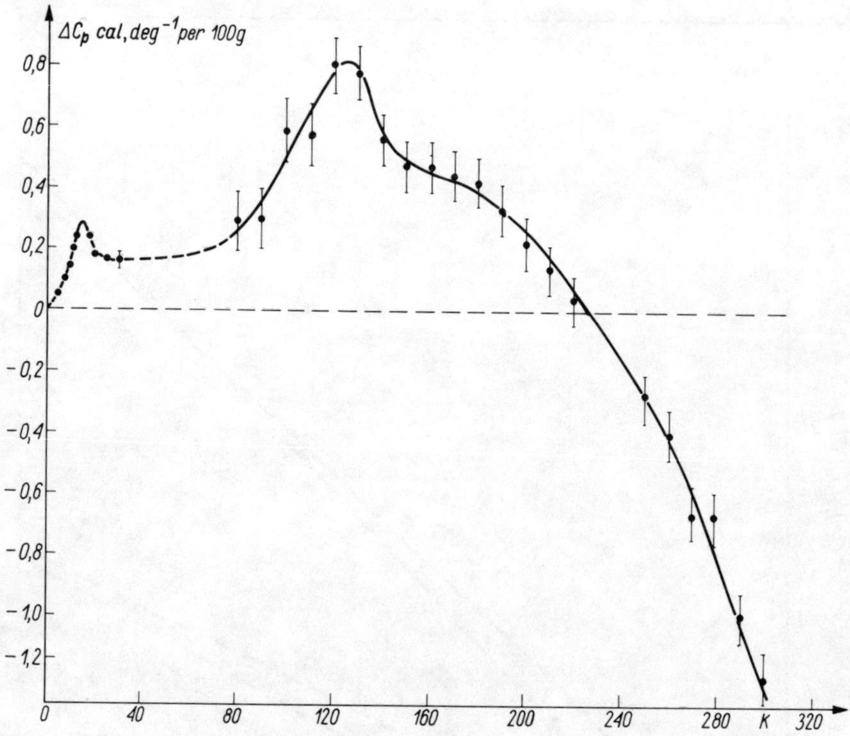

Fig. 10.16. Temperature dependence of the difference of heat capacities of hydrated denaturated randomly coiled and hydrated, native helix collagen ($\Delta C_p = C_{coil} - C_{helix}$). The water content in specimens is 0.35 g H_2O/g protein. According to [14]

hydration layers is, according to the authors, one of the peculiarities of 'intramolecular' melting of the collagen macromolecule.

The details discussed make the understanding of leather properties during wearing easier, as they indicate what is essential in leather behavior.

For the first-order transition the temperature, where $\Delta F = 0$

$$T_m = \frac{\Delta H_r}{\Delta S_r}$$

i.e., melting (shrinkage, or denaturation) temperature is dependent on the enthalpy and entropy quotient of denaturation per residue. For mammalian collagen ΔH_r ranges are of 1000-1500 cal/mole (4200-6300 J/mole) and ΔS_r 4.1-4.9 cal/deg/mole (17.1-20.5 J/K mole). These values are determined by various ways and may slightly differ. Their dependence on pH, cross-linking (i.e., tannage) presence of salts, spatial arrangement and kind of animal is well known.

Effect of salt on phase transitions in collagen

When discussing the meaning of the presence of salts in the medium under conditions close to the isoelectric point, one has to reconsider the lyotropic series. Salts where both components occupy high positions in the lyotropic series, like KSCN, decrease the shrinkage temperature of the sample, while salts from the other end, when present in the solution at low concentration, as, e.g., KCl decrease the T_S and increase it when concentration is high.

Binding of the salts at random coil explains the importance of melting point depression while the mechanism causing decrease of polymer stability due to the salt effect still lacks explanation. Ion concentrations are almost equal on the surface of swollen collagen molecule and inside of it, thus the Donnan effect is small or does not appear. As one may see, however, the decrease of melting point by salts, where both ions are strongly absorbed, e.g., $Ca(SCN)_2$ is greater than if there only one of them is strongly adsorbed, e.g., $CaCl_2$ or KSCN. Most probably the Donnan effect (polyelectrolyte effect) is of only slight when salt concentrations are great, may be due to a shielding action of ions close to the polymer. Under non-isoelectric conditions (e.g., at pH 2-3) the importance of salts in transitions I and II is like the one at the isoelectric point only when the salt concentration is higher than about 0.8 molar. If the concentration is lower, so is the behavior typical of polyelectrolytes:

(1) tendency to decrease the temperature of transitions I and II at very low salt concentrations;

(2) tendency to increase the temperature at higher salt concentrations.

This latter effect may be explained by a decrease of free electrostatic energy of the polyelectrolyte due to shielding of its charge in fixed position by mobile ions. Effect 1 is more difficult to interpret; it is probably connected with the influence of salt on the apparent pK values of the protein functional groups which undergo dissociation. There are arguments, however, speaking against this view; dissociation of the ε-amino group in lysine should be complete at pH $= 2,3$ and increase of apparent dissociation constant of the carboxylic group shall diminish deviation from isoelectric conditions. Analysis of the graphs shown in Fig. 10.12 allows one to state that the occurrence of helical form does not affect the dependence of T_S on salt concentration.

The salt effect on transition III under isoelectric conditions may be characterized in a wide range only for CsCl from among the salts given in Fig. 10.12. Slopes of the curves of transitions III and I for the concentrations chosen are of opposite sign. This apparent paradox may be explained if one assumes the negative value of the heat of fusion ΔH_f through all the range investigated. Increase in the KSCN concentration however causes the helix stabilization at pH 7.4. The helix form is not observed in the presence of KCl. This may be explained by increasing or decreasing activity coefficient of both salts.

Transition III may be difficult to observe, or it may occur at higher temperatures only. The 'crystalline' or helical forms are not observable under non-isoelectric conditions at pH 6 for KSCN and $CaCl_2$. Adsorption of SCN^- ion even in small amounts by a protein of a small positive charge is synonymous with compensation of localised charges and causes insolubility, whereas the adsorption of Ca^{2+} ion increases solubility of the helical form. The mechanisms of salt binding to collagen were discussed recently by Mikhailov (Footnote p. 228).

Meaning of spatial relations

The behavior of collagen discussed here allows us to evaluate one kind of effect on the shrinkage or melting temperatures. Another kind of effect involves the structure of amino acid (peptide) chain, or more exactly, the amino acid sequence in the chain. The first step for full information is to recognize the sequence. The next should be to calculate the energy levels through all chain and primarily energy of the bonds.

Investigations on poly-L-proline and polytripeptides have shown, what was already pointed out by Gustavson [5], that imino acids due to their pyrrolidine ring (and particularly to Hypro content) are responsible to a great extent for the formation and maintaining the collagen helix. The question was whether hydroxyproline, due to the possibility of an additional hydrogen bond of its hydroxyl, does influence it more than proline. This question could not be answered by a straight experiment, as we do not have the possibility of selective exclusion of the hydroxyl from the reaction without influencing other functional groups of the side chains. Only recently Ebert et al. [16] have reported to obtain collagen tosylated at the serine residues in 40-50% (but not at hydroxyproline).

$$
\begin{array}{c}
OH \\
| \\
CH_2 \\
| \\
-HN-CH-CO-
\end{array}
\quad \xrightarrow{\text{Tosylchloride*}} \quad
\begin{array}{c}
O-Tos \\
| \\
CH_2 \\
| \\
-HN-CH-CO-
\end{array}
$$

$$
\xrightarrow[-H_3C-C_6H_4SO_3^{\ominus}]{CH_3-\overset{O}{\overset{\|}{C}}-S^{\ominus}}
\begin{array}{c}
S-OC-CH_3 \\
| \\
CH_2 \\
| \\
-NH-CH-CO-
\end{array}
\quad \xrightarrow[-CH_3COOCH_3]{OH^-,\ CH_3ONa}
\begin{array}{c}
SH \\
| \\
CH_2 \\
| \\
-HN-CH-CO-
\end{array}
$$

*Tosylchloride formula is $CH_3 - \langle \rangle - \overset{O}{\underset{O}{\overset{\|}{\underset{\|}{S}}}} - Cl$

Such a collagen is reported to have T_s higher by about 5°C and to get contracted less during the shrinkage experiment. This is due to an introduction of 1.2-1.4% cystine more than in the native collagen sample, and to the -S-S- crosslinks introduced.

An explanation after many discussions was provided by the production of non-hydroxylated collagen through inactivation of prolyl hydroxylase in the collagen biosynthesis. Thus one may obtain essentially the same protein as collagen except that it contains no hydroxyproline or hydroxylysine, and the corresponding amount of proline and lysine instead. This protein was triple-helical, but its T_D in diluted acetic acid was by about 15° lower than that of the comparable collagen (Fig. 10.17) [17]. It is still uncertain, how hydroxyproline produces this effect. Two models have been recently proposed. One by Traub [18], in which a water molecule forms a hydrogen-bonded bridge between the γ-hydroxyl of hydroxyproline and the carbonyl of the preceeding glycine in the same chain. There is a question, however, whether a water molecule with just two bonds, and apparently exposed to solvent, could make a contribution to the helix stability. In another model proposed by Berg et al. [19] a direct hydrogen bond between the γ-hydroxyl of hydroxyproline and carbonyl of adjacent chain was suggested. This proposal requires however a cis-peptide bond to exist between glycine and proline in the adjacent chain and the presence of a cis-bond in collagen could

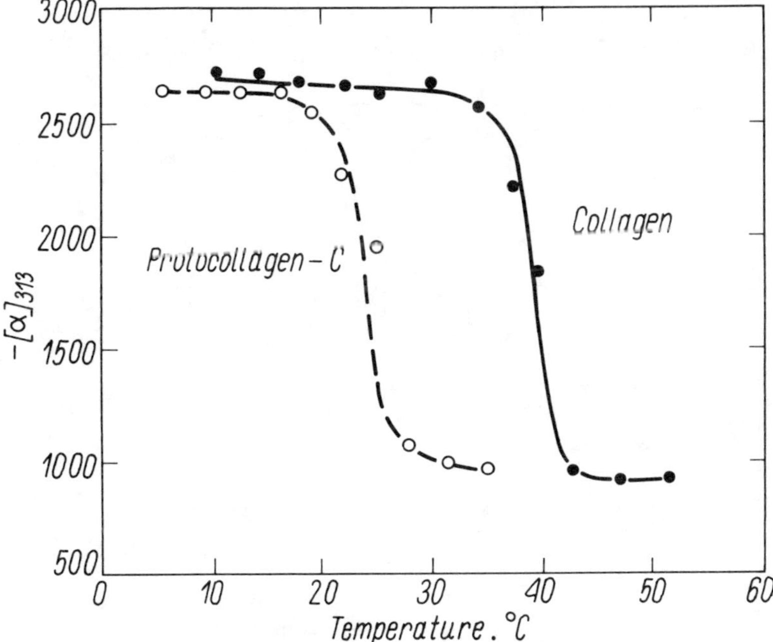

Fig. 10.17. Thermal stability of the triple-helical structure of protocollagen and of collagen. According to [17]

not be confirmed by ^{13}C NMR technique by Torchia et al. [20]. This problem remains open to further investigations. The specific importance of hydroxyproline, however, was recently confirmed by Burjanadze [21] who, analyzing 32 specimens of collagens from various animals, vertebrates and invertebrates with regard to their Pro and Hyp content, was able to demonstrate a very clear correlation between Hyp content and T_D, a less clear correlation between the imino acid sum and T_D, and no correlation between Pro content and T_D, particularly in vertebrate animals (Figs. 10.18 and 10.19). The quoted author indicated that in addition to stabilizing the collagen structure by Hyp content, the number of contiguous proline residues gives added stability since the Hyp content at the third position is equal to the sum of triplets (Gly-Pro-Hyp) + (Gly-X-Hyp). The way in which stabilization through Hyp occurs was not analyzed in his paper, except for a general suggestion the hydrogen bonds occur through water molecule.

Evidence exists that ionic interactions between carboxylic and amino groups are important for triple helix stability. Enthalpy measurements also imply hydrogen bonding and hydrophobic interactions [22]. Intermolecular interactions of the polar and hydrophobic side chains are important for the formation of higher collagen structures [23].

Shrinkage of dry collagen and DTA

In research work it is rather uncommon to test the shrinkage temperature with

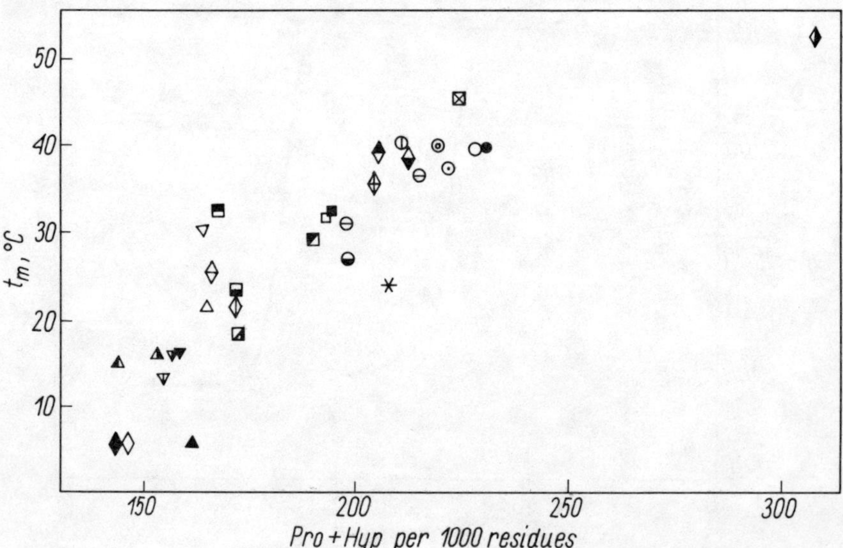

Fig. 10.18. Correlation between t_m and contents of amino acid in various collagens. According to [21]

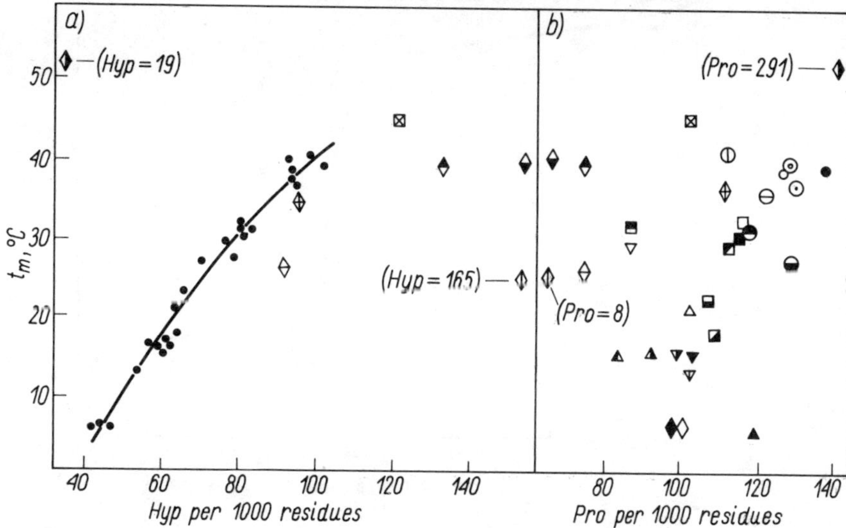

Fig. 10.19. Correlation between t_m and contents of (a) Hyp and (b) Pro in various collagens. Dark circles in (a) designate interstitial collagen of vertebrates. According to [21]

no liquid medium. The results are as a rule related to the sample, immersed in so much liquid that increases in this amount within the limits of 20-30% do not affect the results. The behavior of dry collagen in industrial drying process is important for drying leather in this part of its processing and for the shoe industry as well, because of moist heat setting. As one may see from the previous discussion, basic structural changes are not observable under these conditions. For testing dry or air-dried collagen in order to avoid oxidation heating in neutral gas, e.g., in nitrogen, or in Wood alloy, or in mercury may be applied. As air-dry we define collagen containing the amount of moisture which is in equilibrium with the surrounding atmosphere.

The problem which may be raised here is what happens to collagen when it is heated to temperatures exceeding T_s. At the first stage of heating it loses the bonded water. This, however, is not a simple process and we have already explained the possibilities and known (or suspected) pathways of water binding. Simultaneously, a thermal decomposition (pyrolysis) starts, ending in the burning of the sample. Thermogravimetry permits recording of the weight loss of a sample during heating. This record is usually obtained in the form of a differential curve, in order to obtain better distinguished peaks. Differential thermal analysis is based on the measurement of difference in the temperature increase of the sample, as compared with a standard, using a thermocouple. As the standard, a substance is used whose specific heat remains unchanged in the region investigated. When heat is taken up by the sample in greater amount (e.g., by consumption of the heat of fusion), a distinct deviation in the temperature record

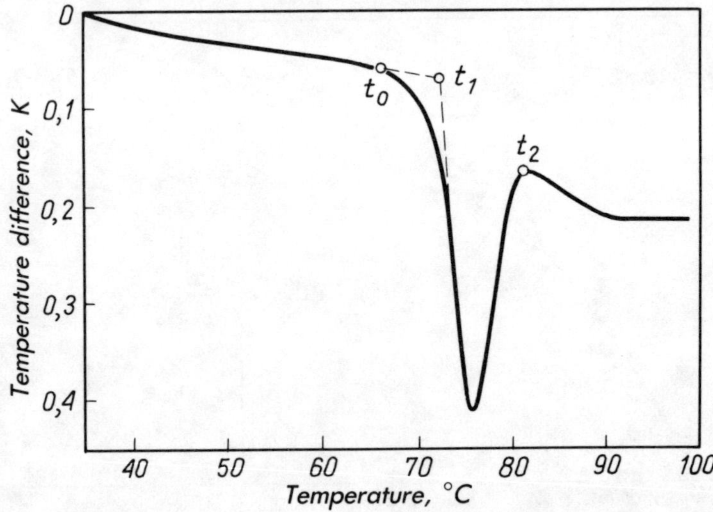

Fig. 10.20. Shrinkage temperature of calfskin, vegetable tanned. Measurement of phase transition by DTA method. t_o—initial temperature, t_1—extrapolated initial temperature, t_2—end temperature.

(Fig. 10.20) is observed. As the initial temperature t_o we consider the temperature at which the slope of the curve starts to decline rapidly. At a temperature where a minimum occurs, the difference in temperature increase is the greatest. As the onset temperature we consider the temperature at which the cross-section of two lines occurs, tangential to the curves of the temperature change. The point thus obtained is a starting moment of substantial changes. From the surface which is limited by the points t_o, extremum on the curve, and t_1, and from the weight of the sample the latent heat of fusion can be calculated.

The whole temperature run for native collagen was demonstrated by Lim and Shamos [24]. It is shown in Fig. 10.21 in normalized (percentage) form and in differential form in Fig. 10.22. A serious weight loss starts at 209°C; probably the residual part of water is liberated there, as it was supposed earlier by Okamoto and Saeki [25]. Very similar results have been obtained by the DTA method from which were determined the endothermic peak at about 209°C and another one at about 290°C (Fig. 10.23) [26]. Then, at about 300°C collagen starts to release heat, i.e., a decomposition reaction overwhelms the breaking of molecular structure by the heat supply, and some volatile destructs appear.

The Russian authors (Mikhailov, Kutyanin) explain the 209°C point in the collagen shrinkage as being a transition between amorphous and viscoelastic state, based on the properties of samples annealed at this temperature and then physically tested. According to them there is no significant difference between collagen and gelatin after such a treatment [27].

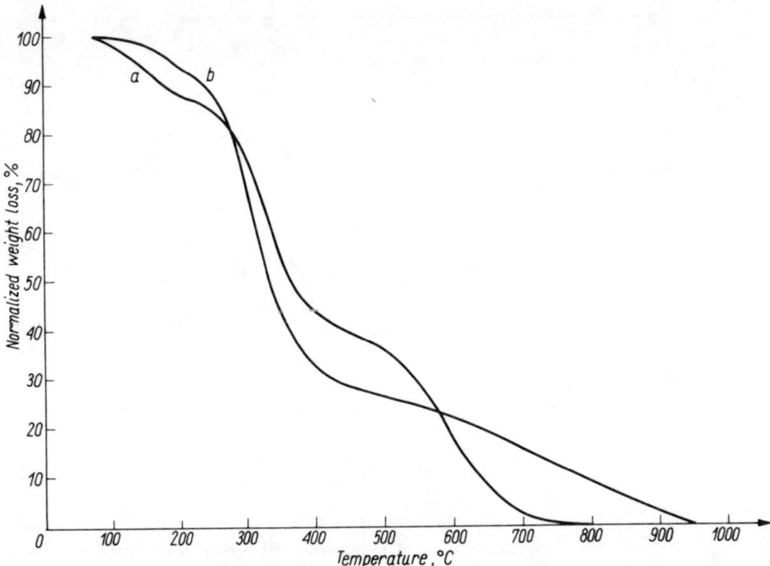

Fig. 10.21. Normalized weight loss of tendon: (a) heated at normal atmospheric pressure in air and (b) heated at about 10μ pressure of Hg as a function of temperature. Heating rate was 3°C/min. According to [24]

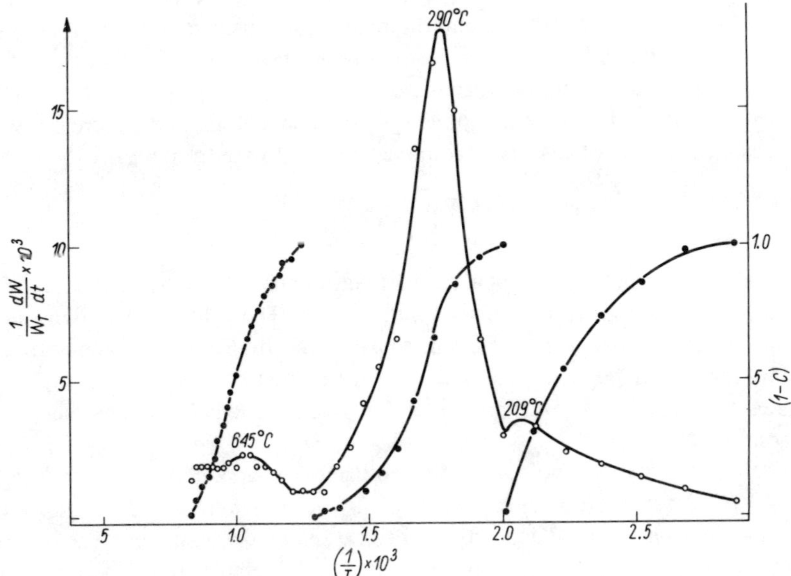

Fig. 10.22. Open circle: time derivative curve of tendon heated in a reduced pressure of 10μ Hg as a function of the inverse of absolute temperature. Solid circle: fraction of the normalized weight remaining as a function of the inverse of absolute temperature. According to [24]

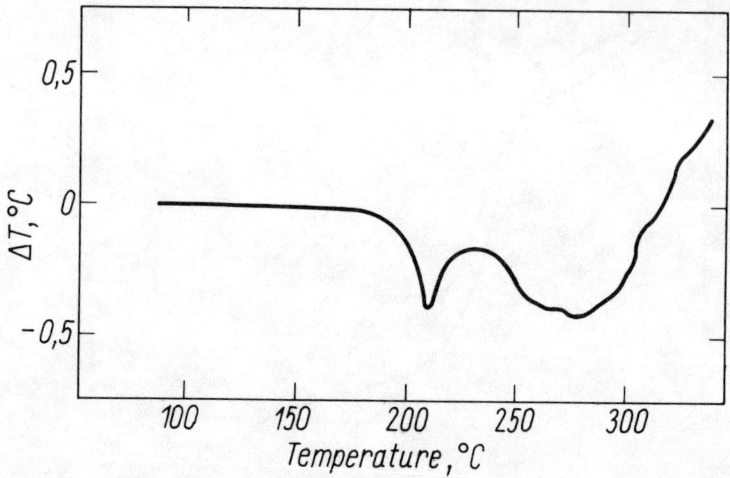

Fig. 10.23. Differential thermal analysis of anhydrous rat tail tendon. Specimen heated at 5°C/min under a nitrogen atmosphere. According to [26]

An integrated approach to the problems of collagen shrinkage and denaturation has been given recently by Harlan and Feairheller [28]. Based on experimental data collected by many authors and on theoretical considerations of Flory, they suggested that the native collagen melting point (taken as equivalent for T_S and T_D, and defined as T_m) in non-reacting solvents is depressed, as results from the theory of a crystalline polymer-solute system. The melting point decreases with increasing amounts of dissolved solute according to the formula

$$\Delta T_m = T_m^0 - T_{.m} = AT_mV_s(1 - \chi V_s)$$

in which T_m^0 is the melting point at a zero solute content, V_s is the volume fraction of the solute for a given polymer—solute system. The value of A is depending on the observed change in the crystalline polymer transition and the sign and magnitude of χ depends on the energy of interaction between the polymer and solute. The maximum T_m depression with a given solute is limited by the extent of solubility.

The intrinsic melting point of native collagen (extrapolated to 'dry' collagen) is 145°C. When the observed values of T_m are other than 145°C, then they indicate the change in collagen crystallites caused by the treatment. The authors interpreted, i.e., the data of Kawamura et al. [10] emphasizing there that the much higher melting point of chrome-tanned collagen indicates crosslinking, and the decrease of this point for vegetable-tanned collagen indicates destabilizing of the helix at low moisture content, probably by competing for hydrogen bonding sites, usually crosslinking the collagen molecule. This suggests that vegetable tannins act as solutes, competing with and blocking the effectiveness of water as solute and thereby providing certain hydrothermal stability (cf. Fig. 10.13).

The energy constant χ in the above equation may be calculated from the formula

$$B = \frac{\chi \, RT}{V_s}$$

where V_s is the molar volume of water. Thus B is an index of curvature in the figure given (Fig. 10.13). Positive B values indicate that the polymer is a poor solvent for water. In conclusion the authors suggest crosslinking reaction between collagen, and chromium or aldehyde, and different kind of reaction (helix destabilizing) for vegetable tannins.

Bowes and Taylor have investigated collagen of chrome leather dried to a constant weight [29]. They determined loss in weight, nitrogen and amino acids using collagen heated to 140, 150, and 170°C for a definite time. The losses at 140 and 150°C were about 20% of weight, at 170°C about 50%. A greater weight loss was observed in methionine, tyrosine, arginine, lysine, serine and threonine. Contrary to common belief these losses were not due to oxidation, because the heating in nitrogen decreased them very insignificantly; pH of the leathers and the kind of tanning agents used has not had an essential influence on the degree of deterioration. In the investigation in question the heating time was 20 days. Collagens of cattle hides and of sheepskins were very alike in their thermal stability in the early heating period. Starting from the second or the third day decomposition of the sheepskin collagen commenced, followed by decomposition of cattle hides (Fig. 10.24a). Water solubility of samples tested suddenly increased in that time, then it declined a little for cattle hides, whereas for sheepskins it stayed on the same level or was still increasing a little (Fig. 10.24b). According to the opinion of the quoted authors, the decomposition mechanism is like the one which occurs during aging of books (leather book binding, furniture, etc.).

Del Pezzo and Fiore [30] tested the properties of leather which after conditioning had been immersed in liquid nitrogen for a period of 1 to 100 minutes at $-200°C$. Afterwards the leather was again conditioned and its stress resistance tested. As they found, storage of the leather at so low a temperature increases a little its stress resistance with time, and decreases its elongation at break. This is best observable in waterproof leathers. Authors explain it by increase of crystallisation degree (ordering of the sample).

Shrinkage temperature as a measure of changes. Shrinkage temperature of

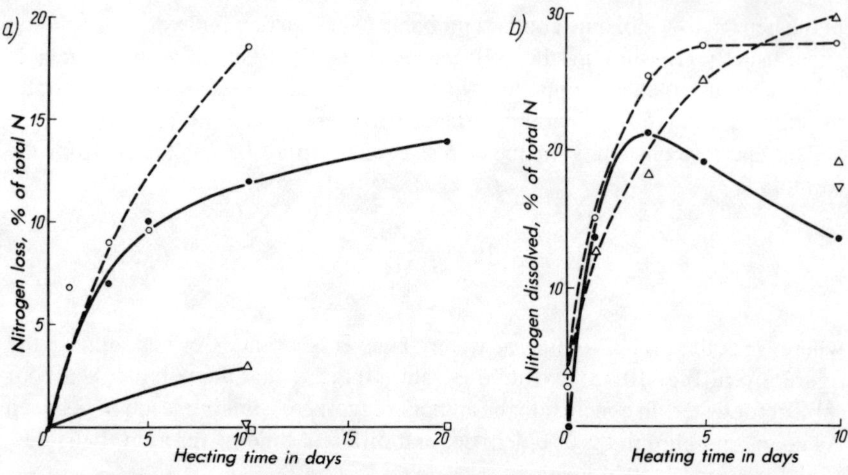

Fig. 10.24. Characteristics of dry heated chrome leathers: (a) nitrogen loss, (b) amount of water-soluble nitrogen in %. Decomposition degree is determined as nitrogen loss. Black circles are for cattle, open—for sheep
●—140°C in air Δ—150°C in air, ○—170°C in air, ▽—150°C in nitrogen, heating time, days. According to [29]

the hides and skins depend on the kind of animal and, as determined by standard method, is (in °C)

Calf	63-65	Cat	60-62	Hare	59-61
Cattle	65-67	Dog	60-62	Whale	61-63
Sheep	58-62	Horse	62-64	Fish from cold seas	33-52
Goat	64-66	Elk,		Fish from warm	
		Reindeer	60-62	seas	49-58
Deer	60-62	Rabbit	60-61		

As one may conclude, the differences between individuals are responsible for a fairly broad scatter of values.

Beamhouse operations influence the shrinkage temperature of the processed material. E.g., T_s for raw cattle hide is 66°C, after liming 56 (50), after deliming 62 (62) and after pickling 56 (60). The T_s is given in parentheses, as measured in baths applied. Such measurement has been done in order to avoid changes, being a result of the reagent washing-out. The T_s of leathers is dependent on the kind and parameters of the tanning process and, first of all, on the kind of tanning agents used. Typical values are shown in Table 10.4.

Testing of changes in soluble collagen under the influence of temperature, inorganic salts and various organic compounds is of great significance for the physical chemistry of collagen, for understanding the thermodynamic and electrochemical properties of this protein. According to the view, based on research done on collagen denaturation and renaturation, there are two steps: the first one

Table 10.4.

Effect of the kind of tanning on T_S [Ref. 16. ch. 5]

Kind of tanning	$\Delta T_s\ K$	
	Laboratory conditions	Industrial process
Basic chrome sulfate	$+30 - +35$	$+20 - +35$
Basic chrome chloride	$+17 - +35$	
Chrome two-bath	$+25 - +30$	—
Vegetable	$- 2 - +24$	$+ 5 - +20$
Formaldehyde	$+15 - +20$	$+ 3 - +10$
Quinone	$+20 - +28$	$+25$
Carbonyl sulfate	$+20 - +25$	—
Mercuric acetate	$+20 - +30$	—
Basic aluminium sulfate	$-10 - 0$	$+10 - +21$
Lignosulfuric acids	$0 - +22$	—
Polymetaphosphoric acids	$- 2 - + 2$	—
Oil tannage	$0 - + 5$	—

leads to the products easily recovering to the triple helix structure. After the second step collagen remains in an unordered state (parent gelatin). One may characterize the first step as that in which drastic change in optical rotatory power occurs as a result of perturbation in helical structure, the second one as that in which gradual disintegration of peptide chain occurs, manifesting itself in a molecular weight decrease. More exact data are available due to Mikhailov [31] and Traub and Piez [32]. The increase of thermal stability in the presence of methanol, ethanol and glycols is well known. This effect is probably due to the increase of thermal energy, which is necessary to break the interamide hydrogen bonds by a decrease of dielectric constant of the medium.

Determination of T_S is an efficient tool of testing in the leather making processes and of their results only if the determination is done in exactly the same way. Experiments in which one uses water-glycerin or water-glycol as a medium are sometimes done; however, they are of no practical use because the reference system is different, and the influence of alcohol on the collagen molecule is not the same as that of water. For thermodynamic reasons one must not accept the correctness of T_S determinations under increased pressure, because in this case the curves showing the phase limits have to be shifted.

Among the tanning agents known only chrome salts may increase T_S over $100°C$, and in these leathers just in a few (published) cases T_S has been determined. Diagnostic importance of this determination is therefore dubious.

Elastin fibers have no T_S. As it has been shown [33], the internal energy of elastin chains does not depend on their conformation. Their retractive force is directly proportional to the absolute temperature.

REFERENCES

1. Zimm, B. H., Bragg, J. K. J. chem. Phys., *31*, 526 (1959)
2. see ref. 1 ch. 2
3. Toyoda, H., Chonan, Y., Imai, T., Matsunaga, A., Kawamura, A. J. Chem. Soc. Japan *18*, 153 (1972)
4. Schubert, B. XIII IULCS Congress Vienna 1973
5. Gustavson, K. H. The Chemistry and Reactivity of Collagen, Acad. Press N.Y. 1956
6. see ref. 25 ch. 8
7. Povarnin, G. Collegium 658 (1914)
8. Flory, P. J. J. Polymer Sci., *49*, 105 (1961)
9. Gorbatshev, A. A., Cimbalenko, A. A., Shkaranda, I. T., Kotov, M. P. Izv. Vys. Uch. Zaved. Leg. Ind., 1, 51 (1971) in Russian
10. Kawamura, A., Wada, K., Wehara, K., Takata, E. XIV IULTCS Congress Barcelona 1975
11. Diebschlag, W., Nocker, W. J. Am. Leath. Chem. Assoc., *73*, 307 (1978)
12. Kanagy, J. R. J. Res. Natl. Bureau Stand., *38*, 119 (1947) and later papers in J. Am. Leath. Chem. Assoc.
13. see ref. 16 ch. 5
14. Andronikashvili, E. L., Mrevishvili, G. M., Japaridze, G. Sh., Sokadze, V. M., Kvavadze, K. A. Biopolymers *15*, 1991 (1976)
15. Sharimanov, Ju. G., Buishvili, L. L., Mrevishvili, G. M. Biofizika *24*, 606 (1979) in Russian
16. Ebert, Ch., Ebert, G., Knip, H. in Protein Crosslinking—Ed. M. Friedman Plenum Press N.Y. 1976
17. Berg, R. A., Prockop, D. J. J. Biol. Chem., *248*, 1175 (1973)
18. Traub, W. Isr. J. Chem., *12*, 435 (1974)
19. Berg, R. A., Kishida, Y., Kobayashi, Y. K., Inouye, K., Tonelli, A. E., Sakibara, S., Prockop, D. J. Biochim. Biophys. Acta *328*, 553 (1973)
20. Torchia, D. A., Lyerla, J. R. Jr., Quattrone, A. J. Biochemistry *14*, 887 (1975)
21. Burjanadze, T. V. Biopolymers *18*, 931 (1975)
22. Finch, A., Gardner, P. J., Ledward, D. A., Menashi, S. Biochim. Biophys. Acta *365*, 400 (1974)
23. Hofmann, H., Fietzek, P. P., Kühn, K. FEBS Letters *89*, 279 (1978)
24. Lim, J. J., Shamos, M. H. Biopolymers *13*, 1791 (1974)
25. Okamoto, Y., Saeki, K. Kolloid—Z., *194*, 124 (1964)
26. Yannas, I. V. J. Macromol. Sci. Revs. Marcromol. Chem., C7 1 (1972)
27. Aiskina, A. Ja., Golubiatnikova, A. T., Kutianin, G. I. Kozh. Obuv. Prom., *18*, 12, 30 (1976) in Russian
28. Harlan, J. W., Feairheller, S. H. in Protein Crosslinking—see ref. 16 this ch.
29. Bowes, J. H., Taylor, J. E. J. Am. Leath. Chem. Assoc., *66*, 96 (1971)
30. Del Pezzo, L., Fiore, U. Cuoio Pelli Mat. Conc., *46*, 338 1970 in Italian
31. see ref. 5 ch. 2.
32. Traub, W., Piez, K. A. Adv. Prot. Chem., *25*, 243 (1971)
33. Hoeve, C. A., Flory, P. J. J. Am. Chem. Soc., *80*, 6523 (1958)

11.

HIDE AS AN ION EXCHANGER AND AS A MOLECULAR SIEVE

Under certain conditions hide behaves like an ion exchanger or a molecular sieve. Crosslinked network of collagen fibers is a polyelectrolyte containing different functional groups, activating at various phases of the leather making process due to the changes in pH and in ionic strength of the medium. As a result of swelling a certain number of water molecules is assembled around the functional groups, and bound to them by various forces. Thus the first contact of a substance approaching collagen molecule is a contact with immobilised water molecules surrounding the collagen functional groups, like it is in ion exchange resins and molecular sieves. The same, yet to a greater extent, is true of the contact between collagen and glycosaminoglycans accompanying collagen. The ability of binding water by hialuronic acid is much greater than that of collagen, and its chemical reactivity is much smaller.

Making a closer approach to the ion-exchange properties of hide in one case or molecular sieve in another, one may point out several analogies. Ionic resins (exchangers) are crosslinked, insoluble polymers, having properties of acids, alkalis or salts. They contain acid or alkaline groups, strongly bonded to the macromolecular backbone, and movable counterions which may be exchanged. The backbones of these resins are obtained by polymerisation or polycondensation. Polymers are most frequently obtained by crosslinking of the links (mers) of divinylbenzene: polycondensates, by crosslinking with formaldehyde.

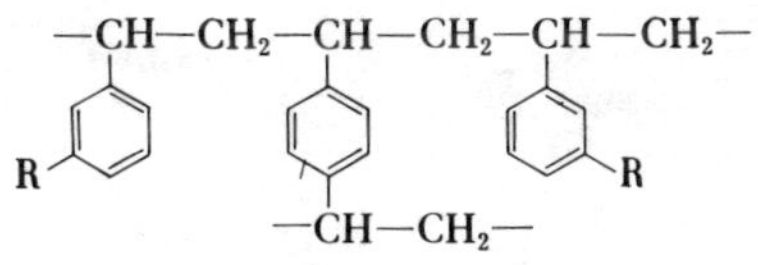

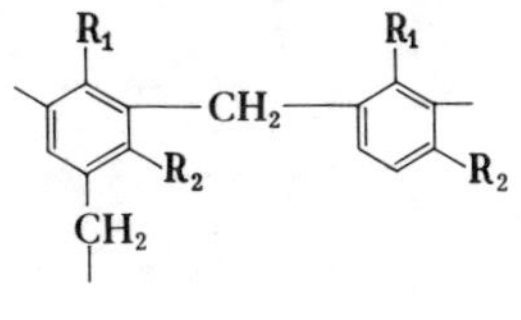

265

As groups R_1 and R_2 can be used acid groups: sulfuric, carboxylic, phosphoric or alkaline: amino or hydroxymethyl groups. The carboxyls, amines, amides and hydroxyls may occur in both exchange resins and collagen; in collagen, however, their number is much smaller.

The resins may be weakly or strongly acidic or alkaline. Synthetic possibilities are very extensive. Often, especially in chemical analysis, derivatives of natural products are used. Among them carboxymethylcellulose (CM), dimethylaminocellulose (DMAE) or derivatives with two functional groups are mostly used. In Fig. 11.1 one may see a schematic comparison of the exchange resins and

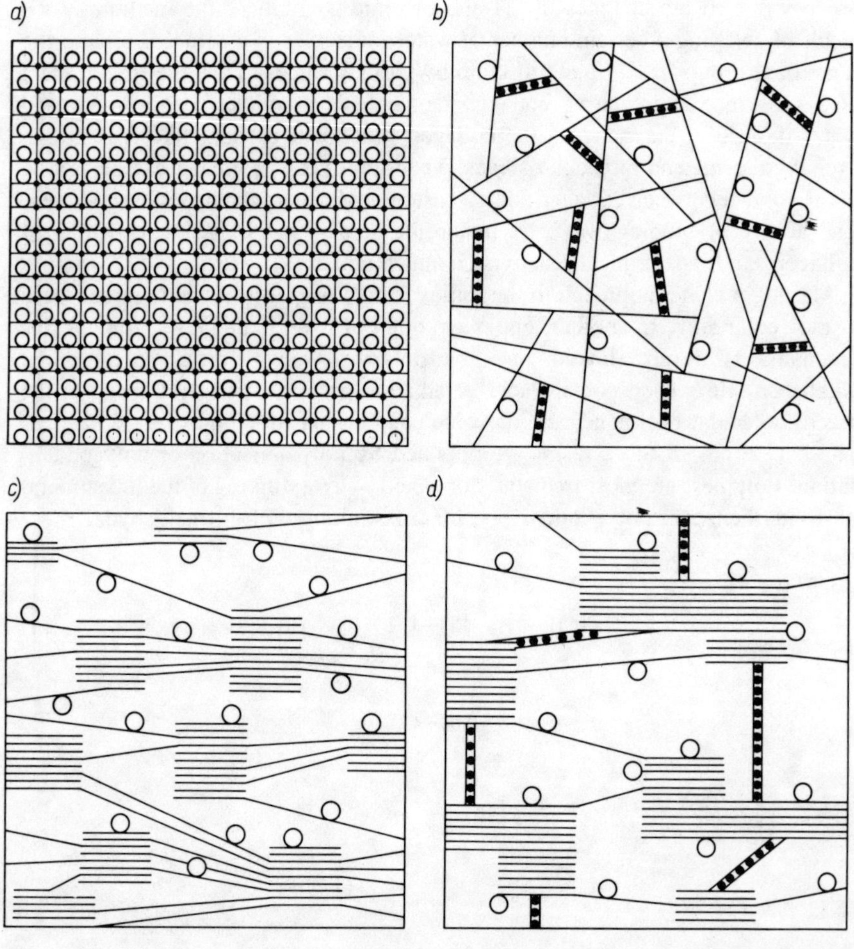

Fig. 11.1. Schematic comparison of microstructures of exchange resins and crosslinked gels: (a) synthetic exchange resin, (b) ion exchange gel, (c) fibrous ion exchange cellulose, (d) ion exchange cellulose, granulated. Full circles and lines connecting are crosslinking bonds, circlesion binding phases.

crosslinked gels serving as molecular sieves.

Ion exchanging resins do not dissolve in water, but they swell in it; macro-molecular network is porous and water penetrates into it. Solvation of functional groups (e.g., sulfonic) occurs along with their total or partial dissociation, which depends on the kind of the groups. Sulfonic groups do ionize completely, whereas carboxylic, to a very small extent. One may assume an existence of solution inside of the wet ionite granule, where the functional groups are the dissolved component. A dissociated sulfonic group consists of an immobilized anion and a free cation. The number of functional groups contained in a defined volume of an ionite determines its exchange capacity. Most frequently one defines the exchange capacity as the number of gram-equivalents per 1 g of dry ionite. The swelling degree of the ionite depends on numerous factors, like, e.g., on the number and kind of functional groups and on the degree of crosslinking. Degree of ionisation of functional groups is depending on the amount of water in the ionite granule according to Ostwald's dilution law.

A similarity between the behavior of ionites and of collagen has been noticed for the first time by Gustavson 1924, who made use of this observation in his investigation. Later on Amberlite IRC-50, partly neutralized by sodium cations, has been found as the most appropriate collagen model. One may use this resin when investigating the reaction between collagen and chrome $^{3+}$ sulfatocom-plexes at various concentrations. This view has been questioned later, although the conclusion concerns the results obtained but not the method itself.

Another group of substances, which are an analytical supplement of the ion exchange resins, are molecular sieves. They do not have active groups to bind dissolved alkalis or acids. Molecular sieves are systems with pores usually about 10^{-8} cm in diameter. This size is comparable with the size of small molecules. Gel filtration method is based on making the solution of a mixture of components to pass through a space (usually column) where a bed of soaked gel is previously placed.

As a gel-forming substance initially starch has been used, later crosslinked dextran (Sephadex), then acrylonitrile gels and agarose gels. Agarose is a poly-saccharide of sea algae, dextran—a polysaccharide produced by glucose pol-ymerases in some bacteria. The 'reverse sieves' principle has been introduced in 1950: Synge and Tiselius have found that bigger molecules are passing quicker through the porous sieve than do the small ones [1]. This is due to the ease of penetration of small ions into the pores of the resin to form more or less stable bonds there. The action of 'reverse sieve' is shown in Fig. 11.2. Gel components, considered as solid phase, slow down the solution flow. A gel-forming substance occupies the greater part of the immediate surrounding of crosslinking bonds. A greater molecule cannot penetrate these regions, whereas the small ones may approach the crosslinking bonds. The speed at which the solute molecules pass the bed is proportional to the average time which they have to stay in the liquid between the gel grains. This average time is given by partition coefficient (in

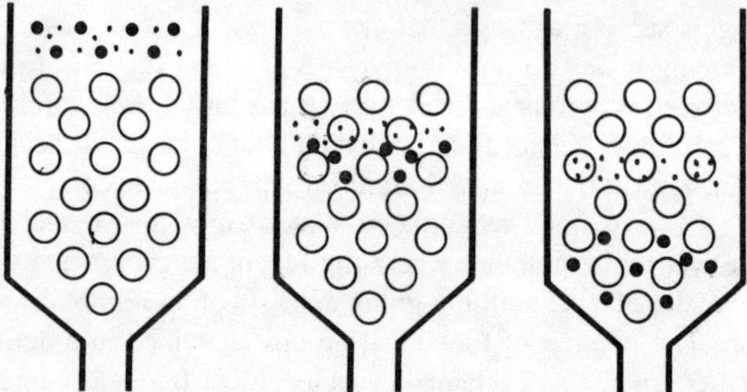

Fig. 11.2. Operating principle of 'reverse sieve.' Separation of smaller and larger molecules. Small molecules penetrate the swollen gel particles and are arrested there (temporarily).

gel chromatograph *sensu stricto* the partition coefficient is depending exclusively on steric effects) of the substance between gel phase and liquid phase. Swelling agent is usually the same solute which brings the solute particles through the gel. In a typical case the gel-forming substance has a great affinity to solvent; it is, however, crosslinked and thus insoluble.

Some substances leave the bed later than it might be expected from their molecular weight. Aromatic compounds are in this group. The cases are known as well, where the homologues having longer chain do have longer effluent time (greater effluent volume) than substances having shorter chain. Aliphatic alcohols belong to this group.

Elastic properties do influence the gel properties. Osmotic pressure in gel is of an order that may change the behavior of proteins. In the common chromatographic gels this pressure is much lower than in the ion exchange resins, where it reaches hundreds of atmospheres (tens of MPa).

Electric charge of the solute does influence its partition on gel. Results of the partition obtained at too small ionic strength are abnormal. Not infrequently the smaller effluent volume of the substance from gel due to a decreased ionic strength has been observed. The molecules of a complex substance pass through the bed only when the ionic strength of the solution is appropriate. Diffusion of charged molecules into pores is impossible in distilled water; however its possibility increases when the buffer concentration increases. Another point is dielectric properties of the medium: gel properties are different than those of the solution. This effect is considered to be relatively independent of ionic strength, when ions having small molecular weight are added to the system, they are dispersed rather regularly in the gel and in the solution as well, and in this way the properties of the system remain unchanged. However, if a charged small molecule, in the presence of ions having opposite charge, comes into contact with gel, which is impermeable for macroion, the small ions are penetrating into the gel structure. Between gel and solution a Donnan potential arises, which is a counterweight for the small ion penetration. This type of Donnan effect may not occur, if the ions in solution do have a differentiated penetration tendency. In case of gels, prepared according to modern technique, having no or almost no charged groups the Donnan effect seems to be a reason for deviation from the chromatographic process, if the ionic strength of the medium is low. The Donnan effect decreases rapidly when ionic strength increases. It is somewhat different in gels than in ion exchange resins, in which it is due to the bound, charged groups in gels and mobile counterions.

Affinity of the majority of the molecules passing through the bed to the substances forming it is markedly lower than their affinity to the solvent. Gel is shielded due to hydration, which is the probable reason for the similarity in behavior of various gels of different structure and chemical composition. This explains why the significant differences in the structure of dissolved substances makes small difference in their chromatographic characteristics: affinity of the solute to water surrounding the gel particles seems to be similar to the affinity of the molecules to unbound water. All these factors responsible for the characteristics of the substance passing through the gel may be applied to the leather during processing. Most frequently we are interested in regular dispersion of small molecules in the hide. This may be done, e.g., by equilibration of the charges in the hide and in the float. Quick binding of small molecules is of interest in surface dyeing.

For tanning agents and dyestuffs the hide is an ion exchange resin, which may work as cation or anion-exchanger, according to its charge. In this case as active groups the carboxyls or amino groups may be considered. Binding of vegetable tannins by hide powder, as it is usually done in classical methods of tannery analysis—is an example, how collagen may be an exchanger for tannins, and molecular sieve—for nontannins.

According to recent investigations [2] the components of fats penetrate into leather to various depth. Poré was able to demonstrate using chromatographic technique that the particular components of fatliquors are fixed in various layers of leather. Thus if an extraction of fat from various layers is carried out, a different composition of the extract results, entirely different from that of the oil applied, although the sum is equivalent to the oil composition used. A chromatography process in leather is also occurring during fatliquoring, in which leather is a carrier.

For fillers, which penetrate the hide without being bound, collagen is a swollen gel into which the molecules enter to be kept in it during drying and shrinkage constricted interfibrillarly, like the vegetable tannins component (cf. Fig. 14.1).

The author could not find whether such investigations on the molecular weight of the fractions of non-tannins have ever been done. Such investigations could confirm or deny this view. Very few papers deal with thermodynamics of single operations of the tannery process: Generally speaking, the rules of data transformation described below can give a general impression about how it is done. The change in the level of free energy ΔF is given by equation:

$$\Delta F = RT \ln \frac{a_2}{a_1}$$

which is fulfilled when a mole of substance at a constant temperature changes its activity a_1 to a_2. If water from the ground state is brought to the state of hydration water of protein molecule, the measure of activity of hydration water against bulk water is $\frac{P}{P_0}$ where P_0 is the vapor pressure over bulk water, and P is the pressure of the vapor, being in equilibrium with the absorbed water (which is always lower). So one may write:

$$\Delta F = RT \ln \frac{P}{P_0}$$

The heat of hydration of protein or another polymer can be calculated from the Clausius-Clapeyron's equation:

$$\frac{d(\Delta H)}{dn} = \frac{RT_1 T_2}{T_2 - T_1} \ln \frac{x_1}{x_2}$$

where x_1 and x_2 are the relative vapor pressures, given at the temperatures T_1 and T_2 the same degree of protein hydration, n is the amount of bound water expressed in moles. The easiest way of determination of $\frac{d(\Delta H)}{dn}$ is to measure the area under the curve, showing this result as a function of n. According to this method Bull has given the values of ΔF_{25} and ΔH for some biopolymers [3]

Table 11.1

ΔF_{25} and ΔH values for some proteins (ref [3] ch. 6)

Polymer	*ΔF_{25}* *25°C*	*$\Delta H/100$ g* *of dry protein*
Nylon, unstretched	254	510
Silk	658	1270
Wool	975	1950
Elastin	1000	2250
Collagen	1650	3640
Gelatin	1532	3800
Albumin of hen egg, native	956	1350
Serum albumin	1034	1450

In this specification we see the strikingly high values of ΔF and ΔH for collagen and gelatin. Using the same formula Mikhailov and Mikaelyan [4] have calculated the dynamic parameters of reaction: pelt-tanning agent (see p. 319)

$$\Delta\mu = RT \ln \frac{D_B}{D_L}$$

where D_B is the equilibrium amount of tanning agent, bound to 1 g of collagen, and D_L—the amount of tanning agent in solution, g/ml. Authors have given data in kcal/mole of tanning agent and, calculating on hide powder, have obtained thermodynamical conditions of reaction between tannins and pelt, leading to leathers:

	$\Delta\mu$		ΔH	
	kcal	kJ	kcal	kJ
for HCHO tannage	-200	-436	$+1000$	$+4180$
for quebracho tannage	-240	-1003	$+150$	$+627$

these values are of comparable order but lower than the values of protein hydration given in Bull's handbook.

The direct calorimetric measurements done in the U.S.S.R. show very insignificant evolution of heat, which is probably due to the parallel running neutralization. The extent of this effect, however, is in accordance with tanning intensity: thermal effect of the covalent binding of formaldehyde is the greatest, of the vegetable tannin, the smallest. Similar experiments on the affinity of tanning agents to hide substance were done a long time ago (1935) by Grassman [5]. As a measure of affinity of tanning agents to pelt he assumed the Freundlich isoterm

$$\frac{x}{m} = ac^{\frac{1}{n}}$$

where c is concentration of the substance in solution, m—the amount of adsorbing substance, x—the amount of substance adsorbed, a and n are constants. In handbooks of physical chemistry it is usually assumed $\frac{x}{m} = y$, or the amount of the adsorbed substance per unit weight of the adsorbant. According to the paper of Grassmann and Bender mentioned, the Freundlich's isoterm describes well the adsorption process (Author's note: there are several new adsorption curves, which may better approach the real process—but insofar as he knows, nobody has investigated it in recent times, perhaps because now we consider this process as very complicated). This curve is an approach to the adsorption from the solution on solid state used for describing the phenomena of ion exchange chromatography.

According to Gustavson tanning is a diffusion process. It has to be emphasised that this may be true only from the point of view of the reaction kinetics. In fact this process is much more complicated. Stather and Laufmann [6], testing the cuts of pelt in the course of its vegetable tanning, found that $B = k \sqrt{t}$, or binding of vegetable tannins is proportional to the square root of time. A better approach as recognized by the same authors, gives the dependence $B = k\log 2t$. This makes the results close to the volume of gel column effluent and molecular weight of the dissolved substance. These authors have used in their investigations tanning agents of molecular weight 1000-2000. In order to compare the result, the distribution of tannins according to their molecular weight and penetration should be determined.

Decrease of the tannins diffusion rate with increase of their way of travel has been discussed for a long time. The following dependence has been accepted since Gustavson has proposed it

$$\Delta m = k_{FT}\, m_{FT} + k_{ST}\, m_{ST}$$

otherwise as he assumed, the apparent weight increase is a function of tannin bound by collagen (m_{FT}) and the amount of tannin in a solution among fibers (m_{ST}), whereas the constant k depends on the kind of tannin used and is determined experimentally. In Fig. 11.3 one may see how the tanning goes when

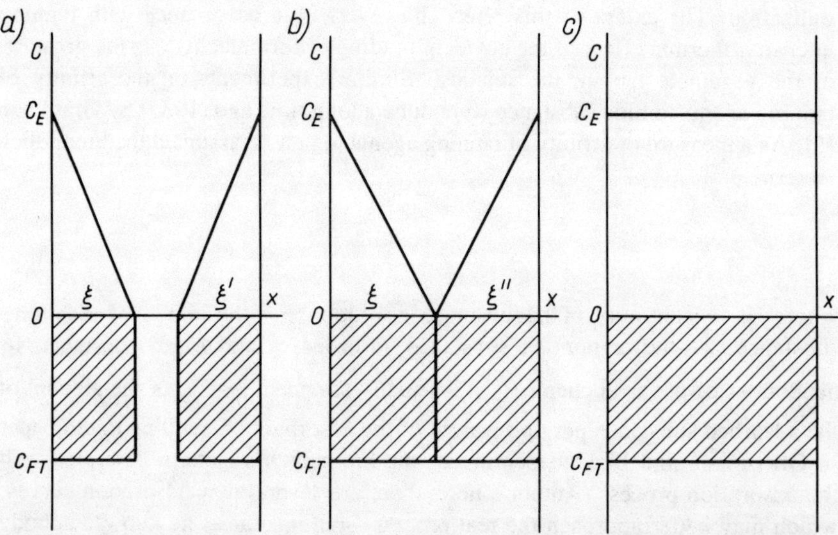

Fig. 11.3. Penetration of vegetable tanning agent into the pelt according to Gustavson (a) during tannage, (b) end of 1st stage of tanning, (c) end of 2nd tanning stage. On the x-axis the distances, measured parallel to the surface, are laid off; OC means concentration, ξ distance between the tanning front and the surface, ξ' distance from grain, ξ'' distance from flesh side. Usually $\xi' < \xi''$.

considered as diffusion of vegtan molecules into the pelt. In the Figure the cuts are demonstrated as they are obtained at various tanning stages.

The fronts are nearing each other until they meet. This is called the end of phase I of tanning (Fig. b). At this very moment the tannin concentration in the solution inside leather drops down from about C_E on the surface to zero at the point where the fronts meet. Then diffusion still occurs until the moment in which roughly the same concentration of tannin in the cut may be observed (the same or very close result one may see in mineral tannage, although it is much quicker).

Diffusion of dyestuffs may be studied in a similar way to the one mentioned above. Otto obtained similar results, investigating penetration of 3-azobenzanthrone into the leather [7]. He controlled the reaction rate changing, e.g., the number of sulfonic groups in the dyestuff. Maybe his efforts to explain the reaction are not valid now, but the deep analogy between tanning, dyeing, and gel chromatography seems obvious.

It can be anticipated that the hide is too complicated and thus not suitable for use in analysis, first because of the difficulties with its purification and homogenization. However, this is not an argument against considering hide or its collagen as a gel containing a certain amount of functional groups. It may help to understand some of its properties. One must not oversimplify the characteristics of hide by saying that, e.g., 'it behaves like an ionite resin.' In another system it may behave like a molecular sieve, and in still another—like chemically inactive bulk polymer.

As it is known from experiment, hide in the soaking process imbibes and binds the more water, the lower the process temperature is. The simplest explanation for it is the increase of the mobility of water molecules, i.e., its enthropy decrease with temperature. A substance or action increasing the distances among molecules has such an effect on unordered system. Increase of ordering, however, is connected with the entropy decrease. There is a temperature point (T_D or T_S) at which the long-range order structure sharply decreases. The problems of water structure have been discussed in Chapters 5.1 and 5.2, as related to hide-water system, in which the structure-making function belongs to the functional groups of hide proteins, and to structure-making ions contained in the system. This effect may be followed on X-ray diffraction patterns, where the band, corresponding to the long-range interactions (7 Å) disappears at a higher temperature. This can also be recognized from the specific heat of the solutions investigated. Its value is higher for the solution containing the structure-making ions, and lower for those containing the 'structure breakers.'

REFERENCES

1. Synge, R. M. L., Tiselius, A. Biochem. J., *46*, 41 (1950)
2. Poré, J. Rev. Techn. Ind. Cuir *71*, 201 (1979)
3. see ref. 3 ch. 6
4. Mikhailov, A. N., Mikelian, I. I. Leder *15*, 77 (1964)
5. Grassmann, E., Bender, R. Collegium *787*, 521 (1935)
ʊ. Stather, F., Laufmann, K. Collegium *786*, 470 (1935)
7. Otto, G. Leder *4*, 1 (1953)

12.

PREPARATION OF COLLAGEN FOR REACTION WITH TANNING AGENTS (BEAMHOUSE OPERATIONS)

'Leather is made in the beamhouse'—this old English saying is at least right in that the operations done in the beamhouse are decisive for the physical form of the fiber network, and thus preparing the raw hide for tanning (section 14.1). Further operations and processes may only modify and supplement the changes caused, and achievements of modern chemistry, particularly in the field of finishing chemicals, may correct the errors of nature itself and of the tanner to a considerable extent.

A typical tanning process starts in the beamhouse. Classical operations of the beamhouse are the following:
(1) Soaking and mechanical removal of scud from hides.
(2) Fleshing, splitting, smoothening of the hides.
(3) Loosening of the hair follicle, of hide fibers and deterioration of epidermis ('liming').
(4) Removal of lime (deliming).
(5) Enzymatic loosening of hide fibers (bating).
(6) Pickling.

The purpose of these operations is to increase the amount of water in the hide to the amount close to that of the 'living' hide, remove foreign bodies and loosen the structure. This loosening makes it easier for the tanning agents, fats, dyestuffs and other substances to penetrate into the hide. In the beamhouse the non-collagenous proteins are removed from the hide, so is its epidermis, hair and globular proteins, melamines, components of cell walls, while the collagen fiber skeleton remains practically untouched.

A present trend in leather making is to combine the processes into single operations, like, e.g., tanning in pickling bath, or enzymatic unhairing, without (or with some) liming. In our treatment this kind of processing can hardly be discussed, as the unit operations are already complicated enough to deserve a separate coverage. In authors' opinions the reader may observe these 'compact' processes by himself and draw empirical conclusions. Maybe the coming sciences of the future will be able to answer some arising questions. One argument here is beyond discussion: the tanner should give some confidence to the scientist, and conversely, the scientist to the tanner's experience and observations.

12.1. Soaking

During soaking the bacteria present on the surface of preserved hide, frequently as spores, are growing very fast. The simplest way to break this growth is to change water after the first few hours of soaking, and/or to add some bactericides not detrimental to the hide. Higher water temperature makes swelling easier, but it makes bacterial growth quicker as well. To chose the proper conditions is the matter of tanner's experience, the condition of hides, temperature of the place and of water used, antiseptics, applied economic calculations, water supply, and of the possibility of the re-use of waste water.

As the parameters of the process there may be considered: float-ratio, and the influence of work done on the material (drum revolving). A generally accepted rule is to use work in the tannery operations, in the form of drum or paddle revolving. This makes the operations shorter by easing penetration of the solutions (floats) to the inside of hides due to their bending and straightening, and also to quicker exchange of the thin solution layer, adhering to the hide, from which the solutes already have diffused to the hide.

The soaking of cured hides has for aim the removal of salt and part of soluble proteins and imparting an adequate plumpness to the hide. This is the result of water returning into the interfibrillar spaces due to which the fibers may slip one against another. Thereby breaking or mechanical damage is avoided in subsequent mechanical operations. Principally the original condition of natural softness is to be restored. However, it is not always reached for the reason of time-consumption and bacterial growth, which may become dangerous. In practice the achieved softness is now still checked by feel only.

Another purpose of soaking is to remove the solids, like sand, stones, parasites and remainings of blood, urine, dung, etc. This needs no justification. Besides, the possibility to damage the machines in later operations is thereby avoided. Impurities make the weight of raw hides false. All dissolved proteins are a very good nutritive medium for the bacteria, molds, etc., The excess of salt, increasing the ionic strength of water adherent to the hides is undesirable as well, because it may change the way of enzyme action in further operations. The dissolved proteins, not washed out, make collagen fibers stick together in the further operations, and thereby inaccessible to the action of the chemicals applied.

In order to prevent bacterial flora growth, soaking may best be done in a way which prevents the bacteria from passing from the first growth phase (lag phase) to the next one. Usually this is done by changing out water and by the use of bactericides or bacteriostatics, mainly derivatives of naphthalene-sulfonic acids and of chlorinated phenols, sulfonated esters of fatty acids, sulfates of fatty alcohols and their mixtures.

Soaking is more easily done for salted hides than for dried ones, where the sticking together of collagen fibers with the matrix is considerable. In this case alkalization of both kinds of hide may be advantageous, as it causes the matrix to form a dispersion. The technologists use numerous ways to facilitate and to accelerate soaking, preferentially by action of the substances that decrease surface tension or by enzyme action.

During soaking of salted hides in the first phase near to the hide and in the hide itself a salt solution is formed, which makes possible a washing-out of soluble proteins. The formerly expressed doubts as to whether or not collagen is lost are now explained: according to our knowledge a certain amount of collagen which is soluble in salt solution is removed in soaking or in the following operations, because it is not a mature, crosslinked collagen, which is more resistant to chemicals. Other salts have been tested for washing-out of proteins, e.g., $CaCl_2$ or $MgSO_4$, in order to check how they influence the process, but they have not been used in practice.

Temperature of the water used for soaking is of influence on the process duration and on bacterial growth. Cold, tap or well water is harder than surface water (rivers, ponds) in certain seasons. This hardness is usually not of negative influence. Well or tap water contains less bacteria, however, and this speaks for its use.

The amount of water used for soaking ranges from about 3:1 to 5:1, as related to hides, when soaking is done in drums, paddle-vats or pits without moving, and 6:1 to 7:1 for dried skins. If the amount of water is too small, the dissolved protein concentration becomes too great, the bacteriostatic and surfactant agents unequally dispersed, and mixing of the contents of the container is more difficult. If the amount of water is too great, the removal of protein is unsatisfactory due to insufficient amount of salt in solution (it shall be about 4-6% in the first stage of soaking), action of bacteriostatics is too weak and the operation too expensive. The solubility of globular proteins during soaking is only partial. Albumins dissolve in water, globulins only in the solutions of salt having lyotropic properties.

The speed of drumming is 4-8 r.p.m. and water penetrates into the tissue, which makes easier the removal of impurities, softening of fibers and action of the compounds used. To control the soaking process properly, it would be recommended to observe the following factors:

—The pH of the solution, easing the swelling of collagen fibers, and so the diffusion of the bath components into hide. The lowest degree of swelling is close to the isoelectric point of hide, i.e., at pH of about 7.

—Presence of salts (including NaCl), contained in the soaking water, as it influences the water structure.

—Surface tension at the hide/water interface, which is mainly depending on the fat content of the raw hide, and on the presence of surfactants in the solution. The highly complicated system made in the soaking operation may not be theoretically described. Water uptake by hide/skin during soaking is inversely proportional to their fat content. This is of particular importance in pigskin processing.

12.2. Fleshing

The fleshing operation does not only consist in the removal of unnecessary fragments of tissue but also excess of water used for soaking and containing salt, soluble proteins, impurities and bacteria are then removed. So this is the operation which may be considered as squeezing away solution from solid.

Fleshing of raw hides done after or during soaking, i.e., after the first soaking, is necessary because:

(1) It imparts to the hides treated fairly uniform thickness by removing local thicker places.

(2) The muscles or fat adhering to the hide decrease locally the results of soaking and liming, and weaken the action of bacteriocides.

(3) It prevents formation of calcium soaps from calcium hydroxyde in the lime float and from fat, particularly it because if improper preservation or too long conservation, free fatty acids have been formed.

12.3. Loosening (opening up of the structure). Depilation

Loosening may be considered as an extension of soaking; its purpose is to separate two structural proteins: keratin and collagen. In the chapter in which the keratin structure was discussed special attention was drawn to its typical feature, i.e., disulfide bonds. At this point both principally differ from each other, and one may obtain both proteins in utilizable form, if separation process is carried out properly. This concerns mainly those kinds of skin in which keratin occurs as wool or bristle. However, keratin contained in the epidermis as well as that situated in the hair follicle is not recovered. Differences between keratin in the form of hair and that being the structural material of epidermis have been discussed in section 3.1.

Methods of hair removing can be divided into two groups:

(1) Methods based on destruction or modification of the epidermis tissue surrounding the hair, so that this may be 'loosened' and mechanically removed.

(2) Methods, in which hair itself is attacked and its structure destroyed. This more drastic method is in practice connected with the use of alkalis (calcium or sodium hydroxide) and of sulfide. This group of the methods includes 'liming' in solution or lime liquor and 'painting' which consists in spreading a layer of paste ('paint') over the flesh side of skin. As a result of such a procedure only the lower part of the hair and the bulb are destroyed, while the upper part may be used after mechanical separation.

The result of both operations of hair removal is shown according to Thorstensen

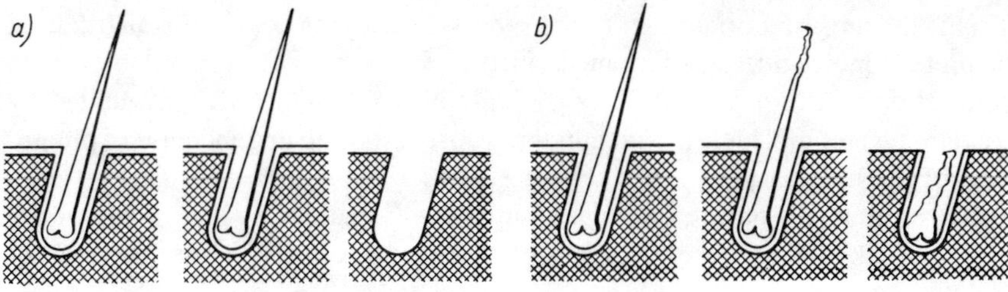

Fig. 12.1. Differences in unhairing: (a) mild, (b) drastic (after Thorstensen). [1]

in Fig. 12.1. Figure 12.1a shows the result of action of unhairing agents used in a mild medium. In this case soft epidermis keratin is removed, and the hair slips out from loose follicle, which remains clean and the surface of dermis as well. Hair is undamaged and may be utilized. In Fig. 12.1b the result of the action of a potent unhairing agent is visible. The hair is destroyed, softened and partly deteriorated. Bulb and epidermal tissue of hair follicle remain in the skin. Both actions mentioned are the extremal ones; in practice the result is always depending on numerous factors.

The choice of the way of removing keratin (unhairing) decides about the necessity or otherwise of subsequent operations of the beamhouse. The following ways of separation of collagen from the remaining hide components are known:

(1) Action of calcium hydroxide without addition of other substances (white lime liquor), or with an addition of some amount of sulfides (sharpened lime liquor). Operation may be done in a bath or by painting.

(2) Action of other alkalis (ammonia, sodium hydroxide) generally not used in the industry with the exception of sodium carbonate used for removing bristles from pigskins (in German Democratic Republic).

(3) Action of calcium hydroxide along with aliphatic amines.

(4) Use of lyotropic substances (not applied in industry).

(5) Thermal action ('scalding') used for removing bristles.

(6) Action of oxidizing agents.

(7) Action of bacteria (sweating).

(8) Action of enzymes.

Liming and painting are the oldest, classical procedures. They are based in principle on action of hydroxylic groups, in the presence of cations, on soaked hide. The reaction between metal hydroxylates and functional groups of proteins may be simply expressed by the formula:

$$\begin{array}{c}-COO^{\ominus}\\-NH_3^{\oplus}\end{array} + MeOH \longrightarrow \begin{array}{c}-COO^{\ominus}\ Me^{\oplus}\\-NH_2\end{array} + H_2O$$

This is the most important part of the process: change in the charge distribution in protein molecule causes changes in the system of hydrogen bonds due to connection of cation to carboxyls by ionic bonds. Cations are surrounded by solvation water molecules, which in turn causes swelling and increase of plumpness of the hide tissue.

Blazej et al. [2] basing on polarographic investigations of lime liquors have proposed the following mechanism of cystine hydrolysis in the presence of NaOH:

$$
\overset{|}{C}H-CH_2-S-S-CH_2-\overset{|}{C}H \longrightarrow
$$
$$
\quad \overset{\curvearrowleft}{OH^{\ominus}}
$$
$$
\longrightarrow \overset{|}{C}H-CH_2-SOH \;+\; {}^{\ominus}S-CH_2-\overset{|}{C}H
$$

According to this view a hydroxyl, having nucleophilic properties induces a bimolecular substitutional reaction S_{N2}. Sodium salt of sulfuric acid thus formed is unstable and a disproportion reaction occurs:

$$
2R-SONa \rightarrow R-SNa + R-SO_2Na
$$
$$
3R-SONa \rightarrow R-SO_3Na + 2R-SNa
$$

and in the presence of sulfides a competitive reaction is:

$$
\overset{|}{C}H-CH_2-SOH \longrightarrow \overset{|}{C}H-C\overset{\displaystyle O}{\underset{\displaystyle H}{\big<}} + H_2S
$$

If pH is > 11, a nucleophilic substitution according to S_{N2} mechanism again takes place:

$$
\overset{|}{C}H-CH_2-S-S-CH_2-\overset{|}{C}H \longrightarrow
$$
$$
\quad S^{2\ominus}
$$
$$
\longrightarrow \overset{|}{C}H-CH_2-S-S^{\ominus} \;+\; S^{\ominus}CH_2-\overset{|}{C}H \longrightarrow
$$
$$
\longrightarrow \overset{|}{C}H-CH_2-S-S^{\ominus} + S^{2\ominus} \longrightarrow \overset{|}{C}H-CH_2S^{\ominus} + S_2^{2\ominus}
$$

If the hides are treated with calcium hydroxide several hours before proper unhairing, the process is delayed or goes wrong. This is a result of the removal of one sulfur atom from disulfide bridge. A very stable monosulfide bond is then formed, as in lanthionine. This reaction has an unknown mechanism. According to Blazej (loc. cit.) two parallel reactions may take place:

$$\overset{|}{CH}-CH_2-S-S-CH_2-\overset{|}{CH}\underset{\diagdown}{\diagup}\begin{array}{l} \overset{|}{CH}-CH_2-S-CH_2-\overset{|}{CH} + S \\[2ex] \overset{|}{C}=CH_2 + CH_2=\overset{|}{C} + H_2S + S \end{array}$$

lantionin is however unstable in alkaline medium. Perhaps it is due to the formation of lantionine sulfoxide as a result of autoxidation. However, this compound has not been found in the medium. Splitting of disulfide bonds when sulfur is bound to peptides occurs at pH 10.6. A transition of disulfides into monosulfides is unknown in organic chemistry; the authors however, do believe it to be an analogue of the formation of cysteine thioether and acrylic acid:

$$\begin{array}{l}NH_2 \\ | \\ CH-CH_2-SH \\ | \\ COOH\end{array} + CH_2=CH-C\underset{OH}{\overset{O}{\diagup}} \longrightarrow \begin{array}{l}NH_2 \\ | \\ CH-CH_2-S-CH_2-CH_2-C\underset{OH}{\overset{O}{\diagup}} \\ | \\ COOH\end{array}$$

Similar to keratin is the behavior of globular proteins having disulfide bonds. Hydrolysis proceeds slowly but more easily if pH of the medium is higher. Penetration of calcium hydroxide into the hide occurs within 12 hours. Binding of Ca^{2+} ion takes more time; it will not be finished even in several days. It may be accelerated by introduction of 'sharpening' agents, which are sulfides, or sometimes cyanides. The action of sulfide, hydrosulfide and cyanide residues has been discussed in section 3.1. Introduction of these compounds makes the ionic composition of the medium more complicated.

The degree of swelling of collagen, in an acid as well as in alkaline medium, depends on the pH, which means that it depends on the dissociation degree of the components applied. Change of pH during liming, e.g., due to addition of chemicals, is shown in Fig. 12.2, according to Thorstensen.

Mellowness of the hide is liming-depending. This feature is less tested, e.g., on the basis of experiments done it could not be discovered as yet which cross-linking bonds and to what degree are destroyed in the process. There are some suppositions but no satisfactory explanations concerning, however, this operation. As one may conclude from more detailed studies done on the gelatin-making process, during prolonged liming in its first stage are broken the covalent

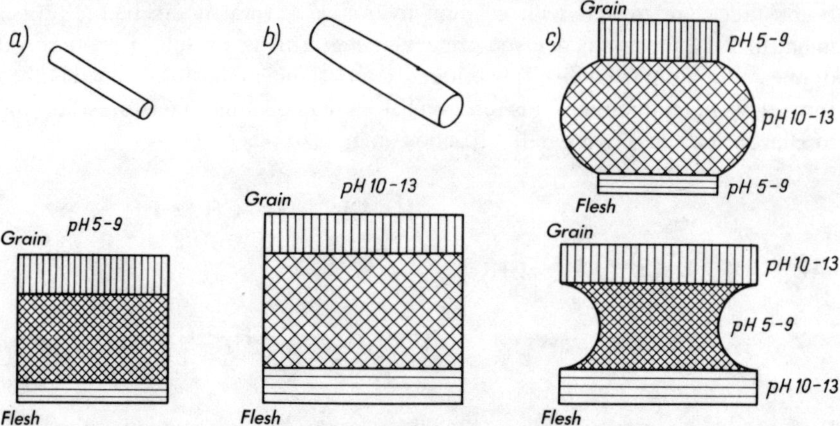

Fig. 12.2. Effect of pH on collagen swelling: unequilibrated pH-values may cause significant stresses: (a) no swelling, (b) equilibrated swelling, (c) unequilibrated swelling [1]

crosslinking bonds, in the next stage a great deal of hydrogen bonds and some peptide bonds. To what an extent the reactions do participate at the first stage has not been estimated. It is to be presumed, however, that all these reactions more or less slowly—start from the beginning of the liming. However, one believes it to be a reason for increased mellowness during liming. As we can tell from practice, the longer is the liming, the more mellow is the leather, i.e., it becomes more plastic and extendible, and less elastic. Therefore, the necessity is obvious, of more prolonged liming of glove- and nappa leather, and short, light liming of sides (boxhides). A longtime (months or weeks) liming makes gelatin from collagen, i.e., its helical structure, at least partly, becomes lost.

It cannot be said with certainty what is the fate of basal membrane collagen during leather making. Since Stirtz [3] has shown its presence in animal hide due to the proper staining method (see Fig. 1.2), only few investigations were done on this subject, however merely from technological point of view and thus without answering the question. Usilov et al. [4] have tested the distribution of the collagen fiber bundles in hide, coming to conclusion that their ends are situated in the grain, lying parallel to the hide surface. In this work no differentiation was made between collagens in particular layers, i.e., if they are formed from collagen I or IV, nor was their composition recognized. Then Zurabjan et al. [5] have tested the properties of chrome tanned leather with grain untouched and light buffed in order to remove the outer layer, together, as they presumed without proof, with basal membrane. In fact fatliquors and fillers have better penetrated the buffed grain. Hence two possibilities may be considered: either the basal membrane is removed in the beamhouse operations, being susceptible to proteolytic enzyme attack, as it results from biochemical investigations (cf.

section 2.2) and very tight network of collagen I fibers forms a barrier against penetration of chemicals, or the membrane remains up to the end of leather processing. Further histochemical and microscopical investigations can give the explanation, which concept is the right one.

No reason is known why the mechanic properties of fiber network are changing in an essential way while the structure remains unchanged (section 1.2, the Grassman's stress-strain curves).

According to old technologists—a subtlety in the 'art of tanning' was the knowledge how to mix the fresh and used lime liquors. In a fresh liquor the swelling is stronger and the opening-up weaker; therefore, this kind of liming is usually done in processing of bottom leathers. The spent liquors, often highly infected with microorganisms *(Stinkäscher)* are better hair-looseners, if the pH is below 11.0 due to the presence of sulfide-ion containing groups, from keratin breakdown. A similar effect may be obtained by addition of methylamine to the lime liquor.

Plumpness of swollen hides for various lime liquors composition has been tested by Toth and Vermes [6]. As a measure of plumpness *(Prallfestigkeit)* has been accepted a sum of reciprocal value of compressibility under defined load and of the resistance of bending. Some results are given in Table 12.1.

The quoted authors have investigated the known phenomenon that looser leather is obtained from too plump pelt than from the moderate plump one. The two curves shown in Fig. 12.3 characterize the dependence of plumpness, loss of protein in pelt and keratin decomposition (which was estimated, e.g., by decrease of sulfur content in the pelt) as function of time and using the same reaction parameters and concentration of sodium sulfide in the liquor.

From Fig. 12.3a it is seen that the degree of protein loss, related to the initial

Table 12.1.

Determination of plumpness index

Property	Butt in good pelt	Butt in weak pelt	Bellies
Resistance to bending kg/cm^2 (10^{-1} MPa)	14.1	5.4	0.2
Reciprocal value of compressibility, kg/cm^2*	$\dfrac{1}{0.11}$	$\dfrac{1}{0.18}$	$\dfrac{1}{0.55}$
Plumpness index	27.2	13.7	2.9

*Testing was done with a load of 1.5 kg/cm^2 (0.15 MPa). Plumpness index is obtained by multiplying the ratio by 1.5.

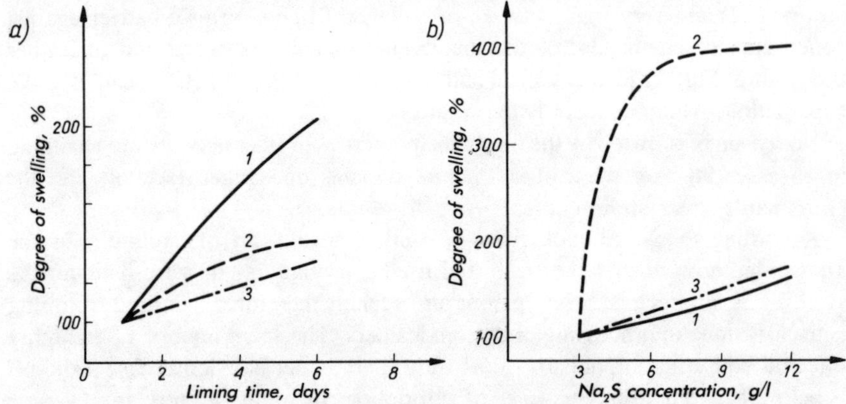

Fig. 12.3. Loss of protein 1, keratin destruction (dissolving) 2 and plumpness 3: (a) as a function of liming time, related to changes observed after one day, (b) as a function of Na$_2$S content determined after a three-day liming.

value, increases faster than the two remaining values, whereas the amount of decomposed keratin is increasing but the rate of this reaction is slowing down with time. The amount of dissolved keratin, however, increases very rapidly, if the concentration of Na$_2$S increases to about 6 g/l, remaining unchanged at still increasing concentration. Protein loss in hide and the plumpness are increasing linearly. According to the authors' opinion the increasing plumpness is a result of increasing loosening up of hair fixation *(Hautaufschluss)* and swelling. This swelling makes more difficult the protein diffusion to the outwards due to stoppage of pores and of interfibrillar spaces. These observations are in accordance with earlier data of slower penetration of the acids to the more plump pelt. Quantitative data, shown in Table 12.2, confirm this argument.

Great amounts of ammonia increasing with time are present in the lime liquors. Reactions occurring have not been investigated by the use of modern methods, however, as a source of this ammonia glutamine and asparagine have to be considered, because the decrease of amide nitrogen content in collagen during liming has for long been known.

Investigating the influence of liming by calcium hydroxide on collagen

Table 12.2.

Influence of the liming time on the grain of leather

Time, days	Final pH	Plumpness index	Protein loss, %	Grain	Touch
1	12.45	15	1.97	full	elastic
8	12.40	30	5.02	loose	soft

Hörmann and Schubert [7] have found decomposition of collagen with prolonged liming time, which manifests itself by the presence of hydroxyproline in solution. Treating such limed collagen with trypsin may solubilize it completely, and treating with chymotrypsin—in 90%.

In a typical liming process certain amount of hairs, particularly short which are soft and thin (underfur), will remain in the hide. Regaining of the hair is very seldom done in processing of cattle hides or calfskins. Liming is mostly done by leaving the hides in lime liquor overnight. Waste water from beamhouse, containing calcium hydroxide, sulfides and products of protein decomposition is highly polluted and difficult to clean or make utilizable. This is why plenty of efforts have been and still are done to replace sulfides in beamhouse with another compound. The difference between liming in liquor and by painting is only in the form of the substances used. Effect on hide is in both cases the same, although swelling during painting occurs just on one side, or if the water present is limited, by draining well is greatly restricted so it may be advisable to add some anti-swelling substance (e.g., $CaCl_2$) to the paint. The same effect may be obtained by using substances that decrease the pH of painting like sodium or calcium hydrosulfide.

The effectiveness of paint depends on concentration and on water content in the hide. The higher is the water content in both of them, the more sharp is its action. Painting of hides just soaked is to be avoided, as it is much better if water content in them may be controlled. Well-done painting makes the hair pull out from its follicle, and it is a little swollen in its lower part due to the action of keratin-decomposing substances. Other painting variants are also known, like: painting from grain side, giving leather of exceptionally soft, delicate grain, as it is mellowed from this side, and painting in drum. In this latter method gradually water is added to the drum and thus the dilution is increased, which gives the intermediate effect between liming and painting. Investigations on drum painting have shown that the increase of the swelling goes on to the amount of float of about 60-80%; then it becomes constant. To maintain swelling of pelt below the degree where its technological properties start to change, one has to keep this volume not over 50% for pure lime liquor, and not over 30%—for the sharpened one. This may be a function of salt concentration in the float [7a]. An advantage of drum painting is a faster penetration of chemicals into the hide by lesser swelling. Under these conditions liming in the internal layers of the hide is easier than by liming in liquor; leather obtained is flexible, softer, with well-filled loose parts. According to Herfeld et al. [8] drum painting makes deliming easier and quicker. Heating of sheepskins which after painting are stacked 'flesh to flesh' to the temperature exceeding the surrounding temperature up to 10°C one may attribute to the vigorous bacteria growth, which occurs in fleece during the process. A revolutionary—as it seems—method of 'liming' in which only highly concentrated (10%) sodium sulfide is used has been proposed by Dorstewitz and Heidemann [9]. This method is part of the technological process which shortens significantly the whole leather-production process and should be applied in connection with other operations. According to the proposal quoted, the hides have to be immersed for 5 minutes in Na_2S solution of a temperature not exceeding 20°C, and then for 1 hour in a 10% Na_2O_2 solution at 25-31°C. Under these circumstances hair and epidermis are totally destroyed, and the rest may easily be separated simply by scudding. Hair bulbs, which are remaining in the hide, are completely bleached, because melanine is decomposed in the peroxide bath. Leather thus unhaired is regularly glossy, plump and very clean.

Methods of shortening of the liming process based on the addition of surfactants to the bath are also known. Decreasing surface tension of the solution makes it easier for the solutions to penetrate into the hides. The consumption of sulfides may be decreased this way. However, the technologists believe them not to be very effective. Of certain importance is the degreasing action of these chemicals and decreasing of their swelling. As a general rule of choice of the liming method one has to consider:

(1) Long pure lime liquors—for soft gloving and garment leathers; short-time ones for sole leathers.

(2) Lime liquors of high sulfide content—for light, flat, strong bound leathers, as for chevreaux.

(3) Mixed types—for remaining leathers.

The effect of temperature on the liming process has been reviewed by Ellement [10]. As one may see from his data, an increase of the temperature increases the liming speed without changing the quality of leather processes. If the temperature of 35°C is exceeded, drastic changes, or even denaturation and splitting of the corium layers may occur. So the limit of 30-32°C is recommended as safe. One has to pay attention to that when temperature increases, the pH-limit for hydrolysis decreases. The liming time increases with temperature when sodium sulfide concentration decreases. (Concentration of CaO is constant and equal to about 15 g/l). The liming process shall be carried out at a constant temperature, which is dependent—for a given leather kind—on sodium sulfide concentration. Other hydroxides, e.g., sodium hydroxide, are hardly used in industry. They are however very often used in research work [6], e.g., for comparison of the effect of concentrations of sodium and calcium hydroxides on collagen swelling. The monographs of Veis [11] and Ward and Courts [12] are devoted to the action of alkalis on collagen. Ammonia solution is a good unhairer; it is not used in industrial processing because hermetic containers have to be used there. The action of ammonia as the unhairing agent is inhibited by calcium hydroxide present in the medium. This is due to its pH, rather than to a complex formation. Optimal unhairing action of ammonia occurs when a 0.6 molar solution is applied. The use of ammonia as an unhairing agent is no longer discussed.

As a marginal work in the unhairing investigation several authors have tried to elaborate the method of testing the hair fixation strength, as a measure of the effectiveness of unhairing. However such method may not be considered as satisfactory as yet.

Heidemann and Yakali [13] investigated the influence of unhairing agents on keratin and on other proteins present in hair and in epidermis. They came to a conclusion that it is possible to bleach hides completely by decomposition of melamines, if short-time bleaching with oxidants is applied (Peracetic or performic acids, chlorine dioxide). These substances are to be used in an acid medium, in concentration of 1-3%. Deliming is not necessary after their use.

Use of amines in lime liquors

In earlier works of American investigators on recycling of lime liquors a conclusion was drawn according to which aliphatic amines are the unhairing agent, because the liquors contained these amines in amounts growing with the time of their use. Actually methylamine, ethylamine and dimethylamine have been found to be the compounds significantly enhancing the unhairing action of calcium hydroxide. Accordingly, over 70 nitrogen-containing compounds have been demonstrated to accelerate unhairing by calcium hydroxide to various extent. Now sometimes lime liquors are used which contain dimethylamine and sodium sulfide, calcium chloride or just dimethylamine in combination with calcium hydroxide itself. Their usage is not explained theoretically, or by isolation of the reaction products. Perhaps the unhairing effect of aliphatic amines is due to their properties as reductors. As it is found, they bleach pelts, which may be explained by a reduction process. The use of amine liming liquors makes purification of sewages easier. Sommerville et al. [14] have investigated unhairing by dimethylamine sulfate in alkaline medium. As alkalising agent sodium hydroxide was used. Applicability of the system used for unhairing has been satisfactory, and two significant remarks have been done, viz.:

(1) Dimethylamine sulfate (DMAS) may be used for unhairing as a main component. In connection with calcium hydroxide this compound is suitable for the processes in which hair saving is required.

(2) Sodium hydroxide was successfully used. This compound was previously considered to be too strong to be suitable in controlled processes.

Further observations of sulfideless unhairing have lead Sommerville to the conclusion that it is possible to apply short liquors (1 : 2) in unhairing without sulfides and a 1 : 1 liquor in a case of liquors containing 3% NaOH, 1% Na_2SO_4, 0.5% NaSH and 1% DMAS, calculated on hide weight. Operation time was 24 hours.

Action of lyotropic agents

Lyotropic agents, discussed in O'Flaherty's manual are not applied in industry.

Their way of action is structure-making of the solution, but they contribute to dissolving of the proteins. Keratin is attacked by these compounds more easily than collagen, thus hair and epidermis may be separated from corium. Urea and sodium chloride affect hide in the same way. Unhairing effect is a property of compounds dissolving protein molecule through transferring hydrogen bonds stabilizing protein molecules on the solute molecules, thus forming protein-solute bonds.

Thermal unhairing

This unhairing (scalding) is used for the removal of bristles from pigskins. Histological changes in pigskins immersed in water heated to 58-60°C for several minutes have been tested by Wang et al. [15]. They came to the conclusion that germinative layer of epidermis is more sensitive to thermal decomposition than others, which contain much more keratin. Short-time scalding does not influence collagen, and one may obtain pigskins of highly desirable properties in processed leather. Scalding requires very exact temperature and time regime to be maintained as a slight change in parameters applied may very easily make the skins useless. Scalding method is not used for other kinds of hides and skins.

A very promising method of grain removal from pigskins is worked out in the Leather Research Institute in Lódź, Poland, by Wojdasiewicz et al. [16]. In this method a limed pigskin is passed very quickly between a pair of rollers, one of them heated to 160-200°C, a contact time is about 0.05-1.0 s. In this process the grain proteins are denatured, however denaturation is very shallow, and a nubuck pigskin is obtained without buffing. Denatured layer dissolves very easily or may be separated spontaneously. In this way several finishing operations are avoided, and the nubuck obtained is particularly soft and of uniform surface.

Oxidative unhairing

This process has been introduced by Rosenbusch [17]. It is based on treatment of hide with a strong oxidiser in a slightly acidic medium. As an oxidising agent ClO_2 is used. It is split off from some compounds in water or can be used in gaseous form

$$5\ NaClO_2 + 4\ HCl \rightarrow 4\ ClO_2 + 5\ NaCl + 2\ H_2O$$
$$4\ keratin-S-S-keratin + 10\ ClO_2 + 4\ H_2O \rightarrow 8\ keratein-SO_3H + Cl_2$$

Chlorine dioxide, when reacting with keratin, splits the disulfide bonds; during this reaction keratinsulfonic acid and free chlorine are formed. Compounds formed are removed due to the drum motion. To maintain the pH of 3.0-3.5, glycolic acid is used. If the pH-value increases sulfuric acid is to be added. Temperature is maintained at a level below 40°C.

Oxydative unhairing takes about 24 hours. When the process is finished, the excess of ClO_2 will be decomposed by the use of sodium thiosulfate, in order to prevent oxidizing of chrome tanning agents (later). This process, replacing all the beamhouse operations: degreasing, pickling and alum pretanning by binding of chlorine to collagen, has some disadvantages, and due to these it is not applied in industry:

(1) It has to be carried out in an acidic medium. Thus all the materials used for construction of the machines in beamhouse are to be replaced by the acid-resistant ones.

(2) It is quite expensive.

(3) Toxic gases are formed.

(4) Leathers obtained are stiff and spready.

Enzymatic unhairing

This is a very ancient process and known as 'sweating' used in primitive tanneries, particularly in processing sheepskins. The sheepskins have been piled up flesh to flesh and left. Operation is done during 1-2 days at 20-25°C, or 1-2 weeks at 8-12°C. The skins are spontaneously heated as a result of the action of proteolytic enzymes which start putrefaction. This process is to be interrupted in due time and hair as well as epidermis removed. It is a bacterial process, in which various kinds of bacteria participate. A major disadvantage of this process is the difficulty in its controlling.

Cathepsins contained in the lysosomes of the cells of the skin are also participating in this process [18], because it is possible to make it under completely sterilized conditions. For autolytic unhairing an optimal pH is about 4. In this 'lysosomal' unhairing probably proteins and glycosaminoglycans are equally attacked. In the products separated from skin, hydroxyproline is found in some amounts, equivalent to 0.3% of skin collagen. This may be a result of attacking of collagen-containing 'lining' of the hair pocket.

Unhairing by the use of commercial enzyme preparations has for long been known: the first technological process was carried out 1910. An advantage of this method is a great simplification of the beamhouse work—difficulties in its controlling are its disadvantages. This process, if not selective enough and without appropriate control, is associated with the risk of collagen destruction due to combined enzyme attack. Contemporary chemical and biological technologies make possible to obtain from plants, microorganisms and animals the enzyme preparations of high degree of specificity. Despite this, enzymatic processes may give some difficulties in the tannery. Enzyme attack (most frequently an enzyme complex) may be very vigorous, so that too much protein may be removed from the skin, and collagen partly decomposed. The resultant hides are then thin and stiff. If the enzyme action is too weak, insufficient amount of protein will be digested, and additional operations necessary, like, e.g., alkaline swelling of hides before pickling (after-liming). The enzymatically unhaired hides as a rule have to be tanned and dyed in different way than hides processed by other techniques. Among the beamhouse operations enzymatic unhairing is perhaps the most discussed and most controversial process. Specific action of proteolytic enzymes and their characteristics have been covered in section 9.3. Commercial enzyme preparations as a rule contain enzyme complexes, activators and buffers, and are adjusted to proper pH.

The methods of enzymatic unhairing are discussed in several papers, e.g., [19]. A special symposium has been devoted to this point during the 8th IULCS Congress at Scheveningen in 1963. However, the enzymatically unhaired hides are considered to be spread, empty, and the economical aspect of this is still unsatisfactory, as a large amount of enzyme has to be used in order to obtain good results. Mold peptidase preparations have also been used. Under these circumstances it is possible to obtain a finished leather, which is very like the classical one. Short 'liming' with the help of NaOH, which is then stopped by addition of $(NH_4)_2SO_4$ makes easier the enzyme penetration through the hide.

According to Felicjaniak [20], who investigated in detail the unhairing activity of pancreatic enzymes, there is a distinct difference in unhairing and proteolytic action of those enzymes. To increase the unhairing activity, the pelt has to be prepared by applying an inorganic chemical before the enzyme is used. The compounds giving optimal results are ammonium chloride, thiocyanate, sodium thiosulphate, and some others. These substances increase the unhairing activity of enzymes, when used in concentration of 1% per pelt weight. Sodium thiosulfate increases proteolytic (not unhairing) activity of the enzyme. Optimal unhairing activity was reached when pH was maintained on the level of 8-9, i.e., somewhat higher than optimal proteolytic activity ($pH_{opt} = 7.5$). For pigskin unhairing some nonionic surfactant addition was found to be advantageous.

The paper of Monsheimer (loc. cit.) is based on the paper of Heidemann and Yakali [13]; according to it the hair follicles are 'lined' not only with soft keratin, but with another protein, rich in glutamine and asparagine. These authors draw attention to the point that the substance which forms this 'lining' consists as well from components of cell membranes, from keratohyalin and pigmented melanosome grains.

The proposed one-step beamhouse process is based on washing-out of salt excess from calfskins and cattle hides, and then on treating of them with proteolytic preparation during 4-6 hours at pH about 11. Because of binding of alkaline substances to collagen, the pH drops then down to the value of 3.5. As no bacterial growth is observed under these conditions, there is no need for any bacteriostatics addition. The quoted paper has a character of preliminary communication—no characteristics of preparations used are given; however, it seems to be a real novelty in the beamhouse practice.

The use of enzymes in the beamhouse, first of all for fur skins, has been investigated by Pfleiderer and Mayer [21]. Using bacterial preparations and mold proteolytic enzymes, they got leathers not only well mellowed and soft, but more resistant to stretch than the ones obtained in the common way. The use of enzymes in question does not damage hairs, whereas globular proteins and 'soft' keratin are digested. A list compiled by these authors concerning applicability of proteolytic enzymes separated from molds and bacteria, is shown in Table 12.3. Proteolytic activity curves for some known, typical enzymes, are shown

Table 12.3.

Some microorganisms which produce proteolytic enzymes

Microorganism	Range of pH	Kind of enzyme	Reaction specificity, kind of use
Aspergillus oryzae	3.0/8.0	mold proteinase	acidic, neutral, alkaline u (soaking) o (unhairing) w (bating) s (swelling)
Asp. satoi *Asp. usami* *Asp. niger*	2.5/3.5	" "	w, o, s
Asp. niger *Asp. flavus*	9.5/11.0	" "	w, o, u
Asp. oryzae *Asp. niger*		" "	w, o, u
Bac. subtilis	10.0/11.0	bacterial proteinase	w, o, u
Bac. cereus *Bac. natto* *Bac. vulgatus* *Bac. mycoides*	6.8/9.8	"	w, o, u, s
Bac. subtilis *Bac. mesentericus* *Bac. polymyxo*	5.8/8.2	"	w, o, u

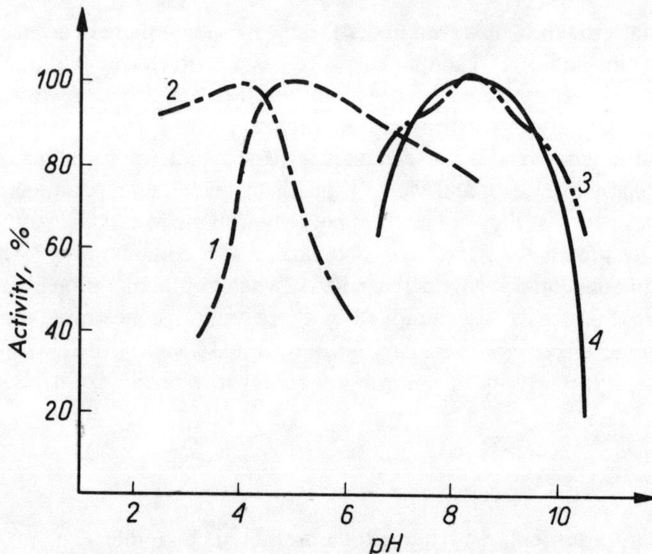

Fig. 12.4. Proteolytic activity of typical preparations according to [21]: 1—papain, gelatin as substrate, 2—pepsin, gelatin as substrate, 3—trypsin, casein as substrate, 4—chymotrypsin, casein as substrate.

in Fig. 12.4. As can be seen from the Figure, papain and pepsin act on gelatin in a 'symmetrical' way: action of papain is strongest at pH $\cong$ 5, weakens rapidly when pH further decreases, and very slow at pH increase. Action of pepsin is the reverse one. This is a result of enzymic activity and pH of functional groups of side chains, which have their places adjacent to the attacked bonds (section 9.3). No significant differences in proteolytic activity of trypsin and chymotrypsin against casein are observed. The curves of proteolytic activity of both enzymes are in fact almost identical.

Liming mechanism has not been elucidated exactly—from the standpoint of collagen fibril structure. Investigations by electron microscope, as well as X-ray analysis did not permit as yet to discover any changes in the fibril as a result of liming, and in a broader sense—separation of collagenous and non-collagenous proteins. Maybe the very detailed investigations on the amino acid sequences in collagen molecule will allow recognition of these changes. Moreover, the limed fibrils give a different picture in the electron microscope, when it is taken immediately after processing: no cross-striation is observed and the swelling is distinctly visible. After successive operations they recover their previous shape (cf. Fig. 3.9).

12.4. Deliming and bating

These processes depend on the unhairing method used. As a rule, they are done in one bath, when hair is removed by liming. According to the technologists,

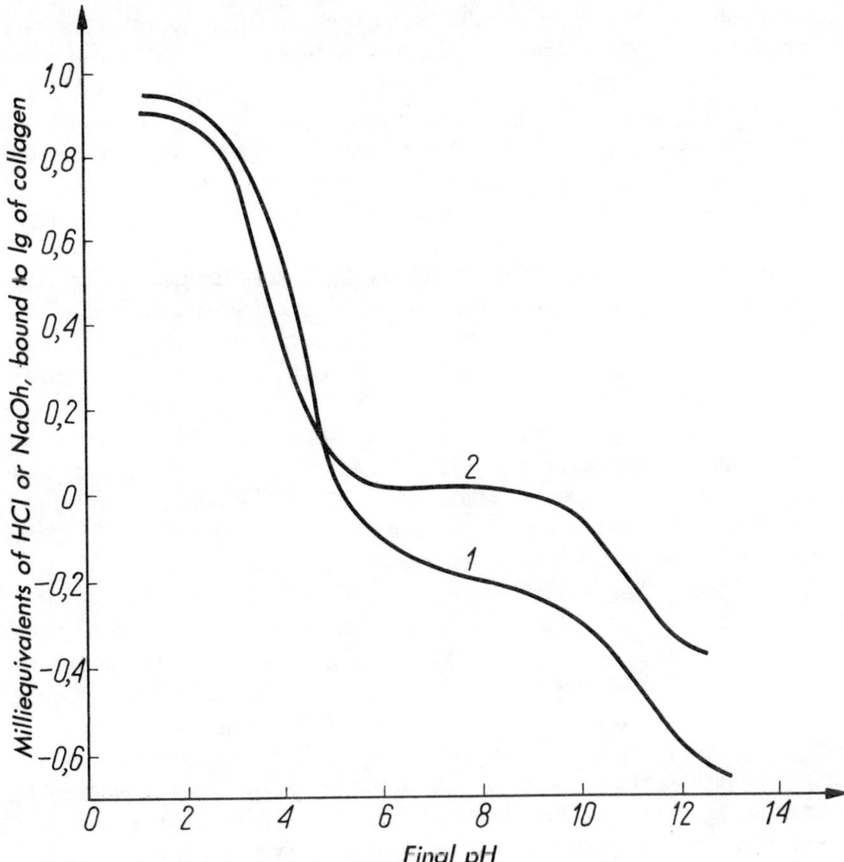

Fig. 12.5. Titration of limed (1) and native (2) collagen.

one can consider this process to be done properly if, beside the removal of calcium ions from pelt—it meets the following requirements:
(1) the pH inside the collagen network is regulated, and the bath is buffered;
(2) the swelling of the pelt is reduced to the expected degree;
(3) the fibers are separated by washing out the matrix;
(4) the products of protein degradation are removed from the pelt.
Points 1 and 2 may be considered as chemical, or physicochemical operations; point 3—as chemical and enzymic operation, point 4—as an enzymic one only.

In our treatment of the process, first of all we will discuss deliming and pH-regulation, and subsequently bating. A certain part of calcium hydroxide, about ⅔ of the amount present in pelt after liming, may be removed by washing out with water. Further washing does not decrease the calcium level. Use of warm water makes possible to remove still a small amount of calcium, but the pelt will be less plump then; mobility of water molecules is increased and so is the accessibility of Ca^{2+} ions. Calcium not removed from pelt at this moment may form the sulfate during tanning. Accordingly, it is necessary to remove Ca^{2+} ions from pelt to a degree which will avoid the formation of 'salt spots' on the leather. Complete removal of calcium is not required, except for deliming of very soft glove leathers. In other sorts of pelt the adjustment of its Ca^{2+} content occurs in the pickling.

Acids and salts are used in deliming, and the aim of the use of acid is to decrease the pH of protein itself. The protein-calcium hydroxide system is buffered by both components. The difference in behavior of native and limed collagen is visible, when comparing two titration curves with acid or base of the native, or limed collagen (Fig. 12.5).

The amount to be added to the pelt can be calculated or found from experiment; at the beginning of the reaction the pH on the surface is very low, but very high inside the pelt. Thus the acid-swelling mechanism operates on the surface, and the alkaline one—inside the pelt. Between these layers there is a hypothetical surface, the pH of which is close to the isoelectric point.

There is no swelling in this plane (see Fig. 12.2). This powerful disturbance may cause breaking of collagen fibril and, in consequence, a danger of loose-grain in the finished leather. Thus the deliming process shall be carried out gradually, slowly, with due regard to the pH inside the pelt. For this purpose buffered systems from slowly dissociating acids and appropriate salts are used. The pH value of the ammonium sulfate solution is about 5. A system containing ammonium salts and calcium hydroxide is of a pH of 7-8 and the solubility of the formed calcium complexes is good under these conditions. A diffusion of the calcium from pelt follows, assuming a minimal swelling level maintained. A further deliming step may be, if necessary, the treatment with hydrochloric acid. Organic acids may be applied as well, particularly lactic acid, which is advisable because of the pH of its solution. Checking of the reaction on the pelt cut during deliming process by using pH-indicator (phenolphthalein or bromo-thymol blue) is a typical and necessary control point. It allows at the same time to control the drumming speed of pelts in order to limit it to the minimum when swelling is considerable or of opposite sign in the particular layers.

Liming and deliming processes may be considered as steps in the removal of fat from hides. This point is of particular importance, when pigskins and sheep-skins are processed, because the fat content of these skins is very high. It is of less importance in processing cattle hides and calfskins. During liming, part of the fat contained in hides or skins is converted into calcium soap. In deliming it is decomposed in a way, which includes conversion of the calcium ion into ammonium-sulfate-calcium complexes, whereas the fatty acids form sodium, or ammonium soaps, soluble at pH about 8. At this pH, usually applied in bating, as well as at the temperature of ca 30°C fat is removed by washing out. To remove fatty acids an addition of surfactants is useful. Non-ionic surfactants, making peptisation of fatty acids easier, have been successfully applied in the U.S. tanneries for this purpose.

Problems of the removal of iron from hide are combined with those of deliming. Iron ions in hide may have their origin either in blood hemoglobin, or come from the water used in processing. Deliming and simultaneous iron ions removal problem is solved by the use of EDTA, as described by Otto [22]. Investigating the tanning action of chrome complexes masked by chelating agents (or rather lack of such action, as mentioned already by Gustavson for oxalatochromate complexes) Otto has checked the action of EDTA-Cr^{3+} complexes on non-delimed pelt, having in mind the formation of calcium and iron complexes. He found that EDTA-Cr^{3+} complexes, penetrating into the limed, lukewarm pelt, do react with Ca and Fe ions, making Cr^{3+} salts free. These Cr-salts have tanning properties. In this case use can be made of the fact that stability of EDTA-Cr^{3+} complexes has an inverse pH-dependence, than that of EDTA-Ca^{2+} (see Fig. 12.6). The EDTA-Cr^{3+} complex is stable below pH = 4.5; at pH = 7 a half of Cr is set free. The EDTA-Ca^{2+} complex is stable at pH = 9; at pH = 7, half of calcium becomes liberated. In this way, putting a limed pelt, rinsed with lukewarm water in a float at pH = ca. 2.5, one has two effects at the same time: deliming and tanning. According to Otto, tanning is very slow at start,

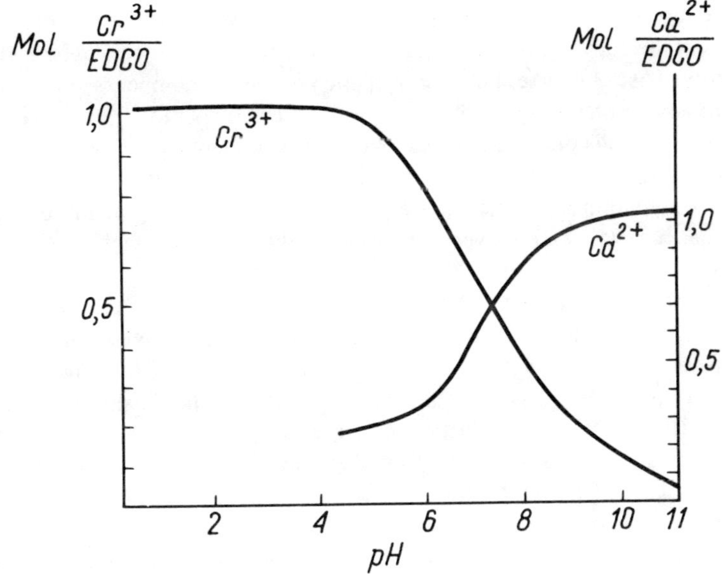

Fig. 12.6. Stability of EDTA chromium and calcium complexes in dependence on pH. The curves show the amount of ions bound. According to [22]

i.e., in grain and flesh layer, then, due to changes in masking, it becomes faster when going deeper into pelt. Use of diluted sodium chloride solution causes deswelling of pelt and lost of its plumpness. This very interesting process, having some features of 'chemical automation,' has not been introduced into industry; maybe some serious difficulties arose during detailed working out, or for economical reasons.

Numerous industrial preparations are known which form complexes or soluble compounds with calcium. They act as buffers and accelerate the deliming process. Typical substances used in tanning technology are the pelt-saving organic acids: formic, acetic, lactic and butyric. Lactic acid anhydride has been used for a long time. It hydrolyzes slowly to yield the acid. Derivatives of glycolic acid, sulfophthalic and other aromatic acids are also used, as well as ammonium salts of organic acids and their mixtures. They form systems of a particularly advantageous influence in leather quality, its smoothness and softness, because they do not cause quick, drastic changes in the swelling degree. Mixtures of acids (e.g., hydrochloric and boric) are used as well as sugars (starch, molasses) which are now rather given up, as acting too slowly and of unequal efficiency.

Bating is done in the same operation with deliming. It means a final removal of non-collagenous hide components, keratin degradation products, globular proteins, elastin and cell structure residues. A condition for successful carrying out of combined operations is avoiding negative action of deliming on the enzymes used for bating.

Since ancient times the enzymes used in bating were in the form of natural products of the body system (dung, urine). 'Sweating' of skins caused by bacteria and autolysis is, of course, but an enzymatic process. A number of preparations which remove proteins and the products of their decomposition from the network of collagen fibers is difficult to be estimated. The amount of enzyme preparation used for bating (an enzyme mixture, as a rule, of known or unknown composition) is small when compared with the weight of pelts. Degreasing action of bating liquors has been known for a very long time, as the natural enzyme sources used, like dung, contained lipases.

Enzymatic processes have to be controlled very carefully. Action of bating baths may be dangerous, as their prolonged action, overdosage of preparations or temperature increase may cause an excessive removal of proteins from the pelt; due to this it may become 'spongy' or empty, and the layers may become separated (loose grain). By insufficient bating, leather may become stiff or even brittle. Collagen is resistant to the proteolytic enzyme activity. If, however, the nonhelical parts of the molecule are removed, an irreversible loosening of the fibril structure may occur. Collagen treated for 90 hours with pronase (which is a mixture of proteolytic enzymes isolated from *Streptomyces griseus*, now frequently used in scientific work) loses about 5% of its weight. Such a treatment is extremly drastic; usually more mild ones are applied.

Table 12.4.

Percentage of calfskin components (as average nitrogen content) and their removal in beamhouse operations

| *Fraction* | *Skin-pelt* | | | | |
	fresh	*soaked*	*limed*	*delimed (Oropon-C)*	*pickled*
Elastin, keratin, nonextractable globular proteins	11.4	13.8	2.2	2.8	2.4
Extractable globular proteins	13.3	9.4	4.2	1.9	1.5
Collagen	73.1	75.0	91.9	93.9	94.3
Nonprotein N	2.2	1.8	1.7	1.4	1.8

Fractionation has been carried out by mild hydrolysis with 0.1 N HCl, and successive precipitation with trichloroacetic acid

Bating affects the components of pelt, making them loose and the collagen fibrils able to shift. This influence is very difficult to control, as no methods in the industry are used which could make possible the immediate process control, allowing to estimate the optimal point. Bating is a short-time process; to test its effectiveness takes much more time than the process itself, so as a rule the visual and manual (feel) evaluation are of decisive importance. To avoid any influence of bating on collagen is impossible; 'if the bating bath does not affect collagen at all, it is not the right one'—says Stather. A typical protein composition of calfskin after particular beamhouse operation is shown in Table 12.4, according to Tancous. According to the Table, liming is crucial for skin protein decomposition; however, still about 2% of globular proteins are removed in bating. Among substances removed at this stage, the residues of tissue are still present, like, e.g., sebaceous glands, mast cells, hair follicle 'linings,' hair, etc.

The fate of elastin in beamhouse operations has been widely discussed. No doubt it is dependent on the liming parameters and on the choice of enzymatic composition of bating bath, however it is not fully understood yet. Taking the elastin properties into consideration, like its elasticity and resistance to chemicals, it seems to be important for leather, whether or not elastin is removed. According to former views, bating does not change elastin, according to newer investigations, 30 to 86% of elastin is removed by the use of pancreatic enzyme preparations. According to Ornes and Roddy [23] this amount is depending on development degree and state of blood vessels. Applying the histological elastin staining technique, these authors, as well as Urban (cf. Fig. 3.6), found elastin fragments in leather.

In bating the muscle tissue becomes destroyed. Non-bated pelt, brought from warm water into the cold water, becomes rough owing to a shrinkage of the hair muscles, still contained in it (*erectores pili*). This effect ('goose-flesh') cannot be observed in a well bated pelt, as McLaughlin has shown in 1929.

The typical decrease of pelt plumpness during bating (deswelling) has not been fully explained. Most simply it would be to consider it as the effect of buffering salts, usually present in the medium; however it is impossible to obtain such an effect just by applying, e.g., ammonia salts alone. One may presume that this deswelling is promoted by the removal of last part of non-collagenous proteins, and thus making water molecules, trapped inside the pelt (in interfibrillar spaces) more mobile. However, this hypothesis would need an experimental confirmation.

Influence of physicochemical parameters (pH, temperature, activators, ionic strength) has been discussed previously, as related to enzyme action (section 9.3).

By addition of surfactants, largely polyglycols and adducts of ethylene oxide alkylphenol and aliphatic alcohols, Pietrzykowski [24] was able to increase the activity of pancreatic enzyme preparations. According to his opinion this is a physical action, based on making active centers accessible due to decreased surface tension, because this is not observed, when crystalline (i.e., pure) trypsin is used. Thus the site of the attack of surfactants is not the protein to be digested, but the structure, surrounding the active center of the enzyme, which is not fully separated from the tissue, or it is shielded by steric hindrance.

12.5. Degreasing

Degreasing is the last step in the beamhouse operations. In the majority of technological processes applied there is no need for degreasing, as the fat contained in raw hide becomes saponified and washed out or just washed out by the use of surfactants.

Pig- and sheepskins may contain after liming still over half of their fat content.

Ineffective and incomplete removal of fat may be the reason for nonuniform tanning and dyeing, and gives rise to grease stains. Fat is contained in hide mainly in its subcutaneous tissue, and between papillar and reticular layers. It mostly occurs in fat cells; therefore, to remove it completely by extraction with solvents is possible only when cell membranes are destroyed in the beamhouse.

Degreasing of raw hides by the use of organic solvents alone is not applied, as it would create significant technological, economical and engineering problem. 'Wet' degreasing is the simplest operation of wringing of pelt, soaked in warm water. Then one may treat the remaining 5-12% fat with emulsifiers and organic solvents, whose task it is to make mobile the relatively large fat molecule. The solvents used should possibly be non-inflammable or at least sparingly inflammable, removable from hide, e.g., by salting out with a salt solution or by centrifuging. Wet degreasing is most frequently based on drumming of wrung-out pelt in 32-35°C together with solvent and emulsifier. The solution of fat in a solvent is then salted out or centrifuged.

For removal of high quantities of inter-fiber fat such as may occur in sheepskin the procedure involves drumming in 10% solvent (e.g., trichlorethylene + 1% surfactant), with the pelts in the 'bated' condition at pH 7-8, when they are less swollen, and the structure more accessible.

However this technique does not remove more than 5% fat, but does equalize the fat distribution over the skin.

In the case of N. Zealand pickled pelts which may contain 25% fat by weight it is common to degrease them in this depleted condition by drumming with 40% kerosene solvent and 1% surfactant. The resultant milky emulsion of grease solution in solvent and water is then washed off with 5% brine solution [25].

REFERENCES

1. Thorstensen, T. C. Practical Leather Technology, 2 ed, Krieger Huntington 1976
2. Blazej, A., Michlik, I., Galatik, A. Kozarstvi *22*, 3 (1972) in Slovak
3. see ref. 4 ch. 1
4. Usilov, V. A., Kutin, V. A., Mikhailov, A. N. Izv. Vys. Uch. Zav. Leg. Prom., *3*, 79 (1966) in Russian
5. see ref. 12 ch. 2
6. Toth, G., Vermes, L. Leder *23*, 48 (1972)
7. Hörmann, H., Schubert, B. Leder *9*, 265 (1958)
7a. Sharphouse. J. H. J. Am. Leath. Chem. Assoc. *74*, 215 (1979)
8. Herfeld, H., Schubert, B., Häusermann, E. Leder *17*, 243 (1966)
9. Dorstewitz, R., Heidemann, E. Leder *26*, 133 (1975)
10. Element, P. G. J. Soc. Leath. Tr. Chem., *48*, 388 (1964)
11. Veis, A. The Macromolecular Chemistry of Gelatin, Acad. Press London 1964
12. Ward, A. G., Courts, A. The Science and Technology of Gelatin, Acad. Press London 1977
13. Heidemann, E., Yakali, T. Leder *22*, 85 (1971)
14. Hetzel, L. V., Sommerville, I. C., Cases, C. J. J. Am. Leath. Assoc., *61*, 536 (1966)
15. Wang, H., DeBoukelaer, F. L., Maynard, M. J. Am. Leath. Chem. Assoc., *49*, 728 (1954)
16. Wojdasiewicz, W., Szumowska, K., Morawiec, R. US Pat. 3951593 1976 and patents in other countries
17. Rosenbusch, K. Leder *17*, 289 (1966)
18. Money, C. A., Scroggie, J. G. J. Soc. Leath. Tr. Chem., *55*, 333 (1971)
19. Monsheimer, R., Pfleiderer, E. Leder *26*, 150 (1975)

20. Felicjaniak, B. XIV IULTCS Congress Barcelona 1975
21. Pfleiderer, E., Meyer, H. G. Leder *23*, 148 (1972)
22. Otto, G. Leder *17*, 285 (1966)
23. Ornes, C. L., Roddy, W. T. J. Am. Leath. Chem. Assoc., *55*, 124 (1960)
24. Pietrzykowski, W. J. Soc. Leath. Tr. Chem., *54*, 135 (1970)
25. Sharphouse, J. H. Leather Technicians Handbook (II ed) Leather Producers Assoc. London 1982

13.

PICKLING

Pickling also belongs to the operations whose aim is to prepare raw hide for tanning. The purpose of this operation is to stop enzymatic bating and to adjust the acidity and pH* of the hide, or its water content for subsequent processes, e.g., storage, chrome or 'dry' vegetable tannage.

To pickle the pelt means to acidify it in such a way as to prevent it from simultaneous swelling under the action of acid. This is usually done by salt addition. The presence of acid in the medium suppresses dissociation of carboxylic groups of collagen side chains. Solution of acid and salt outside the pelt is hypertonic relative to the water contained in the pelt, so it has a higher osmotic tension. This is a reason for the coming over of water molecules from pelt to the solution. Solution in the pelt surrounding becomes less concentrated, and owing to this the electrolytes in it become more dissociated, and diffuse into the interfibrillar spaces in collagen according to the first Fick's law. This process stops when concentration of ions in pelt and in solution becomes equilibrated, and at the same time an interaction occurs between salt cations and collagen carboxyls, and between anions of salts and basic groups of collagen side chains. Ions, remaining inside the pelt keep their solvation water, thus the pickled pelt is hydrophilic and mellow. Inorganic acids, sulfuric and hydrochloric, are used for pickling, the latter only in special cases, or organic acids: formic, lactic, acetic as well as sodium chloride as a swelling suppressant.

Pickling slows down the chrome tanning process, does not allow the tanning agents to be bound to the external pelt layers, as this would stop their deeper penetration. Acid, contained in the pickle, reacts with basic functional groups of collagen side chains, and imparts to it a pH of about 1.5. Thus the basicity of chromium complexes that are in contact with collagen decreases and so does their ability to attach to pelt. This ability is in the technology called astringency. This definition is not very precise, because it is not based on any quantitative determination. It is a measure of affinity of tanning agent to the pelt protein, and so it is a measure of its tendency to be bound by this protein. This affinity is dependent on many factors, and this is why it cannot be expected to be defined in just one way. Only a sum of particular properties may give the information

---

*The definition *acidity* is frequently confused with the pH definition. As *acidity* we understand the amount of ionizable acids contained in solution. During titration the ions are removed and formed anew due to following dissociation when ion concentration decreases, according to Ostwald's dilution law. pH is the measure of the amount of hydrogen ions present at the given moment in solution and may be equal to acidity in an ideal case, when the whole amount of acid is ionized. This may occur in strong acid solution of concentration, approaching to zero. In practice we never meet such a case.

required, what is the astringency of a tanning agent or of a float. This definition is less exact, but more flexible than the classical van't Hoff's affinity definition, used in physical chemistry. It says that the real measure of chemical affinity shall be the maximal molar work utilizable, or electric work, measured during reversible course of reaction.

The first contact of chromium float with leather should occur at pH of the order of 1.5-2.5. The pH of common chromium floats is 2.5-3.5, so the acid used for pickling shall decrease the pH of pelt to the lower value than that of the chromium float. The purpose of the salt, contained in the pickle in concentration of about 10% is to quench the tendency of collagen to swell in an acidic medium. Accordingly, the salt is to be chosen which shall decrease osmotic swelling; that is, such a salt whose a cation and anion have possibly the lowest place in the lyotropic series, where 'structure breaking' is a prevailing feature. This 'breaking' action of salts and acids, manifesting itself by dehydration of pelt, i.e., removal of the structured water from it, makes the start of tanning easier.

The regularity of the acid-to-collagen binding has long been tested, but it could not be expressed by a simple formula. The proposed expressions are of empirical type.

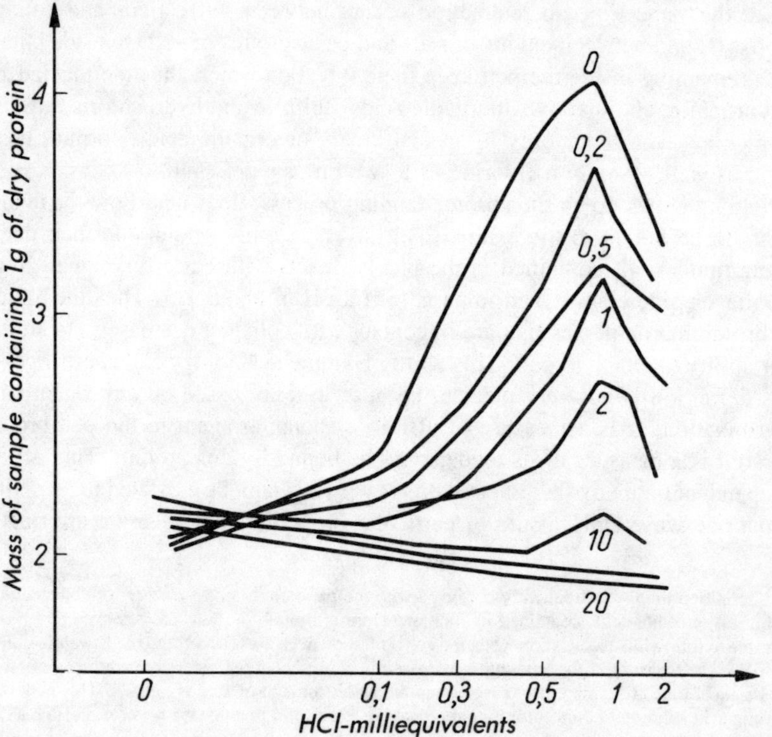

Fig. 13.1. Effect of salt NaCl on the degree of acidic swelling of hide, in mequiv. of NaCl per one gram of collagen.

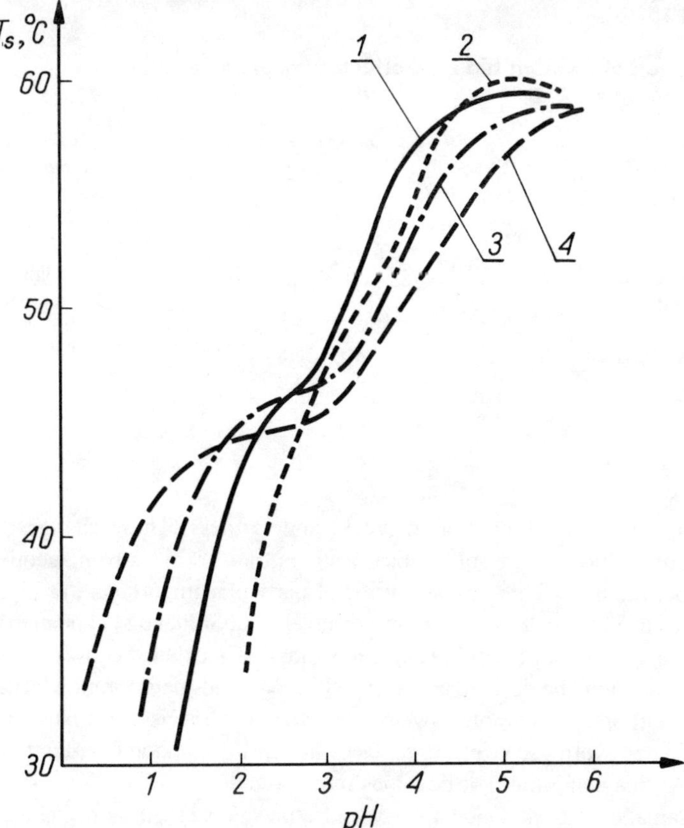

Fig. 13.2. The T_s of pickled pelt in dependence on pH and on the kind of acid used: 1—lactic acid, 2—acetic acid, 3—formic acid, 4—hydrochloric acid.

The effect of salt in acidic medium on pelt swelling is shown in Fig. 13.1. One can see there that swelling may be completely suppressed, and even the water content in the collagen may be decreased by adding appropriate amount of salt. Dehydration is not observed if acetic acid is used as acidifying agent. Dissociation constant of this acid is so low that it is of no influence on dissociation of protein carboxyls, and thus on changes in the system of bonds in native collagen. Fig. 13.2 shows the shrinkage temperatures of pickled pelt in relation to the amount of acid used. The curve for acetic acid is different from those for other acids.

Beside these features the increase of pelt plasticity due to the pickling can be observed, as shown in Table 13.1. In this Table one may see how pressure resistance of the swollen and pickled pelt is increased. However, after swelling the pelt is more elastic, while after pickling it is more plastic.

Prolonged pickling particularly at an increased temperature, may cause irreversible changes in leather, i.e., decrease of its stretch resistance, increase in

Table 13.1.

Deformation of swollen hide, as effected by pickle and acid

Medium	Pressure deformation at kg/cm^2			Deformation after removal of load	
				immediately	after 1 h
	0.05	0.55	1.2	% of initial thickness	
Water	22.6	52.2	56.9	30.7	2.1
0.175 m H$_2$SO$_4$ and 4m NaCl	5.8	32.0	44.5	24.8	19.0
0.05 m HCl	3.4	14.2	17.7	3.2	0.5

the ability to swell, elongation at break, and softness. These changes are advantageous in glove- and double-face leathers, not always advantageous in the shoe upper leathers. This observation is of particular importance, as pickling is sometimes used for pelt preservation. Portavela and Soldevila [1] in an extensive study have shown that during prolonged storage of pickled sheepskins a process occurs, which may be defined as the crosslinking bonds deterioration. According to those authors, an osmotic swelling during pickling is the first step in the process. Then lyotropic swelling replaces in part the osmotic one due to the salt action. At this stage the pelt becomes more soluble.

The damage of a skin due to the acid attack, observed as a change in the swelling curve and T_S after definite storage time is shown in Figs. 13.3 and 13.4. A very interesting, sharp minimum of swelling at pH = 4 appears in the skin exposed to an acid attack. This is accompanied by a shift in T_S maximum, and a very good correlation is observed between these two factors. The authors suggest a shift in the isoelectric point of collagen under conditions of prolonged pickling. They could not find new ionic reactive groups in the pelt treated with prolonged pickling, and associated the topochemical hydrolysis with the subsequent loosening of fibrous structure. They found a very sharp decrease of lastometer strength in leathers obtained from pelts, stored under acid conditions during 34 days as can be seen from the data given below:

	Grain crack		Leather crack	
	load	distension	load	distension
	g	mm	g	mm
Control	33.40 ± 8.70	12.01	34.10 ± 8.16	12.24
After acid attack	29.50 ± 6.29	10.89	30.10 ± 6.19	11.03

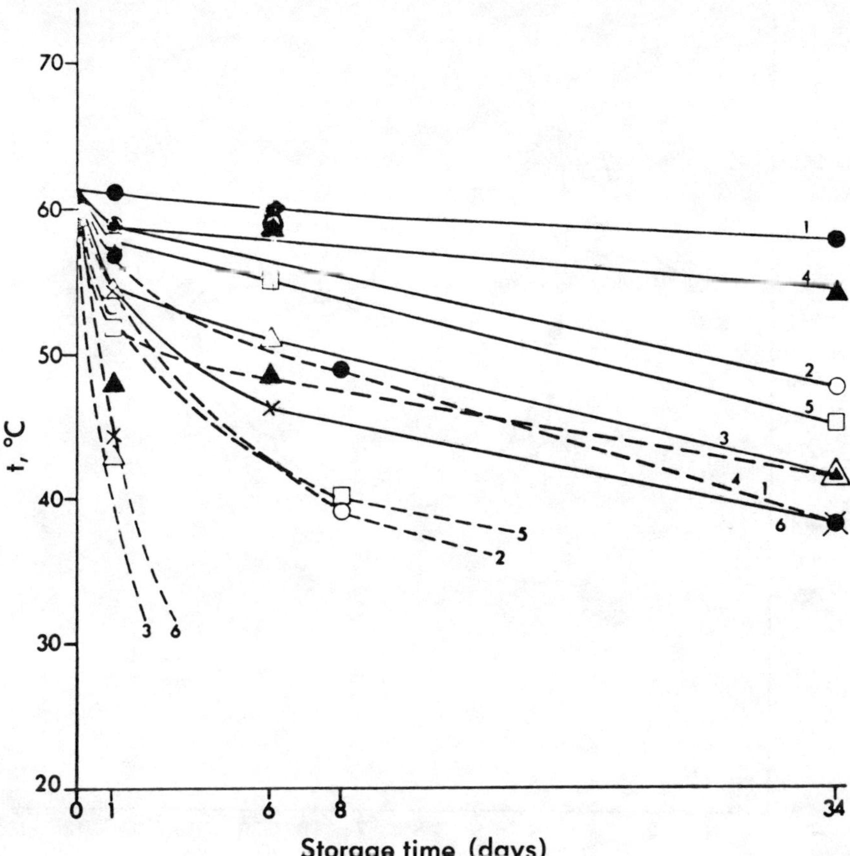

Fig. 13.3. Shrinkage temperature of pelts as a function of pickle and storage conditions according to [1]

Stored at 20°C
● 1% H_2SO_4
○ 3% H_2SO_4
Δ 9% H_2SO_4
NaCl brine density:1,045 g/ml for 20°C
1,090 g/ml for 40°C

Stored at 40° C
▲ 1% H_2SO_4
□ 3% H_2SO_4
X 9% H_2SO_4

For the reason given above, preserving pelts by pickling cannot be considered as safe, unless the moisture content and temperature are kept low.

Alum is sometimes used instead of sulfuric acid in the pickling. Used in the presence of a great NaCl excess it is partly hydrolysed and the basic aluminium sulfate formed then shows a low tanning ability. It acts mostly on the grain side, due to reaction with the alcaline groups of the fibers. Introduction of formaldehyde to the pickling bath, still containing organic acids, causes only slight

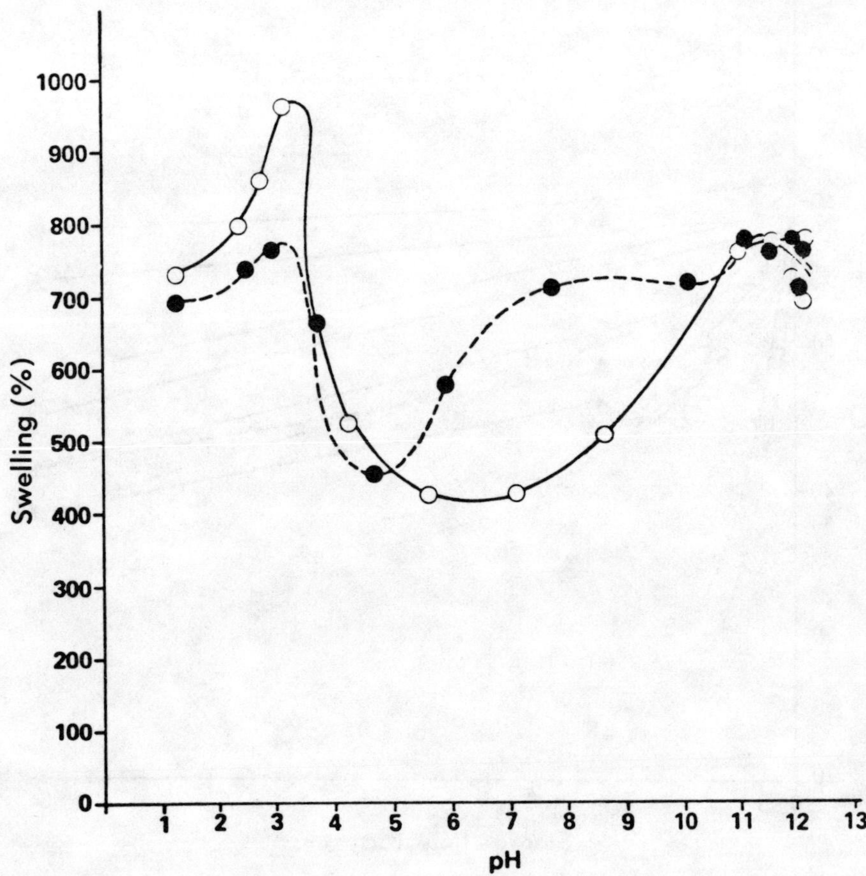

Fig. 13.4. Swelling of pelts at different pH values according to [1]. Open circles—nonpickled pelts, black circles—pickled pelts.

tannage due to a low pH-value, however, this is a process, reinforcing to some degree the very delicate grain of young hides. Other tanning agents for pretanning pelts, may be as well introduced to the pickling. Introduction of organic acids into the pelt may be considered as a method of chrome salt masking, i.e., as an operation which is reverse to topochemical tanning. This method is based on introduction of a tanning agent into pelt as a nonactive molecule. Then it has to be activated, e.g., by increasing the alkalinity. Thus pickling is the preparation of hide for chrome tanning. Under proper conditions it may be omitted, easing in this way tannery procedures. Investigations on tanning without pickling carried out at the Italian Leather Research Institute have shown that such tanning may be done, when chrome alum is applied to sheep- and goatskins. Chrome tanning was carried in these experiments according to Gustavson. The process is based on contacting undelimed pelt with a compound which reacts with the collagen-calcium system, forming in part soluble calcium-complex, in part insoluble collagen complex. The compound introduced must not have tanning ability as it has to penetrate into pelt. EDTA is one of these compounds. Chemical automation of processes of simultaneous deliming and tanning as proposed by Otto [2] has been described in section 12.4. Blazej and Smejkal [3] investigated prolonged action of EDTA on hide tissue. They did not find washing out of non-collagenous proteins, glucoproteids and glycosaminoglycans by the treatment of hide with 0.1 n EDTA-solution at pH 4.5 within 3-30 days. The quoted authors consider the following as main points of the EDTA-action:

(1) metal ion binding,
(2) bacteriostatic action,
(3) formation of chelates with some protein components.

The removal of metals chelated in this way is good; within 30 days 90% of Fe and Al are removed, ca. 85% of Ca and 80% of Mg. This speaks for the formation of Cr-complexes at pH $= 4.5$ which is controversial with the observations of Otto (loc. cit.). The bacteriostatic action is in good agreement with the observation of Yates [4] who found that 5×10^{-3} moles of EDTA decreases markedly the unhairing activity of pronase. An easily observable EDTA action is the removal of hair and epidermis. Thus Blazej and Smejkal (loc. cit.) suggested the formation of the EDTA-noncollagenous protein and EDTA-glycosaminoglycan complexes instead of natural chelates formed by these two macromolecules. Such macromolecular chelates with sufate group participation are supposed to exist in the skin [5].

REFERENCES

1. Portavela, M., Soldevila, M. XIV IULTCS Congress Barcelona 1975
2. see ref. 22 ch. 12
3. Smejkal, P., Blazej, A. XIII IULCS Congress Vienna 1973
4. Yates, R. J. Soc. Leath. Tr. Chem., *56*, 158 (1972)
5. Podrazky, V., Steven, F. S., Jackson, D. S., Weis, J. B., Leibovich, S. J. Biochim. Biophys. Acta *229*, 690 (1971)

14.

COLLAGEN TANNING

14.1. Concept of tanning

Tanning is a process of converting unstable raw hides into leather, with adequate strength properties and resistance to various biological and physical agents. In many languages distinction is made between leather and hides or skins, others have a common term for all of them.

Tanning is a process of introducing a tanning agent into the hide. This is accompanied by introduction of additional crosslinks into collagen, which bind the active groups of the tanning agents to functional groups of the protein. However, not all tanning procedures give rise to clearly defined bonds.

To what extent the resistance to microorganisms is a criterion of tanning remains an open question. It is true that tanned leather is very resistant to putrefaction, although that is largely a matter of humidity. In a hot tropical climate (40-45°C and nearly 100% RH) leather is very strongly attacked by mold already after 2-3 days of exposure. Other changes in the properties of leather occurring in it as a result of tanning are secondary ones, which may but do not have to take place, e.g., the T_s may remain unchanged, the resistance to water penetration may be achieved without introducing crosslinks.

The choice of the type and character of changes to be made in hide or skin in order to correct these properties in leather depends on the producer. The feel of leather may be considered, according to Lollar [1], as a part of its characteristics; it is not, however, independent of some processing stages: hide dehydrated with organic solvents may have a soft, warm and mellow feeling, whereas the dried 'wet blue' is stiff and hard despite being tanned. Thus the feel of leather, a very useful test of an experienced tanner, cannot be considered a tanning criterion. Two major questions arise here, viz.: what is the difference between collagen and other structural proteins, if one considers its capacity for being tanned? and what is the definition of the tanning agent? There is no full and unequivocal answer to these questions but it seems to be easier to answer the second one. The theory of tanning is a subject with a long history. The creators of tanning chemistry have expressed their opinions on that point. The current status in the science makes a unique theory impossible, as too many independent variables would have to be taken into account. The liability of collagen to tanning is the result of an ensemble of its features, whereas none of these features alone is sufficient. The features of this protein most important from the point of view of tanning are:

(1) The triple helical structure with spacings between the functional groups

which make it possible for many molecules to fit in between. The 'pockets' in this structure formed by a great number of glycine residues, and the 'protuberances' mostly in the form of stiff pyrrolidine rings and of side chains will make this structure particularly 'bewhiskered.' Owing to this a molecule of, e.g., a vegetable tannin of low but multipoint reactivity may easily be mechanically kept in the network. Drying after tanning causes shrinkage of interfibrillar spaces and thus keeping the molecules of tanning agent in the way shown in Fig. 14.1.

(2) Numerous, diverse, and variably spaced functional groups of the side chains give the tanning agent molecules the ability to be built in. This applies to some intramolecular bonds of the ester and aldehyde type as well.

(3) Numerous weak hydrogen bonds, oriented along the molecule, may be replaced by hydrogen bridges. The amino hydrogen and carboxyl oxygen participate in them, and an outside molecule, like, e.g., a water molecule. Distances between amino acid residues from adjacent helices do favor these changes because the intramolecular bond is a very long one, its length may attain 3 Å. These are the maximal distances known for hydrogen bonds. It should be emphasized again that only an ensemble of features decides whether or not the tanning process occurs in the particular proteins. This view finds support in a certain liability to tanning observed in other proteins and in some synthetic polymers. The ability to bind tanning agents (lesser than that of collagen) is revealed by keratin, fibroin, gelatin, polyamides and polyvinylpyrrolidone.

Wilson and Merril [2] have formulated the following axiom: 'if the chemical composition of proteins is changed in the way which imparts to it a resistance

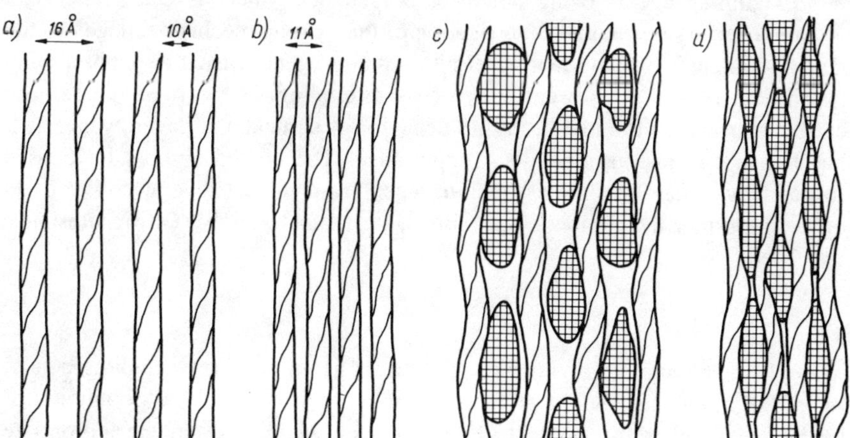

Fig. 14.1. Tanning agent molecules, mechanically involved in the collagen fiber network: (a) untanned, hydrated, (b) untanned, dry, (c) tanned, hydrated, (d) tanned, dry. (Heidemann).

against hydrolysis, it can be considered as tanned, and the material responsible for this change is called the tanning agent.' According to this definition the ability to change the definite feature in a material tanned is indicative of potential tanning properties of an agent. The chemical composition of leather may also be indicative of the probability of obtaining the physical properties required by the user: as one may presume, leather defined as waterproof but having an insufficient amount of fat is not likely to meet the requirement of waterproofness. For a long time efforts were made to support experimentally the view that the reactions between tanning agents and collagen are stoichiometric. According to that view chrome leather is considered as tanned, if it binds 3.38% of Cr_2O_3 in terms of dry weight. However, the available information concerning chromium complexes does not permit us to accept this view; the complexes differ not only in their composition, but also in their behavior with respect to leather. An important factor involved in the binding of the tanning agents is the competition between the particular reactions. According to Mikhailov [3] 'to tan means to treat collagen with substances whose every molecule is able to react with several polar groups of the peptide chains and form between them supplementary bridges resistant to water.' Being most recent this definition should be accepted; though the present author suggests to supplement it as follows: '*tanning is the treatment of collagen in conditions optimal for the reaction with substances, every molecule of which may react with several identical or different polar groups of peptide chains, forming supplementary water-resistant bridges between them, and changing at least one of the essential properties of the protein.*' According to Mikhailov, the reactions in which peptide bonds participate are not included into the tanning reaction. This view can hardly stand up to criticism in the face of the fact of tanning agents being bound to polyamides which have no functional groups in their side chains. Investigation of the tanning mechanism leads to the general conclusion that when a definite tanning agent is used it is not always possible to show an equally definite place of its binding. The work of Wiederhorn et al. [4] in which the highly elastic behavior of shrunken collagen was tested, has led to conclusions of crucial importance. The quoted authors have used formaldehyde, as the simplest tanning agent, and rat tail tendon collagen. In order to estimate the number of crosslinks, a formula for highly elastic substance was used:

$$V = \frac{R_{co}}{R_c} - 1 = \frac{M_{co}}{M_c}(1 + F) - 1$$

where R_c is the number of amino acid residues between two crosslinking bonds, M_c is the molecular weight of the section between bonds, F is the amount of formaldehyde in grams, bound to 1 g of collagen, M_{co} and R_{co} are the appropriate values for non-tanned collagen. For non-tanned collagen it was accepted that

$$R_{co} = \frac{M_{co}}{92.6}$$

this means the average value of one amino acid residue is 92.6. The value of M_{co} was estimated as equal to 55,000 which means that one chain contains two bonds. Putting these values into the above formula we obtain the values for V and M_c. Those values for a formaldehyde concentration of 0.3 and 1.0% in solution for various values of pH are given in Table 14.1.

The quoted authors came to a conclusion that:

(1) Formaldehyde forms 2 to 3.5 crosslinking bonds with respect to one originally present.

(2) The amount of the aldehyde participating in the bonds is only a part of the aldehyde bound by collagen.

(3) Only part (⅓ to ⅛) of amino groups of side chains participate in crosslinking. In the quoted paper [4] a dependence is shown to exist between the shrinkage temperature and the crosslinking bonds in collagen. The results of Wiederhorn were discussed in more detail since they have been obtained with the use of formaldehyde, which is the simplest crosslinking agent. Thus they can be considered as an introduction to this kind of investigations.

About 1950 two tanning theories were formulated, which are close to the contemporary knowledge about collagen and its modifications. Chyzewski [5], having made a thorough analysis of the quantitative relation between collagen and chromium, has accepted the hypothesis according to which in commercially tanned leather one chromium atom is connected to the carboxyl, and binuclear complexes are linked to two groups. The quoted author could not reject at that time the participation of amino groups in chrome tanning; he emphasized, however, that polynuclear copper complexes engaging amino groups as ligands do not have tanning properties. At the same time Batzer [6] expressed a view, according to which tanning is a dehydrative softening of collagen. According to this view the tanning process is a transformation of collagen, softened by tanning agents entered into it. It is based on the exchange of water for tanning agent in places where adsorption in the collagen molecules occurs. Batzer's theory may well agree with current views on binding of a chrome-tanning agent by the collagen carboxyls, and to some extent with the results discussed in section 5.4. According to the author's investigation, there is a change in behavior of water bound to collagen when chrome complexes are introduced. It seems that the majority of water molecules will 'jump over' from the collagen molecule to the chromium complex as the additional ligand.

The number of papers dealing with the influence of various factors on the tanning end effect is huge; this gives evidence of the host of knowledge in tanning chemistry. In this part of the book papers will be discussed which give a synthesis or an approach to a synthetic formulation of information concerning chemistry of the reaction between collagen and the tanning agent or formation of this agent.

Collagen fibers are responsible for leather strength. Addition of the tanning agent 'dilutes' them in a unit volume and decreases the material strength. As the

Table 14.1.

Number of crosslinking bonds in formaldehyde-tanned collagen in relation to pH

Content of HCHO in soln., %	Quantities calculated from formula	pH								
		3	4	4,6	5	6	7	8	9	10
0.3	M_c	—	> 40,000	38,000 ± 7000	16,000 ± 1000	14,000 ± 2000	19,000 ± 1000	20,000 ± 4000	13,000 ± 2000	17,000
	F	—	< 0.0001	0.0001	0.023	0.0054	0.0081	0.011	0.014	0.017
	V	—	< 0.4	0.45	2.5	2.9	1.9	1.8	3.3	2.3
1.0	M_o ~	29,000	15,300	—	16,600 ± 1000	13,300 ± 2000	12,500 ± 1600	15,000 ± 1000	13,000 ± 2400	16,000 ± 2500
	F ~	0.001	0.0023	—	0.0057	0.0089	0.012	0.013	0.017	0.023
	V	0.9	2.6	—	2.3	3.2	3.5	2.7	3.0	2.5

X-ray investigation has shown, it decreases the degree of ordering of these fibers in leather. However, it seems more likely that this is due to mechanical operations to which hide is subjected in the leather making process. The discussions on the tanning reaction mechanisms, very active formerly, have been less during recent years, however, not so much because of exhaustion of the problems, as in view of the increased collagen knowledge. This is a protein, having so many possibilities that a single reaction mechanism can hardly be accepted in definite cases. Discussion of the mechanism of binding of definite types of tanning agents concerns several points:

(1) Kind of bonds which in a given case may come into consideration. The chemical properties of the tanning agent discussed are considered, its amount in leather, the optimal conditions of its binding and the way of removing it.

(2) The isoelectric point of leather, influence of blocking of certain functional groups of collagen or tanning agent or both, or the way that tanning could be made impossible in the presence of the tanning agent in question.

(3) Binding of tanning agent by model substances used instead of collagen having functional groups of one kind only.

(4) Stoichiometric comparison of the number of functional groups of a given kind with the number of functional groups in the tanning agent which have reacted.

(5) Knowledge of molecular weight of the tanning agent, shape of its molecule, arrangement of functional groups and charges in it, which makes possible to evaluate whether it can penetrate into the structure, or it will stay on its surface, binding only some fragments of collagen molecule.

Harlan and Feairheller [7] gave the following scheme of tannages (Table 14.2), according to the functional groups of collagen involved. They also have calculated the number of functional groups available for the reaction. This is a very interesting proposal of the tanning type classification.

Table 14.2.

Functional groups involved in the tanning process

	$\text{Coll} - \overset{\displaystyle O}{\overset{\displaystyle \|}{C}} \diagdown_{\text{OH}}$	$\text{Coll} - NH_2$	$\text{Coll} - \overset{\displaystyle O}{\overset{\displaystyle \|}{C}} - NH\text{-}R$	$\text{Coll} - OH$
Relative number	3	1	19	3
Tannage type	mineral	aldehyde	vegetable	
	chrome	formaldehyde	syntans	
	zirconium	glutaraldehyde		
	alum			
Type of bonds formed	coordinative	covalent	hydrogen	

Considering the contemporary knowledge of collagen chemistry, composition and structure, we can expect a significant progress in the tanning chemistry. It is now possible to investigate the fragments of the collagen molecule and to say exactly which groups of collagen and at what position bind the tanning agent molecule, and by what bond. Such experiments will contribute to elucidation of the tanning process.*

14.2. The leather properties dependent on tanning

Fiber properties

The strength properties of fibers are affected by inter- and intrafibrillar bonds, both of chemical and physicochemical type. They occur only when the fibers are arranged in a parallel way. The cohesion forces between the molecules are significant only inside the fibrils. They operate in the final stages of collagen biosynthesis by connecting single molecules into fibrils. In native, mature collagen the amount of bonds increases only insignificantly, and there is no process of joining molecules into fibrils. Higher-range structure is only statistically ordered, as it may be observed with a microscope on the hide cross-section. Ordering of collagen fibers of that range is observed in tendons, unidirectionally and in layers (planar) in skins of fish and reptiles. Having to do with a network such as occurs in the hide, it is impossible to say if it is reinforced or weakened by the crosslinking. According to older views, tanning was meant to reinforce the hide. However, the investigations on single fibers, started in 1947, have shown that this view is incorrect. These investigations are very difficult and their results not always unequivocal because of their substantial scatter. It is very difficult to carry them throughout all operations in the beamhouse and through the tanning department. An additional difficulty arises from the unequal thickness of leather fibers, and the problems involved in its measurement; according to new sources the strength of those fibers depends on their size. In industrial laboratories the macroscopic testing of physical strength of leather is regularly made. Frequently conclusions are drawn from this testing concerning the influence of particular operations on the quality of the product.

Parameters such as the feel and physical properties of leather are the resultant of all processes contributing to leather production, not to the tanning itself. So Lollar [1] considers the chemical and biological properties of leather as more affected by tanning, than physical ones. From the definition quoted at the beginning of this chapter it follows that tanning takes place only if an external factor (tanning agent) is involved. Accordingly, it is right to expect some 'dilution' effect of native collagen properties. Thus the basic question of Lollar is to what extent except apart from this 'dilution' are the collagen properties undergoing changes during tanning? The chemical reaction itself is not enough

*Cf. note on p. 66

to believe tanning to be the process in which a collagen 'nonwoven' system is formed ready for finishing. Thus it is to be reconsidered whether the changes due to tanning are caused by crosslinking, or by chemical modification of fibers, like the well known acylation, deamination or succinylation.

Effect of tanning on collagen hydration

Tanned collagen as a rule binds less water than the native one. The data given in Table 14.3 demonstrate that there is a general tendency to decrease the water amount bound by leather. This decrease calculated on hide substance is different, being smallest in the aldehyde tanned leathers.

Michlik et al. have shown, using the calorimetric method [8] that the thermal effect of water to collagen binding is smaller in chrome tanned collagen (above 5 cal/g (21 J/g)) than in the native one (32.6 to 38 cal/g (136 to 159 J/g)). In the same paper significant differences in the heat of hydratation between the natural and synthetic footwear materials have been shown; e.g., for Clarino it is 1 cal/g (4.18 J/g). Mikhailov [3] believes it right to calculate the heat of hydration of the system, considering all its components: collagen, tanning agents and fats in appropriate ratio. Using this method one can show that the total effect of hydration of the components is somewhat smaller than that of the product. Tanned collagen remains a hydrophilic substance, even when 'oil tanned' (e.g., chamois leather). The hydrophobization sometimes observed is a phenomenon related to fat introduced in the fatliquoring process, i.e., after tanning, or to the formation of

Table 14.3.

Results of investigation of total absorption of water by collagen and leather of different tannage

Material	Water binding in an atmosphere of 95% RH, %		Hide substance content, %
	after 25 h	*after ca. 200 h (equilibrium)*	
Collagen, acid soluble	33.9	66.0	100.0
Pelt, dehydrated	23.6	50.3	90.0
" " impregnated by silicones	32.0	53.0	84.5
Leather, formaldehyde tanned	20.5	48.1	84.0
" starchdialdehyde tanned	21.1	50.0	77.8
" oil tanned	23.2	45.8	67.5
" chrome tanned (fatliquored)	20.0	42.3	69.0
" " " vegetable retanned	19.5	33.9	53.2
" " " " tanned	18.5	40.5	45.5
Poromerics	1.7	2.2	—

chrome soaps from salts and fat. As a result, the ability of collagen to swell and to bind water decreases drastically. It is dependent on the mutual mobility of the collagen structural elements. This mobility decreases as a result of tanning. The swelling ability, which is close to 200% for native collagen, is 60-75% for the hide tanned with vegetable tannins and about 100% for the chrome tanned one. Thus the decrease of collagen ability to swell may be believed to be one of the consequences of tanning. Tanning not only decreases swelling in the vicinity of the isoelectric point but the swelling in acids as well. Treated with alkalis, chrome or vegetable tanned collagen becomes detanned. Consequently, the collagen thus tanned swells in alkalis, although the formaldehyde-tanned one does not.

It has for long been known that in the water binding process first of all the functional groups of side chains are involved. Recently it was possible to demonstrate by the NMR method that water is bound to native collagen in a different way than to the chrome-tanned one. We can suppose that in this last case water is only in part bound to collagen and the balance is bound to the chromium complex. In the NMR spectrum of oriented collagen (tendon collagen) tanned with chromium complexes a strong band of oriented water molecules is observed [9]. Its changing width and intensity speaks for an additional orientation due to these complexes. This phenomenon is not observed in the formaldehyde-tanned collagen. Investigations done by Kanagy [10] could not demonstrate an essential difference between raw hides and vegetable-tanned leathers (section 17.1). These investigations made a long time ago, by a simple calorimetric method, are still frequently quoted because of their exactness.

Effect of tanning on collagen hydrolysis with acids, alkalis and proteolytic enzymes

Tanning makes collagen more resistant to hydrolysis by acids and enzymes, but not by alkalis. The effect of strong acid on tanned and nontanned collagen has been shown. One may see that strong acid hydrolyses native collagen quickly, but reacts slowly with the chrome formed collagen. (Fig. 14.2).

Tanning with chrome compounds or with tannins does not prevent alkaline collagen hydrolysis. Tanned and nontanned collagen, treated with 0.1 N NaOH at 65°C during 1 hour dissolves to various extents:

Non-tanned	85%
Vegetable tanned	34%
Chrome tanned	29%
Formaldehyde tanned	8%

The results of enzymic hydrolysis of collagen are controversial (Table 14.4), when it was treated with papain at 70°C, thus, after collagen denaturation. In another paper 3 molar urea and 0.1 molar sodium hydrosulfite have been used.

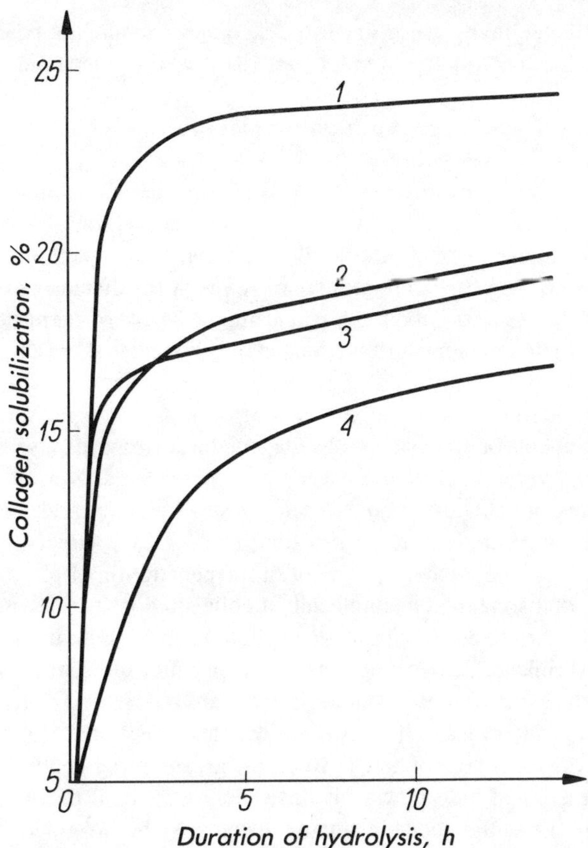

Fig. 14.2. Acidic hydrolysis (solubilization rate) in 0.1 N HCl at 65°C: 1—native collagen, 2—formaldehyde tanned collagen, 3 vegetable tanned collagen, 4—chrome tanned collagen.

Table 14.4.

Effect of tanning on enzymic hydrolysis, expressed in % of dissolved collagen

Collagen	Papain, pH 5.9	Trypsin, pH 8
Nontanned	85	73
Aluminum tanned	38	—
syntans (condensed phenols)	23	—
vegetable	25	29
quinones	—	34
formaldehyde	11	1
chromium salt	2	4

Thus the collagen investigated under these conditions may not be considered as native, and one has to keep in mind some degree of its detanning.

Tanning and structure of collagen molecule

It is impossible to define unambiguously the influence of a tanning agent on the molecule structure. This is so first of all for steric reasons—a formaldehyde molecule has diameter of 3 Å and all its fundamental functions are precisely known. A molecule of the chromium complex is of the diameter of 6-20 Å, that of the vegetable tannin—about 20 Å and more. Many of them have not been isolated and their fundamental functions may be calculated only in approximate way.

Formaldehyde is the most important compound for recognizing the effect of crosslinking agents on the macromolecule structure. However, its way of binding to collagen is very specific and it cannot be considered as a model for other tanning agents. It penetrates into the molecules and joints peptide chains without changing the wide-angle X-ray diffraction pattern of the molecule. It changes, however, its low-angle picture. A proof of the penetration of formaldehyde into structures is that so tanned the molecule of collagen does not decrease its tensile strength after shrinkage. In chrome-tanned collagen, despite its greater thermal resistance, shrinkage causes destruction. According to this, the tanning agents are not regularly distributed in collagen; there are spaces in it, free from tannin. A typical experiment was carried out, in which it was found that the higher the mechanical disintegration of hide powder, the greater its susceptibility to trypsin. If thus disintegrated hide powder is once more tanned, it becomes resistant to trypsin. Tanning with vegetable tanning agents may be imagined as the coating of fibrils with tannin molecules. Forming of such 'shells' is very fast. According to Mikhailov (loc. cit.) the molecules of soluble collagen have increased shrinkage temperature after one minute when treated with vegetable tannings, isolated fiber bundles—in 4-5 minutes, whereas in chrome tanning, 6-24 hours are needed. The process of vegetable tanning of collagen fiber network takes much more time owing to the effects described in Chapter 11.

14.3. Tanning thermodynamics

Investigations of energy levels of the collagen tanned in various ways lead to the conclusion about disturbed collagen structure due to the tanning. As Mikhailov and Mikelyan have found, collagen entropy value increases with the ability of the tanning agent to penetrate inside collagen molecule in this order: formaldehyde > chrome compound > vegetable tannins.

In their paper [11] heat effects have been calculated basing on the thermodynamics of the tanning process. The affinity of a tanning agent to collagen $\Delta\mu$

is expressed by the difference between the chemical potential of the tanning agent in solution (μ_L) and bound (μ_D):

$$\Delta\mu = \mu_L - \mu_D$$

but

$$\Delta\mu = RT \ln \frac{[D]_B}{[D]_L}$$

where R is the gas constant 1.987 cal/K/mole (8.24 J/K/mole), T is the absolute temperature, $[D]_B$ is equilibrium weight of the tanning agent bound by the unit weight of collagen (kg), and $[D]_L$ is the weight of tanning agent in equilibrium with solution. The D value should be multiplied by the activation coefficients; as, however, they are not known, very low concentration was applied presuming that under such circumstance activity coefficient is equal to unity.

The thermal effect ΔH, entropy ΔS and $\Delta\mu$ are related by the equation

$$\Delta\mu = \Delta H - T\Delta S$$

Thus from measurements of affinity $\Delta\mu_1$, and $\Delta\mu_2$ at various temperatures T_1 and T_2, ΔH and the entropy can be calculated. Such calculations done in the paper quoted, based on experimental values, are given in Table 14.5. The thermodynamic parameters of leather formation are gathered in Table 14.6.

Tanning agents are to form crosslinking bonds inside the collagen molecule or between its molecules. The tanning agent may form two or more kinds of bonds which are typical of it. For this to occur, however, two functional groups of collagen have to be at an appropriate distance, and the molecule of the tanning agent has to be arranged in a way, which makes possible to 'find' both these functional groups. Considering the tanning processes it is this situation that is

Table 14.5.

Thermodynamic parameters of selected tanning reactions

Kind and amount of tanning agent bound, given as part of hide powder	$\Delta\mu$ at 25°C kcal/mole (kJ/mole)	ΔH at 25°C kcal/mole (kJ/mole)	ΔS at 25°C kcal/mole (kJ/mole)/K
Formaldehyde, 0.5%	3.3 (13.8)	20.8 (87)	81.0 (338)
Chromium sulfate, 33.2% bas. (Cr_2O_3 6%)	3.5 (14.7)	12.1 (50.6)	52.3 (219)
Quebracho extract, 300% tanning agent	3.4 (14.2)	1.8 (7.5)	19.0 (79.4)

Table 14.6.

General parameters of hide tanning

Parameter	Formaldehyde	Quebracho
Molecular weight	30	1000
Tanning agent bound, %	0.2	70
Typical values: $\Delta\mu$/mole	− 3.0	−3.4
Leather formation per 1 kg of hide powder	− 2.0	−2.4
ΔH kcal/mole (kJ/mole)	+ 15.0 (62.6)	+2.2 (9.2)
ΔH of leather formation, kcal/kg of hide powder (kJ/kg)	+ 10.0 (41.8)	+ 1.5 (6.3)

anticipated. Then the question arises what happens if two or more functional tanning agents are bound at one point, by the use of just one of its possibilities. Then one may assume that the tanning agent forms a kind of prolonged side chain, having certain functional groups and 'seeking' a contact with other. Should this supposition be right, its further consequences are to be expected, usually some changes in collagen, its ability to attach other tanning agents, dyestuffs, fats, surfactants or impregnants. In fact such changes are known and registered but only in few papers are they ascribed to a single-point attachment of the tanning agent. Burton et al. [12] expressed their opinion about participation of chromium tanning agent in attachment of vegetable tannins used then in the process. As these authors believe, great affinity of chrome-tanned leathers to vegetable tannins is due to attaching of vegetable tannins to chromium complexes, rather than to the charged collagen amino groups. Wiederhorn and others [4] investigating the binding of formaldehyde by collagen came to conclusion that in the crosslinking reactions only $\frac{1}{7}$ or $\frac{1}{8}$ of the side-chain amino groups take part, and only a part of the collagen-bound formaldehyde. The amount of formaldehyde bound to collagen at one, or two points, as a function of pH of the reaction is shown in Fig. 14.3. One may see there that one half of aldehyde in neutral medium, more than one half in acidic medium, and less than a half in alkaline medium participates in the crosslinking bonds. The remaining amount of formaldehyde forms hydroxymethylene groups, that is, it is single-point bound. The amount of formaldehyde attaching to collagen is small and smaller than is generally believed, as it makes about 20 mmole (0.6%) in neutral medium, somewhat less in acidic, and more in alkaline. Binding of the dyestuffs as well as binding of fat by leather is different than by hide, owing to an intermediate role of the tanning agents (Sections 18.5 and 19.6).

Tanning and collagen reactivity

The question of influence of the tanning process on the collagen ability to come into chemical reactions is of crucial importance for further operations: dyeing and fatliquoring, as well as neutralization and drying. This influence may be different and depends on which collagen functional groups are involved in the crosslinking reactions, and which remain intact, i.e., ready for other reactions. It should be reminded here the one-point attached tannin, forming an additional side chain with functional group. In order to investigate these phenomena, frequently model substances instead of collagen have been used in tanning chemistry, or the collagen with certain functional groups in side chains blocked or removed. As typical model substances polyamides and polyvinylpyrrolidone have been considered, and ion exchange resins as well. Investigations of this kind were started by Gustavson and continued by others. In the last twenty years homopolymers of amino acids have been synthesized (polyglycine, poly-L-proline, poly-L-lysine, and esters of poly-L-glutamic acid), and regular polymers of tri, tetra-, octa-, and dodecapeptides of definite sequences. To investigate models of polyamide and polyvinylpyrrolidone type is inasmuch wrong, as they have spacial conformation entirely different than collagen. This may lead to erroneous conclusions concerning the place and kinetics of binding of the tanning agents. Such conclusions cannot be transferred to collagen due to structural differences and due to the fact that in the collagen tanning process various competitive, cooperative or independent reactions may occur (Chapter 2). One may believe that regular polypeptides of helical structure able to form hydrogen bonds may be of use for investigations in tanning chemistry. However, as yet they are very expensive, hardly accessible and for the time being used only in biochemistry and in structure investigations.

Among collagen functional groups of greatest importance for tanning reactions are carboxylic and amino groups, whereas in the last group we may consider the more complicated systems of histidine and arginine. Properties of the hy-

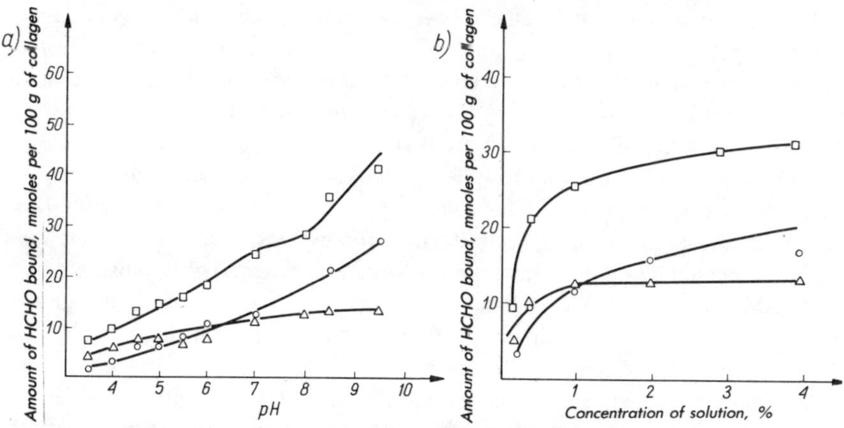

Fig. 14.3. Effect of pH (a) and concentration of formaldehyde in solution (b) on its binding by hide powder: □—total HCHO bound, ○—HCHO one-point bound, i.e., amount of hydroxymethylene groups, △—HCHO two-point bound.

droxyls of hydroxyproline and serine are not much known; the hydroxylysine hydroxyl participates in collagen glycosylation. Functional groups in the side chains of residues of such compounds as aspartic and glutamic acids, histidine, lysine, hydroxylysine and arginine may become protolysed. The pK values for those groups are, respectively, 8.60; 9.67; 9.07; 10.28; 9.67 and 12.6, so are very high, i.e., their acid constants are very low. This is why these groups form very slowly ionic bonds, and pH values of the protolysis are very low for acidic amino acids, and very high for the basic ones. Accordingly, also of great importance are other kinds of bonds between collagen and tanning agent.

As model tanning agents individual compounds of known constitution are used. Among chromium^{3+} complexes it is difficult to find such a compound, sufficiently stable and of a definite composition; thus the tanning with chromium salt models is rare. Aluminium salts or organic compounds are more frequently used. Bifunctional compounds of difluorodinitrophenylsulfonate type (Sanger's reagents) have been introduced to leather chemistry by Zahn and his coworkers [13]. These compounds form crosslinking bonds whose length is dependent on the compound structure. This kind of 'chemical rule' was a very valuable tool in collagen investigations about 20 years ago.

Tanning may be considered as collagen stabilization due to which its reactivity decreases and resistance to microorganisms increases. Leathers are very durable in soil, particularly in dry soil and under water. Among archeological findings one may often see shoes or leather goods, though brittle and weak due to oxidation, but keeping their shape. In the Egypt Museum in Cairo the soldiers' sandals and leather bags are shown from the times of the 17th Dynasty which are about 3500 years old. Only some leather dust scattered around them demonstrates their proceeding deterioration—the shape remains still to be observed by future generations—a testimony of incredible leather durability.

Under certain conditions leather becomes detanned, i.e., the bonds collagen-tanning agent are split. For example detanning and deterioration of chrome-tanned leather may occur in the course of wearing probably due to lactic acid, secreted by sweat glands of the feet. This compound enters the chrome complexes at sites occupied by the collagen carboxyls.

Tanning consist in binding to hide the substances that change its properties in a definite way. Formation of hide-tanning agent bonds is connected with some energetic effect (energy gain). Introduction of various substances into hide may be accompanied by neither energetic nor structural effect. These substances are fillers. Mineral oils, inert inorganic substances like kaolin belong to this group. Some, as fats, may form hydrophobic bonds with the collagen side chains; there the energetic effect is too small to be observed. Others, e.g., dry bentonite, may be maintained mechanically in the collagen network (as bentonite molecules are binding water) increasing their volume without influencing collagen properties. Some inorganic pigments may as well be forced into leather without reaction with it. One may as well fill leather with monomers of resins or with prepolymers, and after thorough polymerization make a kind of replica of fiber network, not connected to collagen, but of some stiffness.

Most important is chrome tannage, as compared with zirconium, aldehyde, sulfochloride, syntan, vegetable and oil tannages. Recently titanium tannage has been reported by Russian scientists and engineers to give very promising results. Combinations of particular kinds, very important from technological viewpoint, are too complicated from the physicochemical one to be described in a brief approach.

REFERENCES

1. Lollar, R. M. in Chemistry and Technology of Leather ed R. O'Flaherty vol. 2, Reinhold N.Y. 1958
2. Wilson, J. A., Merril, H. B. Analysis of Leather, McGraw-Hill N.Y. 1931
3. see ref. 5 ch. 2
4. Wiederhorn, N. M., Reardon, G. V., Browne, A. R. J. Am. Leath. Chem. Assoc., 48, 7 (1953)
5. Chyzewski, E. Physical Chemistry of Tanning, vol. 2, Warsaw 1950 in Polish
6. Batzer, H. Chemiker Z., 76, 397 (1952)
7. see ref. 28 ch. 10
8. Michlik, I., Strabelova, E., Koutny, P., Blazej, A. Kozarstvi 20, 187 (1970) in Slovak
9. see ref. 14 ch. 5
10. Kanagy, J. R. J. Am. Leath. Chem. Assoc., 45, 12 (1950)
11. see ref. 4 ch. 11
12. Burton, D., Reed, R., Wood, M. J. J. Soc. Leath. Tr. Chem., 40, 91 (1956)
13. Zahn, H., Wegerle, D. Leder 5, 121 (1954)

15.

TANNING WITH INORGANIC AGENTS

A vast number of inorganic compounds used to treat collagen increase its shrinkage temperature. Salts of bi- and trivalent metals are in this group. Among them are tungsten and molybdenum salts used for 'staining' of collagen preparations for electron microscope investigations. Crosslinking properties of various salts are shown in Table 15.1, where only maximal T_s values are shown. More detailed data may be found in the original paper [1].

15.1. Chrome tanning

Chrome tanning is the most important tanning method used to obtain light, inexpensive leathers of high thermal and bacterial resistance. To obtain a utilizable product, however, further operations are necessary, like, e.g., dyeing, fatliquoring, possibly retanning and impregnation. A usual procedure is to introduce the Cr^{3+} salts into the hide, adjusted to a pH of about 3 by pickling, and then make the collagen-chromium complex crosslinking reaction to occur. Two-bath chrome tanning initially applied was based on saturation of hide with Cr^{6+} compound, followed by its reducing to Cr^{3+}. This method is no longer used.

Table 15.1.

Effect of some salts on the shrinkage temperature of cattle hide

Salt	T_s °C	Optimal pH	Tanning effect
Control sample	63	—	—
$AgNO_3$	63	7.9	none
$ZnSO_4$	65	7.0	none
$LiCl$	66	8.0-8.4	none
$SnCl_2$	66	4.6	none
$CdSO_4$	67	6.7	none
$Pb(NO_3)_2$	68	6.9	none
$MgSO_4$	68.5	7.5	none
$Cd(C_2H_3O_2)_2$	69.5	7.3	none
$Zn(C_2H_3O_2)_2$	69.5	6.3	none
$ZnCl_2$	70	7 <	borderline
$TiCl_4$	71	6.5-7.8	borderline
$CuSO_4$	73	5.5	borderline
$Au_2(S_2O_3)_3$	73	7 <	borderline
$Hg(C_2H_3O_2)_2$	91	5.2	positive

Chromium complexes participating in the tanning

Chromium cation Cr^{3+} in aqueous solutions occurs as hydrated hexaquochromium ion $Cr(OH_2)_6^{3+}$. The most important property of this ion, from the point of view of tanning chemistry, is that the water held by its ligands can be exchanged for other ions, e.g. acid residues or hydroxyls, which latter can be considered as a case of deprotonation. Due to this deprotonation reaction positive charge of the complex becomes lower.

$$[Cr(H_2O)_5(OH)]^{2+}$$

and this is a complex, having a 33% basicity. The complex having a 100% basicity is the uncharged chromium hydroxide, insoluble in water. Basicity is also a measure of the number of charges of chromium complexes and for more complex compounds it may take not only the values of 33 (⅓), 66 (⅔), 100% (³⁄₃), but intermediate values as well. The chromium compounds used in a tannery are usually of 30-50% basicity, which means that they contain 1.0 to 1.5 hydroxyls per one chromium atom. Lower-basicity components do not attach to collagen molecule, or the bonds formed are very weak. By certain special methods, like, e.g., increase of ionic strength of the solution, the higher-basicity components can be made to precipitate on collagen, but this is not resorted to in tannery practice. Definition of acidity of complexes, as used by some investigators, is a complement of basicity, viz.

$$100\%\text{—basicity } = \text{ acidity.}$$

The trivalent chromium complexes have the coordination number six and form octahedral structures. Their coordination valencies are directed to the corners of regular octahedra. This has been known since classical Werner's investigations on the chemistry of complexes. Hence one may expect seven types of complexes, starting with $[CR(H_2O)_6]^{3+}$ with successive replacement of every water molecule up to $[Cr(OH)_3 X_6]^{3-}$. The hydroxyls may be replaced by other ligands, and thus cationic complexes are followed by anionic ones (three of each) and one uncharged (X is an anion). Not all species of the complexes are isolated and identified. Tanning with cationic and anionic complexes gives different results, as they attack various collagen functional groups. The charge of the main components of chrome liquor may be determined by electrophoresis.

Recent papers of Kawamura and his coworkers [2] have set out a way in investigation of chromium complexes occurring in tanning liquors by electrophoresis technique. They have been able to demonstrate the presence of several chromium complexes in tanning liquors, some of them separable and identified. Similar technique was used by Slabbert [3]. This valuable method is still under investigation, but important conclusions about particular compounds of tanning float may be drawn.

An exhaustive description of the isolated and studied chromium^{3+} coordination compounds may be found in Gmelin's Handbook of Inorganic Chemistry. Immense variability of Cr^{3+} compounds and ability to form various salts is a reason for some scepticism, which is recommended when speaking about characteristics of these compounds derived from tanning chemistry, particularly if an older description is considered. Sometimes they are not chemical individuals, which can be isolated or whose structure and composition are not fully elucidated. In description of structure and properties of chromium Cr^{3+} compounds only opinions of inorganic and complex chemists may be considered as conclusive ones (if at all available in the case at hand).

From the known isomers of hydrated chromium chloride $AgNO_3$ precipitates, respectively, 100, 67 and 33% of chlorine. This suggests the following structures for these isomers:

$[Cr(H_2O)_6]Cl_3$ violet gray
$[Cr(H_2O)_5Cl]Cl_2 \cdot H_2O$ green
$[Cr(H_2O)_4Cl_2]Cl \cdot 2H_2O$ dark green

The question how strong is the bond between water molecules and chromium in water solution can be answered by carrying out an exchange reaction between these ions in ^{18}O-enriched water

$$[Cr(H_2O)_6]^{3+} + H_2{}^{18}O \rightleftharpoons [Cr(H_2O_5)(H_2{}^{18}O)]^{3+} + H_2O$$

The reaction rate for one water molecule is defined by constant k.

$$6k[Cr(H_2O)_6]^{3+}$$

For ionic strength $= 0$ at 27°C the value of k is $2.15 \pm 0.15 \times 10^{-4}\,min^{-1}$. This value, calculated for half-life time of the system, is surprisingly high, $t_{\frac{1}{2}} = 54$ h, whereas for $[Al(H_2O)_6]^{3+}$ $t_{\frac{1}{2}}$ is $10^{-2}s$, and for ions Li^+, Mg^{2+}, Bi^{3+} it is of an order of $10^{-4}s$. Experiments done with $H_2{}^{18}O$ confirm the existence of six water molecules, linked directly to the Cr^{3+} ion in water solution.

As early as 1910 Bjerrum [4] made a classical experiment, which greatly contributed to the understanding of the chemistry of chromium complexes. To a chromium chloride solution NaOH was added in an amount of 0 to 33% of the amount required to bind all the chloride ions by sodium ions; the pH of the solution increased in this experiment from 2.5 to about 4 (Fig. 15.1, curve 1) and after 24 hours it dropped to 3.1-3.2. By titration with 0.1N HCl the pH values decreased according to curve 2. The end pH at titration was about 1.0 for calculated zero basicity. An explanation of this was the acid added during titration (2) and not used up by the hydroxyls which, as it was interpreted by Bjerrum, had to be built into internal sphere of the complex, because they did not react with acid. If the formed basic chromium complexes have been heated or allowed to age for some time, between chromium atoms the oxygen bridges formed, which were highly resistant, remaining untouched even when heated in acid solution. This is called olation, and the oxygen bridges formed in the complexes are defined as μ-bridges. Olation reaction goes in the following way (the unengaged groups and charges are omitted)

$$I \; Cr(H_2O) \underset{\text{acids}}{\overset{\text{bases}}{\rightleftharpoons}} Cr(OH) + H_3O^+$$

$$II \; Cr(OH) + Cr(H_2O) \overset{-H_2O}{\rightleftharpoons} Cr - \overset{H}{\underset{(a)}{O}} \rightarrow Cr \rightleftharpoons Cr \leftarrow \overset{H}{\underset{(b)}{O}} - Cr$$

The forms (a) and (b) are resonant-structures and thus cannot be separated. This is because the ol-bridges are usually depicted by conventional formula Cr-OH-Cr; the Cr-O-Cr angle is usually ca 120°. Kinetics of the reactions preceding olation have been studied. The measured dissociation constant of hexaaquo-chromic cation is

$$K_a = [Cr(H_2O)_5(OH)]^{2+} [H_3O]^+ [Cr(H_2O)_6]^{3+} = 10^{-3.82}$$

which is, in other words, constant of deprotonation reaction

$$[Cr(H_2O_6)]^{3+} + H_2O \underset{k_w}{\overset{k_n}{\rightleftharpoons}} [Cr(H_2O)_5(OH)]^{2+} + H_3O^+$$

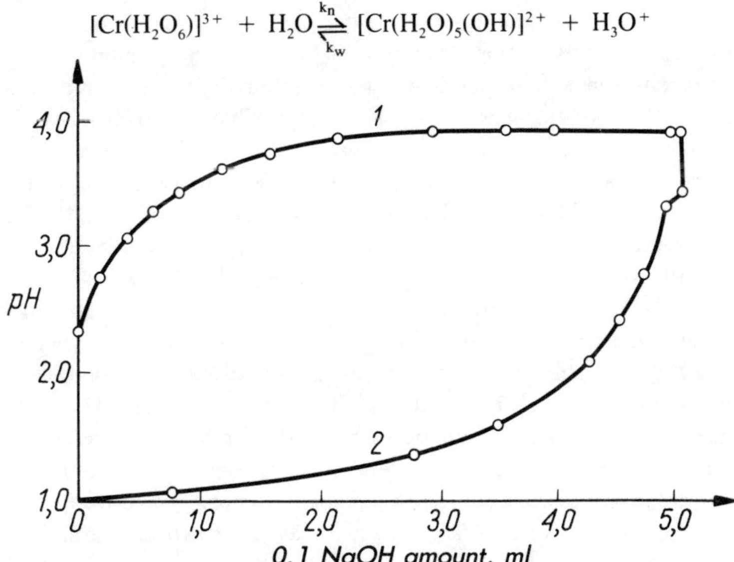

Fig. 15.1. Titration curve of hexaquochromium chloride according to Bjerrum: 1—back titration of NaOH excess immediately after addition of this solution, 2—back titration after 24 hours.

where k_n means rate constant, when going from left to right, and k_w is the rate constant in the opposite direction, whereas $K_a = \dfrac{k_n}{k_w}$. Measurement done by the dielectric relaxation method for fast reactions have shown that k_w is $10^{11} - 10^{12}$ $mol^{-1} \cdot s^{-1}$ for all the processes, in which one of the substrates is the hydrated proton. Thus the rate constant in ionisation reaction $k_n = K_a \times k_w = 10^{7.2}\ s^{-1}$ This reaction, and hysteresis effect in Bjerrum's experiment may be explained only by formation of olated complexes as a result of aging and a low rate of bridge splitting. According to some investigators the lost of a proton may cause oxolation and formation of oxygen bridges. If ionisation and condensation process is going further, its result is a formation of complex with two ol-bridges.

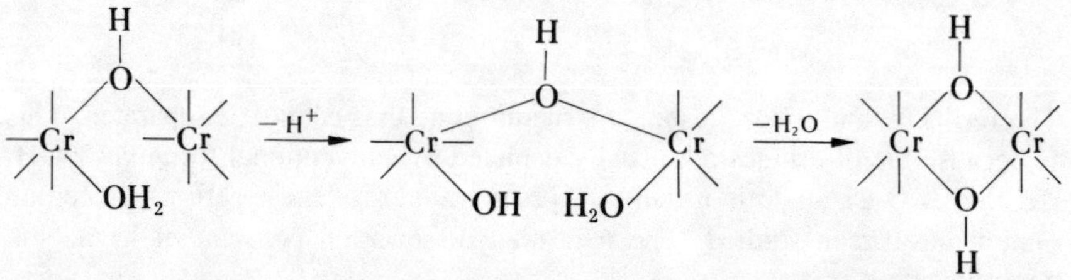

$$
\begin{array}{ccc}
\text{(ol-bridge complex)} & \xrightarrow{-H^+} & \text{(oxo/aquo complex)} \xrightarrow{-H_2O} \text{(di-ol bridge complex)}
\end{array}
$$

Thus some further possibilities arise for the formation of a two-nuclear complex with one oxo- and another ol-bridge, or even two oxo-bridges. Experimental proofs of that kind of transitions, however, are not quite convincing.

Olation process may lead to the formation of tri- and poly-nuclear complexes; however some steric limitations exist in that case. In a complex having two ol-bridges (Fig. 15.2) two octahedra are connected along a common edge, giving approximately planar, four-membered ring consisting of two chromium and two oxygen atoms. The distances among the next possible sites of the coordinative bonds formation (axes 1, 6 and 1', 6') are quite great. The tendency to leave these axes in a position close to the former one, i.e., perpendicular to the ring surface, limits the choice of chelating agents to chelating compounds of appropriate shape and size. Carboxyl and sulfate residues are among such agents. From calculations checked on the model it is obvious that they are almost perfect for the formation of a six-membered, stress-free ring (Fig. 15.2b and c). In complexes containing two ol-bridges the corners most distant from their centers are 4 and 5 as well as 2' and 3'. In fact they are coplanar and cannot be 'stressed' by any of the chelating groups known. They may however take part in the formation of longer chains bound together with diol bridges. Structures proposed in the tanning chemistry literature in which closed rings of, e.g., four Cr-atoms are formed may not be accepted stereochemically. A bond is possible that consists of one ol-bridge and one sulfate group; in this case two conformations may occur: one stiff and one flexible. This last one makes a formation of di-ol bridge possible (Fig. 15.2b). This might be related to the influence of sulfate residues which make olation easier.

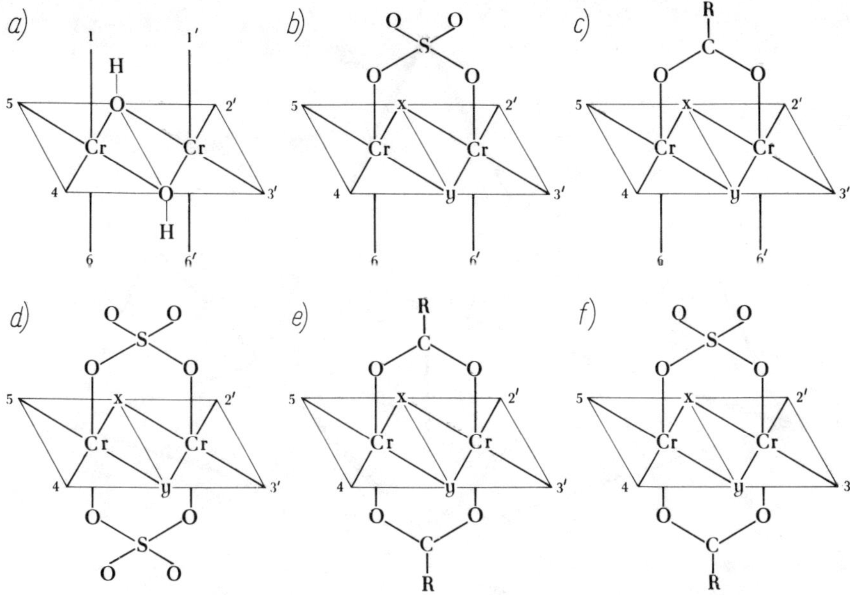

Fig. 15.2. Stereochemical scheme of the two-core complexes of chromium³⁺, containing two ol-bridges. There are further possibilities of forming the next bridge—forming bonds due to sulfate or carboxylate residues (x = −OH or =O, y = −OH or =O, R = H) or alkyl (aryl) residue of the end of the side chain of collagen molecule with a carboxyl group.

If two octahedra are connected with a common wall (not edge), the remaining bonds (4, 5, 6 and 2′, 3′, 6′, 7) change their position in a way shown in Fig. 15.3. Sterically impossible is then to bridge two chromium atoms by additional chelating groups. So distant positions cannot be bridged, which excludes plenty of nonrealistic compounds. Sterically possible however is to connect two chromium atoms through two ol-bridges and at the same time by two bidental ligands (sulfate or carboxylate residues) as it has been shown in Figs. 15.2d, e, f. By forming such bond the other bonds remain approximately coplanar and directed outwards of the metal atoms. This causes a solubility decrease of chromium complexes with increasing dilution of the solution, and also increase of size of chromium complexes.

Kawamura and Wada [5] have applied the electrophoresis technique for separation and identification of three main components of sulfate liquor of 33⅓% basicity. Then Indubala and Ramaswamy [6] using ammoniacal form of Dowex 50W-X12 have isolated from identical liquors as well three complexes, among them a known and already tested [Cr(H₂O)₅SO₄]⁺. The remaining two have been bi-nuclear complexes of a total charge of 2⁺ and a Cr:SO₄ ratio of 2:1. These authors have measured the half-life of the sulfate bridges of complexes with

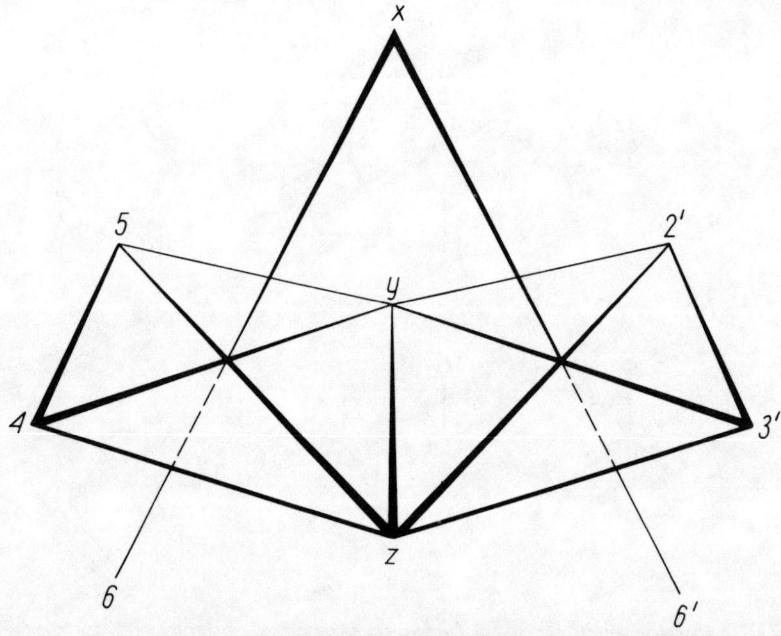

Fig. 15.3. Stereochemistry of a binuclear complex of Cr^{3+}. There is no possibility to form further bridges by two-dentate ligands (notation as in Fig. 15.2).

tracers; they obtained the value of 6-44 hours

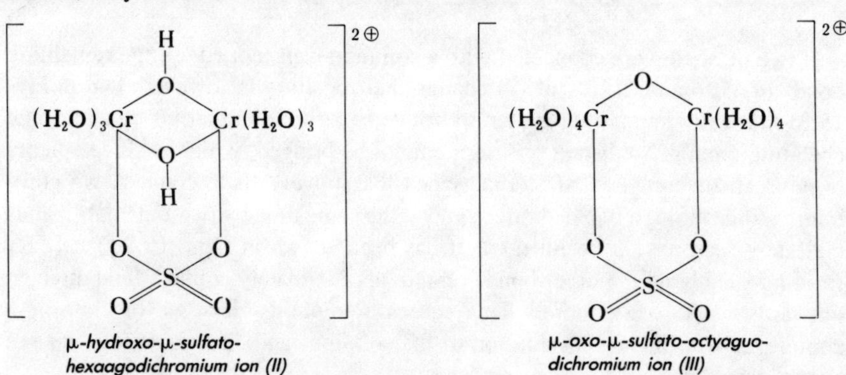

μ-hydroxo-μ-sulfato-
hexaagodichromium ion (II)

μ-oxo-μ-sulfato-octyaguo-
dichromium ion (III)

To solve the problem, whether the chromium atoms are connected with ol-bridges or with oxo-bridges and if the sulfato groups act as mono- or bidentate ligands, is a key to the complex structure, and indirectly—to confirm or deny the current views and theories of binding of chrome complexes to collagen.

Whether the sulfate residues are connected mono- or bidental can be found from the IR-spectrum. Comparing the formula of differently bound sulfate residues one may note a decreasing degree of symmetry

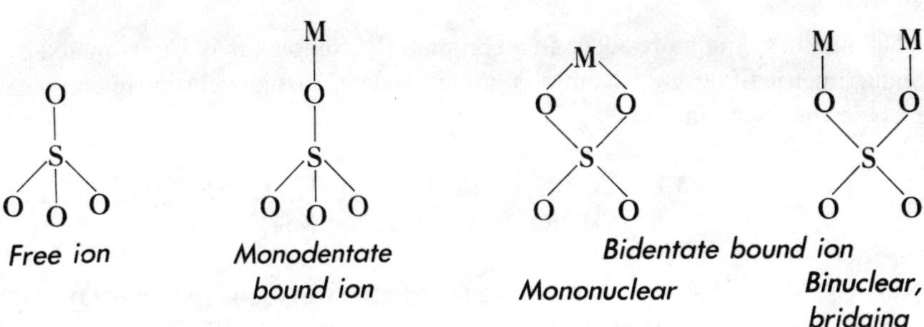

as the number of bonds via oxygen increases. This means the appearance of new bands in the spectrum and from their position and intensity one may conclude about binding of the sulfato group. This method was introduced about 15 years ago and in quoted paper it was used to show that both complexes tested have two nuclei (cores) and have the bridging sulfato groups as they have four absorption bands each in the range of 900-1300 cm^{-1} (Fig. 15.4). Complex II has 2 μ-hydroxo-bridges; complex III—one μ-oxo bridge (see p. 330).

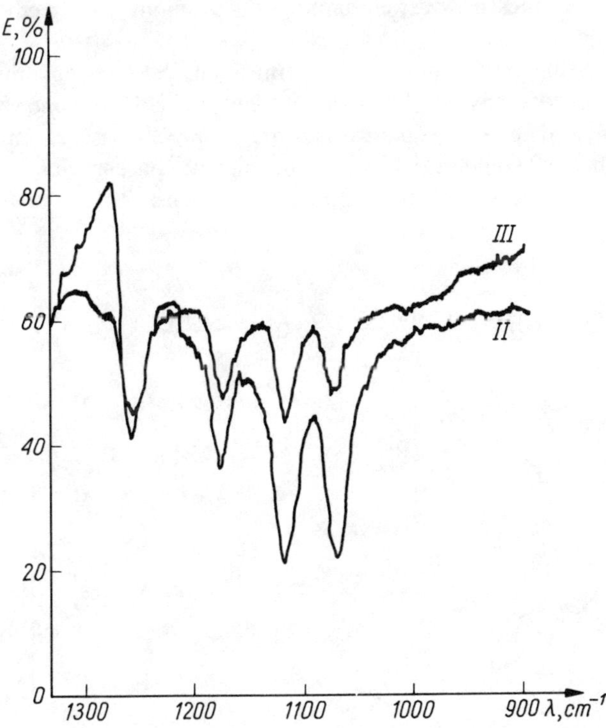

Fig. 15.4. A part of the IR spectrum, showing the difference between binuclear complexes, having μ-hydroxo (II) and oxo- (III) bridges. In both complexes the sulfate ion occurs as a bidental ligand, thus the number of bonds is the same. According to [6]

The number of aquo residues in a chromium^{3+} complex may be estimated by conductimetric titration with ammonium molybdate. Hexaquochromate complex reacts in the following way:

$$7\,Cr(H_2O)_6(NO_3)_3 + 6(NH_4)_6\,Mo_7O_{24} \rightarrow 7(NH_4)_3Cr(MoO_4H)_6 +$$
$$+\ 15\,NH_4NO_3 + 6\,HNO_3 + 18\,H_2O$$

giving Cr : Mo ratio a 1 : 6. This way the ratio 1 : 5.0 was found for $Cr(H_2O)_5SO_4$, 1 : 3.14 for the complex II, and 1 : 3.81 for the complex III, which suggests a ratio of $3H_2O$: Cr and $4\,H_2O$: Cr, as in the compounds of formulas II and III. The recent experiments concerning IR-investigation of chrome complexes, masked with various organic residues could not be believed successful, because no information may be obtained from the spéctra, except that the masking agent obscures the initial spectrum and, in turn, the initial spectrum is changed by the masking agent. However, this publication is worth mentioning (Takimoto [7]).

According to Riess [8] it can be demonstrated how the molecular weight of chromium complexes increases size by thermoelectrical measuring of the difference in vapor pressure over chromium salts solutions. This method seems to be no more valid.

Slabbert, using combination of gel permeation, gel electrophoresis, ion exchange chromatography and Vis spectroscopy, was able to demonstrate that at least 10 ionic and nonionic complexes are present in 33% basic chromium sulfate solution. The six compounds present in the highest concentration, and of determined structure, are shown in the following formulas.

$$10\%\,[Cr(H_2O)_5\,SO_4]^+ \xleftarrow{\;SO_4^{-2}\;} [Cr(H_2O)_6]^{+3}\ (9\%) \xrightarrow{\;OH^-\;} [Cr(H_2O)_5\,OH]^{+2}$$

with further conversions (down the SO_4^{-2} path and the OH^- path) to the following bridged structures:

$$\left[\,O_3S\!-\!O\,\underset{Cr}{}\overset{H}{\underset{O}{O}}\,\underset{Cr}{}\,O\!-\!SO_3\,\right]^{0}\;(11\%)$$

$$\left[\,Cr\,(\mu\text{-}OH)_2\,Cr\,(\mu\text{-}SO_4)\,\right]^{+2}\;(32\%)\quad(\text{after }-SO_4^{-2})$$

$$5\%\left[\,(H_2O)_4\,Cr\,(\mu\text{-}OH)_2\,Cr\,(H_2O)_4\,\right]^{+4}$$

$$20\%\;\left[\,Cr\,(\mu\text{-}OH)_2\,Cr\,(\mu\text{-}OH)_2\,Cr\,\right]_{n}\qquad n = 0,1,2$$

This is an oversimplified mechanism, and no details have been checked as yet, but it gives an idea of the processes occurring in such a solution.

Olation increases the stability constants of the chromium sulfate complexes. There are some exceptions to this rule, namely stability constant of the remaining ligands increases, if two places are occupied by the oxalic acid carboxyls. Probably these irregularities may be a result of an additional resonance involved.

Reactions of chromium complexes (mostly exchange reactions) are characterized by their particular slowness. This feature makes it possible to control the process of making and aging of floats. This also affects certain properties of complexes which are of importance for tannery as, e.g., penetration of them into leather, and binding to collagen, i.e., tanning kinetics. To explain these properties, one has to start with the definition of complex stability, as expressed by stability constant. Chromium compounds are frequently reported as generally stable. One has to distinguish, however, between thermodynamic stability and inertness as an opposite of lability—this latter is related to kinetic aspect of reaction. To give a simple example: as it is known, water is 'stable' as related to its components, it is formed from them with liberation of a great amount of energy. Nonetheless, a mixture of oxygen and hydrogen may be stored for a very long time in darkness at room temperature, as these gases without a catalyst do not react with each other.

The stability constant K_{ML} of a compound formed in a reaction $M + L \rightleftharpoons ML$ is the value of equilibrium constant of this reaction:

$$K_{ML} = \frac{[ML]}{[M][L]}$$

For reaction $M + nL \rightleftharpoons ML_n$ a total stability constant may be defined by

$$\beta_n = \frac{[ML_n]}{[M][L]^n}$$

These constants are determined by activities because their values depend on ionic strength and temperature. If K is great, the compound may be considered as stable. Liberation of energy during its formation defines the formula

$$- \Delta G = 2.303\, RT$$

Such a definition of the stability constant makes it independent of concentrations of the substrates. Stability constants of the attachment reaction of the oxalate residues to chromium ions have been estimated as follows:

$$K_1 = [Cr(C_2O_4)^+]/[Cr^{3+}][C_2O_4^{2-}] = 10^{5.39}$$
$$K_2 = [Cr(C_2O_4)_2]/[Cr(C_2O_4)^+][C_2O_4^{2-}] = 10^{5.17}$$
$$K_3 = [Cr(C_2O_4)_3^{3-}]/[Cr(C_2O_4)_2][C_2O_4^{2-}] = 10^{4.93}$$

and for ethylendiamine (en)

$$K_1 =]Cr(en)^{3+}]/[Cr^{3+}][en] = 10^{16.5}$$
$$K_2 = [Cr(en)_2^{3+}]/[Cr(en)^{3+}][en] = 10^{14.32}$$

the K_3 value is not easy measurable and is about $10^{11}/^{11.5}$. It means that the 1 : 1 chromium diamine complex is about 10^{11} times more stable than the corresponding oxalate complex. These formulas do not contain the values of $[H]^+$ and $[OH]^-$ and thus the stability constant does not depend on pH. It is not quite right, as in calculations dissociation constants of acids and bases K in solution have still to be considered, i.e., to take into consideration the participation of particular reactions: pK_1 for oxalic acid $= 1.13$, $pK_2 = 3.85$ for ethylenediamine $pK_1 = 7.42$ and $pK_2 = 10.14$. It may also happen that in a mixed solution both types of complexes as well as mixed complexes may occur.

In analyzing the reaction of coming of new ligands B into octahedral complexes (Fig. 15.5) one may presume operation of one of two mechanisms:
(a) group B comes in a place of one of the six bound A groups
(b) group B becomes attached, and one of the A groups leaves.

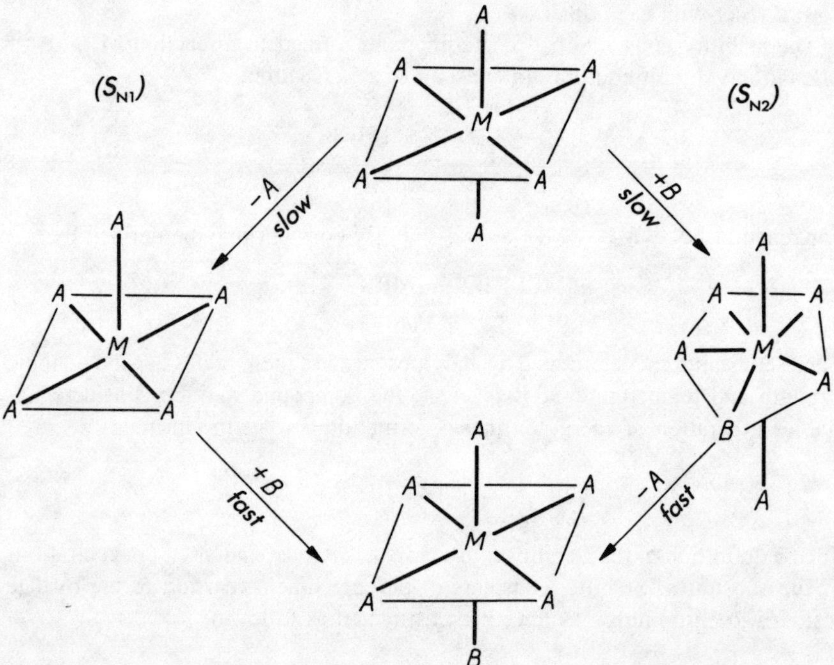

Fig. 15.5. Possible ways and substitution mechanisms in octahedral complex of general formula MA_6. q—quick, s—slow.

The first possibility anticipates an unstable system of five ligands, the second one—a seven-ligand unstable system. The mechanism (a) agrees with nucleophilic substitution S_{N1}, as the elimination of one of the A groups is slow, followed by a quick substitution of group B into the MA system. In mechanism (b) two molecules take part: MA_6 and B, so the reaction rate is limited by the second-order (S_{N2}) nucleophilic transfer, then quick elimination of one of the A groups occurs. Independent of the mechanism type involved, the total activation energy E, in both cases very high for the structures d^3 (see p. 339) it is 2.00 for S_{N1} and 4.26 for S_{N2}. Slowness of substitutions is not synonymous with stability of the Cr^{3+} compounds; however it is a result of their inertness and slow reaction. Stabilities of the compounds, as the values related to the equilibrium constant, may not be related to kinetic phenomena. The series of anions with decreasing affinity to chromium complexes (resorcinol, OH^-, oxalate, lactate, acetate, formate, Cl^-, water) are an oversimplification, valid only in equimolecular concentrations when, according to the law of mass action, a group of low coordination capacity may replace strong coordinating groups.

The rules of coordination of ligands of chromium complexes may be, according to Irving [10], described as follows:

(1) In order to compete successfully with coordinated water molecules in aqueous medium, the oxygen atoms should have an additional negative charge. In other words, the OH^- is a better donor than H-OH. Alcohols and phenols do not enter the complexes, as this process should be carried out in a medium of much higher pH than it is possible for the reason of solubility of Cr^{3+} compounds.

(2) Coordination stability of acidic group is inverse with the acid dissociation constant. This rule is generally right, although the deviations from linearity are great. The reason for such deviations lies in steric hindrances, in the types of bonds and in still other factors.

(3) Amino groups in acidic medium do coordinate chromium only to an insignificant extent, and in neutral medium still much less than do carboxyls. This rule is not synonymous with the differences in stability between amino and carboxyl groups.

(4) The chelate rings formed increase markedly the coordination stability. This is related to 'chelating' effect and in line with complex chemistry as a whole. Second part of the rule, saying that a compound is the more stable, the smaller a chelate ring is, is not confirmed according to current information.

(5) Stability constants of the ligands are decreasing with increasing number of the chromium coordination bonds involved. This rule is based on classical investigations of Bjerrum on thiocyanate chromium complexes and is correct if the steric hindrances do not disturb it.

If compounds, containing functional groups of more than one kind, enter complexes as ligands, they usually take only one coordination place, as, e.g.,

urea, entering chromium^{3+} complex

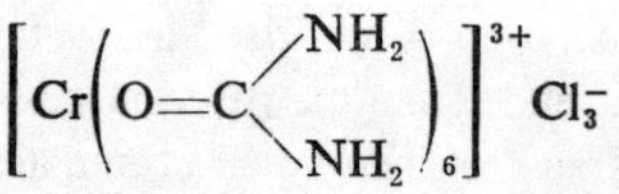

$$\left[Cr\left(O{=}C\underset{NH_2}{\overset{NH_2}{\diagup}}\right)_6\right]^{3+} Cl_3^-$$

The rate of hydration reaction of some types of complexes is $10^{-3}/10^{-6}$ min^{-1} and even 5.5×10^{-7} (half-life over 2 years) for complex

$$[M(H_2O)_5]^+ \; NCS^- + H_2O \rightarrow [M(H_2O)_6]^+ + NCS^-$$

Substitution reaction of OH$^-$ ion seems to be fastened by about one order. But the exchange of sulfate ligand for ionic sulfate as measured by ^{35}S, gives the half-life time $t_{1/2} = 20^h$ at 20°C for monosulfatopentaaquo complex and 30^h for bidental sulfates, as measured at 25°C.

There is geometric isomerism in chromium complexes. Since 1914 two configurations of potassium salt of oxalatochromium complex have been described

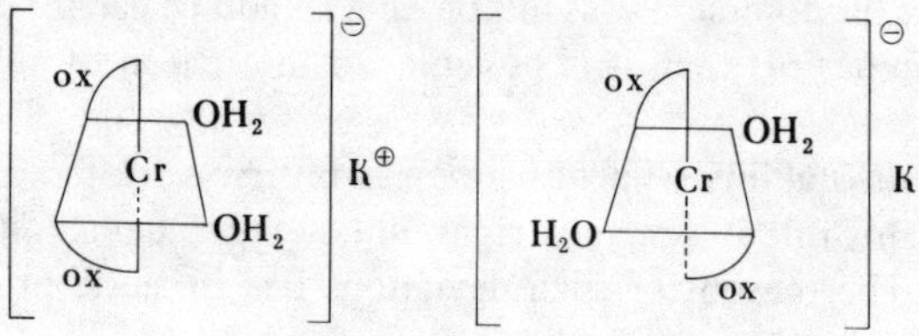

and it is shown, the optical activity of chromium complexes does not result from carbon atom. Trans isomer is less water soluble; so, aqueous solution is richer in cis-isomer. The oxalate ions are very strongly bound and their exchange with ^{13}C-labeled oxalates does not exceed 5% in 6 hours.

The knowledge about chromium complexes, although far from completeness, is quite extensive. However, relating it to tanning chemistry is very difficult, and it seems quite appropriate to quote Albert Einstein here: 'The trouble with chemistry is that it is too difficult for chemists.'

The well known and frequently cited paper of Irving was discussed and in some points questioned by Makarov-Zemlanskii [11]. Unfortunately, this author based his argumentation in many points on older literature, without making any difference between the obsolete and contemporary points of view. What he considered as wrong in applying the knowledge about chromium complexes in solutions to the tanning chemistry was at first the solvation problem. Coordination chemistry usually works with more dilute solutions—about 0.01 mole/l; whereas in tannery we usually have concentrations of 0.1-0.2 mole/l and more. In fact, when concentration of a salt in solution decreases, the activity of water increases with its concentration and this leads to the increased hydration of acidocomplexes. It is known that on dissolving 0.01 mole of potassium chrome alum in water the complex

$$H_3[Cr_4(OH)_3(SO_4)_3(H_2O)_{12}](SO_4)_3$$

is formed; whereas on dissolving 0.15 mole/l one obtains

$$H_2 [Cr_4(OH)_2(SO_4)_4(H_2O)_{12}](SO_4)_2$$

Next point is the hydrolysis of chromium complexes. Although it has been demonstrated that this process, very important for tanning chemistry, as it concerns olation and oxolation, is more exactly explained and the values of hydrolysis constant found are correct for very wide concentration range. The influence of temperature, pH, kind of ions introduced, and presence of organic substances used as masking agents has to be taken into consideration in the range of values used in tanning processes. As a result of the reasoning used, Makarov-Zemlanskii postulates the following trends for future investigation with regard to the conditions of the tanning process:
(1) Properties of concentrated solution (0.1-0.5 mole/l).
(2) Estimation of enthropy and entalpy of complex formation.
(3) Outer-sphere complexes—in order to find out whether they occur in solutions.
(4) Kinetics of formation, composition and structure of homo- and heteronuclear complexes.
(5) Complexes in non-aqueous solutions.
A great deal of such investigations have been done, however, without relating them to tanning chemistry.

The question arises how much the composition of chrome float changes during tanning, i.e., which complexes present there pass into collagen, and which ones remain in the exhausted liquor and may be considered as inert in relation to pelt. Quite an unexpected answer came from the paper of Scroggie and Davis [12], who dealt with recirculation of tanning liquors. To identify the composition of fresh and exhausted liquors, they have used gel filtration on Sephadex, ion exchange chromatography on cellulose acetate and spectrometry methods. According to the authors, no qualitative changes in composition of chrome liquors could be observed, indicating the increase of the amount of any kind of complexes. Thus, all the complexes present in liquor do find their way into pelt (Fig. 15.6). Because the waste water from sammying was also recycled, thus the supposition that some kind of chrome complexes may just penetrate into hide, without being bound, must be rejected.

A constant ionic strength and gradual exchange of chloride ions for sulfate ions throws some light on the importance of anions present in tanning float.

Several papers published in recent years dealt with chromium salt in aquaeous solutions at concentrations close to the ones applied in tanning chemistry.

Menshikov et al. [13] have examined the basicity of Cr^{3+} sulfate using the ion exchange chromatography and spectrophotometric method. According to these authors the amount of the inner-sphere sulfate ions, the pH of the solution, and the polymerization degree of complexes increase with their basicity, although to different extents. Also, the amount of nonionic and anionic complexes increases with basicity, whereas the amount of cationic complexes decreases, although these changes are less expressed. The basicity investigated was from 7 to 46%. Authors concluded that the substantial increase in degree of polymerisation of the complexes and of their reactivity is connected with decrease of SO_4^{2-} content in their inner sphere. These findings are not discussed from a theoretical viewpoint and may be considered as an introductory step in investigation of properties of tanning chromium solutions with methods now available.

The paper of Caminiti et al. [14] deals with a more theoretical problem of the

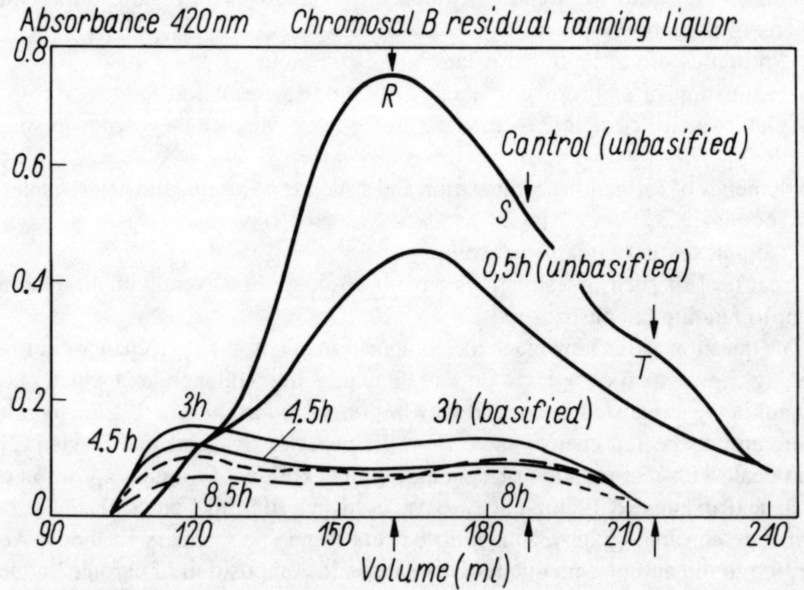

Fig. 15.6. Absorption of tanning liquor at constant wavelength—observation of chrome uptake during process. According to [12]

structuring influence of aqueous solutions of $[Cr(H_2O)_6]Cl_3$ by X-ray method. Authors have tested the reason for the unusual stability of Cr^{3+} hydration complex found and concluded that a satisfactory agreement with experimental data may be achieved with the acceptance of a model in which the hydrated $Cr(H_2O)_6^{3+}$ ions are interacting with a second shell of water molecules. In this model Cr^{3+} hexaaquo ion must be considered as an actual kinetic unit.

By calculation it was impossible to describe the model without an introduction of a second hydration shell. The distance between adjacent water molecules in the first and second hydration shell was found unexpectedly short (2.76 Å), even shorter than the one in ice. Thus, it should mean a very strong interaction between them and a very strong structuring influence. The authors did not touch the very interesting question of exchange of water molecules from first coordination shell for other groups, like, e.g., collagen carboxyls in tanning processes. Their theory, thus, seems to contradict the views about the entering of these groups into chromium tanning complexes.

Color and structure of Cr^{3+} complexes

The Cr^{3+} has an electron structure of the outer M shell defined as d^3 or, more exactly, $3s^2 3p^6 3d^3$. This means that it has three electrons in the last filled subshell d. These three electrons become excited upon absorption of an energy quantum; corresponding to the energy of an electromagnetic wave of a length of 400-700 nm. The electrons are unpaired and their energy levels degenerated. They occupy orbitals d_{xy}, d_{xz} and d_{yz}. These orbitals form clouds of identical electron density in three planes of the coordinate system. These planes lie between the axes and are equidistantly oriented right between them.

Excitation of d^3 electrons by absorbed energy quantum causes their transition to the subsequent free orbitals $d_{x^2-y^2}$ and d_{z^2}. We may show it as follows

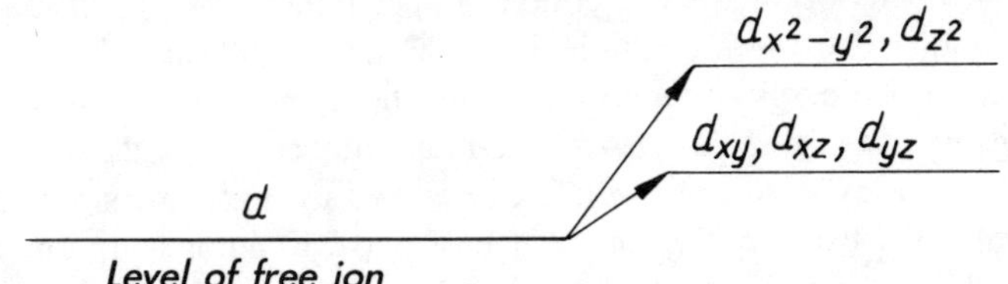

Orbitals $d_{x^2-y^2}$ and d_{z^2} have their energy levels higher than d_{xy}, d_{xz} and d_{yz}, but both are of equal energy. The d_{z^2} orbital concentrates the electron density along the z-axis, with a ring in a plane xy; whereas orbital $d_{x^2-y^2}$ concentrates itself on the x and y axes. Influence of the surroundings on both kinds of orbitals is different.

In the visible light range for complex Cr^{3+} salts, two absorption bands may be expected. This may be accounted for by the ligand field theory, which deals

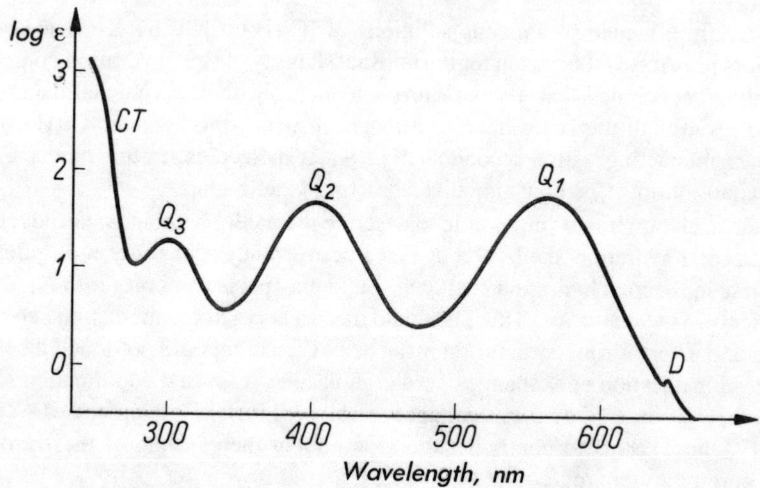

Fig. 15.7. Typical UV and Vis spectrum of Cr^{3+} complex solution.

with mutual reaction of central ion and related groupings. On the assumption of certain simplifications, e.g., whole symmetry of the complex, this theory rejects all *a priori* hypotheses concerning chemical character of the bonds inside the complex and assumes a loss of individual properties of complex forming groups.

The ligand field theory gives explanation of the above picture of electron excitation, which is simple for free ion but more involved in complexes. Owing to the action of electric fields of complex components, the $d_{x^2-y^2}$ orbitals get deformed to become longer or shorter (hybridization) and the overall electric level of atom (term), which is a resultant of the system of electron orbitals, may be split. A typical absorption spectrum of a Cr^{3+} complex is shown in Fig. 15.7. Three weak bonds occur there, defined in spectrometry as Q_1, Q_2 and Q_3 in which the molar extinction coefficients are about 10-100, and the half-widths 1500-2000 cm^{-1}. Energy of the excited states in visible light and in ultraviolet range is by 40-140 kcal/mole (167-585 kJ/mole) higher than in the ground state. Q_3 band is often invisible because it is overlapped by a very strong, sometimes unmeasurable CT band in UV. In a range of 650-750 nm a sharp but weak D band occurs, having molar extinction about 1 and a half-width of about 100-200 cm^{-1}. This band has a vibrational structure. To define and ascribe the bands mentioned is possible by applying the ligand field theory. System of energy levels of d^3 configuration in octahedral symmetry, i.e., for the complexes having coordination number six, is shown in Fig. 15.8. Notation used in this figure is explained in spectroscopy handbooks. Band D—due to transition into state ^2E_g, as it results from its properties, has to be very weak and narrow, and its position is ligand-insensitive. Bands Q arise from the transfer of electrons from π orbitals

t_{2g} to the antibonding orbitals t_{eg}^2. The equilibrium distances in excited states are greater, and therefore the absorption bands are broad. Two basic chromium^{3+} absorption bands are in the vicinity of 420 and 575 nm. The ligands connecting two chromium atoms set up a field of various intensity, which results from their mass and charge. The spectrochemical series

$$I^- < Br^- < Cl^- < NO_3^- < F^- < OH^- < HCOO^- < C_2O_4^{2-} < H_2O$$
$$< \text{glycin} < EDTA < NH_3 < \text{ethylenediamine} < NO_2^-$$

has the property that every ligand, starting from left side, shifts the absorption bands in the direction of shorter wave-length, more than its predecessor. These shifts are small, not exceeding 20 nm and may be disturbed by other influences, exerted, e.g., by lone electron pairs of ligands. Excitation energy by transitions d → d, i.e., from one orbital of subshell d, to the other is very small. This energy is responsible for absorption intensity. The ratio of intensity of both bands and position of their maxima may be easily measured. They account for the color of Cr^{3+} salt solutions, which is a resultant of these two components.

Investigations on spectral curves of chromium compounds used in tannery have been started in 1934 [15]. Then Theis et al. [16] have extended a range of these investigation on complexes formed with chromium salts by acetate, formate, tartrate, citrate, sulfate and other ligands. However, this work was limited to characteristics of the absorption curves without drawing conclusions concerning the structure of complexes. These authors have ordered the acid

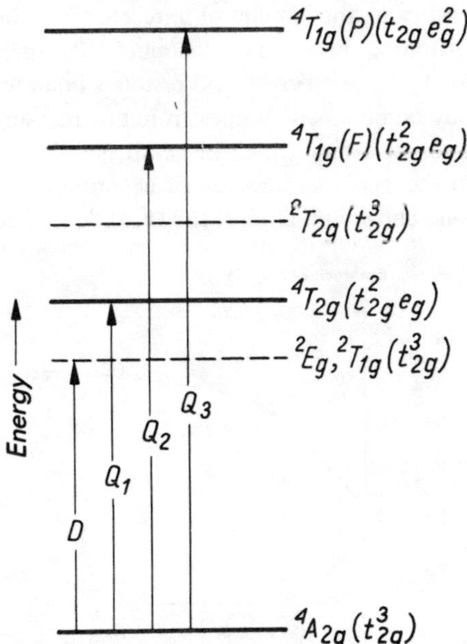

Fig. 15.8. Energy levels of possible electron transitions for d³ configuration of outer electron shell.

anions according to their ability to give changes in extinction in a following, decreasing order: oxalate, glycinate, tartrate, citrate, glycolate, acetate, chloroacetate, formiate. They also found that introduction of sulfuric acid to the medium in a great excess, in a ratio of 600 moles/mole Cr, shifts both peaks towards greater wavelength by about 100 nm, and a small absorption increase occurs (10-20%) only in sector of band of a higher wavelength. Shuttleworth and Sykes [17] have tested changes in the absorption curves of Cr^{3+} salt solutions as influenced by addition of amino acids. According to them, the molar extinction coefficient increases as the amino groups are moving away from carboxyls which indicates building-in of carboxyl groups into complexes, rather than both groups. Thus the stability of the coordinative bond is inversely proportional to the acid dissociation constant.

In the papers dealing with investigations of the phenomena occurring during chrome, and chrome-aluminium tanning Erdmann [18] used the method of spectral analysis and together with other physicochemical methods he applied it for study of the structure of chrome complexes.

To estimate the number of acid residues coordinatively bound to the chromium atom, Erdmann used conductimetric titration and measurement of positions of the absorption maxima, as well as the extinction values at these points. The principle of conductimetric titration in this case is to backtitrate with hydrochloric acid the excess of the sodium salt of an organic acid. Based on the amount of unreacted masking salts, the buffering curves obtained from titration of pure solutions of standard salts, the amount of unreacted masking salts can be determined. From the difference between the amount, determined from the curve and amount of salt added, the number of acid residues bond to chromium with coordinative bond may be obtained. Values from this measurement allow to mark the coordination axes in a way which makes it possible to ascribe the number of acid residues to the maximal increase of absorption. This is the number of acid residues per one chromium atom (Fig. 15.9). The spectral curves, as taken

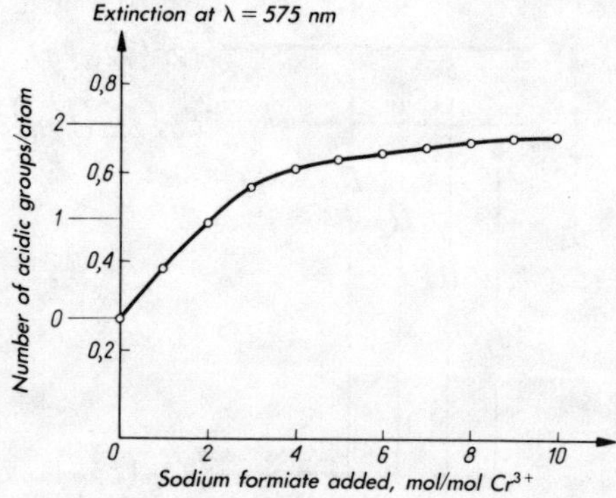

Fig. 15.9. Extinction values at λ = 575 nm and the number of formiate groups per one chromium atom in complex. According to [18]

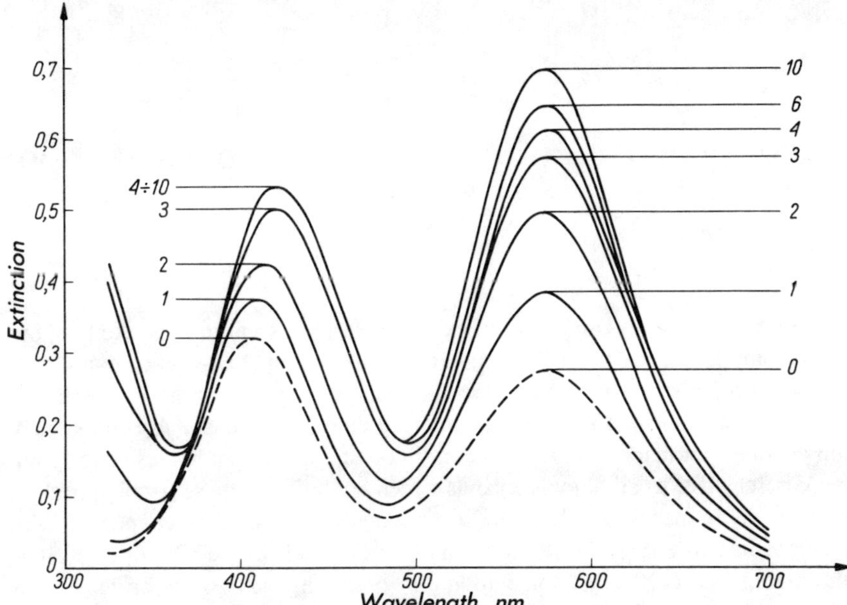

Fig. 15.10. Spectral curves of Cr^{3+} chloride masked with sodium formiate. The Figures refer to the number of formiate moles per one mole of Cr^{3+}. According to [18]

from solutions tested (apparatus record), are shown in Fig. 15.10.

As may be seen there, the addition of 1-3 moles of sodium formiate per one mole of Cr^{3+} will increase extinction in both bands and will shift the 410-nm maximum to 420 nm. Further addition of the formiate (4-10 moles/mol Cr^{3+}) gives only slight increase of 575 nm band; however, the extinction maximum at 420 nm does not increase any more. In Fig. 15.9 one may still see that the amount of the formiate residues, coordinatively bound to the chromium atom does not exceed 1.8. The combined data allowed to write the formulas in which were assumed: the molar ratio of 10:1, three chromium atoms in a molecule, and the results of conductimetric titration; water molecules have been omitted:

green

$$\left[\begin{array}{c} \text{HCOO—Cr} \cdots \text{O=C—O—Cr} \text{------OH------Cr—OOCH} \end{array} \right]^{\oplus} \quad HCOO^{\ominus}$$

violet

One may expect both complexes to exist in solution. An argument like this, but more complicated, carried out in studying the formation of Cr-acetate complexes leads to a conclusion, how to interpret the absorption spectra of these complexes. This is shown in Fig. 15.11. As one may see there, the 420-nm absorption peak almost does not respond to the building-in of acid bridges, whereas the 580-nm peak is sensitive to all anionic ligands which are built-in, these forming bridges as well. This means that building-in of ligands as bridges is connected with changes in two energy levels of the molecule—the lower (580 nm) as well as the higher one (420 nm), and building-in of the ligands not as bridges gives change only to the lower (580 nm) peak.

Figure 15.12a shows building-in of the nonbridging ligands, Fig. 15.12b—partly bridging ligands in their majority giving simple bonds, and Fig. 15.12c—of

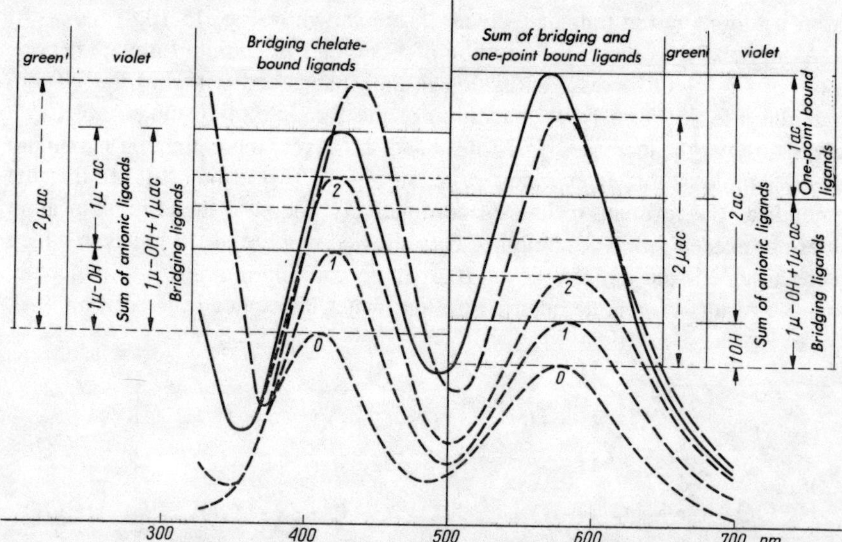

Fig. 15.11. Interpretation of elements contributing to the spectrum of acetate Cr^{3+} complexes in the visible light range. According to [18]. The dashed line refers to the green, binuclear complex of Cr-acetate; the solid line—the violet, containing only μ_{ac} (binding) acidic groups. The ac/Cr molar ratio is equal to two; 0, 1, 2 are curves for chromium chloride, containing 0.1 and 2 moles of NaOH per mole Cr.

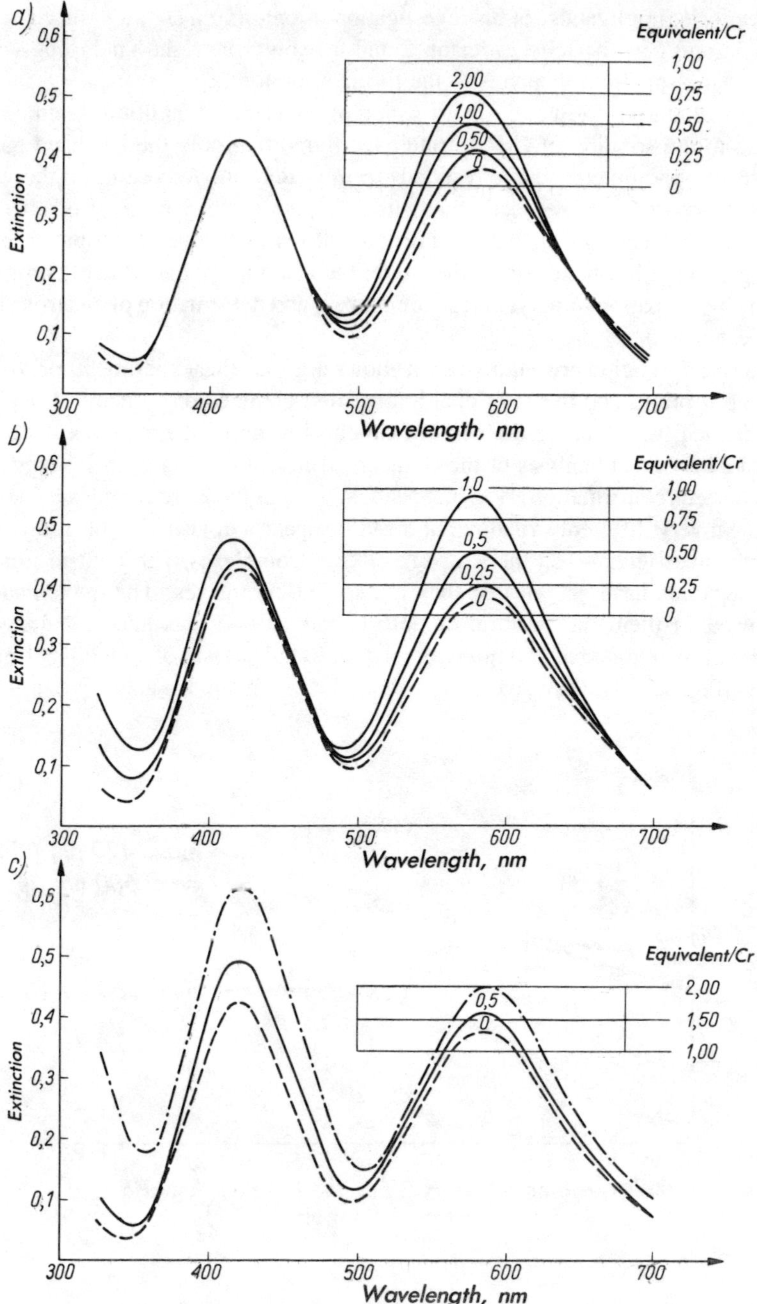

Fig. 15.12. The effect of building-in of ligands of various ability to form bridges into the Cr^{3+} complexes: (a) nonbridging, (b) partly bridging, (c) typical bridging ligands.

typical bridging ligands. In his investigation of chromium sulfates and chlorides of a 33 and 65% basicity Erdmann found that only the right-hand (long-wave) side of the spectrum responds to the formate masking.

Küntzel et al. [19] in their investigation of the effect of addition of aluminum nitrate to the solution of Cr^{3+}-nitrate have found that only the left-hand part of the spectrophotometric curve responds to this addition. According to those authors it is possible to reconcile this with Erdmann's suggestions, if one accepts that not only entering of ligands into complexes as bridges contributes to the changes in the left-hand part of the curve, but also the approaching of chromium atoms by formation of polynuclear complexes, and deformation of electron shell, connected with it.

The interpretation procedure shown allows us to say that experiments on visible rays absorption, and the additional information obtained on a chemical or phys-icochemical way, can be the basis of conclusions about the complex structure, based on an exact analysis of the elements of its color. This is an investigation which needs confirmation on an independent way, as there are complexes known, having a very different structure but the same spectra in visible light. Erdmann's further investigations on the structure of Cr^{3+} complexes with built-in fumaric acid residues have shown that they are spatial complexes. The quoted author followed in them the position of both already tested maxima, and that of a minimum positioned at 360 nm with relation to the amount of sodium hydroxide added (Fig. 15.13). Using the previously described method of analysis of position

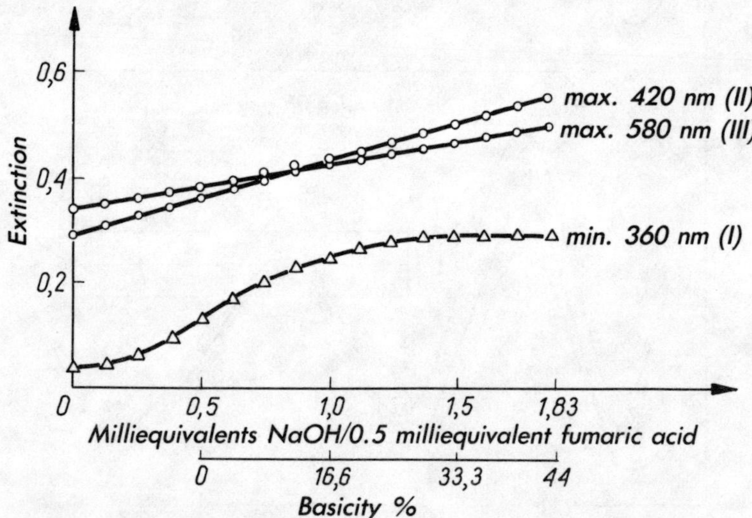

Fig. 15.13. Positions of the extremes of spectrophotometric curves and extinction values at these points vs. molar ratio of NaOH and fumaric acid and basicity of chromium complexes in the system tested [18].

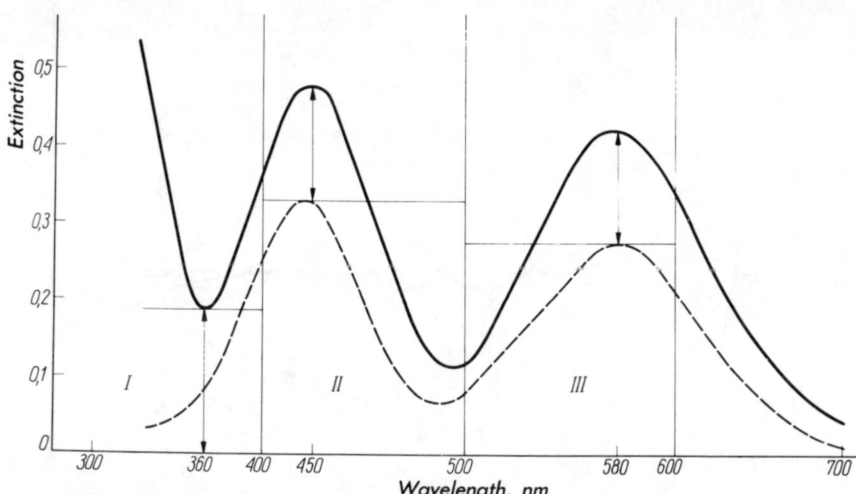

Fig. 15.14. Interpretation of the data from Fig. 15.13. Dependences between the structure and spectrum elements I, II, and III.

and intensity of peaks he ordered them as shown in Fig. 15.14.

Thus investigations of complexes masked by the fumaric acid residues lead to the conclusion that they are spatially crosslinked according to the scheme shown. The systems containing sulfate complexes, obtained by dissolving in the cold of commercial preparations of 33% basicity, revealed a similar behavior. In the spectra of these complexes Erdmann observed changes in the height of the minimum at 360 nm, which points to their crosslinking. According to Erdmann, the spatially crosslinked complexes have the ability of 'filling' the leather. Thus one can say that:

(1) The color of Cr^{3+} compounds depends on the electron shell of this element and, more exactly, on the excitation of the d^3 electrons from M-shell, and not on the presence of certain kinds of bonds in the molecule, as in organic dyestuffs.

(2) The physical elements of color are two absorption bands in the visible light range. The intensity ratio of these two bands, their width and intensity are responsible for the color of the solutions. Change of color may be produced by ligands of chromium complexes which may change the intensity ratio of both peaks (bands) and slightly their position.

(3) Analyzing the intensity ratio and position of both peaks, and making use of additional information of chemical character, some data about the structure of a single complex and its energy state can be obtained.

An interesting effect, yet neither explained well nor utilized by tanning chemistry, is the luminescence of chromium complexes, described by Schläfer et al. [20] (Fig. 15.15).

The luminescence of the Cr^{3+} compounds and its relation to the Cr^{3+} spec-

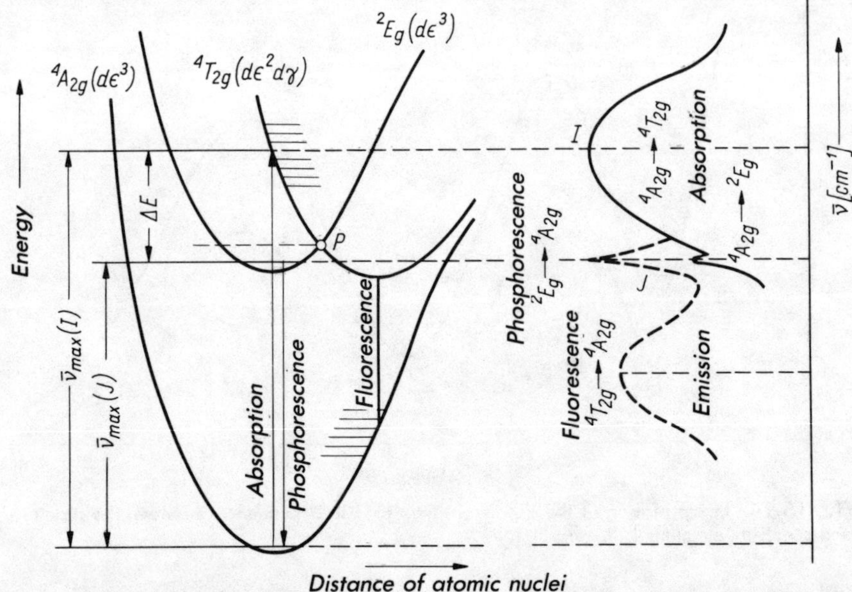

Fig. 15.15. Fluorescence and phosphorescence spectra of chromium complexes. Potential energy of singlet and triplet (excited) states. Energy of maximum is indicated by arrows. At point P there exists greatest probability of transition. Shape of segments of spectrum tested has been shown in the right side of the figure. Markings of energy levels are generally accepted. According to [20]

troscopy are extensively discussed by Lumb [21]. The problem is theoretically difficult as it concerns transitions between various energy levels of excited electrons and their direct or indirect emissions. To understand this well, a detailed knowledge of quantum chemistry is required. In brief, it can be said that luminescence, closely related to visible spectra, is strongly dependent on the presence of other ions in the medium. Thus, for the time being, investigations in this field, related to tanning chemistry, have not been carried out yet. The existing knowledge is based mainly on the results of investigation of solids containing Cr^{3+}, like ruby crystals, where Cr^{3+} is a dopant, or others.

Tanning complexes and chrome tanning theories

Chrome tanning consists in building in carboxylic groups of collagen as ligands into the chromium complexes. It was Gustavson who first recognized this fact [22]. It was shown that if carboxylic groups are blocked by methylation, then collagen binds less chromium and its T_s is lower. An amount of 1% of Cr_2O_3 in leather is, however, enough (of course, in the form of complexes) to form an amount of crosslinking bonds, which would essentially affect the T_s, appearance and properties of hide.

The residues of organic acids are built in very strongly as ligands of the chromium complexes. This relates primarily to dicarboxylic (oxalic, tartaric) and monocarboxylic (formic) acids. Oxalic complexes are very stable; they make binding of chromium by the hide carboxyls difficult. These groups which ionize at a proper pH, do react with coordinative chromium compounds to enter them as ligands. Unionized carboxyls do not react there. They dissociate at pH 2 to 4, and then the rate of their entering the complexes is the greatest.

The ligand exchange in chromium complexes is the fundamental principle of chromium tannage theory as formulated by Gustavson, Kuntzel and Shuttleworth. Essential for this theory is the acceptance of the chromium^{3+}-collagen carboxylic group complex as the most stable among the complexes known. Masking ligands, being bound weaker, give way to collagen carboxyls.

As already mentioned the half-life of chromium complexes is particularly long, reaching 6-40 hours, which as compared to other complex-forming metals is by several orders of magnitude greater (e.g., the half-life of Al-complexes is about 10^{-2} s). A fundamental condition for the reaction is to maintain proper parameters of the medium.

Hydroxyl ions have an even greater affinity to chromium than collagen carboxyls. They are capable of expelling the carboxyls from the complex, or of detanning the leather. In other words, chrome-tanned leather is not resistant to alkalis, assuming of course a sufficiently long time and adequate concentration. Chrome tanning may be defined as the control of four competitive reactions:
(1) entering hydroxylic groups into the complex to give it the required charge;
(2) binding of the anion in the first sphere of the complex, in order to make it susceptible to the masking;
(3) the masking reaction: entering of the organic acid residue into the complex instead of the inorganic one;
(4) reaction of the protein carboxyl group with the chromium complex.
Crosslinking could be believed to be the fifth reaction, i.e., reaction 4 proceeding when the uni-point binding has already been formed. Neither kinetics nor thermodynamics of this reaction has been tested separately. It is presumably slow; probably it still proceeds during the piling of the leather after tanning.

In discussing the chrome-tanning process it is presumed that the chromium ion is a reaction center. The hydroxylic group has a tendency to penetrate to the inner sphere of the complex. This penetration is quite quick and accompanied by formation of μ-hydroxo bridges. Other ions present in the medium, e.g., sulfate ion, enter the complex as well, and the reaction proceeds according to the competition principle. Positions in the complex are occupied by ions of the greatest complexing capability. The entering ion influences the stability of the complex and prevents, or at least retards, precipitation of chromium hydroxide by a pH increase. Thus the coordinating ion masks the chromium solution and retards its reaction with hide. The chromium complex of a high masking degree has a low tanning ability. Such reaction types are investigated in tanning chemistry.

A technological question of precipitation of chromium tanning complexes in leather was investigated by Babich and Mikhailov [23]. As some clusters of chromium complexes may be observed on tanned collagen fibrils with the electron microscope, the authors dealt with the screening effect of these clusters, concluding that they may, in certain cases, act as protection against sticking together of the fibrils. This effect depends on fatliquoring which may decide which of these functions prevails, due to the dispersing action of fats. The chromium salts precipitated on fibers, and forming clusters may reinforce the mechanical strength of leather. Thus it is recommended to fatliquor the pelt before tanning, if soft, plump leather is to be obtained. Equally important is the drying of tanned leather under strenuous conditions, as this results in olation of chromium complexes and prevents sticking together of the fibers. According to the quoted authors topochemical olation of chromium complexes decreases the mechanical strength of leather, whereas in leathers in which the complexes are less olated, the screen effect is quenched, whereby the mechanical strength is increased.

Another chrome-tanning theory was proposed by Lasek [24]. In that theory great stress is laid on the behavior of the sulfate ion taking part in the formation of polynuclear complexes from the binuclear ones, with previously built-in collagen carboxyls. Depending on the basicity of the medium tanning occurs in the inner or outer sphere; this last process is in fact precipitation of basic salts. This theory has been strongly criticized [25]—some of his suggestions, however, were recently confirmed [23, 26].

Masking agents enter complexes, if they are sufficiently ionized at low pH. When the organic acid used for masking is weak, a higher pH facilitates its reaction, as its ionization increases. The increase of pH of the tanning liquor is a reason for the growth of basicity of chromium complexes, and their oxolation as an aging effect (deprotonation), whereas the coordination affinity of the sulfate residues remains unchanged. In this reaction step the masking agent remains coordinated with chromium, the reactivity of protein increases, however, and the carboxyls enter the complex. The sulfate residue may in the case of increasing activity of both carboxyls be removed whence crosslinking occurs.

According to Harlan and Feairheller [26] the following reaction types can take place as the carboxyl ion attached to collagen enters the complexes:

(1) Entering two carboxyl ions into the same chromium complex.

(2) Olation with elimination of water and formation of a linkage between two complexes

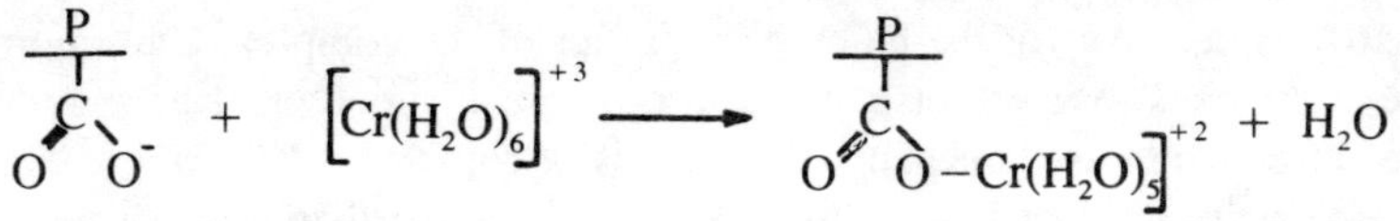

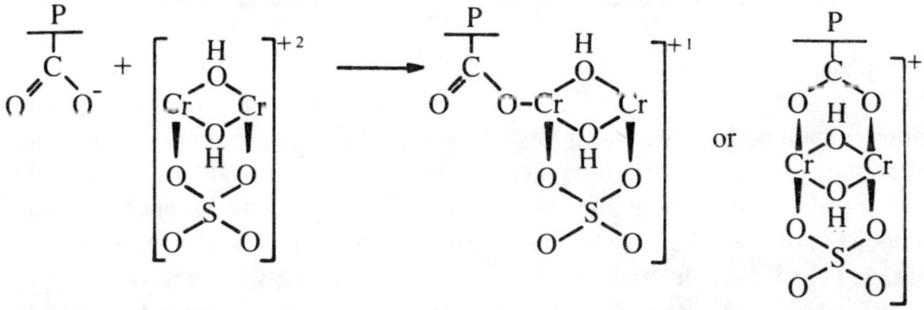

Increasing the alkalinity of the reaction mixture favors the olation reaction. When the reactions proceed and multinuclear complexes are formed with ol-bridges, hydronium ions are released and highly stable oxolate bridges are formed.

COMPLEXING   OLATION   OXOLATION

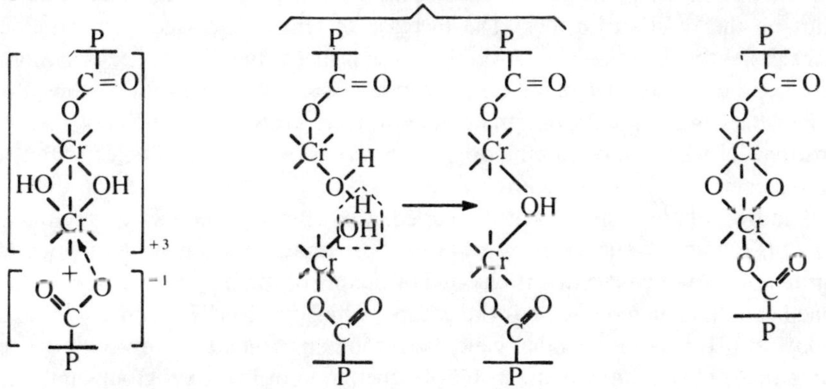

Note: in these formulas the coordinate bonds are marked with a dotted line in order to differentiate them from valence bonds.

It has been shown that bidentate sulfate groups remain in the final complexes after leather curing and dyeing. Their apparent significance for improving stability of the complexes has been recognized.

Similar coordinative complexes are involved in other tannage procedures.

The decrease of complex susceptibility to addition of alkalis, sodium carbonate or ammonia is a sign of masking. The content of masking salts is usually expressed in moles per Cr gram-atom. This ratio is called the masking degree.

This quantity is decisive for the tanning properties of the complex. Chromium carbonate complexes, as yet not isolated, are of great importance, as sodium carbonate is frequently used to change the basicity of complexes in solution. The increase of basicity produced by sodium carbonate is lower than it might result from stoichiometric calculations. This means that during neutralization with sodium carbonate not just one but more mechanisms are involved.

Variable parameters of chrome tanning

Chrome tanning is now a major procedure of leather making applied to all kinds of leather produced by industry. There are plenty of monographs, handbooks and specifications, devoted to this subject and many research papers. A technologist, however, has to remember that all tanning parameters are operating together, and the industrial process is a resultant of many variables.

Diffusion of chromium salts into leather. In processes, preceding tanning collagen fibers have to be separated, and the interfibrillar substance removed. In that way they are more easily accessible to tanning agents. The tanning agents in turn penetrate the hide more readily if they are of low molecular weight. This value increases with growing pH. This is true of the pH of float and pelt. The penetration rate depends on the rate of binding of the tanning agent by protein, thus on the reaction kinetics. The increase of temperature and concentration accelerates the diffusion rate. A similar effect is that of the float and pelts motion. The reproductivity of tanning results can be assumed, if one observes just the parameters indicated above: stratigraphic (layer) analysis is their only confirmation. Distribution of chromium in the layers should be equilibrated within a few percents of the relative error.

Binding of tanning agent to collagen. Collagen is positively charged at a low pH (about 2). The tanning agent easily penetrates, but it is not easily bound at this pH. An increase of basicity and of the pH of the liquor increases on the one hand dissociation of carboxylic groups, which reaches 75% at pH = 3, and 100% at pH = 5; on the other hand the chromium complexes grow at the same time in size. This growth is a result of penetration of hydroxyl groups into the complexes. The amount of μ-hydroxo bridges also increases the degree of oxolation while the charge of the tanning agent molecule decreases. These processes decrease the stability of the solution and increase the tendency of complexes to precipitate (astringency). In other words, the complexes increase in size to a degree when they can crosslink two collagen chains. A further pH increase is already hazardous, as olation can cause precipitates to form.

Composition of the complexes. The more stable are the complexes, the lesser is their capacity to attach to collagen. For this reason the steric arrangement of the complex is as important as the strength of the chromium-ligand bonds. The more masking groups in the complex, the weaker and slower is its binding to collagen.

Presence of neutral salts. Sodium chloride acts as a structure breaker, increasing the velocity of water molecules motion in solution. If sodium sulfate is present in the medium, it decreases the amount of Cr_2O_3 bound by collagen from chromium sulfate. High concentration of the sodium salt prevents precipitation of chromium sulfate from a solution of high basicity. This property was used in the investigation of tanning with highly basic sulfates [27].

There is a problem which though it has lost some of its importance in more modern tanneries is still encountered in those applying older technologies or working in the countries where the supply of the proper chemicals is limited. It consists in the reduction of bichromates by glucose or, even worse, by molasses. The course of this reaction depends on temperature, concentration, molasses composition and on the liquor storage time. However, it must be apparent to the tanner that in using this process he never knows which were the masking agents, products of glucose or sucrose decomposition, which formed chromium complexes, particularly if it is noted that an average molasses composition is 50% sucrose, partly inverted, and more than 10% of ash, mostly potassium carbonate. This composition varies so as usually untypical molasses is supplied to tanneries. This problem was considered in many older publications, but the analytical methods used there were insufficient to give an answer to the question, what exactly happens during chrome liquor preparation. Quite recently Takenouchi [28] investigated the reduction of biochromate by glucose, and was able to separate eight chromium complexes by their partitioning on SP-Sephadex C 25. He suggested the presence of sulfato, formiato, sulfatoformiato, oxalato, hexaquo and hydroxo complexes in solution, obtained under exactly controlled conditions. It must be assumed that preparation of chrome liquors by reduction of bichromate with molasses or with technical glucose is not fully reproducible for many reasons; moreover, it depends strongly on the reaction parameters and liquor storage time, because complex formation very slowly reaches equilibrium. The good results of tanning which may be obtained in this way cannot be denied, but this process, when uncontrolled, may be the reason for multiple technological problems.

Takenouchi suggested that the complexes obtained have varying abilities to enter into reaction with collagen. This is inconsistent with the recirculation experiments of Davis and Scroggie [12], however, those authors used more uniform preparations, obtained commercially by inorganic reduction.

Aging of floats and tanning time. Every type of float has its optimal conditions of use. The best is a 4-5 day float; old floats react slower with collagen. In research, floats aged for several months are often used to make chromium complexes in solution very stable.

The tanning time is controlled by the relationship between diffusion and binding. During tanning a float composition changes and so does the pH and the charges of complexes. These changes may be drastic. They can be prevented by using dry strongly masked chromium preparations in such a way that the ligands of the complexes are slowly let free, the diffusion and binding being thereby controlled. This is connected with adjustment of the float amount; tanning is usually started with no float, just with the amount of water contained in the pelt which is released due to the osmotic processes and the dissolving of the masked complexes, which are then slowly decomposing.

Chrome tanning has thus to be carried out in a way allowing the chrome complexes to penetrate into leather and then to be bound to it. A strong tanning agent of great astringency, attaching quickly to leather, seals it for penetration of other tanning agent molecules. A complex, which is not easily bonded to collagen, penetrates well into leather but it does not, or only weakly, attach to leather and may be easily removed, e.g., by washing with water. The tanning of leather with pH not adjusted, e.g., due to the uncorrectly done pickling, may be accompanied by another effect, viz.: chromium salts penetrate easily from the surface of leather, having pH of about 2, to the inside of leather, where the pH is of the order of 5-6, i.e., a value at which the tanning agent binds easily to collagen and is deposited there. Then, improperly tanned leathers are obtained, in which the internal layers contain more chromium than the external ones. The medium layer of leather is then darker. The point in tanning is to find a resultant of these two extremes. The classical way is to change the pH of the float by introducing hydroxyls or carbonate ions into the system. Building-in OH groups into the inner complex sphere causes olation and an increase of the complex ability to become bound to the collagen functional groups. Thus, after penetration chromium binding occurs due to increased basicity of the solution. No objection can be raised against this process from the theoretical point of view. In practice, however, alkalization of the float, due to addition of sodium carbonate, changes the pH of the medium very drastically in view of the limited rate of increasing the basicity of the chromium complex. The variation of the pH of the float during tanning with time is shown in Fig. 15.16. The pH value jumps visible in this Figure are of short duration; however, they may be harmful for the leather not tanned yet. This effect may be avoided by the use of 'selfbasifying' floats.

15.2. Tannage with zirconium salts

The zirconium tannage process, being initially a technological curiosity, gained in importance during the years 1940-1950, due to the discovery of new, rich zirconium resources in many countries. Zirconium tannage yields a white, full and light leather resistant to abrasion and light, having the touch and appearance (except color) of chrome leather. Quadrivalent zirconium, a weak electropositive metal, forms salts that yield many and different complexes. They are difficult to obtain in crystalline form, and from the point of view of tanning chemistry more important are the properties of Zr^{4+} ions and their bonding with other ions in aqueous solutions. This problem was reviewed by Hock [29]. His consider-

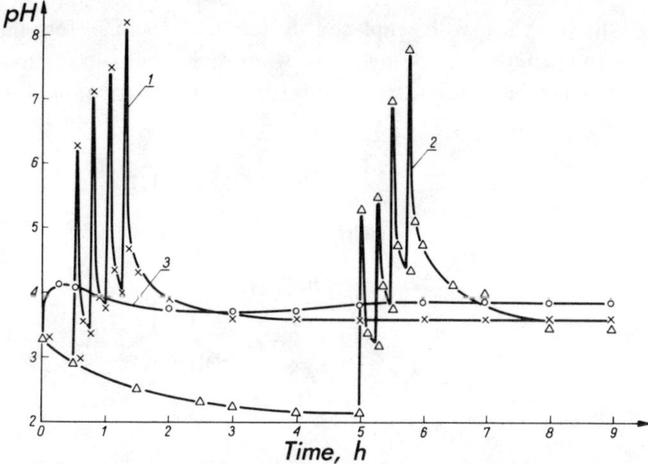

Fig. 15.16. Basification and self-basification of chromium liquors: 1—early basification, 2—late basification, 3—self-basification.

ations are rather consistent with those of some Soviet investigators [30]. In brief, the coordination number of zirconium is eight, with oxygen contact at the antiprism corners. Water molecules participate in achieving the coordination number eight. New bonds of zirconium ions in solution may be expected, if the oxygen atoms entering into the reaction are more reactive towards zirconium than those of water. Sulfate ions, etc., are bonded with zirconium through their oxygen atoms; the zirconyl group (Zr = O) occurring in crystalline compounds does not seem to appear in solution. Direct bonding between zirconium ions does not occur in aqueous solution. Zirconium-containing ions tend to assume a tetrameric core structure consisting of four zirconium atoms at the corners of the cube, where each zirconium ion is linked to its two adjacent zirconium ions by oxygen atoms of two hydroxyl groups forming double ol-bridges; one below and the other above the square. This is in accordance with Metelkin et al. [31]. Those authors suppose that the zirconium complexes used in tanning have a structure of the type defined above where single hydroxyls may be replaced by the acid residues forming spatial systems of a number of charges that stem from molar ratio, e.g., four, when four hydroxyls are replaced by carbonate residues (molecular ratio 1:1)

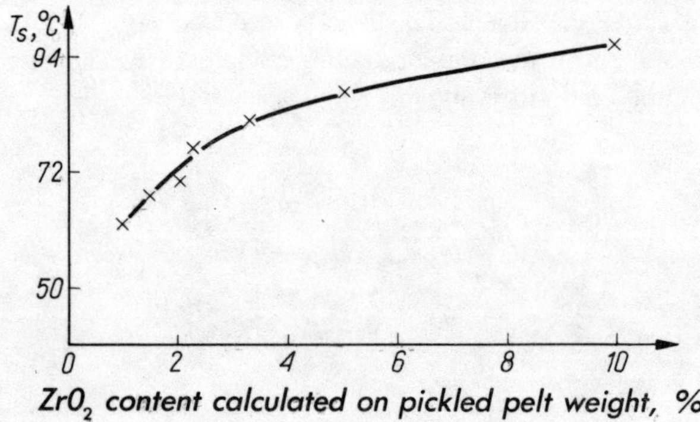

Formation of insoluble zirconium compounds starts when NaOH is added already at pH = 1.5. The behavior of the zirconium tanning agent toward masking substances is different, compared with that of chromium tanning agents. Monocarboxylic acids do not affect the precipitate formation; hydroxy acids, however, have a masking effect: after alkali addition to the solution, containing a masking agent in a ration of 1 mole per 1 mole of ZrO_2, no precipitate is formed. Potassium sulfophthalate, masking very efficiently the chromium tanning agents, does not mask zirconium salts. A comparison of two commercial preparations, one of zirconium and one of chromium, permits evaluation of the differences in their susceptibility to masking. From Table 15.2 it is obvious that chrome tannage and zirconium tannage must be done under different conditions.

Zirconium tannage is characterized by the data shown in Fig. 15.17. One can see there that T_s increases as the zirconium content (ZrO_2) increases to 10%.

Fig. 15.17. Dependence of T_s on the amount of Zr bound by leather.

Table 15.2.

Susceptibility to the masking of zirconium (Blancorol ZB) and of chromium (Chromosal B) tanning agents with organic acids

	Blancorol ZB, precipitation			*Chromosal B, precipitation*		
	initial pH	*addition 0.2 n NaOH, ml*	*final pH*	*initial pH*	*addition 0.2n NaOH, ml*	*final pH*
Formic acid	1.55	6	1.70	2.35	28	5.7
Acetic acid	1.50	10	1.92	2.40	32	5.7
Oxalic acid	1.20	30	2.08	1.90	47	5.2
Citric acid	1.30	—	—	1.75	32	5.7
Tartaric acid	1.35	—	—	1.90	61	5.4
Potassium sulfophthalate	1.55	47	3.65	1.90	—	—
*Masking agent added, mole/mole ZrO_2 or Cr_2O_3	1.70	14	2.50	2.90	14	5.2

*Potentiometric titration

The final T_S value is then 96-97°C. Thus zirconium leather does not withstand the boiling test. A recommended test, which permits evaluation of zirconium tannage, is the observation of a leather strip immersed in water of 70-75°C for one minute. If the medium layer does not gelatinize, the tanning has been done correctly. The dependence of T_S and of the amount of zirconium bound on the float basicity, as calculated on hide substance, is shown in Fig. 15.18. As one can see, the T_3 decreases and the amount of Zr bound to the hide protein increases, though to a smaller extent, with growing basicity.

Zirconium tanning is quick; it takes from two hours for sheep- and goatskins to about 10 hours for thick sole leather.

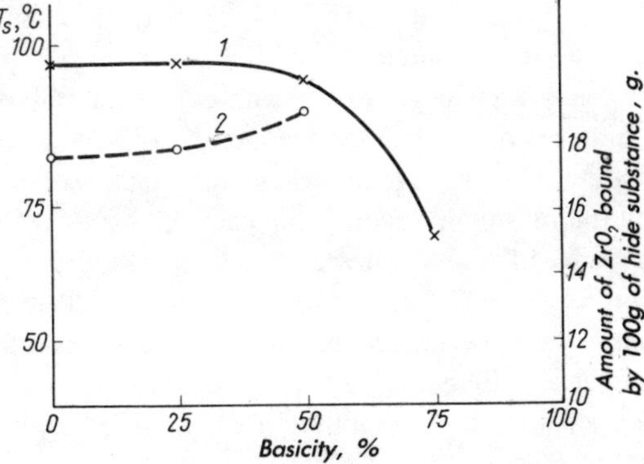

Fig. 15.18. Effect of the float basicity on T_S (1) and on the amount of ZrO_2 bound (2) calculated in terms of the hide substance.

The mechanism of zirconium tannage differs from that of chrome tannage. The conclusions of the authors investigating the problem differ in some points, but they may be generalized the way Hock [29] did it. One of the most important remarks of the quoted author is that a 'Detailed examination of the three recent theories of the chemistry of zirconium tannage suggests that some areas of difference are the result of too literal quantitative acceptance of experimental tannage data obtained at relatively high pH values when additional separate deposition of some form of hydrous zirconium oxide, not chemically bound to collagen, must have been possible.'

Investigations done on pre-treated collagen (acetylated, deaminated, chromium-pretanned, etc.) have shown the correctness of the quoted opinion. The point of view accepted previously, according to which the amino groups of the collagen side chains are involved in zirconium tannage is right, not being, however, the only one. It has been experimentally demonstrated that cationic, anionic and neutral zirconium complexes can coexist in solution, which explains the above statement. No quantitative information on this subject exists. It may only be indicated that the concentration of added cations (Na^+, NH_4^+) and anions (Cl^-, SO_4^{2-}) influences the degree of hydrolysis and formation of complexes, in which SO_4^{2-} ions are linked directly to zirconium. The suggestion of Williams-Wynn that zirconium chloride tannage depends on the availability of carboxyl groups must be accepted, because it is supported by very strong experimental evidence. A 'multipoint attachment' of zirconium to collagen is the most probable explanation of the experimental findings. At least three mechanisms of zirconium-to-collagen binding have to be considered:

(1) Polar binding of anionic sites of the zirconium complexes to amino groups.
(2) Polar binding of cationic sites of the complexes to carboxyl groups.
(3) Covalent binding of neutral sites and the oxygen atoms of nonpolar carboxyl groups of collagen, without participation of amino and imino groups.

Some other covalent bonds between oxygen atoms of collagen and modified hydrous zirconium oxide, precipitated when the pH of the solution is high enough, cannot be excluded. Such bonds exist between zirconium salts and other biopolymers, e.g., cellulose or starch.

It has not been investigated yet to what an extent the polymerization degree of zirconium complexes may affect the amount of collagen which can be bonded to collagen in tanning. However, there exists a relation between these two factors, and a certain optimum value, specific for each kind of complexes present in solution and for each of the above named mechanisms of reaction, is suspected.

The suggestion of three key mechanisms of zirconium binding explains the phenomena observed in experiments the interested reader is referred to. In brief, it concerns the effect of sulfate ions to promote the binding of neutral and anionic complexes by citrate ion, the precipitation of modified hydrous oxide in the absence of these ions, etc.

Lebedev et al. [32] have observed the cross-striation patterns of collagen fibrils with an electron microscope after zirconium tannage, in native and modified (deaminated, deguanidized, acetylated) collagen. The density increase of dark strips is the greatest in unmodified, zirconium-tanned collagen. This increase is explained by the formation of zirconium salt 'cuticles' around the fibrils. The presence of the 'cuticles' on the modified collagen may be considered as an argument in the discussion on the existence of different tanning mechanisms. The authors observed a shrinkage of the cross-striation repeating cycle of about 15-16% in relation to native collagen (from 64 to 52.5-55 nm). This indicates a strong interaction between collagen and zirconium salts used and the existence of a tanning mechanism, independent of the 'cuticle' formation.

There is a discrepancy between the observations of Ranganathan and Reed (decrease of zirconium fixation, masked with citrate) [33] and Williams-Wynn (increased fixation) [34]. According to Hock [29] there must have been a difference in the media applied, causing promoted anionic species formation in the observations described in these two papers. In that of Williams-Wynn the results obtained have been high due to high hydrous oxide precipitation in the blank sample, and the prevention of this effect by citrate addition in the parallel experiment. As a whole, the significance of precipitation of hydrated zirconium oxide, easily occurring when the pH of the tanning processes is increased to about 3, makes difficult the evaluation of various results and probably is responsible for differences obtained by the authors mentioned. On the other hand, the appropriate pH of optimal zirconium fixation, lying at about 1.5 is very low in relation to collagen and undesirable for further treatment of zirconium-tanned hides, making thereby the masking essential.

The importance of the steric factor is still to be mentioned, although it has not been investigated; the relatively large molecules of zirconium complexes must fit into the spaces between collagen chains in order to be able to react.

Practical application of zirconium tannage is discussed by Zorn et al. [35]. They also pointed out the use of zirconium for retanning. It is recommended, if unextensible, tough, white leathers are required, resistant to aging, while maintaining the properties of mineral tanned ones. Zirconium tanning agents also counteract the shrinkage (wrinkling) of grain, making leather more spready, than the chrome-tanned one, moderately filled but having a distinct, uniform grain. Combined retanning with zirconium participation is frequently applied. Many technological data can be found in the monograph by Metelkin et al [31].

15.3. Aluminum tannage

Aluminum tannage is one of the oldest tannage methods. It was widely used for upper, upholstery, glove leathers and furs. Chrome tanning has eliminated aluminum tanning, leaving it only as a method for production of glove leathers and in combined methods. Tanning with aluminum salts gives white, soft and extensible leathers. An old name 'tawing' has been used for this process, as distinguished from tanning, which was understood as vegetable tanning. Tawing consisted of treating the skins with potassium alum in the form of paste including sodium chloride, egg yolk, flour and water. Aluminum-tanned leathers are sensitive to water and high temperatures. These are the main reasons why it has been replaced by newer methods that give more resistant products. This way of tanning is at the same time pickling with sodium chloride and sulfate residues from alum, and fatliquoring with egg yolk.

Aluminum-tanned leather has a shrinkage temperature range of 75-85°C, depending on the method used. Its water resistance is unsatisfactory, since even cold water slowly removes aluminum compounds from the leather. So far there is no general theory of aluminum tannage. Supposedly, it is related to chromium tannage, i.e., it involves the incorporation of collagen carboxyls into the complexes. However, these complexes are much less stable than those of chromium; and thus their binding to collagen is much weaker. This is the reason why aluminum salts are used in contemporary tanning almost exclusively in combined tannage processes.

Solutions of aluminum sulfate give an acidic reaction, even if cold prepared. Their heating causes a further pH decrease. After cooling of the solution the pH returns almost to its initial value. According to Chambard and Grall [36] the difference in behavior between chromium and aluminum during hydrolysis consists in that Al^{3+} cannot bind sulfate residues inside the complexes formed:

$$Al_2(H_2O)_{12}(SO_4)_3 \rightleftharpoons 2H^+ + SO_4^{2-} + [Al_2(OH)_2(H_2O)_{10}]^{4+} + 2SO_4^{2-}$$

Under the conditions of anionic chromium complexes formation it is impossible to obtain similar aluminum complexes. The SO_4^{2-} ion does not enter the complex, if the concentration is equal to or smaller than 0.3 mole/l. This explains why under the same conditions solutions of chromium salts of higher basicity are obtained as compared with the aluminum ones. The critical precipitation point and pH of aluminum sulfate solutions is affected by inorganic salts, present in the solution. Sodium sulfate and sodium chloride at low concentrations cause pH increase. In equimolar solutions sulfates have a stronger effect than chlorides. At the same time one can easily see that the pH of potassium alum solution is much lower than that of aluminum sulfate solution. However, if a solution of potassium sulfate is added to this solution, it results in a further pH increase, rather than in a decrease, as it should be expected from alum formation. In solutions of pure aluminum sulfate, independent of its basicity, no complexes are formed containing anions in their inner spheres. These are not formed in a mixture of aluminum and potassium sulfates of zero basicity. When the basicity of aluminum sulfate is 22-33%, a complex of the type

$$\left[\text{Al}_2 \begin{array}{l} (\text{OH})_2 \\ (\text{SO}_4) \\ (\text{H}_2\text{O})_8 \end{array} \right]^{2+} \text{SO}_4^{2-}$$

is formed.

A condition for proper aluminum tanning is the presence of neutral salts preventing collagen swelling in acidic medium. Usually NaCl is added in an amount of $\frac{1}{3}$rd to $\frac{1}{4}$th of the aluminum amount.

A detailed study of the masking action of organic acids on aluminum sulfate and chloride was done by Simoncini et al. [37]. It was a part of the ARCA-MONTEDISON project and led to the development of an aluminum tanning agent, applicable under industrial conditions. In the second part of the work the authors described a tanning method based on the use of aluminum complex containing citric acid or EDTA as the complexing agent and giving very good technological results. Vegetable tannins and glutaraldehyde are recommended for pretanning and retanning in this process.

In aluminum compounds olation and oxolation may also occur. However, concluding from a slight increase of the resistance to acids of the aged aluminum complexes, the degree of olation is lower than in chromium complexes, and the kind of complexes with oxygen bridges differ to some extent, as it does not raise the acidity of the medium, but only water is produced, viz.:

$$\left[(\text{H}_2\text{O})_5 \cdot \text{Al} \begin{array}{c} \text{H} \\ \text{O} \\ \\ \text{O} \\ \text{H} \end{array} \text{Al} \cdot (\text{H}_2\text{O})_5 \right]^{4\oplus} \longrightarrow \left[(\text{H}_2\text{O})_5 \cdot \text{Al}-\text{O}-\text{Al} (\text{H}_2\text{O})_5 \right]^{4\oplus} + \text{H}_2\text{O}$$

It is more difficult to obtain basic aluminum complexes than analogous chromium complexes. According to Küntzel [38] a relation exists between concentration of the solution and its critical basicity, i.e., the basicity at which a basic deposit will be precipitated. The higher the concentration, the higher is the critical basicity of the complexes, and at concentrations of 5.1, 8, 13.3 and 24% one can obtain a basicity of 30, 40, 50 and 60, respectively. Temperature increase makes it possible to obtain a solution of aluminum sulfate of higher basicity. The values quoted relate to 20°C.

In older literature there are suggestions according to which the basic sulfates are a mixture of highly basic two-nuclear complexes of 83% basicity and sulfate of 0% basicity. The last salt serves as dispersing medium for the highly basic complex. The existence of such a system is possible only in the presence of sodium sulfate, which is considered to be a solvent—perhaps due to its ability to immobilize water molecules around it and thus to make the whole system more stable. This theory, as far as the present author is aware, has not been confirmed by more detailed investigations by spectroscopic or other methods and thus cannot be assumed to be proved.

Neutral salts, added to aluminum sulfate solution cause various changes in it. Investigation of these changes is difficult, because the compounds involved are of varying stability and the method of electron spectroscopy are of no help here, as it was described for chromium. Aluminum complexes are known in which residues of organic acids occur as ligands: oxalates, tartrates, etc. They attach the aluminum complexes at pH 4-5, complexes formed do not have tanning properties. Aluminum complexes are characterized by marked instability of their inner sphere.

According to Williams-Wynn [34] the anions can be arranged in the following order according to their increasing ability to form complexes with aluminum: $NO_3^- < Cl^- < SO_4^{2-} <$ formiate $<$ acetate $<$ succinate $<$ tartrate $<$ lactate $<$ malonate $<$ citrate $<$ oxalate.

The masking of chromium salts with organic acids occurs when they are present in great excess. Tartrate complexes are formed in a 1:1 ratio. However, as it seems, citrates and lactates do not form complexes of constant composition with aluminum. Results of experimental tanning (comparison of T_s) have been shown in Table 15.3. When comparing the properties of leathers obtained it was found that the most stable leathers are obtained when formiate, acetate and lactate are used. Comparing the touch of the leathers led to the conclusion that the more the protein carboxyls are ionized, the more easily collagen enters aluminum complexes. This speaks for the improvement of the wearing properties of leather with the increase of the pH in the tanning process. If the complexing agent is too strong, it will not be displaced by carboxyls and the skin will remain untanned. Similar conclusions have been drawn by Küntzel and Rizk [39]. They investigated the formation of gels in aluminum salt solutions and found two reasons for it: aging of the solutions and the presence of basic salts; the reversible gels formed turn into sols on heating.

Table 15.3.

T_s of leathers tanned with masked Al-salts (T_s of the pelt $= 60°C$)

Tanning agent	Without masking	6 eq./g atom Al		3 eq./g atom Al		
		formiate	acetate	lactate	tartrate	citrate
Aluminum sulfate	63	75	74	77	62	62
Potassium alum	65	74	78	75	62	62
Sodium aluminate	63	75	75	76	62	63

As a rule, aluminum tanning is done in floats of zero basicity at high concentrations in the presence of sodium chloride to prevent swelling, because the pH of the float is low, being about 2.5 to 3.5. In this process, no washing is used. Aluminum-tanned leathers are more resistant to hydrolysis after aging. It has been an old rule that aluminum-tanned leathers are to be finished three months after tanning.

The use of aluminum salts for tanning is attractive because of their accessibility and low price. However, to use them some concessions are to be made. Namely, either the tanning process must be time and work consuming, or the preparation of salts must be appropriate, which means, as a rule, that they have to be used together with other agents or components, like the 'ready to use' preparations.

The classical pure aluminum tannage gave very fine glove or fancy goods leathers of very limited wearing durability.

15.4. Tanning with iron salts

Tanning with trivalent iron salts was a method evoking interest in the times of chromium scarcity, e.g., in Germany during both World Wars. One may demonstrate numerous similarities between Fe^{3+} and Cr^{3+} salts in their ability to form complexes and coordinative bonds. However, the leathers obtained by industrial tanning with Fe^{3+} salts are thin, brittle and do not resist aging. Formerly it was considered that the lack of resistance to aging is due to the decomposition of iron salts under the influence of organic matter, however later it has been shown that the reason is both oxidation and weak binding of the complex to hide as well as too acidic a character of the complexes formed. When the pH value of tanning lies in the range of 1.8-3.0, leathers are obtained with T_s from 65 to 91°C. Among the iron complexes masking compounds citric acid has given the best results. Using citrate complexes at pH 3.0-5.4 leathers have been obtained containing up to 28% of Fe_2O_3. Some scientists believed that the catalytic effect of hydrolyzed iron salts on collagen oxidation systems is the principal reason for irresistancy of iron-tanned leathers. A similar view has been expressed by Krzywicki [40] who has for years been an adherent of iron tanning. He emphasized that despite some tendencies to accelerated aging the obtained leather may be of good quality, provided that:
(1) The iron salts are not too diluted and the solutions are not heated.
(2) Oxidative agents are not used in excess and chosen properly.
(3) The basicity of salts is high and constant.
As oxidant $K_2Cr_2O_7$ or CrO_3 were recommended for upper leathers, and $KMnO_4$ for sole leathers. In this last case an addition of sodium silicate (water glass) was recommended at the end of the process to precipitate insoluble iron silicate, which gives some roughness to leather, protecting the user against slipping.

15.5. Titanium tannage

In the past decade titanium $^{4+}$ salts have been proposed as tanning agent. According to the reports of Soviet scientists and technologists who have elaborated the detailed procedures ammonium-titanium sulfate $TiO_2 \cdot SO_4(NH_4)_2SO_4 \cdot 2H_2O$ (being a stable, water-soluble white salt) has been preferably used [41]. This salt may be used alone, or in a mixture with chromium and/or zirconium compounds. It is applicable for both light and heavy leathers. The T_s of titanium-tanned leathers is up to 100°C. The tanning time for heavy leathers in drums is 6-9 hours. The quoted authors recommend using 2-3% of tanning salt (calculated on TiO_2), and subsequently a zirconium tanning (3-6% as ZrO_2) during 20-24 hours. Urotropin (2%) and sodium sulfate (2%) have been used for neutralizing, then retanning with syntan was applied. The obtained leather is said to be flexible, light colored, with very high abrasion resistance and compactness. Very similar estimates have been made by Strakhov [42].

The tanning mechanism, despite many publications dealing with the process, is not clear, nor the structure of tanning agent and the way it acts are reported. However, Roman et al. [43], studying the binding of titanium salt to ion exchange resins concluded that it attaches preferentially to the amino and imino groups, to a smaller extent to sulfo groups and hydroxyls; anionic complexes prevail in solution. The titanium compound is better attached to the functional groups, if 0.3% of citric acid is added, supposedly masking the complex. Generally titanium tannage is done at acid pH and organic acids, i.e., citric, tartaric and lactic, are recommended as masking agents and to increase the pH; acetic, formic and oxalic acids are ineffective.

Numerous publications concerned with technology of titanium tannage and combined tannages with titanium salt have appeared in the Soviet literature. It would be too early to give any evaluation of this process, as it has been introduced in one country only. However, it may arouse interest of tanning in countries where titanium ores and titanium salts are abundant.

15.6. Combined tannages

Tremendous work has been done in the field of combined tannage. Its very convincing technological aim is to obtain leather most suitable for its intended application—by the cheapest and easiest means. Combinations of mineral tannages as well as of mineral, vegetable and aldehyde tannages are known and applied widely. Crown leather, Danish leather and Dongola leather are the old examples of 'fine art' tanning. The combined tannage processes are mainly based on experience and tradition—now assisted by technical knowledge and, last but not least, by the need of utilizing some chemicals and useless by-products from other industries.

The theories related to these processes have been developed, but in the present author's opinion the combined processes are too complicated to be explained in a monograph like the present one. More additional studies are required to provide some clarity about the phenomena occurring in such processes, but even then they would lack certainty to allow their full acceptation, unless a detailed study of every parameter would prove its real values. This in turn would require an extra monograph on every specific topic. Accordingly, we are not going to discuss such processes, and we refer the interested reader to the relevant literature.

REFERENCES

1. Chakravorty, H. P., Nursten, H. E. J. Soc. Leath. Tr. Chem., *42*, 2 (1958)
2. see ref. 10 ch. 10
3. Slabbert, N. P. XVI IULTCS Congress Versailles 1979
4. Bjerrum, N. Z. Phys. Chem., *73*, 724 (1910)
5. Kawamura, A., Wada, K. J. Am. Leath. Chem. Assoc., *62*, 612 (1967)
6. Indubala, S., Ramasvamy, D. J. Inorg. Nucl. Chem. Letters *35*, 2055 (1973)
7. Takimoto, T. Hikaku Kagaku *23*, 131 (1977) in Japanese
8. Riess, J. W. Leder *16*, 49 (1961)
9. Slabbert, N. P. XIV IULTCS Congress Barcelona 1975
10. Irving, H. M. N. H. J. Soc. Leath. Tr. Chem., *58*, 51 (1974)
11. Makarov-Zemliansky, Ja. Ja. Kozh. Obuv. Prom., *19*, 4, 35 (1977) in Russian
12. Davis, M. H., Scroggie, J. G. XVI IULTCS Congress Versailles 1979
13. Menshikov, B. I., Strakhov, I. P., Makarov-Zemliansky, Ja. Ja. Kozh. Obuv. Prom., *20*, 12, 35 (1978) in Russian
14. Caminiti, R., Licheri, G., Piccaluga, G., Pinna, G. J. chem. Phys., *65*, 3134 (1976)
15. Elöd, E., Schachovsky, Th. Collegium *772*, 484 (1934)
16. Serfass, E. J., Wilson, C. D., Theis, E. R. J. Am. Leath. Chem. Assoc., *44*, 647 (1949)
17. Shuttleworth, S. C., Sykes, R. L. J. Am. Leath. Chem. Assoc., *54*, 259 (1959)
18. Erdmann, H. Leder *17*, 10 (1966)
19. Küntzel, A., Riess, J. W., Rizk, S. Leder *17*, 69 (1966)
20. Schläfer, H. L., Gausmann, H., Witzke, H. J. chem. Phys., *46*, 1423 (1967)
21. Lumb, M. D. (ed) Luminescence Spectroscopy, Acad. Press London 1978
22. Gustavson, K. H. J. Am. Leath. Chem. Assoc., *19*, 446 (1924)
23. Babich, N. P., Mikhailov, A. N. Kozh. Obuv. Prom., *18*, 1, 40 (1976) in Russian
24. Lasek, W. J. Am. Leath. Chem. Assoc., *62*, 487 (1967)
25. Küntzel, A. Leder *18*, 253 (1967)
26. see ref. 28 ch. 10
27. see ref. 10 and 11 this ch.
28. Takenouchi, K. J. Am. Leath. Chem. Assoc., *75*, 150 (1980)
29. Hock, A. L. J. Soc. Leath. Tr. Chem., *59*, 181 (1975)
30. Babich, I. Ja., Shapilskaya, A. Ja. Kozh. Obuv. Prom., *12*, 1, 26 (1970) in Russian
31. Metelkin, A. J., Kolesnikova, N. J., Kuzmina, E. V. Zirconium tannage, Moscow 1972 in Russian
32. Lebedev, O. P., Strakhov, I. P., Baramboim, N. K. Izv. Vys. Uch. Zaved. Leg. Prom., *4*, 49 (1975) in Russian

33. Ranganathan, R., Reed, R. Bull. Centr. Leather Res. Inst. Madras *9*, 202 (1962)
34. Williams-Wynn, D. A. J. Soc. Leath. Tr. Chem., *53*, 64 (1969)
35. Zorn, B., Rieger, I., Schmid, H. Leder *16*, 73 (1965)
36. Chambard, P., Grall, F. Bull. AFCIC *10*, 88 (1948)
37. Simoncini, A., Manzo, G., Fedele, C. Cuoio Pelli Mat. Conc., *54*, 439 and 711 (1978) in Italian
38. Küntzel, A. Colloquiumber. d. T. H. Darmstadt *3*, 27 (1947)
39. Küntzel, A., Rizk, S. Leder *13*, 101 (1962)
40. Krzywicki, E. Tanning Technology, 2 ed. Warsaw 1947 in Polish
41. Metelkin, A. N., Rusakova, N. T. Kozh. Obuv. Prom., *20*, 6, 9 (1978) in Russian
42. Strakhov, I. P. (ed) Chemistry and Technology of Leather and Fur 3rd ed., Moscow 1979 in Russian
43. Roman, A. S., Gnesina, N. M., Konopelkina, L. V. Iz. Vys. Uch. Zav. Leg. Prom., *6*, 61 (1974) in Russian

16.

ALDEHYDE TANNAGE

Aldehyde tannage is probably the most investigated process, especially formaldehyde tannage. The reasons for the interest aroused by this kind of tanning among scientists are discussed in section 14.1. Among the tanning agents used, aldehydes have the smallest molecules. Being well-defined compounds with one or two functional groups of the same kind, and thus binding collagen in the same way, they are frequently used as model tanning agents. Extensive investigations of tanning properties of aldehydes, carried out by Gustavson [1], have led to the conclusion principally that those aldehydes have tanning properties which are not attached to an aromatic ring, or the aldehyde groups are not combined with a carbon chain having more than four carbon atoms and polar groups.

The reactions of formaldehyde, glyoxal, glutaraldehyde and starch dialdehyde are tested in greater detail as they are used in commercial tanning.

Tanning with formaldehyde

Formaldehyde as a tanning agent has been known for a very long time; the campfire smoke contains it in large amounts. Presumably, tanning was observed for the first time when meat together with skin were smoked. Probably the only tanning gas precisely known, it allows, what is also important in research, to tan without any contact with the bath. Using DCDO (deuterated formaldehyde) protons are not introduced into collagen. Formaldehyde will be very easily bonded to the other proteins, not exclusively to collagen. It also is attached to the cell membrane components, denaturing them, and thus it has for long been used as a drastic disinfectant.

The amount of aldehyde being attached to the hide is small ranging, according to various sources, from 0.2 to 2%. Part of this aldehyde is connected to protein, part remains in the hide unbound. One can consider it as mechanically involved in the collagen structure, however, without meeting functional groups capable of binding it.

Determination of formaldehyde in leather is difficult and a great difficulty is encountered in separating free aldehyde from the bound one. Numerous methods are described, mainly based on washing out or mechanical squeezing of unbound aldehyde. Bound aldehyde is usually determined by distillation from over the sample acidified with sulfuric acid. However, part of aldehyde remains irreversibly bound.

Concentration of aldehyde and the pH of the solution are significant for its binding by the protein. Time is of little influence, as this reaction is quick and it is easily completed in 12-18 hours. The effects of concentration and of the pH of the bath on the amount of aldehyde bound in 24 hours is shown in Fig. 16.1, and the effect of pH on T_s of formaldehyde treated collagen in Fig. 16.2. From the first Figure it is seen that the amount of aldehyde bound increases with concentration. This is a peculiar property, because the amount of tanning agent bonded, as a rule, does not depend on concentration. Most investigations carried out on formaldehyde tannage have been performed out at low or moderate concentrations. This is why we have very scarce information about the groups involved, if the operations are done at high concentrations. The possibility of formation of formaldehyde oligomeres in the tanning process cannot be excluded. On the basis of the curves shown in the figures it is clear why it is important to estimate exactly the amount of aldehyde chemically bound and of that remaining in the leather due to physical forces. The rate of binding to collagen is greatest in the first hours of reaction; then it decreases to stop completely after

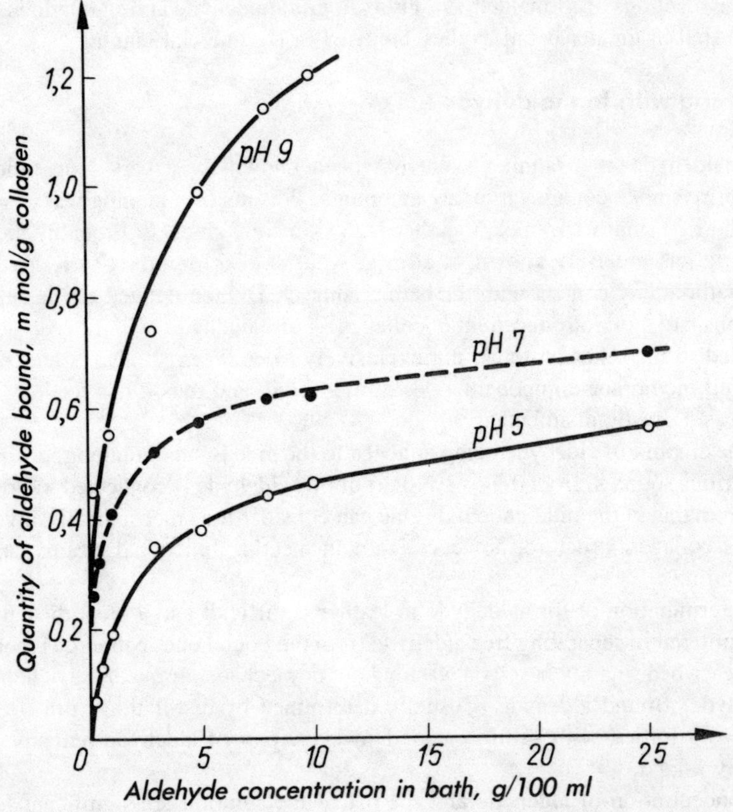

Fig. 16.1. Effect of aldehyde concentration in the bath and of the pH on the amount of HCHO bound by hide powder.

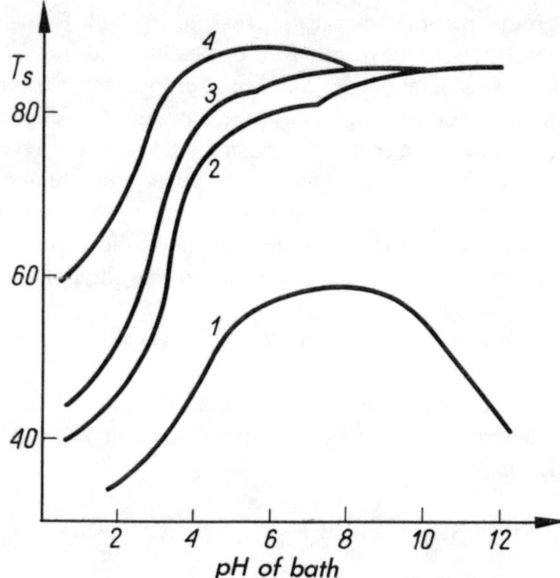

Fig. 16.2. The T_s of untanned (1) and formaldehyde tanned collagen as a function of the pH dependence of the bath containing: (2)—0.5% of HCHO, (3)—1.0%, (4)—1.5%.

12-18 hours. The binding of formaldehyde by casein, however, takes several days. Casein tanned this way was one of the first plastics developed about fifty years ago (Galalith).

The effect of pH on aldehyde binding seems to be similar for all proteins. The amount of aldehyde bound increases with pH until pH reaches values at which testing becomes difficult for analytical reasons. At low pH the amount of bound aldehyde increases, if conditions are applied which can be considered as dehydrating: high concentration, acidity, ionic strength and high temperature. The T_s of the formaldehyde treated collagen is dependent on formaldehyde concentration and on the pH, whereas T_s maxima are shifting towards higher temperatures and lower pH values as the aldehyde concentration increases. According to the paper by Wiederhorn et al. [2], three introduced crosslinking bonds occur per 55,000 molecular weight units, as the result of formaldehyde crosslinking. That means that out of every 7th-8th amino group of the collagen side chain one is able to form crosslinking bonds. Much more bonds of this kind are formed, if previously shrunk collagen is used for tanning. Apparently small amounts of crosslinking bonds, which are necessary to increase the shrinkage temperature, indicate that this change is not only due to the formation of additional crosslinking bonds, but also to some other processes taking place, which cannot be defined now.

The amino groups participating in the crosslinking with formaldehyde have already been mentioned. In the reactions of free amino acids the α- and ω-amino groups come into consideration. In collagen α-groups are involved in peptide bonds. Accordingly, their participation in the binding of aldehyde is unlikely. From polarimetric measurements it has been found that free lysine binds one aldehyde group to each of its amino groups; however, the ε-amino group reacts first. The difference between the ionization constant and the binding rate of both groups is not great. The reaction mechanism in acidic medium is probably identical with the mechanism operating during formol titration:

$$R-NH_3^{\oplus} + CH_2O \longrightarrow R-NH-CH_2OH + H^{\oplus}$$

Subsequently the reaction probably occurs with the formation of a bridge between OH and the amide group as it has been observed in casein and in model systems over a wide pH range:

$$R-NH-CH_2OH + NH_2-CO-R \rightarrow R-NH-CH_2-NH-CO-R + H_2O$$

Such a bond is formed, e.g., in the proline-acetamide system. Formation of methylene bridges in the Mannich type reactions has been previously observed in protein systems. The reaction was primarily carried out in alkaline or slightly acidic medium. Investigations done later allowed us to assume that this reaction may be carried out in a very acidic medium as well. The amino group gives with formaldehyde a stable hydroxymethylene derivative. This derivative, however, may still bind one aldehyde group, yielding an unstable compound, or it may react with amido-, guanido-, imidozolo-, indolo- and phenolic groups. It was believed that these reactions can contribute to the crosslinking by methylene bridges. According to Feairheller et al. [3] collagen crosslinking cannot proceed in this way, as in the collagen hydrolyzate, previously subjected to this reaction, the crosslinking products could not be isolated, though Mannich's reaction may occur under formaldehyde tanning conditions:

$$>NH + H_2C{=}O + H-\overset{|}{\underset{|}{C}}- \longrightarrow >N-CH_2-\overset{|}{\underset{|}{C}}- + H_2O$$

The Mannich reaction during aldehyde tanning would require participation of active hydrogen. It has not as yet been discovered what hydrogen atom it is. If, as we now presume, it is the one at nitrogen in the peptide bond, the question cannot be answered why collagen binds aldehydes more easily than other proteins, despite that 20% of its peptide bonds do not contain hydrogen (those from imino acids). Then still the particular helical structure may be of importance because of the distances in it but this is a supposition only. Investigations on

Mannich's reaction between collagen and malonic acid, which is proved to be the best source of active hydrogen, led to the discovery [4] that the product obtained, which is enriched in carboxylic groups as compared with the starting one, reacts more easily with the chromium and zirconium tanning complexes. Mannich condensation itself gave compounds, having T_S close to that obtained under the given conditions due to the action of formaldehyde (various reaction pH values were applied). Subsequent chrome tanning gave a product containing about 10% Cr_2O_3 more, which is not very convincing because of the great dispersion of the results.

There is no doubt that the site to which formaldehyde is attached is the ω-amino groups of the side chains. The former theory of participation of the peptide group in aldehyde binding, represented still by Mikhailov [5], has not found experimental support; although, according to Gustavson [6] deaminated collagen binds formaldehyde. However, in this reaction the T_S does not increase, nor is this preparation more resistant to trypsin (under appropriate conditions, like, e.g., in water-free medium, catalysis, increased temperature, the hydrogen of the peptide bond may be substituted in polyamide by formaldehyde). Prolonged washing of the formaldehyde-tanned collagen with water causes a decrease of its aldehyde content until the value of almost 0.2% is reached—without a T_S decrease.

The amount of aldehydes attached to collagen in tanning, according to Filachione [7], calculated on protein weight, equals:

for formaldehyde	1.4-2.2%
for glyoxal	3.2-3.7%
for malonic aldehyde	2.5-4.4%
for starch dialdehyde	8.5%

Recently Lange and Pauckner [8] investigated among others the influence of experimental conditions on formaldehyde binding, and found a strong dependence of the amount bound on the pH and concentration (increasing with a pH increase). Their data are in agreement with those of Filachione. A very sharp increase of the aldehyde uptake results from pH increase to 9, as well as a T_S increase. If the amount of formaldehyde added is too low (0.5%), it becomes bound chiefly in the external layers of the hide. This is particularly well observable at high pH values of about 9. The T_S of the middle layer is then low. Two typical plots are reproduced from the quoted paper: the total formaldehyde uptake vs. pH (Fig. 16.3) and T_S of the particular layers vs. pH (Fig. 16.4). These observations may be considered as an introductory report on the kinetics of formaldehyde binding to the pelt. There may be some uncertainties, however, as a formaldehyde determination method of Highberger and Reitch was used which is relatively old (1938) and not very accurate [9].

Various data concerning aldehyde tanning have been collected by Gustavson [1]. He investigated the so-called Ewald's reaction. Collagen tanned with for-

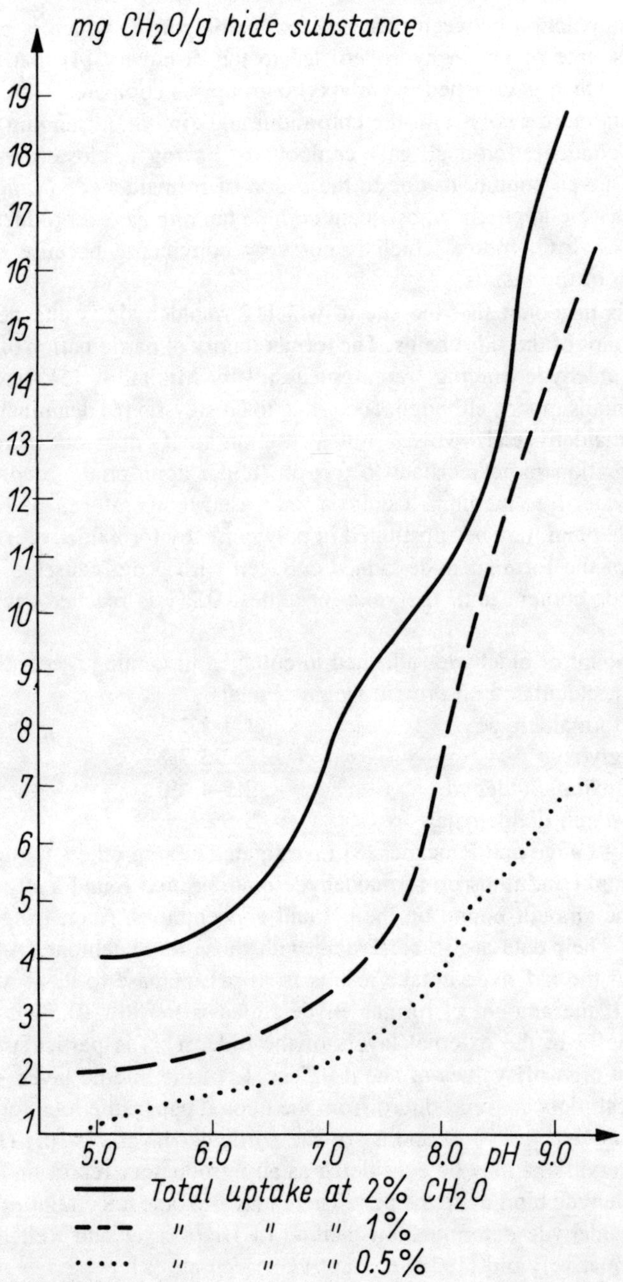

Fig. 16.3. Formaldehyde uptake vs. its concentration in solution and pH of bath. According to [8] with permission

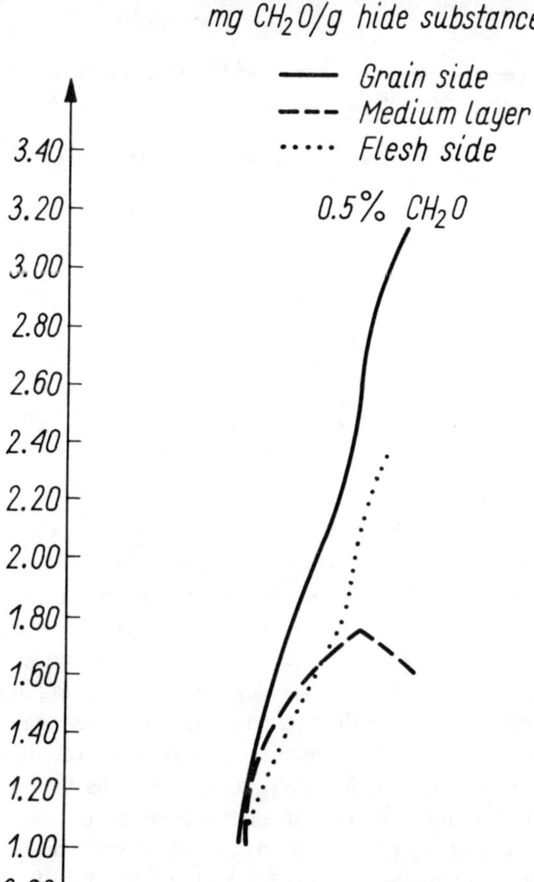

Fig. 16.4. Formaldehyde retention in the particular hide layers vs. pH at constant concentration. According to [8] with permission

maldehyde and then shrunk recovers almost fully its previous length after cooling. The quoted author paid attention to the good results obtained by tanning of hides in organic solvents with glyoxal. Using glyoxal for tanning in chloroform, benzene or toluene a very good yellow-colored leather may be obtained. Its T_s is several degrees higher than for one tanned with glyoxal in water, which is supposed by Gustavson to be due to the lower value of dielectric constant of the organic solvent. Tanning with glyoxal gives according to this investigation satisfactory results despite that it is not a crosslinking compound and has just one active group. This reaction is still unexplained.

Table 16.1.

Content of single- and twopoint attached formaldehyde in relation to the number of shrinkage operations performed

Material	Total HCHO content, mmole/100 g protein	HCHO content as hydroxymethy-lene groups, %	crosslinking %	ΔTs K
Pelt tanned with formalde-hyde in an amount of 1% on pelt weight	21.02	38.63	61.37	36
After 1 shrinkage	17.77	46.48	53.52	34
After 2 shrinkages	11.34	54.01	45.59	22
After 3 shrinkages	8.50	62.50	37.50	20
After 4 shrinkages	8.20	63.40	36.60	20

An experiment in which the crosslinking and single point bound formaldehyde are to be differentiated has been undertaken by Abd Alla et al. [10]. They based it on the determination of a number of hydroxymethylene groups present in collagen, and on the total amount of bound formaldehyde. The results of investigations, characterizing the changes in collagen shrunken several times, are shown in Table 16.1. In the Table one can see the decreasing amount of crosslinking bonds, despite that crosslinking partly remains after heating; the amount of unipoint-attached formaldehyde increases. These observations may allow one to get a better knowledge of the reaction of methylene bridges crosslinking. Simoncini et al. [11] have found that more formaldehyde may be attached to hide in the presence of inorganic salts. According to those authors the above is due to the fact that protelysis is facilitated by the salt anion. Its cation attaches to the collagen side chain carboxyl by one valence.

Glutaraldehyde tannage

Glutaraldehyde forms probably semiacetal bonds with hydroxyls of hydroxyproline, hydroxylysine and serine. With phenols it yields insoluble compounds, so it cannot be used with vegetable tannins.

or

$$
\begin{array}{c}
\diagdown \\
\diagup\!\!\!\!\!-\text{OH} \quad + \quad
\begin{matrix}
\text{O} \\
\parallel \\
\text{CH} \\
| \\
(\text{CH}_2)_3 \\
| \\
\text{CH} \\
\parallel \\
\text{O}
\end{matrix}
\quad \longrightarrow \quad
\begin{matrix}
\text{OH} \\
| \\
-\text{O}-\text{CH} \\
| \\
(\text{CH}_2)_3 \\
| \\
\text{CH} \\
\parallel \\
\text{O}
\end{matrix}
\end{array}
$$

With amino groups it may react in three ways viz.:

$$
\begin{array}{c}
\diagdown\!\!-\text{NH}_2 \\
\\
\diagup\!\!-\text{NH}_2
\end{array}
\; + \;
\begin{matrix}
\text{O} \\
\parallel \\
\text{CH} \\
| \\
(\text{CH}_2)_3 \\
| \\
\text{CH} \\
\parallel \\
\text{O}
\end{matrix}
\; + \;
\begin{array}{c}
\text{H}_2\text{N}-\!\diagup \\
\\
\text{H}_2\text{N}-\!\diagdown
\end{array}
\; \longrightarrow \;
\begin{array}{c}
\diagdown\!\!-\text{NH}-\text{CH}-\text{NH}-\!\diagup \\
| \\
(\text{CH}_2)_3 \\
| \\
\diagup\!\!-\text{NH}-\text{CH}-\text{NH}-\!\diagdown
\end{array}
\; + \; 2\,\text{H}_2\text{O}
$$

or,

$$-\text{NH}_2 \;+\; \text{O}=\text{CH(CH}_2)_3\text{CH}=\text{O} \;+\; \text{H}_2\text{N} \;\rightarrow\; -\text{N} \;=\; \text{CH(CH}_2)_3\text{CH}=\text{N}- \;+\; 2\text{H}_2\text{O}$$

or,

$$-\text{NH}_2 \;+\; \text{O}=\text{CH(CH}_2)_3\text{CH}=\text{O} \;+\; \text{H}_2\text{N}- \;\rightarrow\; -\text{NH}-\underset{\text{OH}}{\text{CH}}\text{(CH}_2)_3\underset{\text{OH}}{\text{CH}}-\text{HN}-$$

From stoichiometric calculations it follows that if the pelt would bind aldehyde by lysine ε-aminogroups only, it would bind only 0.7% of aldehyde by weight. The actual amount, however, is greater, which points to the participation of other collagen groups, probably hydroxyls. That is in accordance with the suggestions already made.

The use of glutaraldehyde in tannage involves two important questions, in what form does it react and why is it considered to decrease the leather strength when used in leather making. The first question has been answered by Gillett and Gull [12]. Using pure glutaraldehyde solutions they proved by UV spectroscopy (which was later confirmed by Heidemann and Bresler [13] by the NMR technique) that glutaraldehyde in 25-50% aqueous solutions is usually oligomerized (3-5 molecules), when stored under conventional conditions giving cyclized oligomeres and unsaturated α, β-aldehydes. Oligomerization may be prevented by adding alcohols, or storage at low temperatures (−20°C). What form is tanning-active remains still to be found. Gillett-Gull believe that this is the monomer form, whereas according to commercial experience oligomerization is of little or no influence at all. Recent experiments with chemically modified glutaraldehyde [14] give no indication in this respect, nor is it indicated what is the nature of this modification, although the tanning experiments are promising.

Glutaraldehyde used for tanning gives the leather a perspiration resistance, washing resistance, better fullness and density. The decrease of strength at break has not been elucidated, though it has been confirmed repeatedly. One can expect deep changes in the molecule, perhaps breaking of some peptide bonds by the transfer of the bonds to the aldehyde group—that is, however, the present author's supposition only.

Investigating the problem of leather tear strength and elongation at break decrease due to the glutaraldehyde tannage Keller and Heidemann [15] concluded that the strength decreases with thickness increase. This fact may be generalized, as it concerns not only glutaraldehyde tannage. The same is observed in chrome-tanned leather as well as in acetone dried one (Fig. 16.5), thus the quoted authors conclude that this is more a matter of changes in thickness, and it is most pronounced in glutaraldehyde-tanned leathers perhaps as a result of the greatest

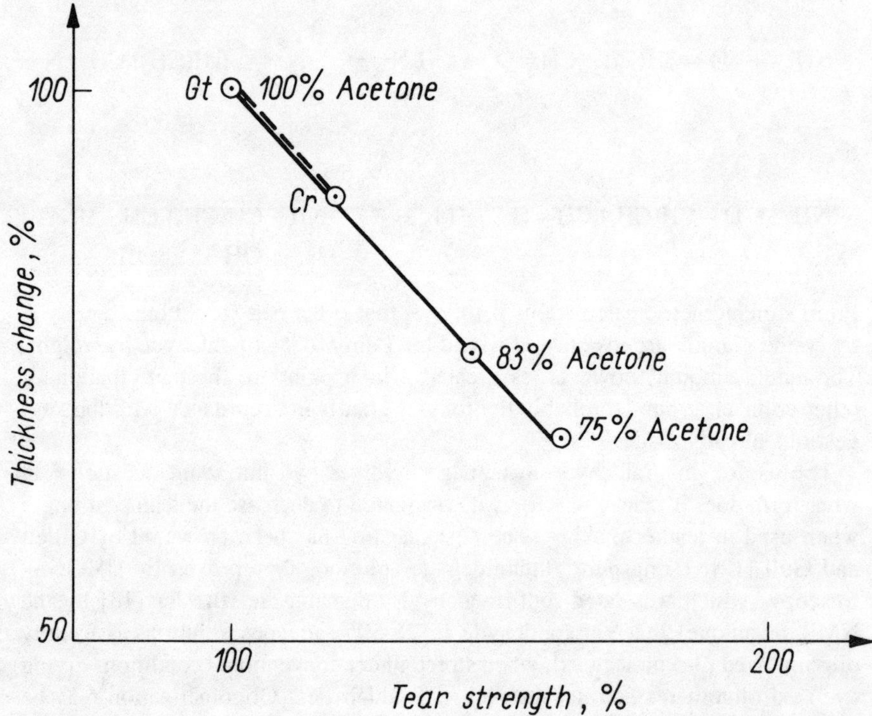

Fig. 16.5. Apparent changes in the tear strength due to tanning and dehydration (acetone): Gt = glutaraldehyde-tanned leather, Cr—chrome-tanned leather. According to [15] with permission

thickness increase occurring in this type of tannage.

Feairheller [16] recently questioned the opinion that the strength of leather decreases due to glutaraldehyde tannage as it has not been satisfactorily statistically proved.

The exhaustion rate of glutaraldehyde from the bath is greater than that of formaldehyde and of glyoxal, particularly at pH about 8 (Fig. 16.6), and the lower the pH the slower the exhaustion.

Leberfinger et al. [17] have investigated the chemistry and kinetics of glutaraldehyde binding. Balancing the amount of aldehyde which may be attached to the functional groups of various kinds of side chains they came to the conclusion that ω-amino groups may bind 0.93-1.86 g/100 g collagen, hydroxyls—3.9-7.8/100 g collagen, depending on whether aldehyde is single- or two-point bound. Pelt may, however, attach up to 21% of glutaraldehyde. This has made the authors convinced that glutaraldehyde must be polymerized, which has actually been proved in the recent years. Monomeric glutaraldehyde occurs in water solutions in very small amounts. When examining the binding of glutaraldehyde to chrome-tanned leather the authors found that it will be attached in smaller amounts than to pelt, and that the reaction rate is higher. At pH = 2 chrome

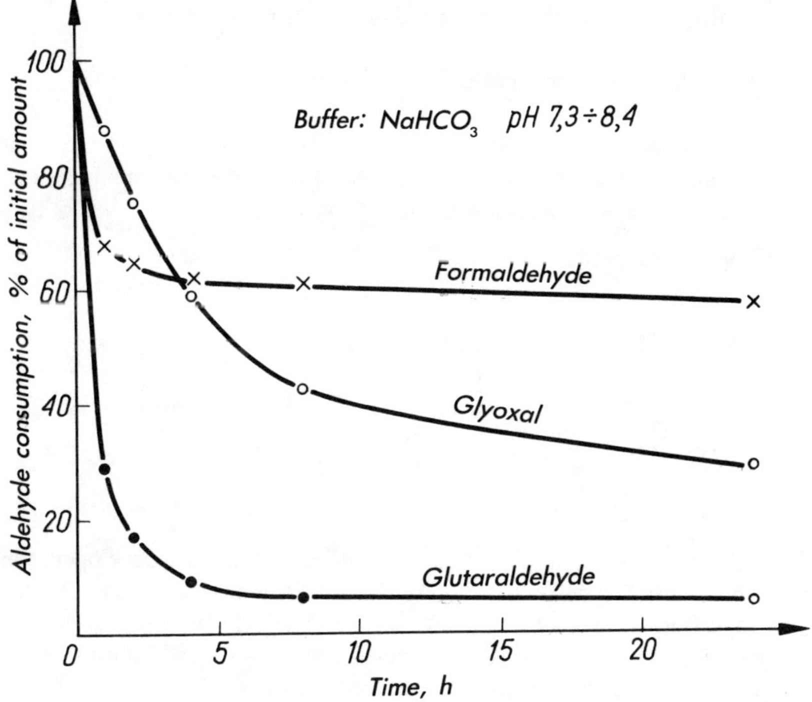

Fig. 16.6. Exhaustion rate of aldehyde from bath.

tanned leather binds only 0.75% from a 3% solution, and the equilibrium is established in about 5 hours. At higher pH values the amount of aldehyde bound to chrome leather is still smaller.

Glutaraldehyde is most often used for pretanning of skins to be chrome-tanned. It binds better to pelt than to chrome-tanned leather. When applied as an additive it increases the perspiration resistance. In leathers containing some glutaraldehyde, the amount of substance removed by washing is lower. Of special interest is the ability to increase perspiration resistance of insole and glove leathers.

According to more recent data [18] azomethine bonds are of special importance in binding dialdehydes to collagen. These bonds of chromphore character are responsible for the yellow color of leather tanned with dialdehydes. In classical organic chemistry these bonds are considered unstable, which does not seem justified, as under conditions of acidic hydrolysis the azomethine bond (Schiff's base) between collagen and glutaraldehyde remains intact.

Now we know that the reduction of Schiff's base gives a stable bond:

$$-CH_2-N=CH- \xrightarrow{H+} -CH_2-HN-CH_2-$$

Collagen destructs containing such bonds have been isolated [19].

Tanning with starch and cellulose aldehydes

In the United States investigations have been conducted jointly on the glutaraldehyde and starch dialdehyde tannings agents [20]. The latter preparation is obtained by oxidation of starch with periodic acid.

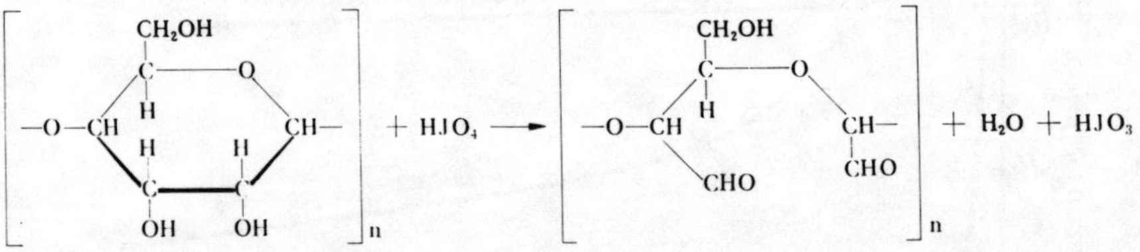

$$
\begin{bmatrix}
\begin{array}{c}
CH_2OH \\
| \\
-O-CH \overset{\displaystyle C}{\underset{\displaystyle C}{}} CH- \\
OH \quad OH
\end{array}
\end{bmatrix}_n
+ HJO_4 \longrightarrow
\begin{bmatrix}
\begin{array}{c}
CH_2OH \\
| \\
-O-CH \quad CH- \\
CHO \quad CHO
\end{array}
\end{bmatrix}_n
+ H_2O + HJO_3
$$

After a simple method of recovering periodic acid has been developed, this oxidant became accessible in the U.S.A. The process may be controlled and thus starch of various degree of oxidation may be obtained. The product is water insoluble. To dissolve it, one can use pressure heating at 120°C during 30 min (D-Starch A) or heat it with sodium carbonate (D-Starch T). In Fig. 16.7 the tanning rate is evaluated in dependence on pH in terms of T_s changes. As it is seen, the tanning process proceeds best in a basic medium, and its rate is fairly high. The shrinkage temperature assumes a practically constant value after 6-24 hours. According to the investigations mentioned, the use of dialdehyde, solubilized with sodium carbonate, is more advantageous because of its stabile pH-value. In this case only slight pH changes are possible. It is a rule that the higher the oxidation degree, the higher the T_s of leather obtained. Leathers obtained in this tanning process are soft and stretchy.

Nayudamma et al. [21] investigated in detail the binding of starch dialdehyde to collagen, to modified collagen and to amino acids. They found that for the binding reaction pH = 7.2 is optimal. Collagen attaches then 9.7% of aldehyde. Binding, like in the cases of other aldehydes, proceeds better in the presence of 0.1 gram-equivalents of sodium acetate added per 100 g of protein. The amount of dialdehyde bound increases then by 75%. Investigation of modified collagen has shown that basic collagen groups are primarily responsible for this reaction. Deaminated collagen attaches only ⅓rd of the amount of aldehyde bound by the native one, and deguanidinated collagen binds 20% less. Color measurements done on the amino acids which have already reacted with aldehyde show the most intense browning reaction with lysine and glutamine. Infrared spectrum of the starchdialdehydeglycine compound has a strong band at 1650 cm^{-1}, for which, according to the quoted authors, the formation of Schiff's base is responsible under the reaction conditions. This is indicated by the results obtained

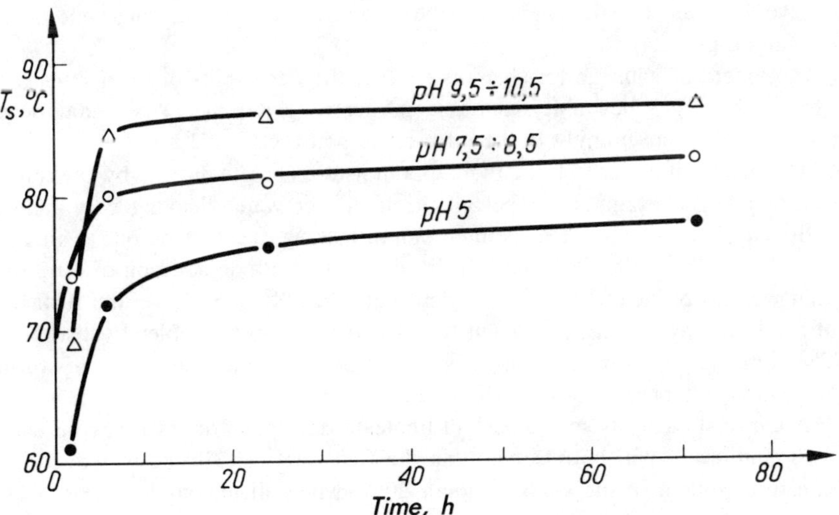

Fig. 16.7. Increase of T_s during tanning of calfskin with starch dialdehyde A vs. the process pH.

by other authors, and discussed in the paper quoted.

Cellulose dialdehyde has been utilized as well, and the results obtained were compared to those for starch dialdehyde [22]. Cellulose dialdehyde consists of 1000-2000 dialdehyde units, connected by a β-glycosidic bond in position 1,4. In tanning tests the T_s increase obtained varied from 5-6°C at pH-5, to 25-28°C at pH = 10. The tanning took about 6 hours.

Decrease of aldehyde polymerization degree improves but insignificantly its tanning ability. Because of the similarity of leathers obtained in this process to the glyoxal-tanned ones, and the occurrence of glyoxal in floats, the quoted authors came to conclusion that glyoxal, as well as other small-molecule aldehydes which are degradation products of the modified biopolymer in question, is the substantial active tanning factor there.

Oil tannage

Oil (fat) tannage is a very old way of imparting the properties of finished leather to skins. Discovered probably in the East it correlates with the Eastern sense of comfort and quality. Oil-tanned leathers are light, soft, air-permeable, and resistant to washing. As tanning agent unsaturated glycerides are used.

The favored ones are found in cod-liver oil, which are found to have the most suitable type of unsaturated fatty acids in their composition. These may have one up to six double bonds in the aliphatic chain, but 15% should have at least four to give the necessary reaction products from oxidation and polymerization to give the characteristic 'chamois' leathering effect under the normal conditions of tanning [23, 24].

In modern oil tannage for chamois leather, the flesh splits of sheepskins are used as having a desirable open fiber structure. After the usual beamhouse processes they are brought to the iso-electric point, e.g., pH 5.0.

This makes it easier to bring them to a moisture content of 50% by pressing, samming or squeezing, to expel all the interfiber water, leaving only damp, hydrated fiber structure. This is important in making the skins porous to air and oil. The skins are then 'stocked' or dry drummed with an addition of 40% (of their weight) of the cod-liver oil, which should readily spread over the surfaces of the fibers by interfacial tension forces and be almost completely absorbed. The skins may then be hung up (if stocked) or drummed in a current of warm air. The resultant reactions are shown in Fig. 16.8.

Additional catalysts such as cobalt linoleate (the 'paint driers') may be used with caution. It will be observed that the first effect should be to reduce the moisture content of the skins to about 20%, when all unbound water is lost,

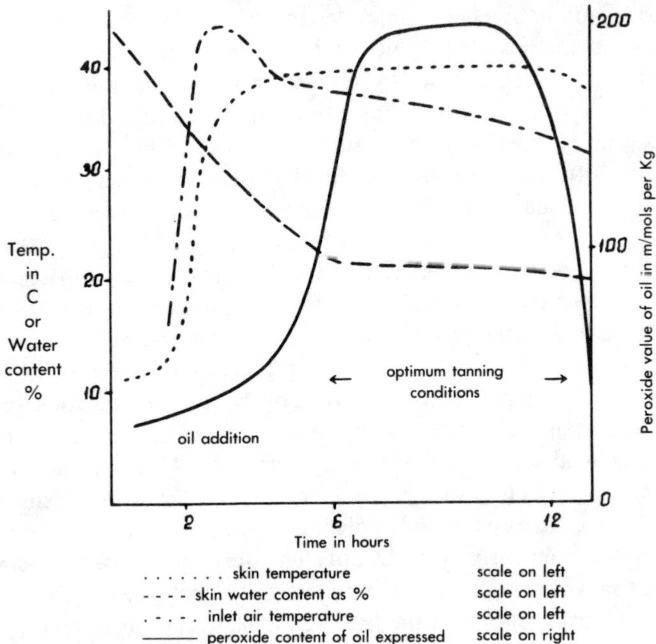

Fig. 16.8. Control factors during course of chamois tannage in a drum fitted with air blast. Courtesy J. H. Sharphouse

which aids oil absorption into the fibers, increases the gross porosity and hence the internal surface exposed to the air. The temperature is now held at the maximum permissible (40°C), with no further drying. Oxidation reactions of the oil now occur with exothermic liberation of heat, increase of peroxide value of the absorbed oil, acrid fume liberation, yellowing of color and leathering or tanning of the skins. After some hours these reactions are completed as shown by the drop in exothermicity, peroxide value, fume liberation, etc. Traces of free water do catalyze the reaction.

Surplus oil is removed by warm damp pressing, (Moellon or Degras) and then washing with warm solutions of sodium carbonate in water at pH 8-9 to saponify the greases. This surplus oil may be marketed as such or sulfited, but it is significant that it has no longer any tanning properties.

Of the 40% oil offered only 5-7% remains fixed to the fibers. Attempts to reduce the amount of oil offered result in thinner harder leather unless aided by additional aldehyde or sulphonyl chloride tannage.

Modern systems remove the surplus oil by solvent degreasing systems which eliminate effluent problems caused by oil wash liquors, and eliminate drying and staking occasioned by aqueous scouring. However the resultant degreased leather is waterproof unless treated with alkaline surfactants. The resultant chamois leather is very soft, very stretchy, rapidly absorbs 600% water, which can be readily 'wrung out' and makes it very suitable for window cleaning.

Elsinger [25] provides a schematic picture of the chemical processes which may be involved, showing the mechanism of the breakdown of the oil and its potential hydrolysis, oxidation and polymerization products. It must be emphasized that all these potential reactions are very dependent on reaction conditions of temperature, humidity, etc., but most significantly on the presence of the hydrated protein. The latter reacts with some of the nascent oxidation products which does not occur with simple oxidation of the oil in the absence of the skin and results in an oxidized oil with no tanning properties. Aldehydes are produced and have a tanning effect, however such by products as acrolein, aldehyde fatty acids, or hydroxy compounds readily polymerize under the conditions of tannage and such polymers are difficult to identify precisely by analytical techniques. Their nascent monomers formed during tannage (probably when the peroxide values are high) may react with the hydrated collagen, via carbonyl, carboxyl or hydroxy groups, giving bonds ranging from covalent to dipole. Subsequent tanning may lead to polymerization or condensation on to these 'fixed monomers,' so that the fibers become coated with fixed fatty acid residues, which account for the softness of chamois leather, as opposed to simple aldehyde tannages. Nevertheless the rise in shrinkage temperature, and the pronounced Ewald's effect, whereby the wet-heat shrunk chamois leather recovers its area on cooling, implies that the prime tannage is of the aldehyde type. A drop in the iso-electric point is characteristic of aldehyde tannages and in fact there is a large drop after chamois tannage, i.e., the leather has enhanced alkali binding power, which is well-known to the chamois scourer. This is largely due to free carboxyl acid groups in the fixed oil breakdown products on the fiber.

When neutralized with alkali they may behave as fixed soap tannages, accounting for the high water absorption and washing properties.

Chamois leather behaves like aldehyde leather on retannage with either chrome salts or vegetable tannins, provided these fixed residues are taken into account.

Pretannage of the pelt with small amounts of formaldehyde or glutaraldehyde is often done in practice to (a) make the preliminary squeezing out of surplus water from the pelt easier, or (b) to safeguard the skin against accidental overheating by raising its initial temperature by 7-10°C. This does not significantly alter the rate of subsequent oil oxidation, but often reduces the amount and color of the 'fixed oil' after alkali washing. The use of aliphatic sulphonyl chlorides (e.g., Immergan—B.A.S.F.) mixed with the cod-liver oil also gives good consistency of tannage, particularly when using natural fish oils of doubtful constitution.

Chamois leather when adjusted to suitable pHs shows poor affinity for anionic dyes but good affinity for basic, cationic dyes and strong affinity for reactive triazinyl dyes, which may fix also on oxidized oil residues. It is remarkable that such oil tannage should give chamois leathers, whose main use is as wash leathers where rapid water absorption and non-greasiness are essential. It is also one of the few commercial tannages not done in a float of water containing the tanning material in solution, like all the classical considerations of aqueous solutions, but involving the reaction of the simple hydrated collagen with tanning materials in gaseous or hydrophobic form. The use of other aldehydes in gaseous form is also possible by such a system.

However there is no record of experiments to use other insoluble, potentially collagen reactive, materials by this system of application to hides or skins although there are now many interesting possibilities of using new chemicals by such techniques.

A new method of oil tannage was proposed recently by Lange and Pauckner [8]. It is but a formaldehyde or glutaraldehyde pretannage, with subsequent cod-liver oil treatment. In such a process a fast entering of cod-liver results and the T_s achieved is higher than in the usual process. The method is mainly of technological interest. However, important observations on the pH effect on the formaldehyde and glutaraldehyde binding are made there (see p. 371).

REFERENCES

1. Gustavson, K. H. The Chemistry of Tanning Processes, Acad. Press N.Y. 1956
2. Wiederhorn, N. M., Reardon, G. V., Browne, A. R. J. Am. Leath. Chem. Assoc., *48*, 7 (1953)
3. Feairheller, S. H., Taylor, M. M., Gruber, H. A., Mellon, E. F., Filachione, E. M. J. Polymer Sci. Part C *24*, 163 (1968)
4. Feairheller, S. H., Taylor, M. M., Filachione, E. M. J. Am. Leath. Chem. Assoc., *62*, 408 (1967)
5. see ref. 5 ch. 2
6. Gustavson, K. H. Leder *13*, 233 (1962)
7. Filachione, E. M. Leder *11*, 141 (1960)
8. Lange, J., Pauckner, W. Leder *30*, 152 (1979)
9. Highberger, J. H., Reitsch, C. E. J. Am. Leath. Chem. Assoc., *33*, 341 (1938)
10. Abd Alla, M. F., Nayudamma, Y., Jayraman, K. S. Leather Sci., *18*, 218 (1971)
11. Simoncini, A., Manzo, G., Ummarino, G. Cuoio Pelli Mat. Conc., *47*, 113 (1971) in Italian
12. Gillet, R., Gull, K. Histochemie *30*, 162 (1972)
13. Heidemann, E., Bresler, H. Leder *25*, 229 (1974)
14. Alford, P. M. J. Am. Leath. Chem. Assoc., *73*, 250 (1978)
15. Keller, Ch., Heidemann, E. Leder *27*, 176 (1976)

16. Feairheller, S. H. XVI IULTCS Congress Versailles 1979
17. Leberfinger, R., Landbeck, F., Matschkal, H. Leder *22*, 271 (1971)
18. Bienkiewicz, K. J., Grzegorzewska, U., Kosiński, M., Pawlak, J. XIV IULTCS Congress Barcelona 1975
19. Bailey, A. J., Robins, S. P., Balian, G. Nature *251*, 105 (1974)
20. Filachione, E. M. Leder *11*, 141 (1960)
21. Nayudamma, Y., Thomas Joseph, K., Bose, S. M. J. Am. Leath. Chem. Assoc., *56*, 548 (1961)
22. Kontio, P., Harva, O., Tuomarla, J. J. Am. Leath. Chem. Assoc., *60*, 48 (1965)
23. see ref. 25 ch. 12
24. see ref. 5 ch. 14
25. Elsinger, P. Leder *15*, 289 (1964)

17.

TANNING WITH VEGETABLE AND SYNTHETIC TANNINS

Classical tanning chemistry arose from investigations on the binding of vegetable tanning agents to hide collagen, from efforts to accelerate this naturally very slow process and to eliminate its disadvantages. A real difficulty consisted in preparing the extracts for own use, in setting their concentrations and technical parameters, because the final product was very sensitive to changes in such parameters. But as late as 1931 S. Borsuk, a Polish technologist was still of the opinion that 'chrome leather shows much greater heat conductance than the vegetable one, being therefore too hot in summer and too cold in winter. For this reason it is unsuitable for soldiers and tourists.' Thus it was not so much for the quality of vegetable tanned leathers, but for their price, resulting from the slowness of the process, and the increasing demand for upper leathers that the vegetable tanning has been almost completely abandoned and replaced by combined and chrome tannage. The development of technology of synthetic resins and of cellulose industry made it possible to eliminate heavy leather, so that this kind of sole now becomes a luxury. This product, however, has its unquestionable advantages for which it will not be completely eliminated.

17.1. Vegetable tannins and tanning with their use

Vegetable tannins are products of plants. The plants have a particular ability to synthesize aromatic compounds. These compounds participate in the fundamental metabolism of plants and in several of their 'life cycles,' like, e.g., synthesis of terpenes, steroids, karotenoids and xantophils. The shikimic acid cycle is of essential importance for the formation of tannins. The compounds formed in this cycle include two groups:
(1) aromatic amino acids (phenylalanine, tyrosine, tryptophan),
(2) aromatic phenolic and phenylpropenic acids.
We shall deal with the second group only, as tannins are derivatives of phenolic acids, and lignins—of alcohols and phenylpropenic acids. Flavonoids, the coloring components of flowers, are of similar structure. The synthesis of aromatic compounds in plants is directly connected with carbohydrate katabolism, which leads to the formation of shikimic acid. As substrates in this reaction serve phosphoenolpyruvate, one of the end products of glycolysis, and erythroso-4-phosphate, which is an intermediate in hexose regeneration in the pentosophosphate cycle. These compounds participate in condensation due to the action of

transaldolase (EC 2.2.1.2). The product of this reaction, sedoheptulose-7-phosphate, is cyclized with splitting off a water molecule, by the action of dehydroquinate dehydratase (EC 4.2.1.10) giving dehydroquinone and then quinoic or dehydroshikimic acid. This acid by action of shikimate dehydrogenase (EC 1.1.1.25) yields shikimic acid.

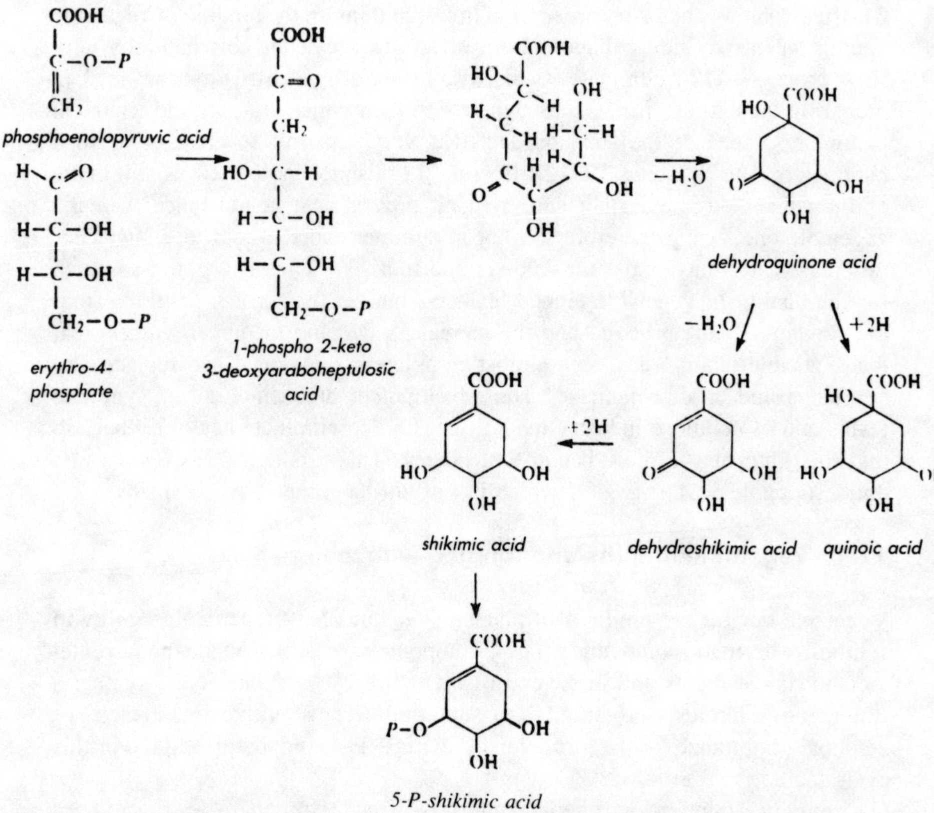

$$
\begin{array}{l}
\text{COOH} \\
|\\
\text{C}-\text{O}-P \\
\|\\
\text{CH}_2
\end{array}
\qquad
\textit{phosphoenolopyruvic acid}
$$

$$
\begin{array}{l}
\text{H}-\text{C}{=}\text{O} \\
|\\
\text{H}-\text{C}-\text{OH} \\
|\\
\text{H}-\text{C}-\text{OH} \\
|\\
\text{CH}_2-\text{O}-P
\end{array}
$$

erythro-4-phosphate

$$
\begin{array}{l}
\text{COOH} \\
|\\
\text{C}{=}\text{O} \\
|\\
\text{CH}_2 \\
|\\
\text{HO}-\text{C}-\text{H} \\
|\\
\text{H}-\text{C}-\text{OH} \\
|\\
\text{H}-\text{C}-\text{OH} \\
|\\
\text{CH}_2-\text{O}-P
\end{array}
$$

1-phospho 2-keto 3-deoxyaraboheptulosic acid

dehydroquinone acid

shikimic acid   *dehydroshikimic acid*   *quinoic acid*

$-H_2O$ $+2H$ $+2H$

5-P-shikimic acid

Phosphorylated shikimic acid is very active; its condensation with phosphoenolpyruvate leads to prephenic acid, and by transamination, to tyrosine or phenylalanine or, still through successive reactions—to tryptophane. The shikimic, quinoic and p-hydroxybenzoic acids have a group name of phenolic acids. They may be considered as parent substances of tannins, like phenylpropenic (cinnamic) acid is that of lignins (see below). Of course, appropriate enzymes participate in these reactions. As an example, the relationship between some aromatic compounds is shown with participation of gallic acid—one of the important components of tannins:

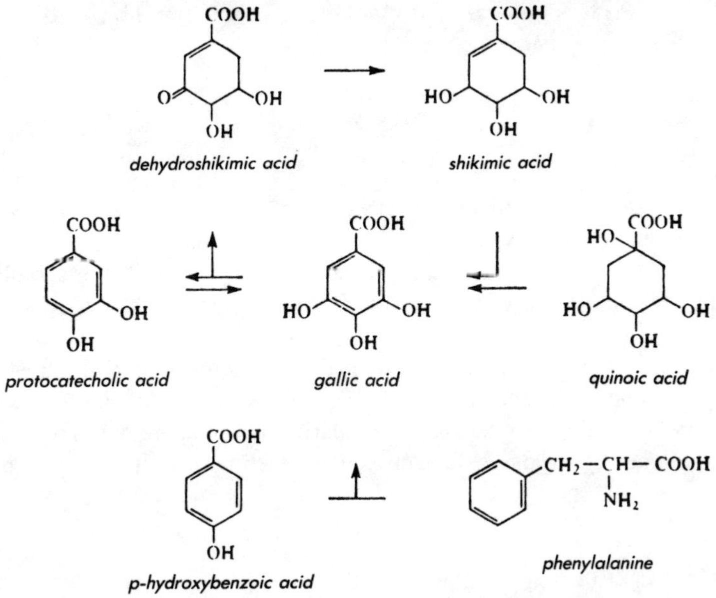

COOH

dehydroshikimic acid

COOH

shikimic acid

COOH

protocatecholic acid

COOH

gallic acid

COOH

quinoic acid

COOH

p-hydroxybenzoic acid

$CH_2-CH-COOH$
NH₂

phenylalanine

Gallic acid participates in many condensation reactions from which tannins result. They are substances containing phenolic and carboxylic groups, and are widely distributed in the plant world, although, as a rule, in low concentrations. Their physiological function in plants remains a mystery. Perhaps they act as inhibitors and/or controllers of certain enzyme reactions, though this seems to be too little for their amount and distribution. According to other viewpoints, they participate as diphenols in redox reaction in plants, serving as participants of redox cycles [1]. These processes proceed with the participation of reductases (group 1), e.g., p-diphenyloxydase (EC 1.10.3.2) or catechol oxidase (EC 1.10.3.1). This last enzyme participates in the catechol to o-quinone oxydation.

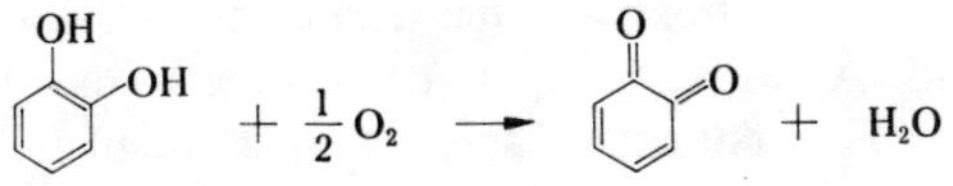

$$OH \quad + \frac{1}{2} O_2 \longrightarrow O \quad + H_2O$$

Typical feature of the group of polyphenol oxidases is the reductibility of quinones by ascorbic acid and enzymes, e.g., by glucose-6-phosphate dehydrogenase (EC 1.1.1.49) in the presence of NADP (Nicotinamide dinucleotide phosphate). Putting AH_2 as substrate, one may write the following reaction scheme:

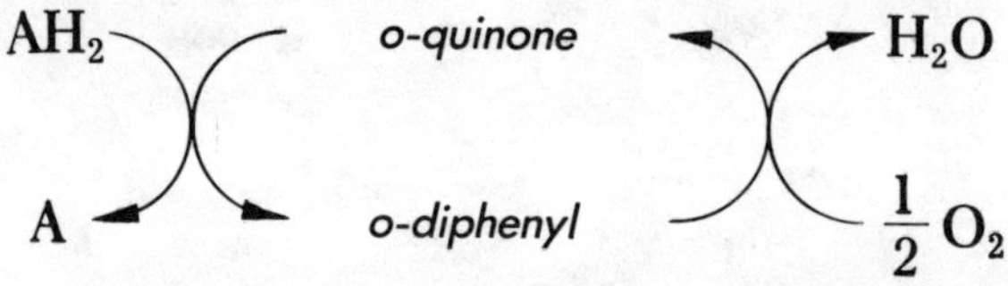

$$AH_2 \quad o\text{-}quinone \quad H_2O$$
$$A \quad o\text{-}diphenyl \quad \tfrac{1}{2}O_2$$

In such a system small amounts of catechol or proper quinone may repeatedly undergo reduction and oxydation. These compounds may also act as hydrogen carriers and the whole system catalyzes the oxygen uptake, being a biological analog of Wieland's pallad system. They form the 'breathing model' of the living cell.

Systems with phenol oxydase participate in the browning of the freshly cut surfaces of apples, potatoes or other vegetables in air. In plants containing polyphenol oxydase, o-diphenols, particularly catechols, are oxidized to appropriate quinones by oxydases; the subsequent reactions are already spontaneous:

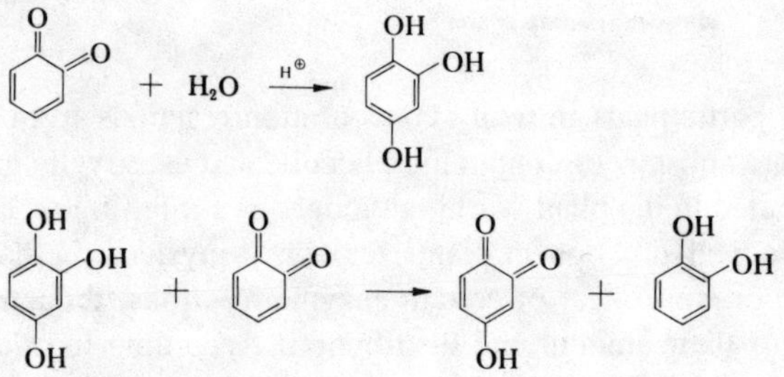

Thus a reaction follows, which is a transition reaction in obtaining compounds of melanine type by polymerization (Section 1.1). These transitions are already a step towards formation of tannin systems. The chemistry of vegetable tannins did not explain, however, fully why the tannins are formed by various plants in various amounts. As we know, more tannins are formed in plants growing in tropical or subtropical climates. Tannins may accumulate in all parts of the plant. They are mixtures of a vast number of substances. In a tanning extract of one part of tannin-giving plant, e.g., from bark, one may find 80 or more compounds, e.g., by paper chromatography. A part of them are non-tannins, i.e., substances not binding collagen. However, in the description of tanning processes one cannot forget these substances as they participate in the process, even if indirectly: they fill leather and stimulate the tanning process. Sugars, salts, acids and phenols are among those substances. It is very difficult to separate completely tannins and nontannins, as some nontannins may, under certain conditions, give tannins, e.g., phenols, whereas others, like sugars become part of the tannin entering into a greater molecule. Over 400 chemical individuals of the polyvalent phenols type have been described, which have or may acquire some tanning properties. Inclusion in this group of substances is not only a matter of function, which they may perform, but also of the size and shape of molecules, their structure and biogenesis. In principle, this group includes substances of molecular weight from 500 to 3000. However, in extracts from tannin-producing plants, substances with a smaller molecule may also be present. It has not been clarified yet, just how they do affect the tanning mechanism. Surely they (e.g., gallic acid) may more easily penetrate to the active collagen centers than can bigger molecules. Whether and to what extent small molecules of such a 'pretannin' opens the way for tannin is not yet clear. In this sense phenols are the substances responsible for tanning, but if phenol has to have a tanning capacity, it should contain the proper number of functional groups and form a molecule big enough to fit in among collagen chains. Gustavson [2] estimated the lowest molecular weight of the molecule at about 500. Such a size has, e.g., the system of three gallic acid residues attached to one glucose molecule. The upper limit of molecular weight is the size at which it is difficult or impossible for the molecule to penetrate into leather. The size of a vegetable tanning agent is decisive for its usefulness. To control this size, dispersing agents may be applied, if it should be decreased, or protecting colloids, if the size is to be maintained. Besides dispersing agents the pH and the presence of salts, i.e., ionic strength, are also important.

As molecule size controlling factors some auxiliary syntans may be applied (see p. 410), lignosulfonic extracts (Section 17.4) or sulfonation of tannins. Their advantage as dispersing agents is that they do not affect the chemical and technological properties of vegetable tannins. Their action is limited only to the change of the tannin molecule size.

The classification of vegetable tannins into hydrolizing and condensed ones proposed in 1920 by Freudenberg is used to date. The main difference between the two groups lies in their behavior toward hydrolyzing substances, first of all acids, then enzymes. Hydrolizing tannins, having a polyester structure, easily hydrolyze to the respective sugar or polyhydric alcohol and polyhydric phenol with a carboxyl group. As concerns the hydrolysis products these tannins may be classified into gallotannins, derivatives of gallic acid, and ellagitannins, derivatives of ellagic, i.e., 3,3'-4,4'-5.5'-hexahydroxy-1,1'-dicarboxyldiphenolic acid usually isolated in the form of stable dilactone. Usually gallotannins are esters of gallic acid or other phenolocarboxylic acids, and sugars or polyhydric alcohols.

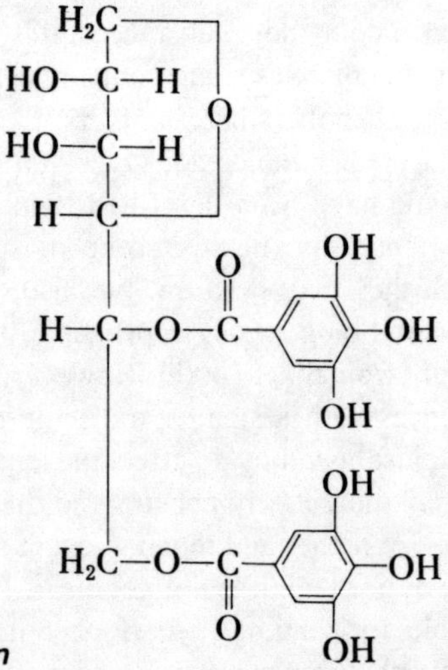

Acertannin

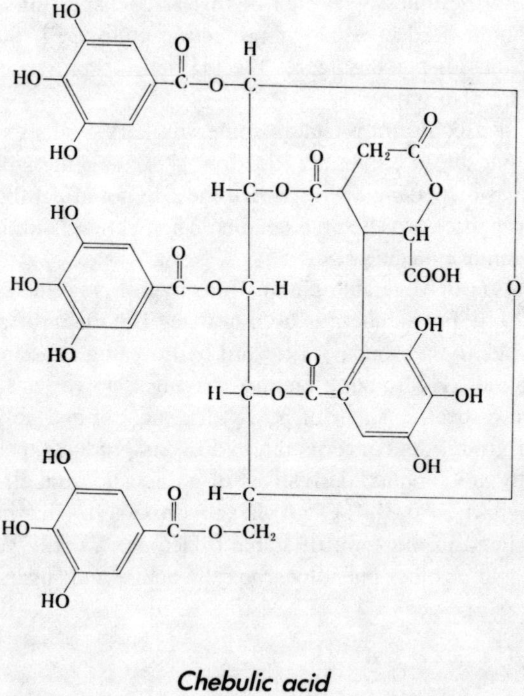

Chebulic acid

Tannins, being elagic acid derivatives, have a structure of the following type:

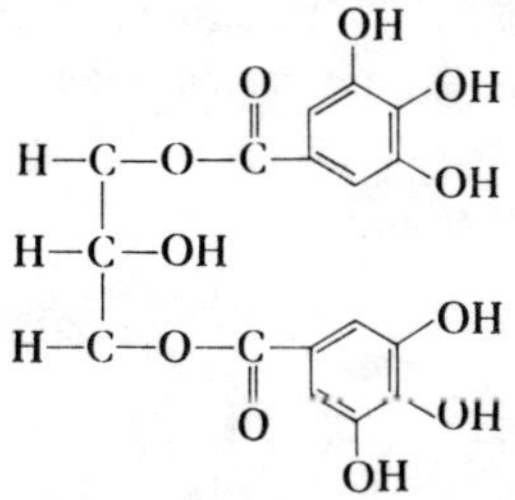

The structure of many tannins in this group has not as yet been clarified and they have been characterized on the basis of the destruction products.

In the hydrolyzing tannins group there are many gallic acid derivatives: hexahydroxy diphenic, chebulic, dehydrodigallic and others. As a rule they may be derived from gallic acid by splitting off a water molecule and/or CO_2, followed by condensation. These transitions in plants are the result of enzymatic process. The following relationships occur [3] between both tannin types:

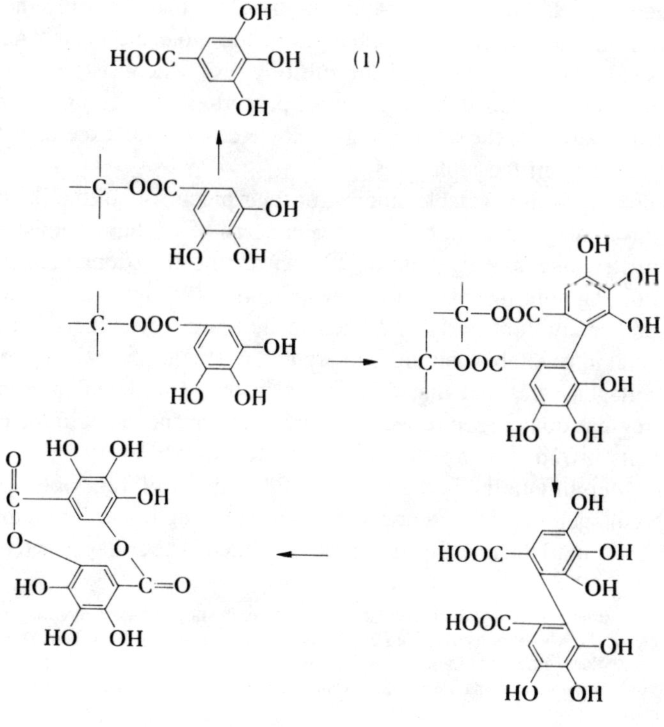

(1)

(2)

(3)

The second component is usually D-glucose, but other sugars may occur as well.

Condensed tannins are not decomposed by acids; they gradually polymerize, becoming phlobaphens—insoluble derivatives. Condensed tannins of catechol type, known as flavone tannins, are catechol derivatives.

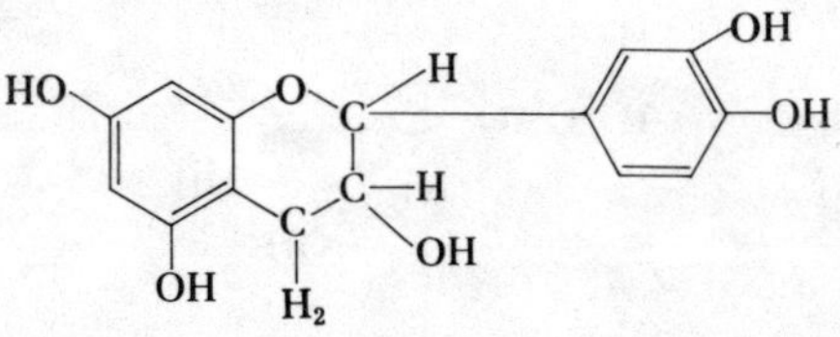

Tanning extracts contain entire sets of compounds polymerized to various degree. Some properties of the particular tanning extracts are discussed, e.g., by Thorstensen [4].

Sulfiting as applied to the condensing tannins consists in treating their solutions with a mixture of sodium sulfite and hydrosulfite. A part of the bonds in tannins is then split. Sulfiting may be graduated as its degree depends on the amount of the reagent used. Initially phlobaphenes dissolve, then not only the size of the tannin molecule decreases, but changes in the molecule occur. According to Endress [5] a substantial part of the sulfiting process consists in converting phlobaphenes into substances having tanning properties. If sulfiting is too intense the tanning properties of the solution may be lowered due to an excessive decrease of the size of the sulfited molecules.

Papers dealing with vegetable tannins, or their precursors of the flavone type and derivatives of gallic acid, are now the concern of organic chemistry and of stereochemistry, like investigations of collagen are mostly a domain of medicine. However, the tanning mechanism is insufficiently investigated, despite its importance and many controversies related to it. Many researchers believe that most essential in vegetable tanning is to protect, or shield, the existing molecular form of collagen fiber and that this occurs in the 'unordered' regions*. The 'ordered' regions do not require such shields, as the arrangement of the molecule itself protects it from water penetration and bacteria.

The first step in tanning is the binding of hydroxyls of vegetable tannins to the active collagen centers. The next step—the binding of further tannin molecules continues until the interfibrillar spaces are filled. The collagen active centers

*Unordered and ordered regions are obsolete definitions from the point of view of collagen structure, as we now know; the whole molecule is ordered in its own way. It would be perhaps better to speak of polar and nonpolar regions. In this sense the expressions used may be consistent with experimental results. The only possible explanation of 'unordering' is the deformation of the helical molecule by shifting it apart by great tannin molecules.

which react with vegetable tannins are various functional groups of its side chains and according to the opinion of some scientists, peptide bonds as well. Osmotic swelling in the protein surrounding makes active centers more accessible when in water solution. This stage of the process ends when collagen has absorbed half of its weight of vegetable tannins. The difference between chromium and vegetable tanning becomes striking here, since 3% of the chromium tanning agent is sufficient to form stable, difficult to destroy bonds between the tanning agent and collagen.

Further steps of the vegetable tanning process are already very slow; the old-type processes require even two years to be finalized, i.e., to achieve equilibrium.

It is not clear what are the active centers of collagen in relation to vegetable tannins.

The extensive discussion on the general mechanism of vegetable tanning between Gustavson and Shuttleworth continued through many years. Gustavson and his group supposed that the basic collagen groups are rapidly bound to the polyphenols of tannin, particularly those that are condensed. Those bonds, which should be stronger than the hydrogen bonds, perhaps due to covalent forces, impart to the leather thermal stability.

Shuttleworth suggested that the tannin molecule is bound to collagen at many points by hydrogen bonds, the interchain bridging effect being responsible for the T_s rise. Covalent bonding and formation of a quinone-like structure do not take place in vegetable tanning.

The above theories have been summarized by Sugano [6]. Endres [5] has expressed the following view regarding the formation of hydrogen bonds between phenolic groups and the peptide bond oxygens:

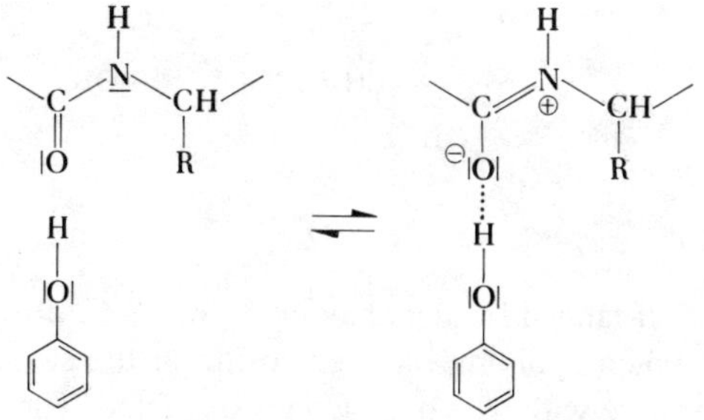

These bonds are considered to be reversible and may be split by the action of alkalis or polar solvents. The existence of bridges of this kind requires affinity of polyphenols to polyamide, which should increase with the number of hydroxyls and the strength of the tanning effect. Affinity of phenols to collagen is a necessary, though not sufficient condition for their tanning capacity. Many substances may be bound to leather, but during this binding no tanning effect occurs (resorcinol, hydroquinone). According to Endres tanning consists in crosslinking of many peptide chains by the same polyphenol molecule, as it is shown in the formula, where we can see that a molecule of the tanning agent has to be of appropriate size and have an appropriate number of hydroxyls:

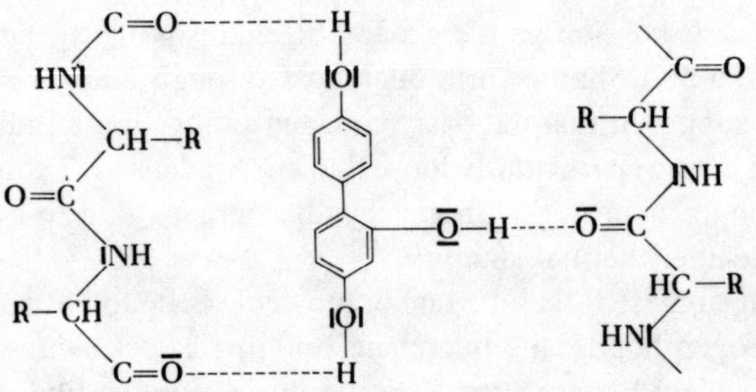

Testing a derivative of pentahydroxystilbene (piceatannol) belonging to the flavanol group

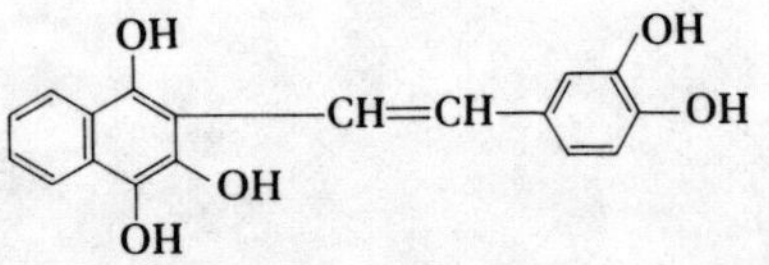

Endres obtained well-tanned leathers having T_s of 88°C; however, he obtained no tanning effect when applying monoglucoside of the same compound. This is explained by the ease with which aglucon is oxydated. The use of piceatannol in an oxygen-free medium did not result in a T_s increase. In the presence of oxygen in piceatannol, like in other flavanols, quinone systems appear which are bound irreversibly to pelt. The quoted author confirmed this reaction mechanism introducing piceatannol into pelt in the absence of oxygen and then treating it with an oxidizing solution. Thus, after the first tanning step the T_s failed to increase. A T_s increase occurred, however, after step 2, when the piceatannol derivatives could no longer be washed out of the leather. Accordingly, Endres demonstrated that between collagen and flavonol-type tannins hydrogen bridges are formed, if the size of the molecule makes them possible to enter into the collagen structure; however, these bridges alone are too weak to consider their formation as a tanning effect. Tanning consists in oxidizing of piceatannol to its quinone, and then an addition of free collagen amino groups in the 1,4 position via covalent bonds. Additionally, these condensed aggregates of tannin molecules constitute fillers bonded by hydrogen bonds. These bonds are reversible whereas covalently attached piceatannol may not be removed, even by washing out with dimethylformamid or alkalis. Glycosides, containing piceatannol as aglucon, do not have a tanning capacity. In Fig. 17.1 we see how Endres et al. [7] have studied the tanning process with condensed tannin (catechol, cyanidendiole, piceatannol). The action of tannin on pelt resulted in an increase in T_s by 5-7°C at pH 5-6. During the next few hours the T_s value increased by 2-3°C due to further transfer of tannin molecules to the fibrils; subsequently, no more T_s increase was observed. The experiment was done in the absence of oxygen.

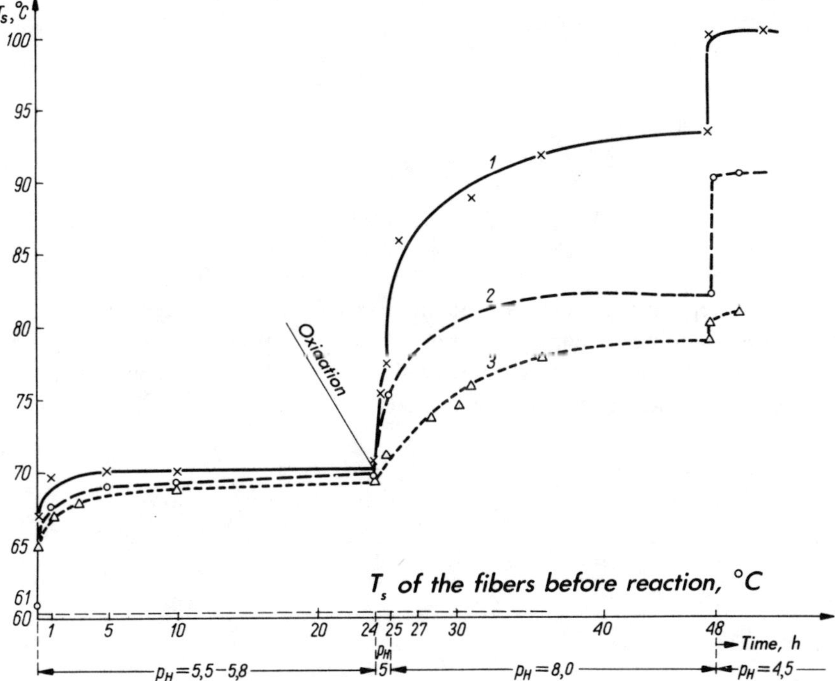

Fig. 17.1. Two-step tanning with vegetable tannins. The first step is carried out in the absence of oxygen, the second one in the presence of an oxidizing agent: (1)—piceatannol, (2)—catechine, (3)—cyanidendiol. According to [7]

Accordingly, the entire T_S increase effect is due to the crosslinking by hydrogen bonds. Addition of an oxidizing agent, e.g., $K_3[Fe(CN)]_6$, causes a marked T_S increase of 6-15°C. This is, however, a slower reaction than the first step. This step corresponds probably to the already discussed addition of the collagen amino groups at position 1,4 to the quinoid compounds. The greatest T_S increase occurs at pH 8-8.5. At this pH value the hydrogen bridges do not play an important part. If, however, the pH is decreased to the value of 4-4.5, the phenolic hydroxyls (still remaining in the tannin molecule) have to form new, strong bonds, with resultant giving further T_S increase. This was observed indeed and ΔT_S was 5-10°C. The T_S end value for piceatannol was 99°C, for catechol 90°C, for cyanidendiol 81°C.

Further investigations on tanning done with the use of isolated chemical species (chebulic and chebulinic acids) have led to the conclusion that the formation of hydrogen bonds increases the value of T_S by $\leqslant 10.5$°C under the same conditions as used for tanning with the mimosa, quebracho or fir extracts which give a T_S increase of 20-25°C. The formation of these bonds increases by oxidation during piling, among others due to the slow, but constant influence of the air oxygen.

According to Endres et al. (loc. cit.), quinones are bound by the collagen amino groups in the following way:

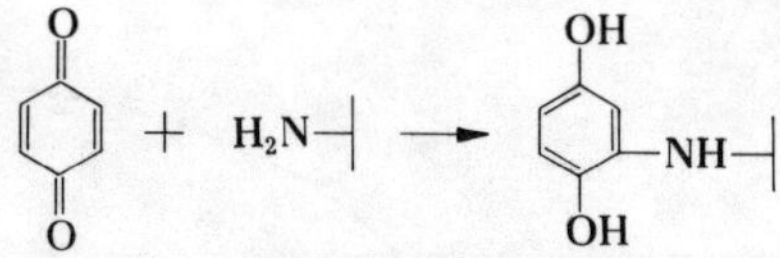

In the first step, as a result of nucleophilic attack on quinone, ring aromatization occurs and binding of collagen to it through the amino group. The redox potential of quinone decreases by about 250 mV, due to which the compound obtained will be easily oxidized by the next quinone molecule.

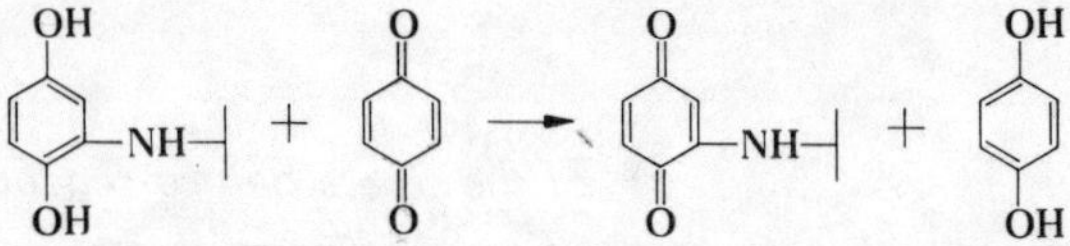

The amino quinone formed is again bound by the collagen amino group, to yield an aromatic compound which is in turn oxidized by the next quinone molecule, so that as the end effect two hydroquinone molecules and a 'crosslinking quinone' one are formed.

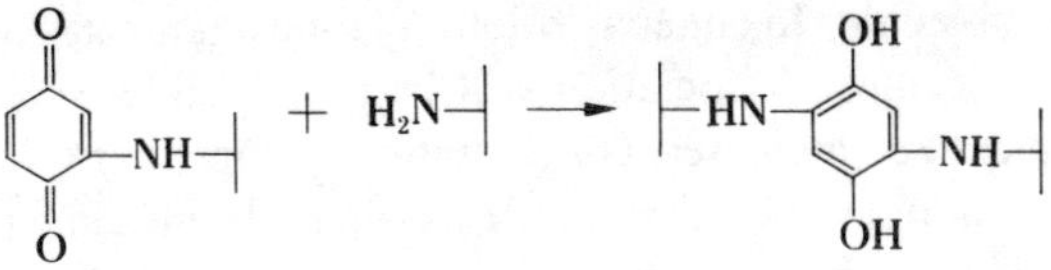

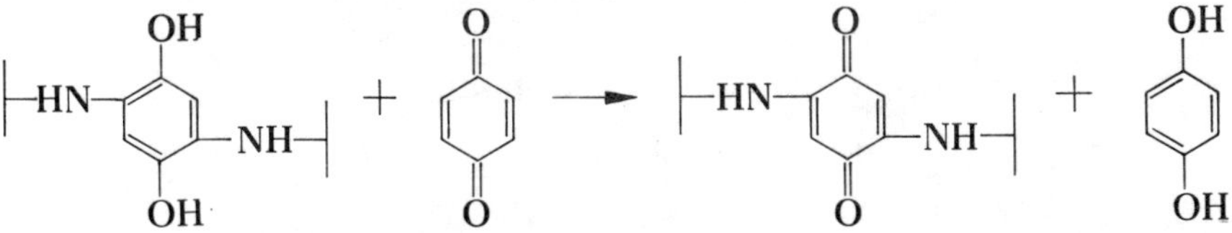

The products of this type were separated from the reaction mixture. The reaction kinetics is pH-dependent; in acidic or strongly alkaline (pH = 10) medium the T_S increases rapidly at the process start, then it maintains the level reached or drops slightly. In neutral solutions the T_S initially grows very quickly, then the slope of the curve declines, but the reaction remains incomplete for a long time. The quoted authors [8] investigated the change of absorption in ultraviolet and visible light, which occurs in the tannin solution with time. Changes in the electronic spectrum allow it to follow the course of the quinone into hydroquinone conversion: by this method it has been found that there is no great difference in the behavior of solutions having pH 6 and 8. Thus the authors confirm the theory of the two-step process of quinone tanning: primary tanning consists of the adding of collagen amino groups at 1 and 4 positions to quinone ring or to the 2,6-naphthoquinone ring, and secondary tanning in the action of the condensed hydroquinones and in the formation of hydrogen bonds. This occurs in the interfibrillar spaces, because the polymerized tannins are unable to penetrate into the molecule. Primary tanning occurs in an acidic or neutral medium; in a weak alkaline solution it easily transforms into the secondary process.

Intrafibrillar tanning gives a larger T_S increase than the interfibrillar one: in the primary tanning quinone dimers and trimers give a greater T_S increase than monomers. Intrafibrillar tanning, as testing of leathers has shown, gives a more spready and brittle material than does the interfibrillar tanning, which contributes to a good filling of the interfibrillar spaces.

A common property of the components of vegetable tanning extracts is their hygroscopicity. They are well miscible with water to form polydisperse solutions, partly of a colloidal type. They dissolve well in organic solvents containing oxygen in their molecule. In aqueous solution labile aggregates are formed from nonassociated molecules. Association will be promoted by such factors as: increased molecule size, increased concentration, temperature decrease, pH decrease and addition of neutral salts. Labile tannin molecules in aqueous solutions have approximately a spherical shape. They contain 200-600% water, calculated on dry weight, their diameter being 14-40 Å.

Besides colloidal particles deposits of 6-13μ particle diameter occur commonly in vegetable tannins. Sulfiting, i.e., addition of sulfite groups to the quinone systems, decreases these deposit particles. Sulfonic group, attaching itself to the ring, where the quinone oxygen occurs, has a low degree of dissociation:

$$
\begin{array}{ccc}
& \text{CH--CH} & \\
\text{R--C} & & \text{CH} \\
& \text{CH=C} & \\
& \text{OH} &
\end{array}
\longrightarrow
\begin{array}{ccc}
& \text{CH--CH} & \\
\text{R--C} & & \text{CH} \\
& \text{CH}_2\text{--C} & \\
& \text{O} &
\end{array}
+\ \text{NaHSO}_3 \longrightarrow
$$

$$
\longrightarrow
\begin{array}{ccc}
& \text{CH--CH} & \\
\text{R--C} & & \text{CH} \\
& \text{CH}_2\text{--C} & \\
& \text{OH} \quad \text{SO}_3\text{Na} &
\end{array}
$$

As a result of the formation of quinone type carbonyls, darkening of phenols occurs when exposed to light. Many quinones have good tanning ability, however, there is no positive correlation between quinone system content and tanning ability of the substance. This is probably due to the formation of strong quinhydrone systems. In the case of quinone systems in view of specific intermolecular interaction donor-acceptor complexes or π complexes are formed as a result of action of bonding forces between electronic systems of saturated and unsaturated molecules. Usually the reaction is stoichiometric. Quinhydrones, products of the reaction between quinone and hydroquinone or between their derivatives, may serve as an example of a π-complex. These compounds are intensely colored, e.g., quinhydrone is dark green. The tendency of the quinone component to bind its partner increases as the electron cloud density decreases. Aromatic compounds of decreased density of electron cloud in the ring containing strong electronegative substituents may be used as the quinone component, e.g., tricyanobenzene or trinitrobenzene. Increased electron charge density in the ring favors its binding to quinone. Suitable complex compounds are formed with special preference by these aromatic compounds, in which the substituents present activate the aromatic ring for electrophilic substitution: phenolic, methyl or amino groups. Compounds of a structure close to quinones, as maleic or phthalic anhydrides, have the ability to form such complexes as well. The existence of some intermolecular compounds may be explained by the transfer of molecular charge from the donor to acceptor (Fig. 17.2). Very similar systems occur in aromatic polynuclear compounds such as tetrahydroxyquinhydrone and its reduction products

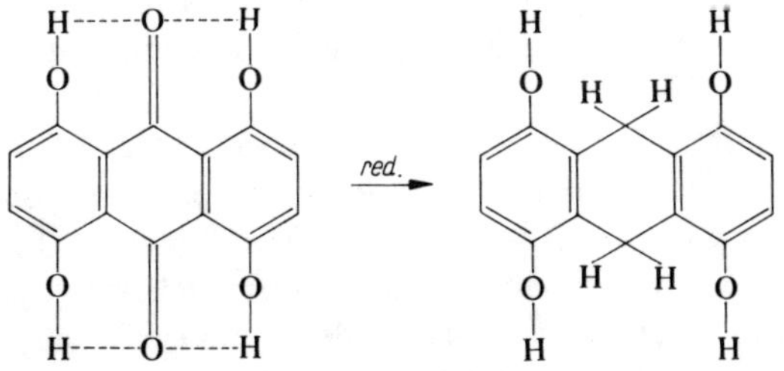

As a study of their affinity to collagen has shown, compound 2 is the one which reacts much more readily with collagen. This rule, however, is not always true:

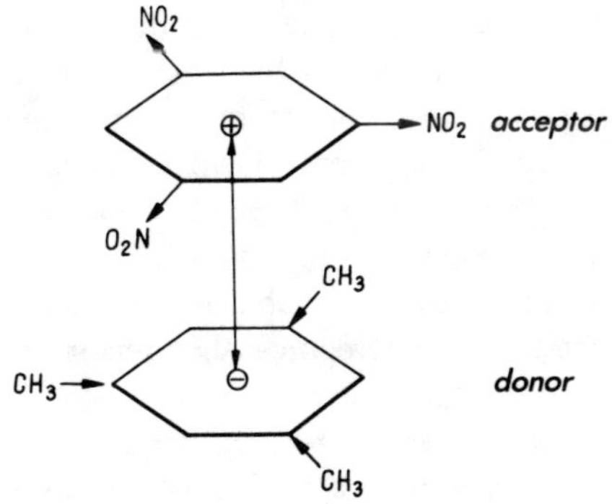

Fig. 17.2. A typical example of charge transfer complex (CT complex).

unoxidized 2,6-dihydroxynaphthalene does increase the collagen T_s by 11°C, after conversion to the respective quinone, by 30°C. Catechol has no tanning properties except when it is condensed. The quinone systems may either attach phenolic groups or act on them as directing substituents.

The hydroxyl groups occurring in the tanning agents behave like very weak acids (pK = 8-13). They contain, however, some centers of much greater activity (pK = 3.2-3.4).

Susceptibility to the action of light or oxidizers, manifesting itself in a change of color of the solution, is typical of o-quinones. A more general property of vegetable tannins is their liability to oxidation, oligomerization and formation of π-complexes.

The tannin extracts give a precipitate with gelatin solution. This is one of the old methods of checking the tanning ability of the extract. In such a precipitate 1-4 g of tannin is given per 1 g of gelatin. From this it can be concluded that tanning of collagen occurs on its surface. Collagen takes up less tanning agent, since in a gelatin solution the collagen destruction products are more easily accessible. This method belongs to the classical ones in tanning chemistry, and may be applied for testing the stability of collagen-tannin bonds, and of the pH effect on the rate and extent of tannin binding. Formerly it was used for fractioning the tanning extract components. On the phase boundary between the gelatin gel and the tannin solution a strong membrane is formed, which remains after heating of the gel to its melting point. The strength of this membrane (film) may be considered to be the measure of tanning ability of the extract, of course if the film is obtained under standardized conditions. The films obtained are very thin and permeable for substances of molecular weight smaller than 300. Unequal dispersion of vegetable tannins in collagen is confirmed by irreversibility of the structure destruction due to shrinkage and by X-ray data, according to which the high-angle collagen reflexes are not changed due to vegetable tanning. The background is, however, much darker in these diffractograms, which is the result of introduction of a great amount of amorphous molecules into the structure. According to this [9], the tanning agent attached to collagen deforms the helical structure of the latter extending it by a large molecule or shrinking it due to the formation of crosslinking bonds (cf. Fig. 14.1).

According to Mikhailov [10], vegetable tannins do not penetrate into the tendon collagen which has no micropores: they precipitate on its surface, like the large tannin molecules, which do not penetrate into the leather. The accessibility of elements of the collagen fiber network is, among others, limited by the degree of swelling. How alike are the curves of swelling and sorption of tannins can be seen in Fig. 17.3 according to Gustavson.

Those curves almost coincide in the acid range; in the alkaline range, however, the maximum of tanning agents binding is shifted to the left with respect to the swelling maximum. This shift is most probably related to suppression of dissociation of phenolic groups.

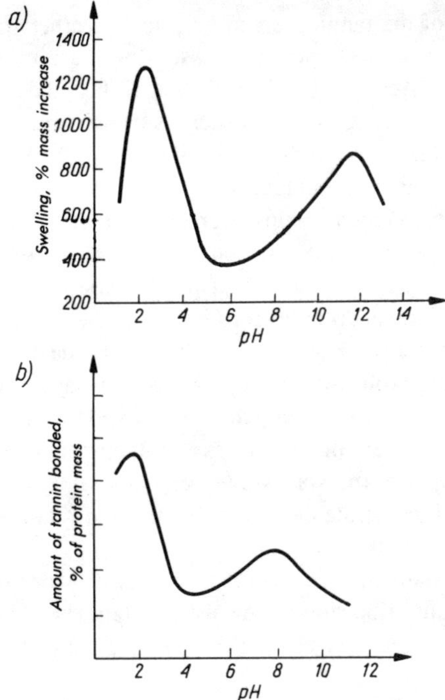

Fig. 17.3. Effect of tanning process pH on (a) collagen swelling, (b) sorption of tannins by collagen.

Not all vegetable tannins contained in the tan-liquor will be bonded to collagen with the same strength: a part may be removed by washing with water, and some by washing with a dilute solution of sodium hydroxide. The ratio of strongly absorbed tannins to those which are water-removed is a conventional measure of the stability of their binding. The cognitive value of this property is insofar low, as the binding strength is to a considerable extent parameter-depending, i.e., it depends on concentration, bath index, time, pH, rH and the tanning temperature, drying level or conditions of extraction. The washability limits may vary for the same tannin from 20 to 90%. Non-crosslinking, singlepoint-attached tannins, as well as those nonbonded but mechanically entangled in the structure are an additional factor in leather, which reacts with the fats, dyestuffs and impregnants introduced. This concerns not only vegetable tannins, but also syntans, aldehyde and inorganic tanning agents. The estimation of the bond stability and usefulness of a tanning agent requires finding out how pelt prepared in a definite way would behave toward the substance under test. The shrinkage temperature of vegetable-tanned leathers may give some preliminary information about the tannin tested: condensing tannins increase it by 20-30°C, hydrolyzing ones by 5-20°C. Identification is usually complicated because tanning mixtures are usually used in leather processing. A significant dependence exists between the T_s and the amount of the tanning agent present in leather (cf. Fig. 17.4).

The vegetable tanning process can be assessed by checking the product resistance against enzymes. According to Lipschitz et al. [11] the lower the tanning pH is, the less potent proteolytic enzymes such as trypsin and papain do attack the vegetable tanned collagen. According to their evaluation, 3.75 is the optimal pH value for vegetable tanning as judged from the enzyme attack.

The blocking of basic collagen groups decreases markedly its ability to bind vegetable tannins (by 30% and more), which points to their participation in the vegetable tanning process. In vegetable-tanned leathers sometimes case-hardening may be observed. The medium layer of leather is then horny and untanned. This layer swells strongly in 10% acetic acid. It is easily destroyed by bacteria, which results in blistering (splitting) during extensive wearing. Case-hardening occurs when the tanning agents are bound too rapidly and too strongly to collagen on the leather surface, preventing the access of tannins to deeper layers. A countermeasure is to apply at the start slowly and weakly reacting synthetic or combined tannins, which penetrate deep into the pelt and saturate its most active centers.

Another type of case-hardening is osmotic removal of water from the pelt. It occurs when high concentration liquors are used in tanning. Then tannins still concentrate on the phase boundary closing the passages in pelt. This is accom-

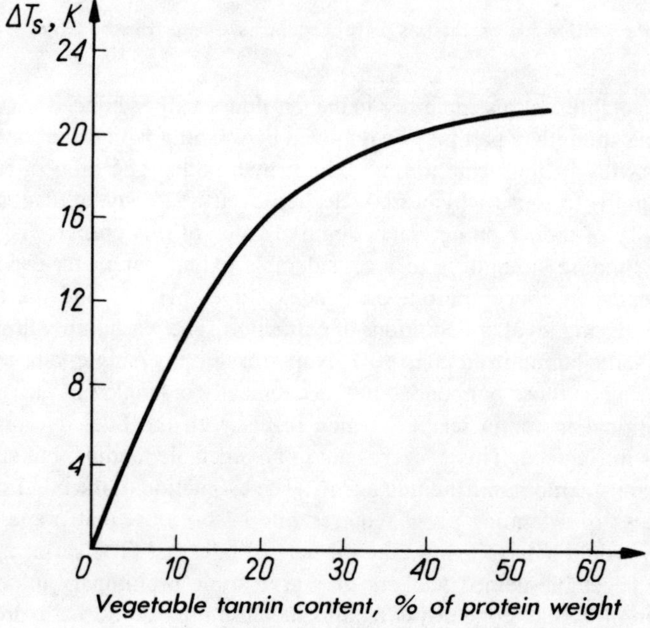

Fig. 17.4. Variation of T_s of vegetable tanned leather on the amount of tannin bound.

panied by an increase of pH of the medium. This effect has been explained by Heidemann [12] who investigated the influence of liquor concentration and the kind of tannins used. When concentrated extracts of vegetable tannins are used a decrease of pelt volume during tannage is noted. This effect is particularly pronounced, when quebracho, wattle and fir bark are used: it is smallest in sulfited quebracho or chestnut extract. This effect corresponds directly to the removal of water due to osmosis. The mass balance of the system is shown in Fig. 17.5. As one can see there, the loss in weight due to the loss in water content is set off by the weight increase due to binding of tannins. When too concentrated solutions of a large-molecule tannin are used, the removal of water immobilizes them in the pelt. They cannot exceed some depth of penetration, and the medium layer remains untanned. The volume of the tanned layer decreases by about ⅓. The appearance of osmotic case-hardening may be reversible when the tanning solution used is diluted. Then the osmotic pressure of the solution decreases, water returns to the leather and makes the tannin molecules mobile. According to the author cited above, case-hardening is not only de-

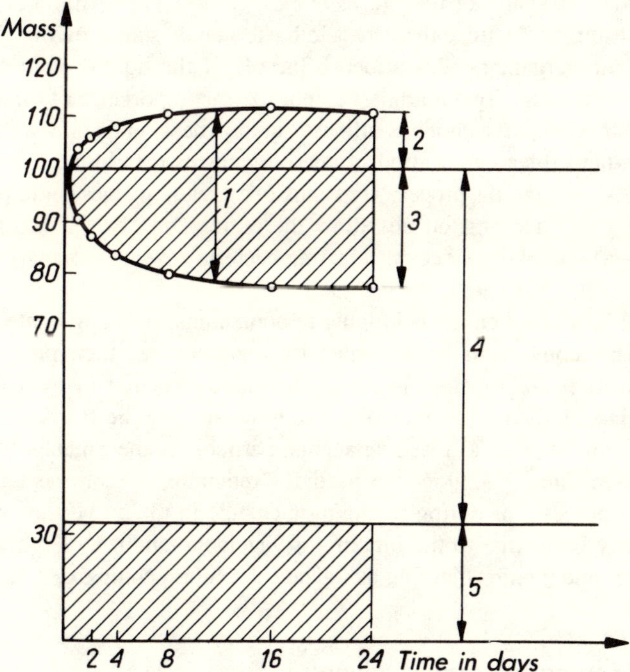

Fig. 17.5. Relationship between the binding of vegetable tannin and water loss. Tanning without agitation with mimosa extract of d = 1.091 (The data have been calculated on 100 g of pelt): (1)—tannin bound, (2)—mass gain, (3)—water loss, (4)—water content in pelt, (5)—dry substance of pelt. Bayer's information

pendent on tannin concentration but also on the hide structure, water content in pelt, temperature, liming and possibly on the way in which the pretannage has been done. Heidemann *(loc. cit.)*, investigating the water content in the internal zone, has demonstrated that this view is justified. Mention has already been made of the nontannins, substances which are part of the tannery extracts, but have no possibility of temporary or permanent tanning (e.g., inorganic salts). However, the effect of the nontannins in vegetable, syntan or combined tannage processes is essential, and it cannot be unambiguously defined because of its variability. Some examples can be discussed, emphasis being laid on the fact that no general rule can be drawn up, as the system is too complicated.

The importance of inorganic salts for tanning consists mainly in that they influence osmotic pressure of the liquor and in this way they may contribute to water removal and case-hardening due to concentration of tannins in the external leather layers. They may as well buffer the liquor, influencing its pH and ionic strength, and thus may cause diffusion processes. The content of inorganic salts in tannins is low, not exceeding several percent of the dry substance.

Sugars, polysaccharides and their derivatives are a very important component of the liquors, because of their fermentation ability. The medium is of pH 3-5, thus yeast, molds, lactic and acetic acid bacteria may start growing. Metabolites of these microorganisms do influence the pH of the liquor, solubility and aggregation of tannins. Technologists emphasize the importance of these side processes, as they impart adequate fullness, mellowness and grain softness to the vegetable-tanned leather. Action of microorganisms mentioned, being very distinct in a slow tanning process, has still one meaning: these microorganisms, controlling the medium, do not allow the putrefacting bacteria to grow there. This is easy to explain, since putrefacting bacteria develop sparingly in an acidic medium.

Small-molecule phenols, belonging to nontannins, may aggregate, to become tanins. This conversion is facilitated by oxidation, as then phenols become quinones. Their aggregation is crucial for the conversion of these compounds into tannins, as then molecules of appropriate size can be formed.

Nontannins may act as masking agents, particularly their metabolites (organic acids) when tanning liquors are used for retanning of chrome leathers. In a combined process, or during retanning a change in the properties of chromium leather may occur due to the binding of vegetable tannin by leather, or due to nontannins and tannins entering as ligands chromium complexes, already bound to leather.

17.2. Tanning in organic solvents

Tanning in an organic solvent was applied in some pilot plants. The first patents related to this process were given in 1910 and 1911. Pelt dehydration with alcohol or acetone was suggested there, and then the use of acetone as solvent for vegetable tannins. Roddy [13] observed that quebracho in acetone solution penetrates through heavy leather already after 48 hours. However, the use of acetone is not economically advisable because of the high volatility of this solvent, and the related losses. In 1945 Ushakoff took a patent [14] for the tanning of acetone-dehydrated leathers, subsequently saturated with alcoholic tannin solution. Organic and inorganic tanning agents have been suggested in the patent. The solvent is removed and tanning agent combines with leather immersed in water forming adequate bonds. The method was called acetone method and was applied in some American and Canadian tanneries. The process is said to be inexpensive due to solvent recovery. It should be realized that no tanning occurs in the organic solvent. This process consists in the introduction of the tanning agent into pelt, where tanning follows as soon as water is added. In a paper by Weber [15] a review of previous experiments on nonaqueous tanning processes is given. He reports that in a factory in Czechoslovakia syntans from polyphenol group were applied in a denatured alcohol solution. The solvent is recovered in the factory. Alcohol as a reaction medium was introduced after pickling and formaldehyde pretannage. The process of obtaining sole leathers, requiring a total of 60 days, was shortened to four days. Tanning alone takes only three hours. In the method described condensation products of polyphenols are used which in the course of the reaction are converted into quinones which have tanning ability much more pronounced than hydroquinone. The syntan Kortan A 30, produced in Czechoslovakia by oxydation and condensation of polyhydric phenols, has a molecular weight of about 1500 and does not dissolve in water, but is soluble in ethyl alcohol. Leather obtained in this way has a low water content (10% when air dried), lower ash, fat and washable components contents. It shows good wearing properties. According to the quoted author a further progress in this process was expected, however, no reports have appeared to date.

17.3 Quinone tannage

Quinone tannage is usually separately discussed as a specific tanning process. It is of some theoretical importance. The tanning agent involved is a compound of small molecular size, containing functional groups of one kind only. In the times of chromium shortage (e.g., in France during World War II) quinone was used in the pretanning process. According to Lasserre [16] this method was sometimes used in France for obtaining calf boxes of very subtle grain. This tanning process is conducted in buffered solution, since on pH increase quinone solutions become dark and so does leather. The properties of the anion of the buffering compound affect the value of pH at which quinone-to-collagen binding occurs: for the borate buffer the optimal binding pH is about 5, for the phosphate one 7-8. These optima are very sharp, the slopes of both sides of the curve being very steep. Quinone reacts with the amino groups to form aminohydroquinone or 2,4-aminoquinone (section 17.1). The tanning process takes 24-48 hours. Quinone, owing to its structure and molecular size, is an excellent crosslinking agent. The T_s of quinone tanned leathers may be as high as 90°C. The amount of quinone bound is high, 20% and more, particularly at optimal pH. The proteolytic resistance of the quinone-tanned leather is very high.

17.4. Syntan tannage

Synthetic tanning agents, or syntans, have been primarily produced with the aim to substitute vegetable tannins, to accelerate production and to make it cheaper. Introduction of new functional groups, which do not occur in natural vegetable products, such as sulfonic groups, opened new prospects for tanning. Syntans are frequently used in retanning in the production of chrome-tanned leathers. Leather gains new properties distinguishing it from conventional chrome leather.

Syntans may be classified into replacement and auxiliary ones. Replacement syntans may be used as straight tanning agent, the auxiliary ones can be considered as substances assisting the tanning process. Among replacement syntans there are those able to replace other tanning agents, and those of special destination. Special syntans impart some properties to leather which cannot be obtained by the use of other chemicals.

The size of syntan molecules affects the tanning effect, as is the case with other tanning agents. Of practical importance are compounds of molecular weight of 800 or a little less, which corresponds to 2-4 aromatic rings. Syntans with greater molecules do not tan at all or tan only superficially. Syntans are mixtures of homologs or related compounds rather than chemical species. The average molecular weight is of analogous importance here.

Bonds between the aromatic rings in syntans are essential for their properties. For example, extension of a molecule by a $-CH_2-$member weakens the tanning ability of a syntan. The replacement of such a bridge by a sulfonyl one increases the tanning and improves the resistance of the tanned leather to light. Resistance to light is also increased, when a sulfonamide sequence $-SO_2-NH-$occurs in the molecule. Light-colored though of poor resistance to light are the leathers tanned with a syntan in which the aromatic rings are connected via a bridge of urea and formaldehyde residues ($-CH_2-NH-CO-NH-CH_2-$). The bridging groups can be put in the following order of increasing light resistance:

$$-CH_2- \; < \; -CHR- \; < \; -CH_2-NH-CO-NH-CH_2- \; < \; -CR_2- \; < \; -CO_2- \; < \; -SO_2-NH-$$

Apart from the bridges connecting the aromatic rings significant is the amount, kind and position of the functional groups built into these rings. As a rule only hydroxyls, sulfo groups and amino groups have been applied. The kind of functional groups and their distribution in the syntan molecule are responsible for the syntan solubility. Thus the factors characterizing a syntan are: the number and kind of rings in the molecule, the number and kind of functional groups and the molecular weight. Syntans may contain chromium components as well (in Poland Rotanin CR). As already mentioned, syntans produced in the industry generally are not chemical individua. Separation of the technical products is economically unsound, so generally one speaks of an average of the properties mentioned. However, in research work frequently individual compounds are used, as, e.g., by Reich [17].

The hydroxyls contained in synthetic tannins are attached to the aromatic rings, hence they have a phenolic character. Their ability to form hydrogen bridges with amino groups of the collagen side chains or possibly with peptide bonds of the main chain is almost identical to that of vegetable tannins. Their amount and position may be of equal importance. Synthetic tannins, containing dihydric phenols, show a greater tanning ability than those containing no hydroxyls attached to the benzene ring or only one. It is known that derivatives of β-naphthol are better tanning agents than those of α-naphthol. This can be shown on the example of two isomers of dihydroxy-9-phenylxanthene. The 3.6-isomer (I) is a very good tanning agent, giving $\Delta T_s = 25°C$, whereas the 2,7 isomer gives almost no T_s increase, ΔT_s being only 5°C.

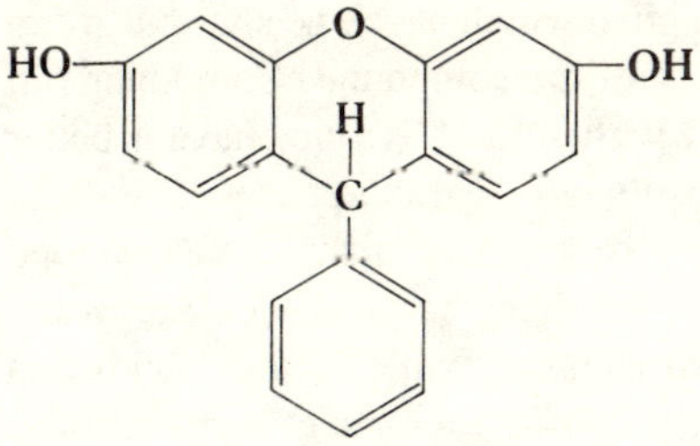

I 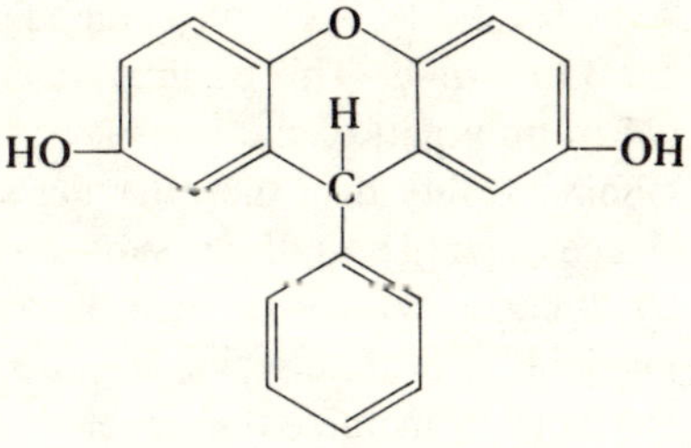   II

A difference in the resonance interaction can be seen when the hydroxyls are in the para- or in meta-position with respect to the carbon with the methine group. Of importance in this case is the charge distribution density in the ring. Regular distribution of hydroxyls in the molecule is particularly advantageous as then the maximum possible amount of conjugated bonds can be formed. The importance of these groups lies less in the formation of bonds between collagen and the tanning agent than in influencing the properties of the molecule itself.

However, it is desirable to introduce them into the molecule in the greatest possible amounts.

The presence of other substitutents in the ring influences the reactivity of the particular groups as well. For instance, the hydrocarbon substituents decrease the tanning ability of syntans. Cresolic compounds are poorer tanning agents as compared with phenolic ones. Sulfo and carboxylic groups in the orto and para position improve the tanning ability of syntans, whereas they decrease this ability when in the meta position. As a rule hydroxylic groups do not occur in the syntan side chains. However, they occur in the side chains of lignosulfuric acids, as groups imparting an alcoholic character to the molecule. They do not influence its tanning properties, but improve its water solubility.

The sulfonic group, as substituent in the aromatic ring, is strongly dissociated, and gives an acidic character to the compound. Due to the mesomeric effect it affects the charge distribution density. This effect occurs, when the electrons of the substituent lone pair become conjugated with the syntan double-bond π-electrons or those of the aromatic ring sextet. This effect is defined as $+M$ when the substituent gives the electrons to the residue, or $-M$ when it attracts them. These effects cause typical changes in the organic compound molecule: shortening the bonds, and changes in the dipole moment values. If a substituent is strongly negative, then the lone electron pairs are shifted in the opposite to the polarization direction. Without discussing the effect of the particular substituents on the behavior of an aromatic compound one can say that the amount of sulfo groups in the compound adversely affects its tanning ability but increases its water solubility. As a rule compounds are applied which have one sulfonic group per 3 to 4 rings. This ensures water solubility of the compound. Compounds in which only part of the rings are substituted with sulfonic groups have a better tanning ability than those having sulfo substituents distributed regularly. Substances having no sulfonic substituents at all, like, e.g., novolaks, are best bound by collagen. However, it is impossible to use them because they are water insoluble. Phenolic resins are also dispersed using aromatic sulfonic acids and products from mixed sulfonated and nonsulfonated monomers condensation. From observations of various authors it follows that the sulfonic group makes penetration of the tanning agent into leather easier and quicker, probably due to electrostatic interaction between the sulfo groups and basic collagen groups. A strong bond of complicated character is formed between protein and sulfo group of the following hypothetical form for long accepted without proof:

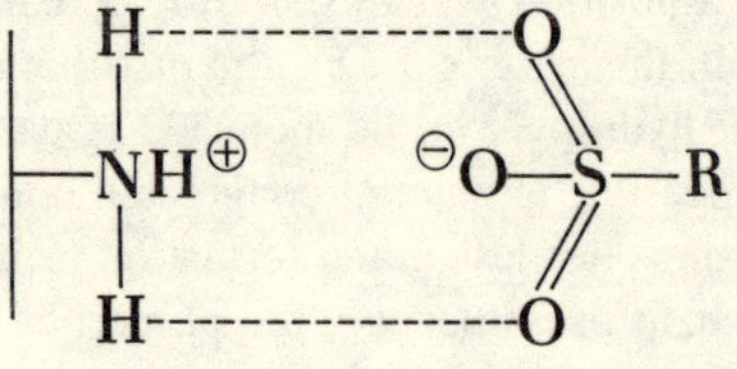

$$
\begin{array}{ccccc}
 & H\text{-----------------}O & & \\
 & | & & \parallel \\
-\!\!&NH^{\oplus} & {}^{\ominus}O\!-\!S\!-\!R \\
 & | & & \parallel \\
 & H\text{-----------------}O & &
\end{array}
$$

Table 17.1.

Washing-out acids from collagen as related to time

	Amount of acid removed by washing with water after		
	3 days %	12 days %	30 days %
HCl	99.6	100	—
H_2SO_4	86.0	98.2	100
HO$_3$S–⬡⬡–CH$_2$–⬡⬡–SO$_3$H	11.0	18.9	19.0

This is a semipolar bond accompanied by two hydrogen bonds. The strength of this bond allows us to compare the washing-out of three acids from collagen, viz., hydrochloric, sulfuric, and 2,2-dinaphthylmethane-6,6-disulfonic (see Table 17.1).

The amount of condensed aromatic rings is also of importance for imparting to the compound the tanning capacity; two- and three-ring substances are better tanning agents than those with one ring. Amino groups, occurring sometimes in syntans, may act as hydrophiles, however, their action is limited to the acidic medium only. If the amino groups are the only functional groups they participate in the binding of collagen. They probably form hydrogen bonds with oxygen from the peptide group. The solubility of syntans containing only amino groups decreases with the acidity of its solution, whereas under the same conditions its ability to bind to hide increases. The possibility of syntans being bound to the peptide chains, as proposed by Reich, is shown in Fig. 17.6. As it can be seen there the symmetric or asymmetric distribution of functional groups in the syntan, the repulsion of charges of the same sign, stiffening effect due to two-point attachment of the syntan molecule to one chain and the effect of chain expansion play an important role in their binding.

Among synthetic tannins of special importance are chromium salts, complexed

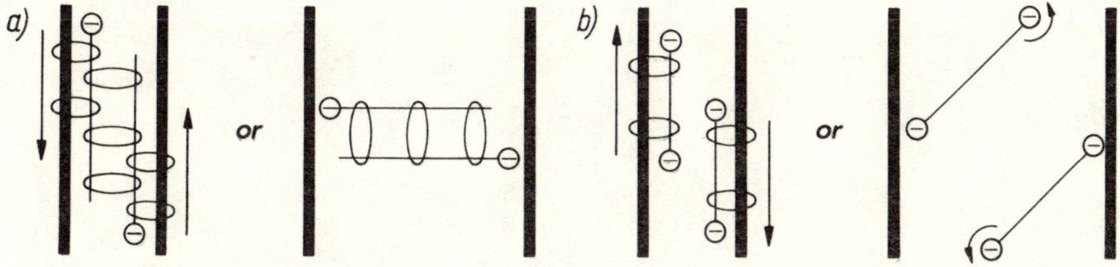

Fig. 17.6. Result of various effects of syntans or their components: (a) stiffening combined with ordering of the unsymmetrical sulfonated compound by aggregation, (b) elimination of the stiffening effect due to the repulsion of symmetrically sulfonated molecules.

with condensation products of benzene and naphthalene derivatives. The already mentioned Rotanin CR is a product of this kind. It contains 12-13% of Cr_2O_3 and derivatives of naphthalenesulfonic, hydroxybenzenesulfonic and o-phthalic acids. It is a tanning agent which may be applied alone; it gives very bright leather of T_S about 100°C, of properties and characteristics similar to those of chromium leathers [18].

Given above are some data concerning the properties of principal functional groups, when attached to aromatic syntans used as replacement products. Numerous compounds are applicable for tanning together with others. They positively influence leather properties, though they do not have their own tanning ability. The already mentioned nontannins belong to this group, playing an anologous role to that in vegetable tanning, as well as dispersing agents, bleaching agents and resin tannins. Substances belonging to this group are called auxiliary syntans. Discussion of this group of substances would lead us to macromolecular products, produced by industry. Their number is very large, and in some classes of the substances almost any preparation—starting with novolacs, through diene and acrylic resins—may be applied for filling, tightening or reinforcing leather. Thus the significance and use of auxiliary syntans are not discussed here, only lignosulfonic acids will be briefly outlined. These substances are the main solid components of sulfite liquors, a waste product of cellulose manufacture. They are not replacement syntans, but they may serve as a raw material for syntans or be used directly. They are produced in great amount by the paper pulp industry, which was the reason why they aroused interest. The structure of the main components of the sulfite liquor is very close to that of condensed vegetable tannins. This is why investigations are carried out aimed at utilizing these wastes as fully valuable tannins.

In the structure of lignine a guaiacol grouping occurs, though the polymeric structure has not been fully recognized yet. They are phenylpropane derivatives linked by ether bonds; coniferyl alcohol, a phenylpropane derivative as well, may be given by way of example.

The lignines, as we believe now, probably reinforce the cell walls and protect them against physical and chemical influences.

In sulfite liquors degradation products of these polymers may be found: guaiacol, protocatechic acid, 4-n-propylguajacol, 4-allylguaiacol and others. In the course of lignine decomposition the methoxy group attached to the ring generally remains intact, whereas other bridges between the propane chains, or between the propane chain and the ring are split. Calcium hydrosulfite is bound to the molecule at the site, where the ether bridge was broken. The functional groups present in the chain are decisive for the way this attachment occurs. With hydroxyls ester-type compounds are formed:

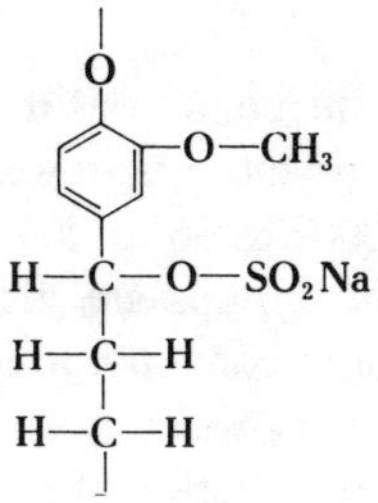

In the case of double bonds sulfonic acids are obtained:

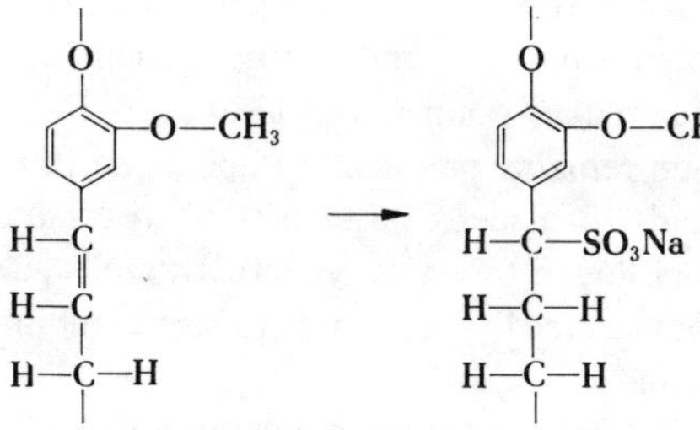

In the case of carbonyl group addition products are formed:

$$\text{(organic structure with aldehyde group)} \xrightarrow[]{NaHSO_3} \text{(organic structure with } NaO_2S\text{—O—C—OH group)}$$

The molecule size depends on the conditions under which the sulfite liquors are obtained. Usually substances of molecular weight below 3000 participate in about 37%, of MW 3000-10,000, in 36 to 26%, of MW higher than 10,000 in 40-47%. As a rule, more sulfo groups are present in compounds of a lower molecular weight. Sulfite liquors contain about 50% of calcium salt of ligno-sulfonic acids as calculated per dry weight. Various methods are used to purify the spent sulfite liquor, e.g., precipitation of inorganic compounds with sulfuric acid and sodium sulfate, or by fermentation (ethanol or yeast fermentation). After fermentation the liquors (brews) are suitable for leather production as by fermentation they are freed from sugar, organic acids and in part of inorganic ones. The knowledge of the reaction mechanism of lignosulfonic acids with collagen may be summarized as follows [18]:

(1) All sulfo groups are capable of reacting with basic collagen groups; on the part of collagen about 90% of the basic groups enter into reaction.

(2) Collagen deamination only insignificantly influences the binding of ligno-sulfonic acids (this makes point 1 a little unclear).

(3) The T_s of collagen remains practically unchanged in this reaction.

(4) Lignosulfonic acids displace a large part of vegetable tannins from their compounds with collagen (not vice versa), whereas auxiliary syntans, containing sulfo groups, are displaced from their combinations with collagen by vegetable tannins.

Based on older literature data one can conclude that ionic bonds are formed between collagen and lignosulfonic acids. These bonds are still strengthened by additional factors such as, e.g., ionization of the side chain groups. As the T_s does not change, one can presume that there is no crosslinking (which is not necessarily correct).

Tanning with sulfonyl chlorides is a process developed in Germany during World War II. In this process a mixture of sulfochlorinated paraffins, having

15-30 carbons in the chain, served as tanning agents. In the reaction of binding to collagen, hydrogen chloride is split off, and the sulfonic group forms an ionic bond with the amino group of the collagen side chain. Probably collagen reacts with sulfonylchlorides through its ε-amino groups of lysine.

$$R-NH_2 + R'SO_2Cl \rightarrow R-NH-SO_2-R' + HCl$$

The reaction proceeds in the presence of bases, whose role is dual: neutralization of the hydrogen chloride and suppression of the amino groups dissociation. Usually sodium carbonate or hydrogen carbonate is used. Assuming a 4.02% lysine content in collagen one should expect the amount of 0.89% sulfur to be bound, if the reaction course were as shown above, or 1.78 of sulfur, if two amino hydrogens were substituted in lysine. Actually 3.7% of sulfur becomes bound. Based on stoichiometric considerations one can presume that only hydroxyls of hydroxyproline and of serine may be responsible for binding of sulfur in such an amount. Sites, where sulfur is bound, should be pH-dependent: in acidic medium sulfur is bound by hydroxyls, which do not undergo dissociation (hydroxyproline) and by ε-amino groups which up to pH 8-9 are dissociated.

A certain part of sulfonylchlorides becomes only mechanically associated to collagen, acting occasionally as a lubricating agent, imparting softness and mellowness to leather. The T_S increase as a result of this processing is only 2-3°C. This is because no crosslinking was considered to occur in this process. The optimal pH of the processing is 7-8 or more. The leather obtained has a particularly high tensile strength, its stretch resistance exceeds 2-3 times that of chrome leather; its elongation at break is also high. Significant resistance of sulfonylchloride tanned, greased leather makes it suitable for the products exposed to harsh climatic conditions (harness). However, this leather is hardly used now, being replaced by plastics of equal resistance. Recently in India several papers were devoted to technological aspects of sulfonylchloride tannage [19].

17.5. Retannage

Tanned leathers, particularly the chrome-tanned ones, are as a rule once more treated with the tanning agent, in order to impart to them the properties required. The technological purpose of retannage is not only a rational use of the already bonded chrome tanning agent, but equally binding other substances to it, equilibration of the consistency (touch) by filling looser parts, or strengthening the grain and increasing or decreasing, as required, the ability to attach the finishing agents. Preparation of leather for mechanical processing is another purpose of retannage.

The retannage principle, based on the distribution of charges on the leather surface, can be presented as follows:

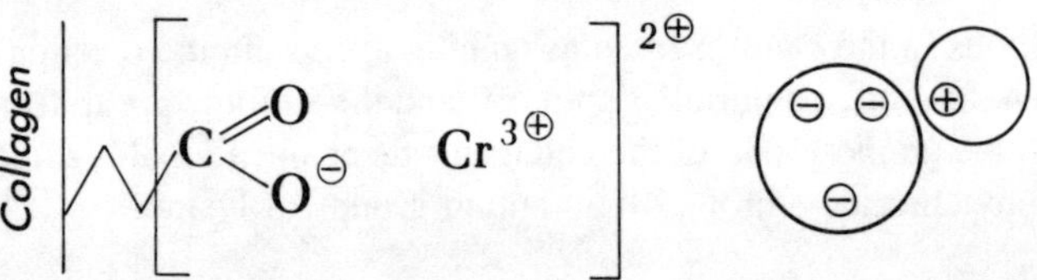

Thus proper retanning of chrome leather by an anionic tanning agent not only weakens the positive charge of the leather-tanning agent system, but it may also make it negative. In the retanning process a negatively charged group attaches to the Cr^{3+} complex that is one-point linked to collagen side chain carboxyl. This makes it possible to attach to collagen a molecule of another substance. This may be, e.g., a tanning agent of some other characteristics, or a highly charged molecule. In this way one can have, e.g., basic dyestuffs bonded to chrome-tanned leather.

Various tanning agents are used for retannage, inclusive of chromium compounds. Retanning with chrome tanning agents makes the leather soft. Vegetable and synthetic tannins also fill leather well in its weaker parts (bellies). In order to fill leather in its weak parts, resin syntans can be used whose molecular weight is suitably chosen to make them enter into the loose part only, but not into the tight one.

By proper choice of the tanning agents for retannage the color of leather can also be changed; bright leather may be obtained, if special syntans, sumach extract or zirconium compounds are used.

Numerous substances, e.g., dyestuffs and fatliquoring compounds, show strong affinity to leather. They attach quickly and strongly to the functional groups in leather or to chromium complexes. This reacting mode makes it difficult for a dyestuff or fat to penetrate into leather. Accordingly, retanning is usually done before dyeing and fatliquoring. Owing to this affinity of dyestuffs and fats increases and leather is more stable. Retanning after neutralization or in a new float instead of neutralization may be done as well. Usually a short treatment time and temperature up to 40°C are required. The tanning of pelt imported, e.g., in the form of wet blues or kips as well as second step of combined tannage inclusive of filling, loading and fixation can be considered as retannage.

Vegetable leathers are filled and loaded by introducing the tanning agent aggregates, which are not leather-bound in interfibrillar spaces. Excessive filling of leather is undesirable as then it becomes too stiff and brittle, and its resistance to abrasion is lowered. It is generally accepted that abrasion resistance is greatest when the content of wash-out is about 20%.

It is advisable to use tanning agents of a fairly low molecular weight, as those of great molecular weight cannot penetrate satisfactorily. The liquor used should contain more tanning agent than the previously used one, in order to prevent washing-out of the already fixed tanning agent. Another method is also used: filling of leather with a dry, pulverized tanning agents, with some addition of surfactants to the extract in order to decrease the surface tension of water contained in leather.

Some additives, apparently only loading the leather, actually perform more functions. Magnesium sulfate causes, e.g., precipitation of insoluble deposits from tannins. Owing to this the tannins may be better fixed in leather. This results in a brighter leather. Generally speaking, fixation of tannin in leather is due to its transformation into an insoluble substance. In the old tanning methods, this was done by precipitation of the condensing tannins inside leather. Later antimony, titanium or tin salts have come to be used in order to speed up the process and make it easier. These additives have been given up for both health safety and economic reasons. Nowadays solutions of soluble proteins, gelatin or casein, are introduced into leather. With tannins they form insoluble precipitates. As fillers urea, dicyandiamide resins or other condensation products are used.

Vegetable or syntan retanning of chrome leather has an aim, among others, to modify its properties. During chrome tanning collagen carboxyls enter the complexes. Other collagen groups, mainly ε-amino groups of lysine and imidazol histidine groups, remain unbonded. The ability to bind a vegetable tannin by chrome-crosslinked leathers is not decreased, yet it may increase as compared to pelts. If, e.g., pelt binds irreversibly 35% of vegetable tannin, the same pelt, after chrome tanning, will bind 61%. To give such an effect several factors have to combine, among others the change of the isoelectric point of the leather, deshielding of part of the basic groups by fixing the existing pores by chromium complex crosslinking bond, and finally the possibility of functional groups of vegetable tannins building into chromium complexes as ligands.

Combining the properties of vegetable and chrome tanned leathers gives the following advantages:
(1) Acceleration of the process with regard to quick vegetable tanning methods.
(2) Increasing the stability and fullness of the product.
(3) Significant increase of abrasion resistance of sole leathers chrome-vegetable tanned as compared with that of vegetable-tanned ones.

Vegetable retanning of light leathers, particularly of suedes, is aimed to decrease the binding rate between leather and the acidic dyestuff, and thus to obtain a more uniform coloring. Syntan retanning is also associated with the isoelectric point increase and, if a proper composition of bleaching syntans is used, for leather brightening.

Formaldehyde has a retanning property as well. It crosslinks the amino groups and some dyestuffs, which gives a T_S rise. Chrome retanning of vegetable-tanned leather is also used. The vegetable-tanned leather binds less acids than pelt. Vegetable tanning, engaging amino groups, makes it more difficult to bind acids, formed by chromium salt hydrolysis, and decreases the amount of chromium bonded by leather probably due to filling of interfibrillar spaces.

Recently Lokanandan and Ranganathan [20] have shown by measuring the friction coefficient between two grain surfaces of leather, that retannage increases the surface smoothness. They were able to demonstrate this difference in the scanning electron microscope picture: the surface of retanned leather did not have the small lesions present on the unretanned. This effect is not observable on buffed samples.

It is difficult to have a general view on the finishing processes as a whole, because too many fields of chemistry give their contributions to it and offer their best. At the same time, however, the manufacturers keep very carefully their production secrets, offering only a ready preparation plus recommendations for use.

REFERENCES

1. Kaczkowski, J. Plant Biochemistry, Warsaw 1980 in Polish
2. see ref. 1 ch. 16
3. Schmitt, O. Th., Mayer, W. Angew. Chemie *68*, 109 (1956)
4. see ref. 1 ch. 12
5. Endres, H. Leder *15*, 102 (1964)
6. Sugano, E. J. Agric. Chem., Japan *19*, 201 (1974) in Japanese
7. Stadler, P., Endres, H. Leder *14*, 175 (1963)
8. Endres, H., Stadler, P., El Sissi, A. A. Leder *15*, 102 (1964)
9. see ref. 17 ch. 1
10. see ref. 5 ch. 2
11. Lipschitz, P., Kremen, S., Lollar, R. M. J. Am. Leath. Chem. Assoc., *44*, 371 (1949)
12. Heidemann, E. Leder *11*, 99 (1960)
13. Roddy, W. T. J. Am. Leath. Chem. Assoc., *38*, 184 (1943)
14. Ushakoff, A. E. Brit. Pat. 565065
15. Weber, H. Kozarstvi *16*, 168 (1966) in Czech
16. Lasserre, R. Bull. AFCIC *21*, 222 (1959)
17. Reich, G. Ges. Abh. Inst. Freiberg *16*, 40 (1960)
18. Kraft, J., Rodziewicz, O. Synthetic Tannins, Warsaw 1971 in Polish
19. Olivannan, M. S., Nayudamma, Y. Leather Sci., *25* (1978)
20. Lokanandan, B., Ranganathan, N. Leder 31, 125 (1980)

18.

DYEING OF LEATHER

Leather dyeing is a transition process between tanning and finishing, as differences in the chemistry of attaching of tanning agent and dyestuff may be very slight.

From the viewpoint of dyeing, leather is a specific material. Just the very penetration of a dyestuff into a tridimensional network of collagen fibers is different than the penetration into fabrics or nonwoven. Certain kinds of leather need not be dyed through; it is enough to dye them on the surface. This decreases the consumption of a dyestuff, which is usually an expensive material. In this case it is important to know how penetration of this dyestuff proceeds and the way its use has to be observed very exactly. The processes done affect the chemical properties of leather. The result of dyeing is not only depending on the tanning agent and method used: reactions with fat, surfactant and water should also be taken into consideration. According to Thorstensen [1] the binding of dyestuff is a resultant of many facts, and the use of a dyestuff is a combination of knowledge, technology and art. In this chapter the principles of visual evaluation of color will be discussed, and then basic problems of dyestuffs binding by leather.

18.1. Principles of knowledge about color

Color is always connected with light. It is a physical as well as physiological phenomenon. It depends not only on optical properties of a given substance and on the kind of illumination, but on the eyesight as well, on human mind and its reaction to light, as to a stimulus. A majority of substances do not emit light; however, from the light, incident on them, a part becomes absorbed, part reflected and part transmitted. Individual components may vary in dependence on wavelength of the incident radiation (this is the principle of absorption spectroscopy).

Transmitted and reflected light have different properties, compared with the incident one. Chemically the process of perception and comparison of the colors is rather complex and highly interesting. It works in the system: eye lens-optic nerve-brain cells.

Possibilities of remission determination and the optics of films

Directed reflection may be expressed in terms of Fresnel's formulas given in the handbooks of physics. Values of the diffused reflection have been determined by Kubelka and Munk [2]. Their paper has formed fundamentals of the film optics.

417

Monochromatic index of directed reflection is a function of optical density, n, and of absorption coefficient x. Remission is a quantity, which is an analogy of reflection index in the case of diffused reflection. Let us assume that we have a film with totally covering power, or a film which does not allow the incident light to come through it to the background, or layer, (film) which, when its thickness increases, does not increase the remission value. In this case R is remission of such a film, characterized by the absorption index K and dispersion index S. The weakening of a light beam, passing through an elementary (unit) layer dt (Fig. 18.1) depends on the absorption (I = Kidt) and dispersion (II Sidt). The beam of light, passing through layer III of thickness dt is weakened by Kjdt in layer III, and by Sjdt in layer IV. The dispersion effect changes the direction of the light beam in layers I and IV. Generalizing, a change in the light intensity beam di by passing through layer dt is

$$di = -(K + S)idt + Sjdt$$

and because di = $-[(I) + (II)] + IV$
the change in the 'backbeam' is

$$dj = -K(+ S)jdt + Sidt$$

According to the solution of Kubelka-Munk, the R value which satisfies this equation, is connected with K and S as follows:

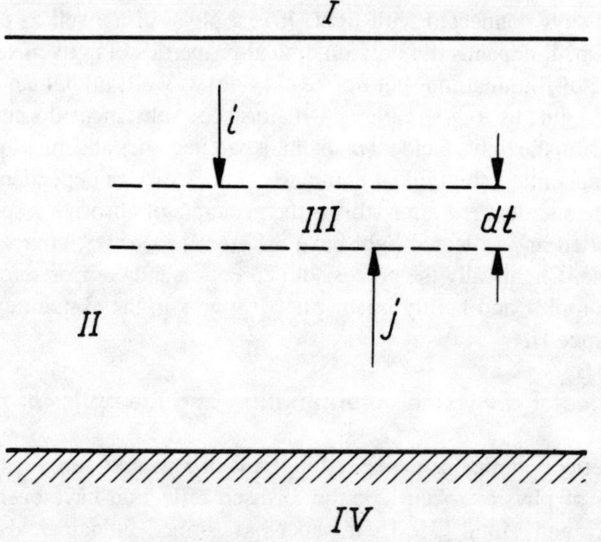

Fig. 18.1. Absorption and dispersion in the colored layer: i—incident light, j—reflected light, I—air layer, II—pigmented layer, III—elementary (unit) layer (dt), IV—background.

$$R_\infty = 1 + \frac{K}{S} - \sqrt{\frac{K^2}{S^2} + 2\frac{K}{S}}$$

or

$$\frac{K}{S} = \frac{(1 - R_\infty)^2}{2R_\infty}$$

Remission as a function of the layer (film) thickness (x)

$$R = \frac{(R_g - R_\infty)\left[R_\infty - R_\infty'\left(R_g - \frac{1}{R_\infty}\right)\right] e^{Sx\left(\frac{1}{R_\infty} - R_\infty\right)}}{R_g - R_\infty - \left(R_g - \frac{1}{R_\infty}\right) e^{Sx\left(\frac{1}{R_\infty} - R_\infty\right)}}$$

where R_g is background remission. This value can be found by measurements.

Typical pictures of remission spectra for white light are shown in Fig. 18.2. They are obtained with a spectroreflectometers, i.e., simple spectrophotometers, or the recording models but equipped with additional facility to record the intensity of light reflected by a sample.

In reflectometric apparatuses designed for measurement of directed reflection with elimination of the diffusion component, and angle of incidence α is equal to the observation angle β. This coefficient is directly measured in relation to the compensated intensity of the incident light. Remission, however, is measured indirectly, in relation to the reflecting ability of the surface of a standard substance. Such a standard, or disperser, is used mostly as powder, pressed or deposited on a surface. These are solids, of $K \to 0$. From the formula of Kubelka-Munk, presuming $R_g = 0$, it results that:

$$R_\infty = \frac{S_x}{S_x + 1}$$

This means that substances of a high dispersion coefficient S are remitting almost completely at low film thickness. The incident light is disperse reflected according to Lambert's law.

A good disperser should have a good covering power. It may be calculated from the above formula after transformation:

$$S_x = \frac{R_\infty}{1 - R_\infty}$$

Substances with optimal covering power have a refractive index for visible light between 1.4 and 1.9 and a particle diameter d: $\lambda < d < 10\lambda$ (λ -wavelength of incident light). Then the dispersion coefficient does not depend on wavelength.

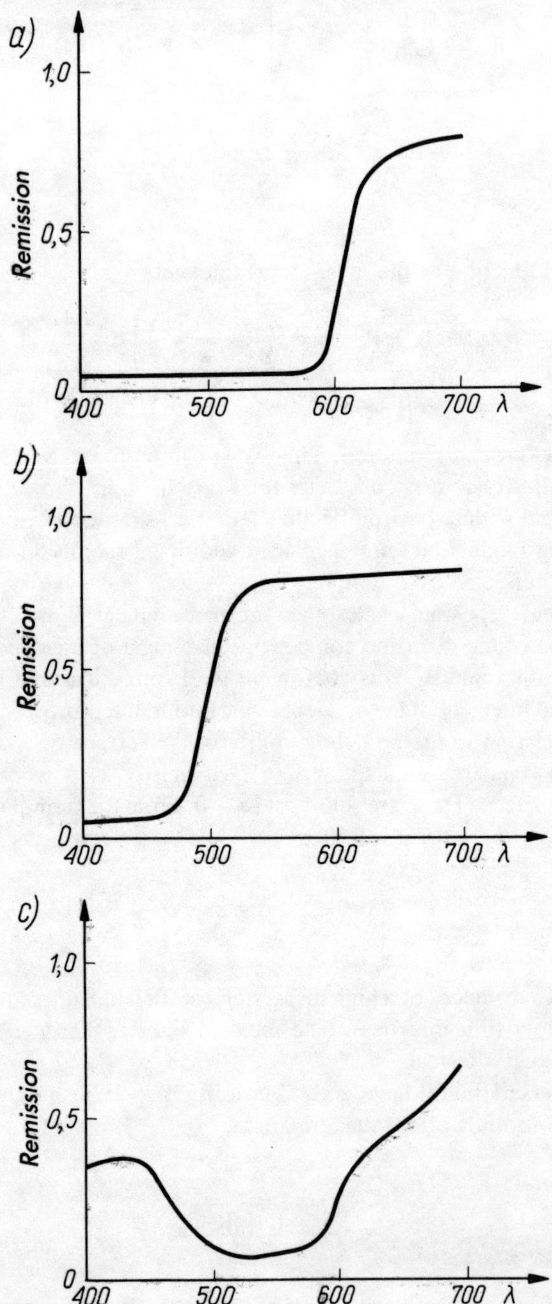

Fig. 18.2. Remission of white light by: (a) red, (b) yellow, (c) purple bodies.

Good disperser, meeting these requirements, is characterized by equienergetic remission spectrum close to 1.00 (cf. Fig. 18.3a).

As standards most frequently MgO, $MgCO_3$, $BaSO_4$, Al_2O_3, pulverized quartz, etc., are applied. Worsening remission (Fig. 18.3b, c) speaks for the growing participation of black color.

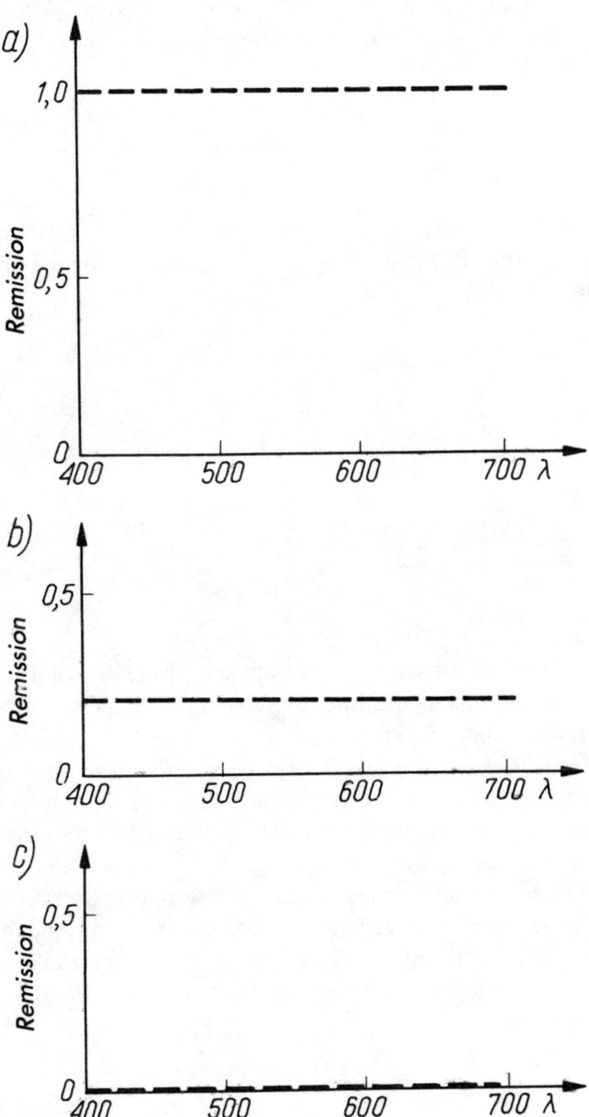

Fig. 18.3. Remission of white light by: (a) white, (b) gray, (c) black bodies.

Remission is often measured quantitatively. In order to measure it, the directional component is eliminated by matting the surface of the sample tested, or by grinding the substance tested with disperser properly chosen. Direct reflection is accompanied by polarization of light, thus the use, e.g., of two crossed polarizing prisms eliminates this component fully.

According to Kortüm and Schreyer [3], remission values in Kubelka-Munk equations are directly proportional to concentration c (in mole fractions) of the absorbing substance

$$\frac{K}{S} = \frac{(1 - R_\infty)^2}{2R_\infty} = \frac{\varepsilon c}{0.4343\,S}$$

where ε is molar absorptivity coefficient. Both the absorption coefficient K and the dispersion coefficient S are additive values, so for multicomponent systems the following equations are valid:

$$K = c_1 K_1 + c_2 K_2 + \ldots + c_n K_n$$

$$S = c_1 S_1 + c_2 S_2 + \ldots + c_n S_n$$

thus

$$\frac{K}{S} = \frac{(1 - R_\infty)^2}{2R_\infty} = c_1 \frac{K_1}{S_1} + c_2 \frac{K_2}{S_2} + \ldots c_n \frac{K_n}{S_n}$$

Evaluation of the dyeing intensity is frequently done by remission measurements. Development of colorimetry has increased, among others, possibilities of instrumental programming of colors, or composition methods, making it possible to obtain an exactly determined color of expected optical properties.

In this Institute the ATR (Attenuated Total Reflection, see below) method was used for testing the coats (films) on leather [4]. In this paper a simple apparatus was used (revolving spray gun) for coating leather samples in a reproductible way. Pigments and dyestuffs tested in this work were characterized by absolute and relative transparency of the coats, made of standardized solutions. Absolute transparency P_{BP} is a difference of absorption in a given wavelength range from $\lambda_1 = 400$ to $\lambda_2 = 700$ nm as determined for total coverage (R_2) and the first uniform coverage (R_1):

$$P_{BP} = \int_{\lambda_1}^{\lambda_2} f(R_2)d\lambda - \int_{\lambda_1}^{\lambda_2} f(R_1)d\lambda$$

and relative transparency is product of P_{BP} and of the first member of the equation on its right side.

In Fig. 18.4 typical examples of coated surface are given, as observed in the paper quoted. One may see there, how the spectrum of reflected light changes when transparent, nontransparent or partly transparent coats are applied. Using this method one may observe the results of applying each coating layer, determine

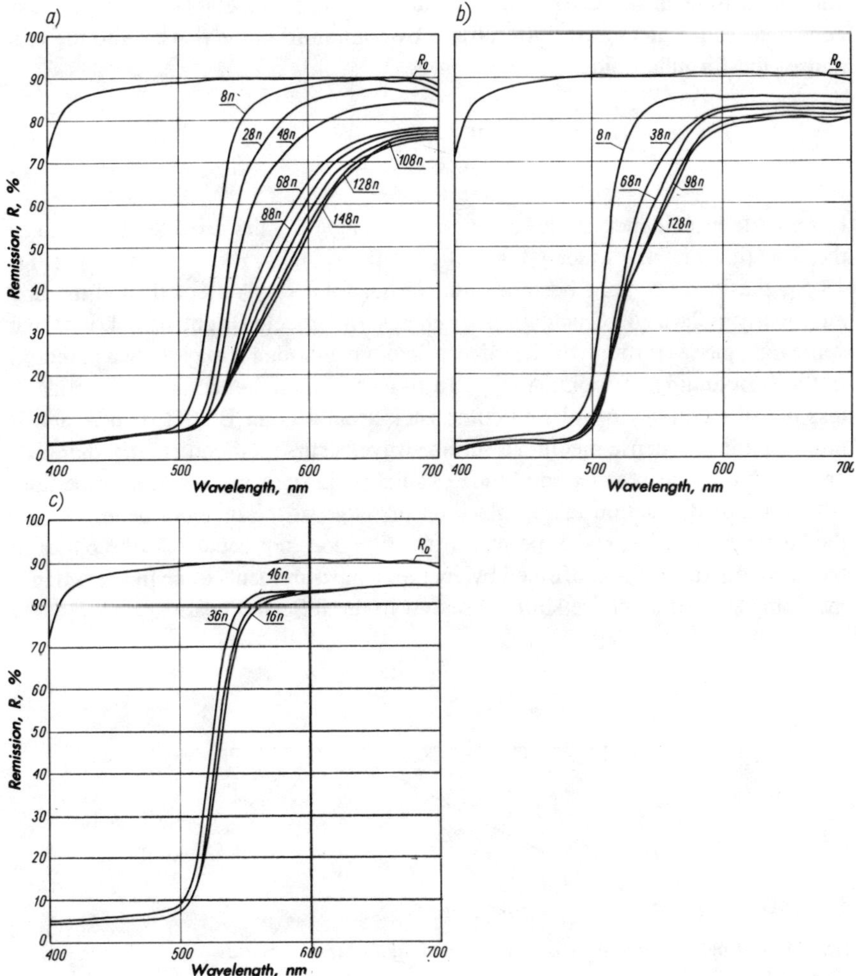

Fig. 18.4. Remission of a yellow coat applied on chrome leather: (a) transparent coat, (b) semi-transparent coat, (c) nontransparent coat, n—number of coats, R_o—remission of background. According to [4]

the differences between pigments or dyestuffs and check whether the compositions claimed as transparent are really so.

The recently used ATR technique makes the characterization of the surface tested easier over a wide range of wavelength in ultraviolet, visible and infrared spectra. The principle of this method is the measurement of intensity of the light, reflected by the medium of a lower optical density than the one from which light is coming and to which it becomes reflected. If the refractive index of monochromatic light in optically denser medium n_1 (e.g., in glass) and in the less dense one (e.g., air) n_0, as light refracts by coming from the denser into the less dense, the formula holds:

$$\frac{\sin \beta}{\sin \alpha} = \frac{n_1}{n_0}$$

For a certain incidence angle, called critical angle α_c, the reflected beam takes the direction of the surface ($\beta = 90°$, $\sin \beta = 1$). Is α greater than α_c (Fig. 18.5), the incident light becomes almost quantitatively reflected, and thereby elliptically polarized. Calculation of energy of reflected light has shown that some light passes through the less dense medium although it may seem a paradox. In the surroundings of point A (Fig. 18.6) part of radiation comes into optically less dense medium, in order to come back around point B. Between A and B points in the less dense medium a surface wave occurs, perpendicularly damped, having sharply decreasing amplitude. Radiation in optically less dense medium is a result of diffraction taking place on the edge of the incident beam. Inside the beam, between A and B points, diffraction does not occur. On the edges of the beam diffraction is confirmed by energetic measurements done in a less dense medium, as well as phase shift, observed in the reflected beam.

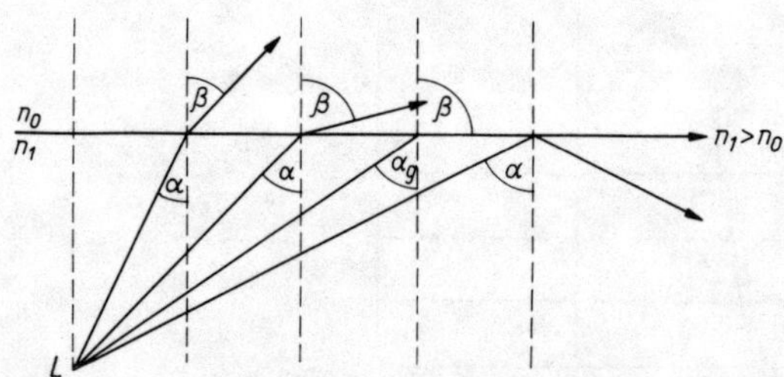

Fig. 18.5. Limiting angle α_g when the beam of light passes from a denser medium to one of lower density ($n_1 > n_0$).

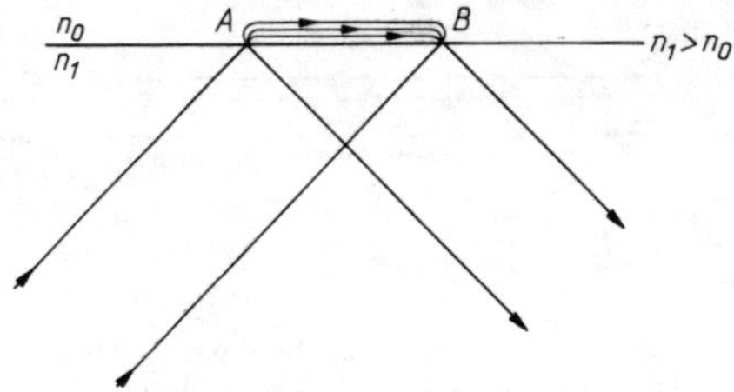

Fig. 18.6. Effect of diffraction of the light beam on the total reflection.

In Fig. 18.7 IR-spectrum of dibutylphthalate is shown as an example of the ATR method for substance identification. Another modification of this method is multiple reflexion. In this method the light coming many times from one medium to another becomes more damped, which makes it possible, e.g., to

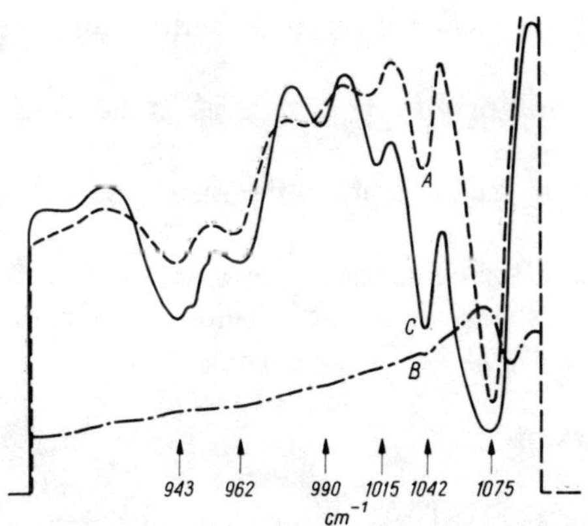

Fig. 18.7. IR spectrum of dibutylphthalate: A—transmission spectrum, B—conventional spectrum in reflected light, C—ATR spectrum obtained by the use of AgCl crystal.

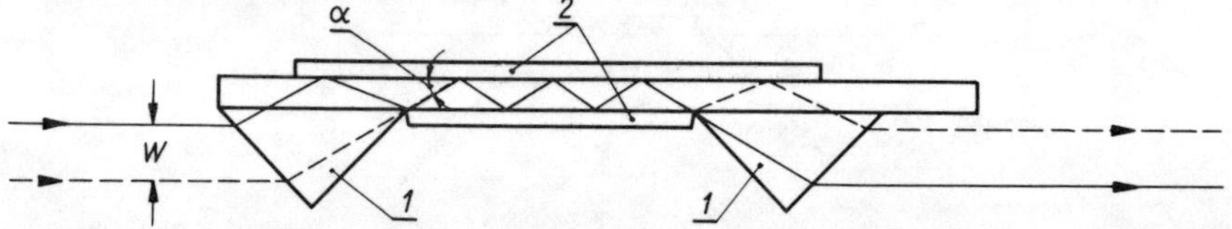

Fig. 18.8. Principle of obtaining an ATR spectrum by the use of two prisms (multiple reflection); A plate of the same material as the crystals is placed between samples. The width of the parallel beam (W) remains unchanged, however, the picture is reversed. 1—prisms, 2—samples, α—incidence angle.

determine substances present in the film at very low concentrations (Fig. 18.8).

The film (layer) is characterized by selective absorption ability for certain wavelengths. This property makes its surface, consisting from many randomly distributed surfaces, to reflect more radiation in certain directions than in others. This phenomenon was described by Wright as early as 1900. Results of his observations may be summarized as follows:

(1) Intensity of the reflected radiation is not symmetrical in relation to the angle of incidence α and to the observation angle β. It is not in accordance with Lambert's cosine law and with Helmholtz's rule of shift. (According to cosine law the intensity of reflected radiation is proportional to the ratio of cosine of the angle of incidence and observation angle. The shift rule says that parallel light beam, passing through a film of equal thickness, having refractive index different than that of the surroundings, becomes only parallel shifted);

(2) Cosine law applies only for matt surfaces, if the angles α and β are small enough;

(3) Are the angles α and β great, so the matt surfaces may give high 'gloss peaks.'

As a rule, intensity of the light reflected by a matt surface in a definite direction increases in a way, speaking for participation of directional reflection, if the azimuthal angle is 180° and $\alpha = \beta$.

Gloss measurement

Gloss measurement is based on determination of the angle of distribution of intensity of the reflected light by means of a goniometric reflectometer. Its working principle is shown in Fig. 18.9. At constant incident angle the reflection angle is measured, as related to the incident beam. The observation angle is varying from zero to 90°. First, the intensities are determined for white standard, then for sample tested (Fig. 18.10). Gloss value is determined by the use of three conventional values: brightness (remission, R), height h and sharpness s. Brightness characterizes capability of the internal diffusion of the reflected light, height is the reflection ability, and sharpness—a measure of surface smoothness.

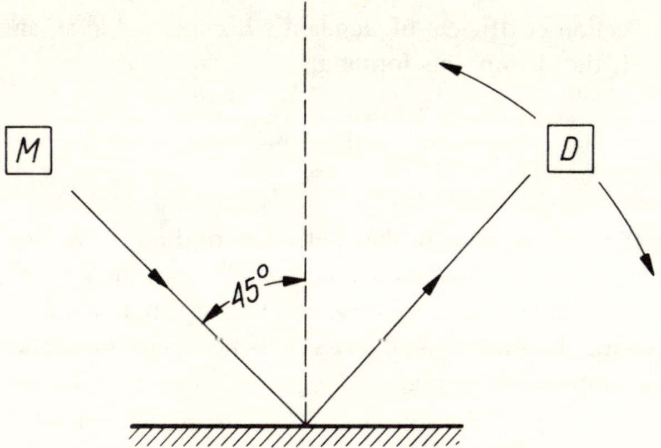

Fig. 18.9. Light path and principle of goniometric reflectometer: light source (M), receiver (D).

Remission is determined by this equation:

$$R_o = \frac{u_o}{s_o}$$

where u_o is the reflection coefficient of sample's background at an angle 0°, and

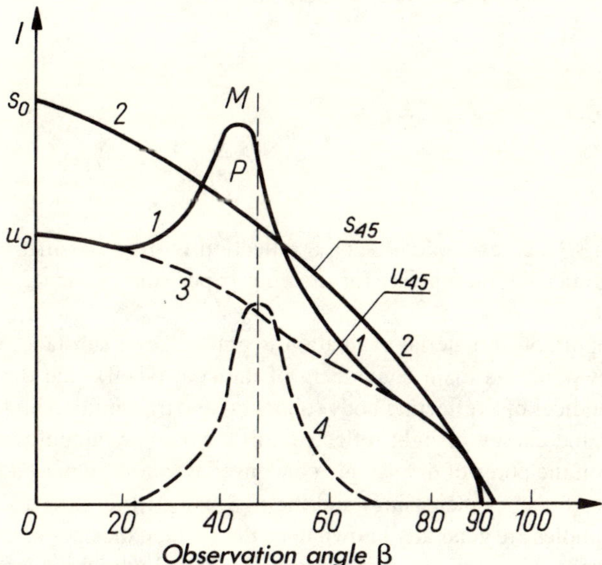

Fig. 18.10. Angular distribution of intensities of the light reflected by the white standard (s_0, s_{45}, 90°), M is the maximal reflection point. Curve 4 presents the difference between the background reflection and sample reflection.

s_o is the reflection coefficient of standard's background at an angle 0°.
Height is calculated from this formula:

$$h = \frac{p - u_{45}}{s_{45}}$$

where p is the highest value of the coefficient of direct reflection of the sample determined, u_{45} and s_{45} are the coefficients of diffuse reflection of background of sample and standard, respectively, at an observation angle $\beta = 45°$.

To determine the sharpness of gloss, it is necessary to calculate the relative surface F area by formula (Fig. 18.10):

$$F = \frac{P - u}{s} = \frac{P}{s} - R; \quad \frac{u}{s} = R$$

where P is surface under curve 1 (d_o, P, 90°), s—surface area under curve 2 (s_o, s_{45}, 90°) and u is the surface under curve 3 of the matt sample background (u_o, u_{45}, 90°). Width of the gloss b, i.e., degree of its diffusion, is:

$$b = \frac{F}{h}$$

and sharpness δ is its reciprocal value of it.

$$\delta = \frac{1}{b} = \frac{h}{F} = \frac{\dfrac{P}{s_{45}} - R}{\dfrac{P}{s} - R}$$

In Fig. 18.11 an example of such a reflection is shown as measured with an angle photometer (goniometer) for drawing paper (the incidence angle shown by arrows).

Very important in material evaluation is grain size of substance, being from optical viewpoint the main component of the coat (shell), and the ratio of the refractive indices of a reflecting body (pigment) and transmitting body (polymer). Evaluating the energy of light reflected still the sample humidity is to be considered. If in the pores of dry sample condensed water has replaced air, the ratio n_o decreases (as n_o is increasing) and the K/S value in formula 18.8 increases.

Moist samples are generally known to be darker than the dry ones: their spectra in visible light, however, do change more, the more hydrophilic is the substance tested. Influence of hydration and drying on spectrum of pulverized Cr_2O_3, 'diluted' with calcium fluoride is shown in Fig. 18.12. As one may see, the

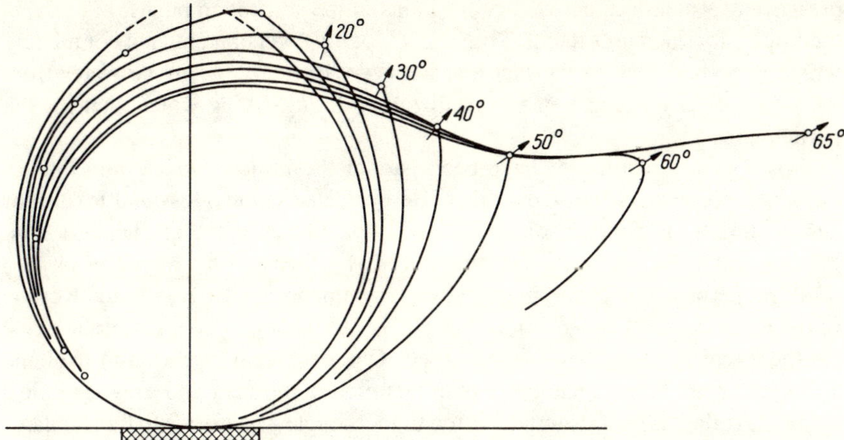

Fig. 18.11. Reflection intensity of white light from drawing paper vs. angle of incidence.

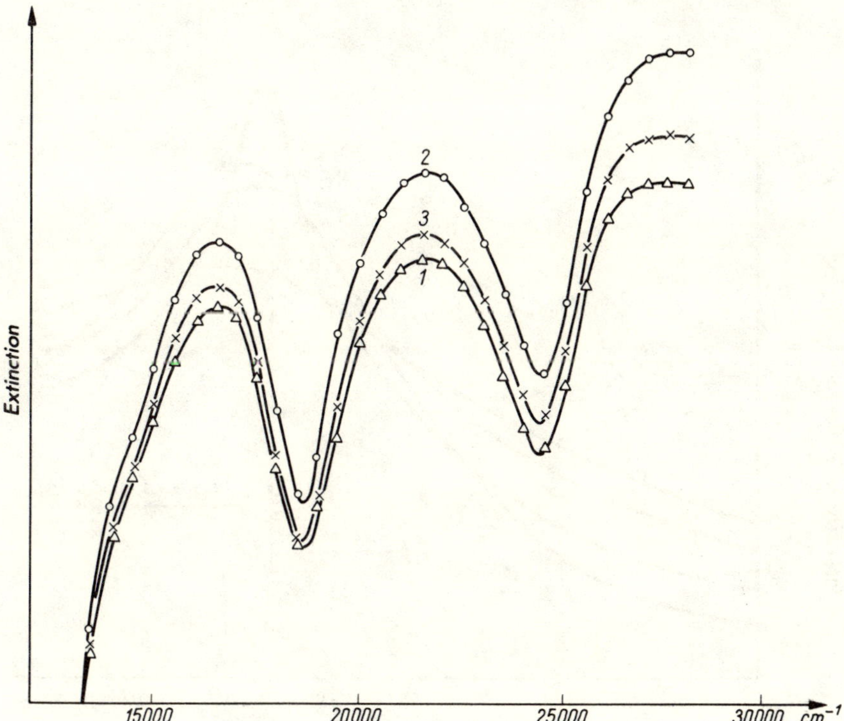

Fig. 18.12. Differential spectrum of light reflected by a mixture of Cr_2O_3 and CaF_2 (CaF_2 used as reference): 1—waterfree sample. 2—sample after 76 hours in a water-vapor saturated atmosphere, 3—the same sample as under 2 after 3 days over P_2O_5 in vacuum.

presence of water does not change positions of the absorption bands, but affects their intensity (increases them). This effect is partly reversible, but not completely which is perhaps due to the fact that it is very difficult to remove water from the pores, or else because of recrystallization during drying which changes the particle size.

Troy's investigations [5] have been part of the project, whose aim was to introduce poromerics into the market. He evaluated color, gloss and texture of new materials, in comparison to the good box calf accepted as standard. Assuming that color is the easiest reproduced, he focussed his attention on two other properties. For gloss investigation a goniophotometer was used. Results of the measurements are shown in Fig. 18.13. In this Figure a logarithmic scale for the intensity of reflected light is used. One has to pay attention to the band width and peak height. Sample 1 is in this picture too matt and gray; sample 2 is too gray; the gloss of material 3 is too high. Sample 4 conforms to the standard.

Of importance is the width of the absorption band. It is related to the texture of the material. Texture cannot be exactly defined by visual impressions but it is rather transferred to abstract ideas. In individual cases microphotography

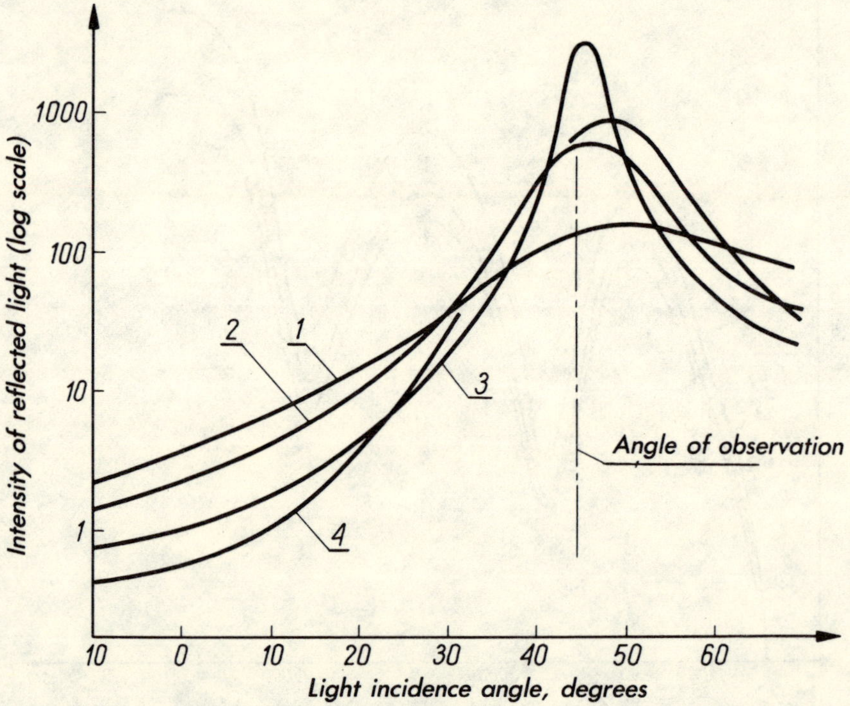

Fig. 18.13. Goniophotometry of shoe upper materials—poromerics of different finish. Curve 4 is for good material and for boxcalf. According to [5]

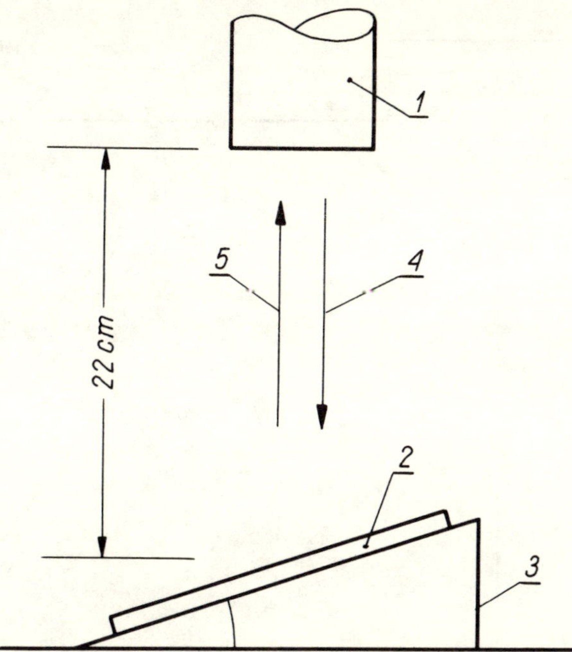

Fig. 18.14. Microphotographs of material surface set used in testing of poromerics: 1—objective with vertical light, 2—sample, 3—its inclination angle (variable), 4—incident light, 5—reflected (measured) light. According to [5]

technique can be used as shown in Fig. 18.14. Thus the surface microstructure can be investigated, e.g., by comparing photographs of box calf grain and the grain embossed on the surface to be tested by applying various exposure times. This is a very good procedure in a case of coarse embossed grain, but is not always suitable to find out the difference, if the grain is subtle. In a photograph taken in a 50-80 magnitude one may spot some 'sparkling' points which impart to the material more gloss than is intended. Such points may be found by change in the exposure time, or in the inclination angle of the sample in relation to the objective. A naked eye may see two points separately if they are spaced 50 μm apart; under very advantageous circumstances this distance may decrease to 10 μm. This is a limit of useful texture testing, because it is of no use for the microscope to apply magnifications showing smaller details, as they will have no more influence on the texture. In Fig. 18.15 one may see what happens if one observes a surface from the distance usually used for reading. Then there is a difference in the point of view of about 15° between the right and the left eye.

If the light beam is narrow, or the texture of the surface is very subtle, there

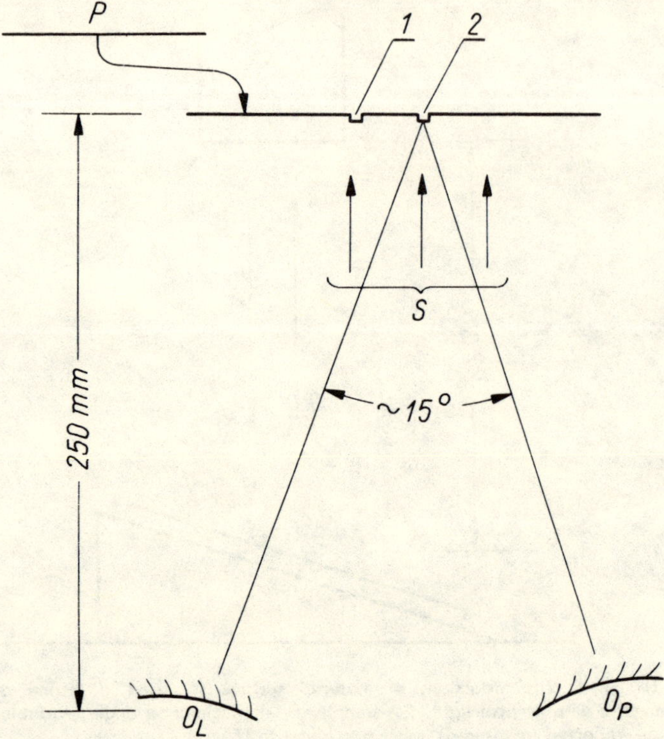

Fig. 18.15. Two-eye texture observation from a distance of 25 cm. The mat point 1 on the surface can be seen with two eyes, the glossy point 2 illuminated from above—with one eye only.

may be some differences in seeing with the left and the right eye. In such cases certain impressions of the depth of the picture occurs. This partly explains such an impression of some depth of the structure, as it may be observed in properly finished box calf.

Color and its determination

In perception of color by man's eye an infinite amount of theoretically possible stimuli may give only a limited amount of impressions. Thus identical visual impressions exist, for which different stimuli are responsible (Fig. 18.16).

Distribution of remission of some simple and composed color is schematically shown in Fig. 18.17. We refer to colors as ideal 'cut' off the white light spectrum, if they have remission curves of approximate shape as shown in Fig. 18.17. A brilliant color is the one having the greatest brightness among all possible,

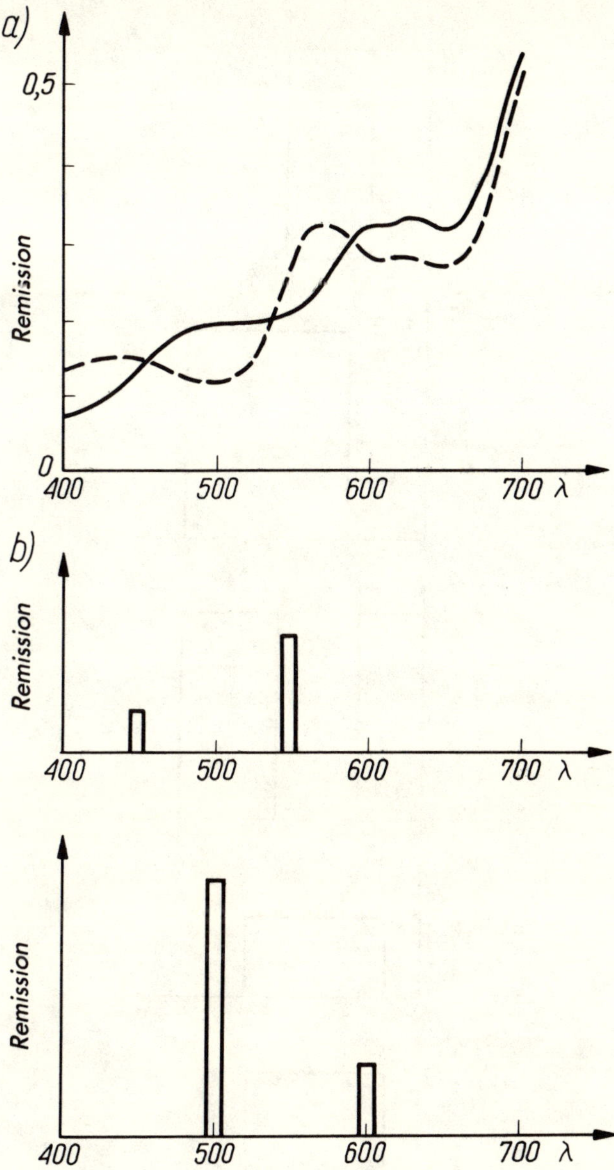

Fig. 18.16. Colors visually equal: (a) real and (b) schematic examples.

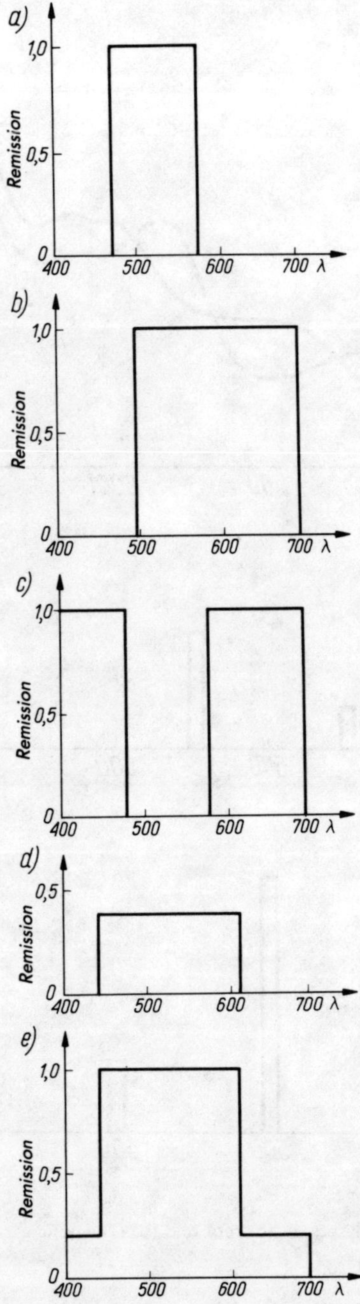

Fig. 18.17. Schematic distribution of remission of various colors: (a) green, brilliant, (b) yellow, optimal, (c) purple, optimal, (d) green, optimal, darkened, (e) green, optimal, lightened.

whereas the remaining part of the spectrum has a vanishing value. Every ideal color is brilliant, but not vice versa. In the case of a brilliant color the perpendicular lines in the spectrum are in points of supplementary colors. Two last examples (Fig. 18.17d and e) are of lightened or darkened shades. The narrowing of the range where remission has the value of unity (broadening of the range, where the remission $= 0$), more than in a brilliant color, makes the saturation increase at the beginning, together with its darkening until complete blackness.

Practical equivalents of these samples are much more difficult and complex. Introduced in 1931 by the International Commission on Illumination (*Commission Internationale d l'Eclairage*) the classification system of colors is still a basis for color systems. In some countries, it is a basis of the standardized measuring system. It is based on the principle of color synthesis, making it possible to obtain all real colors from the three basic ones. Conventionally, the basics accepted are: red R, green G, and blue B colors mixed on equienergetic levels, giving the white color as a sum. In Fig. 18.18 a scheme of the apparatus is shown, in which the observer may compare two colors: the tested one and that obtained by mixing of the basic colors; intensities are so adjusted, they may be considered equal.

So color is identified by three measured values, defined exactly by wavelength and intensity. Of course, this identity may concern only physiological, sensual impression, and not their physical identity, as they may be obtained by mixing the light of other wavelengths.

In 1964 the wavelength of three basic colors has been exactly fixed. The standard visual stimuli are: 645.2 nm (red), 526.3 nm (green) and 444.4 nm (blue). The CIE system is based on the color triangle (Fig. 18.19a). Three basic colors in the spectrum R, G, B are placed as points on a plane. Triangle corners thus obtained are their chromaticities. In order to show a mixture of two basic colors, e.g., R and G, we connect our points, and side RC shows all possibilities

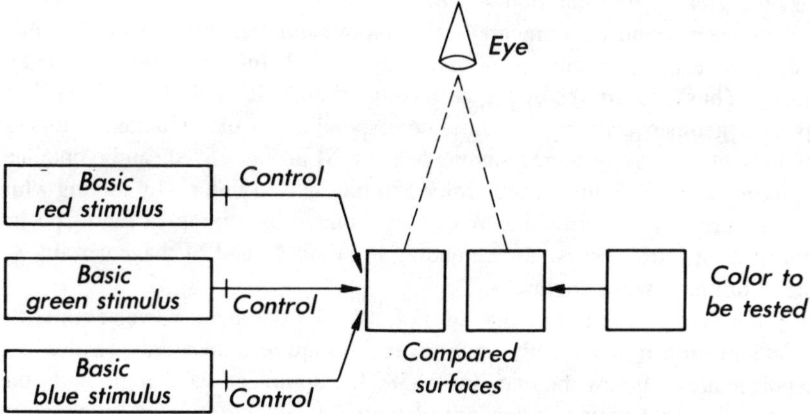

Fig. 18.18. Determination of color by addition of three basis colors.

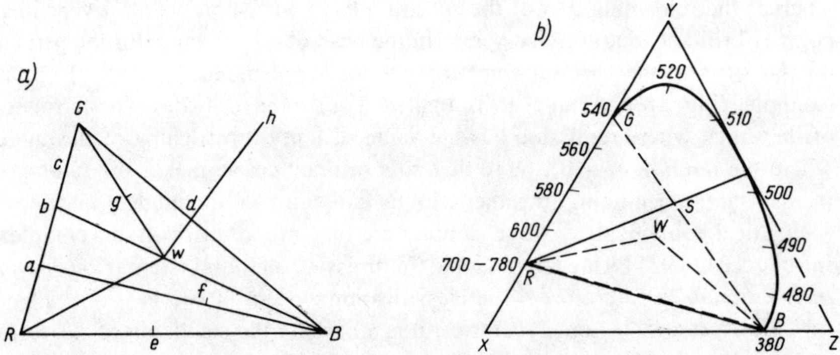

Fig. 18.19. The color triangle (a) and supplementary color triangle (b).

given by mixing these two colors. Both the remaining triangle sides have the same sense. The place of a given color on the appropriate point on the triangle side depends on the proportion in which both components participate in the color, e.g., the greater participation of component R in a color, the closer it lies to the corner. Points a, b and c represent the colors: orange, yellow and yellow-green. On the GB line analogous transitions are obtained between green and yellow (d), whereas on the BR line—the purple hue which is absent in real spectrum. All colors being a mixture of two colors from the R, G, B triad are lying on the periphery of RGB triangle. A mixture of any three colors is formed by finding first the 'resultant' of two colors, e.g., point a and then by drawing a straight line, corresponding to all a and b ratios, containing variable amounts of e. An example of such a mixture is point f. A particular case of such a mixture is the white color at W point. Connecting this point with R, G, B sides straight lines are obtained, being a measure of saturation with a given color. The closer to the peak, the greater the saturation will be.

It has been found experimentally that more saturated (brilliant) colors may exist, than, e.g., a mixture of G and B at point d; full saturation is noted at point h. Thus one arrives at the new color triangle R G B (Fib. 18.19b). It depicts a geometric place of points, corresponding to the saturated colors of visible light. The color corresponding to point M in Fig. 18.20 can be obtained by mixing white W with spectral color 520 nm, and M′ color—by mixing white W with purple A. In turn white W can be obtained by mixing M′ with spectral color 540. Spectral colors, corresponding to points M and M′ have certain, so-called dominant wavelengths.

For the area of spectral colors—in Fig. 18.20 a dominant wavelength is the one, which in a mixture with white gives the required, unsaturated color. For the purple area—below the line 700-W-380 the dominant wavelength is the one of spectral colors which, when mixed with a given unsaturated color, gives white.

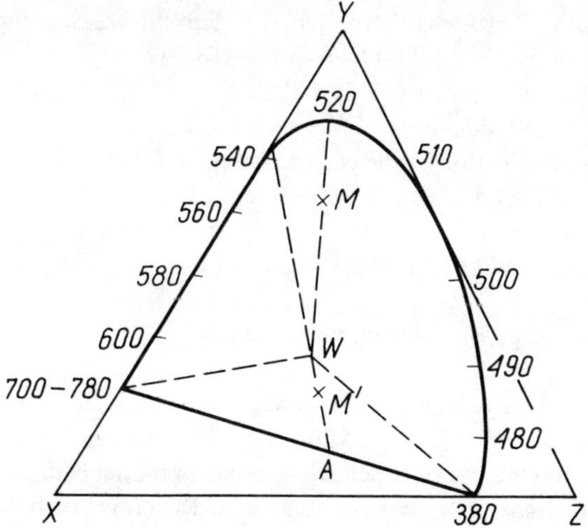

Fig. 18.20. Dominant wavelength and colorimetric purity.

In the evaluation of identity of colors a criterion of saturation, or purity, is applied. This is equivalent to a ratio of the monochromatic component to the total intensity. If the color M has brightness of B_M, and the brightness of spectral color is B_{520}, the colorimetric purity is given by the expression:

$$p = \frac{B_{520}}{B_M}$$

Using these two definitions—wavelength and degree of saturation—one may exactly define the place of a given color on the surface of the color triangle. In many national standards for color measurement two systems of numerical expression of colors are in force—a regular one and a supplemental, which is applicable, if the observation angle of a given colored surface does not exceed 4°. A reason for introduction of the supplementary system is the differences in color perception when at the center or when on the periphery of the seeing field of the eye.

In the CIE system it is accepted to express the chromaticity (defining 'colorimetric purity' and 'dominant wavelength of the light' as it would be received by the eye of a 'standard observer' under specified conditions) is expressed by components of basic stimuli X, Y, Z. It is sufficient to give any two out of three values, as $X + Y + Z = 1$. This whole system is described above. It is not free of disadvantages. They can be eliminated by using other much more involved systems, e.g., Munsell color atlas, etc.

Colors of nontransparent physical bodies are more complex in character, than the spectral colors. They depend, e.g., on the kind of illumination. As a rule, none of the illumination source is equienergetic (the same amount of energy, carried with waves of every wave length). For this reason the standardized illumination sources are used (shown in Fig. 18.21).

Illumination C corresponds to a daylight dispersed equally by the cloudy skies (position of sun and thickness of the cloud layer may be of importance): A corresponds to illumination by the electric lamp, and B—to plain sunlight. These values are characterized in national standards. Illumination standards have their places in the spectrum of black body radiation: A corresponds to temperature of 2848 K, B—ca. 4800 K, and C—ca. 6500 K.

Conventional ways of color comparison

Light incident at the eye's retina results in a reaction of decomposition of visual purple (also called rhodopsin and erythropsin). This substance, of a great, protein-type molecule, having dyestuff behavior, exists in retinal cells. Decomposed by the influence of light, it becomes regenerated. Reactions of both its decomposition and regeneration occur in a definite time. Thus the effect of light does not

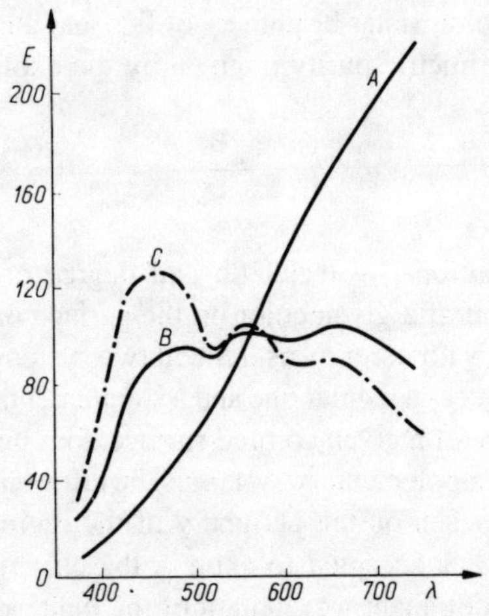

Fig. 18.21. Energy distribution of various light sources A, B and C.

disappear at once, remaining for about 0.1 s. This makes the pictures, following quick one after another, to be perceived as continuous, like in cinematographic procedure.

In order to discuss more in detail the eyesight impressions, one may assume that all of them are colored. The noncolored effects may be defined as those occurring in the range of gray, i.e., from black to white through all shades of gray. These shades may be arranged in a decreasing order of gray, if the evaluation conditions remain unchanged. As the contrasts in gray color may be compared with contrasts in the perception of colors by the eye, the gray scale is introduced as a standard which allows to compare the contrasts between two colored surfaces. Comparison of colors by means of gray scale may be carried out when necessity of color change exists, or if a color is transferred from one surface to another (usually white) by bringing these two surfaces into contact. Colors are compared with one of two gray scales. Each scale contains five pairs of standards (usually cardboard colored white to dark-gray) of contrast between them growing in geometric progression. In a scale used for evaluation of change of surface it is 0, 1.5, 3, 6, 12 units; for the scale used for evaluation of the degree of change in white—0, 4, 8, 16 and 32 units. These units, called NBS-units (National Bureau of Standards), are equal 'five times the smallest observable difference under best observation conditions.' The gray scale for evaluation of the degree of change in white, has a contrast zero (fifth degree), when both the surfaces compared are equally white. The gray scale for evaluation of change in the color of surface has a contrast zero (5th degree), when two surfaces to be compared are not differing in an observable degree. Usually detailed description of the use of both gray scales has to be included into national standards, together with the purchase sources of the standardized and accepted scales.

Choice of the proper contrast, corresponding to a given degree of the gray scale, depends to some extent on fitness of the observer's eye and may be subject to some fluctuations. In order to make it easier, and to standardize, at least to some extent, the procedure after which the gray scale is used, the standards of the so-called blue scale have been introduced. Both these systems combined are used preferentially to determine contrast, being a result of the color change, rather than for determination of differences in the colors. Blue scale standards are made of eight pieces of cloth, dyed to blue, with dyestuffs of various resistance to light. Changes occurring in these pieces of cloth under the effect of a definite amount of light are corresponding with the degrees of contrast in the gray scale (the light used should be of the dispersed daylight C-type). This effect may be obtained, e.g., by illumination of standards with xenon lamp of defined strength and coming from a definite distance like in Xenotest apparatus. As a basic change in standard No. 3 by 1° of the gray scale is accepted, i.e,. contrast equal to four. Resistance of blue standards toward light grows from 1 to 8. Use of the blue scale has also been standardized by the national standards.

Principles of color mixing (matching)

Color science contains some freedom, coming from essential difference between subjective perception of color impression by human eye and spectrometry—i.e., science related to the emission and absorption of electromagnetic radiation of various wavelength—visible radiation included. The mixing of colors does not change their fundamental characteristics based on spectrum of reflected or dispersed light, but it results in changed visual impression. In order to evaluate this change properly, it should be remembered that the human eye can differentiate between two points, if they are seen at an angle greater than $1'$ ($0.291\ 10^{-5}$ rad) (under particularly favorable conditions the value of this angle may be slight less). Also, differentiation between similar colors should be kept in mind, i.e., colors adjacent in the CIE triangle. This has been classified by MacAdam [6]. Based on the tested sensibility of a group of (randomly chosen) persons to particular colors, he worked out a collection of ellipses, inside which the color differences have been below the discernibility limits. Thus the colors inside the ellipses of the CIE triangle gave identical visual impressions. As seen from Fig. 18.22, where the ellipses in tenfold magnification have been shown, the surface

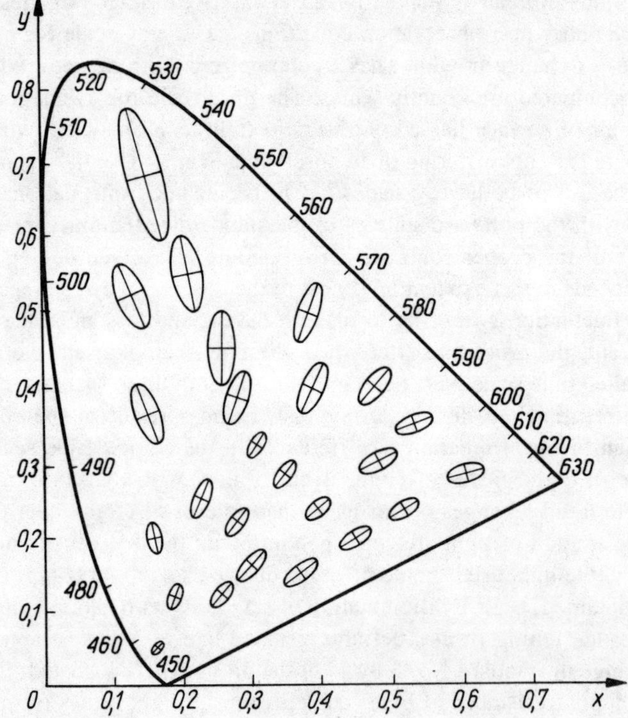

Fig. 18.22. MacAdam's ellipses in the CIE triangle. Tenfold value of normal declination on the color plate.

of ellipses is varying: human eye is less sensitive in the green range, more sensitive in the blue one. This investigation allows to accept proper tolerances in color mixing.

Thus we have demonstrated that the color of light beams mixture depends on the color of its components, rather than on its spectral composition.

From this rule the following conclusions may be drawn:

(1) two separate, equal colors a and b mixed in pairs with two other equal colors c and d, will give mixtures of identical color. Resultant colors: a + c is identical to b + d (unless a chemical reaction occurs),

(2) two identical colors a and b, subtracted correspondingly from two identical mixtures of colors c and d will give two groups of components c − a and d − b of identical color,

(3) if a radiation unit a has the same color as the unit of another radiation b, every amount c of units of radiation a gives identical color, as amount d (if d = c) of units of radiation b.

In other words, arithmetic rules are applicable to colors mixing and, because we are not speaking about spectra evaluation, very many replacements may be done.

Practical rules of color matching

Following points in color evaluation may be considered:

(1) color as a mixture of colored light
(2) color as the illumination effect
(3) color as a result of mixing of transparent dyeing substances
(4) dyeing through coating with pigments and dyes
(5) other dyeing methods.

Only point 4 will be discussed in this chapter, as the others have less relevance to our subject.

Color properties of transparent substances may be easily foreseen and determined, based on their characteristic absorption curves and transmission. Their visual evaluation cannot be a base for determination of those properties. The group of nontransparent substances, having greatest practical use, is less exactly characterized, as these substances are to the least extent governed by the rules of spectrometry. From the viewpoint of the properties of colored substances the following systems are possible: particle suspended in a liquid may be colored and transparent, colored and nontransparent, uncolored and nontransparent, but suspended in a colored liquid.

Two kinds of nontransparent particles, used for dyestuff productions, may be used: pigments (colored solids) and lacquers, nontransparent solids adsorbing fine particles of transparent or nontransparent colored substances.

The degree of reflection on the phase boundaries, i.e., on the pigment particle surface, depends on the ratio of refractive indices at the pigment liquid boundary. If the difference between their values is great, a sharp, nonselective reflection occurs at the phase boundary. The light entering the particle penetrating its surface becomes selectively reflected, due to which a color appears. This color will be weakened, if the ratio of the refractive indices is great. Some liquid media used for color production change optical properties of films drying: in most cases their refractive index increases. The refractive index in particles is always greater than in liquids. Thus when it increases in liquid, the ratio will decrease (tending to unity), and the color will then be reinforced. An exception is water. It has a refractive index greater than air, so during evaporation of water the ratio of refractive indices increases and the color becomes weaker. In a case of white colors it is welcome, if the particles have possibly great nonselective reflection; it is the stronger, the greater the ratio of refractive indices.

Optical transmittance of a color depends significantly on the degree at which light is coming through the layer (film) and reaching the background, being at the same time the bottom surface of liquid. The amount of light transmitted by the color film depends on the size and concentration of its particles and on the thickness of the film. This transmittance still depends on the color of the background. Black background will absorb the remaining part of the radiation, which passed through the color film, whereas white background makes the nonselective reflection almost quantitative. If the background is colored, a selective differentiation occurs, as related to particular wavelengths, because a part of radiation that comes through the colored layer is reflected more intensely than the remaining one (comp. p. 419).

The value expressing the effect of background on color of the reflected light is called the covering power of a pigment. Conditions of its application have to be standardized in such a case, and it has to be taken into account that light reflected by white background comes twice through the dye layer, which makes color saturation much stronger.

Lower covering power and color saturation depend on the film thickness: the thicker it is, the more chances in absorption or reflection exist. However, certain differences are noted, e.g., when paint is applied onto white or black background. If the background is white, so the increase of the thickness of film of low dyestuff concentration layer may give greater saturation of color than if the concentration is greater. On black background the increase of film thickness gives an effect comparable to the effect of concentration rise (cf. Fig. 18.4). Very low covering power is typical of the pigments whose particles are transparent. Their selective action is based on transmission, rather than on reflection of light. This power slightly increases if the ratio of refractive indices between molecules and the liquid is increased. In pigments of very low covering power very essential is the effect of background. However, if the pigment layer is very thick, then even on the white background its brilliance will decrease due to strong absorption of light in the film itself, and the color may seem black.

When the nontransparent and white particles are suspended in a colored liquid, then the greater the difference between the refractive indices, or the greater the concentration of suspension in colored liquid, the more light becomes reflected. To obtain greater color saturation, the color can be applied onto the white background, rather than to increase pigment concentration in it, as thus the light path in the reflective absorbing layer will become extended, if, of course, such an extention is possible, and if it will not be replaced by extinction. When discussing the coloring effect one has to consider always the kind of surface, its smoothness, gloss, etc. Addition of particles of a white or black pigment to a given color changes its saturation and brightness of the mixture. Accordingly, the black or white particles have to change the impression given by the mixture. The color of particles added is not visible in mixtures of this kind. Similar effect will be achieved in water painting, when the color is diluted with more water. This is so because action of this color is to subtract a definite component from white light. Colored particles are then more dissipated, and the color of the pigment and its covering power decrease.

Addition of a white pigment to the given dye increases concentration of suspended particles in liquid. The covering power of this mixture increases, however this effect may quench the light of high color saturation, coming from deeper layers, thus decreasing total saturation. Increase of color saturation may, however, occur, if the primary color is very dark, as then the chances increase for reflection of light on pigment particle and on its way back to the observer.

In Fig. 18.23 in the CIE graph the mixing effect is shown for white pigment

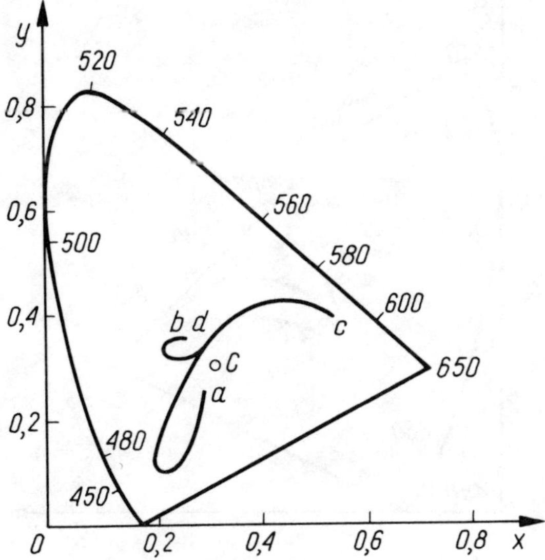

Fig. 18.23. Mixing (matching) of ultramarine (a) viridian green (b) and cadmium orange (c) with zinc white (d). C—white point.

with three dyes: a) blue, b) green, c) orange. White color of the pigment is marked d, close to the white light source (C).

When mixing very finely divided white pigment with a black one, a bluish coloring of the mixture may be observed. This effect can be explained in the same way, as the blue sky color: blue and violet components of white light are more dispersed than the others, as the degree of dispersion increases strongly with decreasing wavelength.

The effect of mixing of two colors depends on a number of factors like: absorption of light in all dyes, optical activity of the dye, light absorption in the dyed layer, transparency of both dyes, external surface of film, or kind and color of substrate. Sometimes it is possible to apply the absorption additivity law to such a mixture, but this is, however, not always possible. In Fig. 18.24 the matching effects of several known dyes are shown. None of the lines, however, is straight line. This means that the laws of spectrometry, generally well obeyed by solutions, are not directly applicable in a case of reflection on a colored surface.

18.2. Dyestuffs used for leather dyeing

Collagen contains functional groups of various properties. Part of them are bound in the tanning process, in which a choice may be made which of them are to be

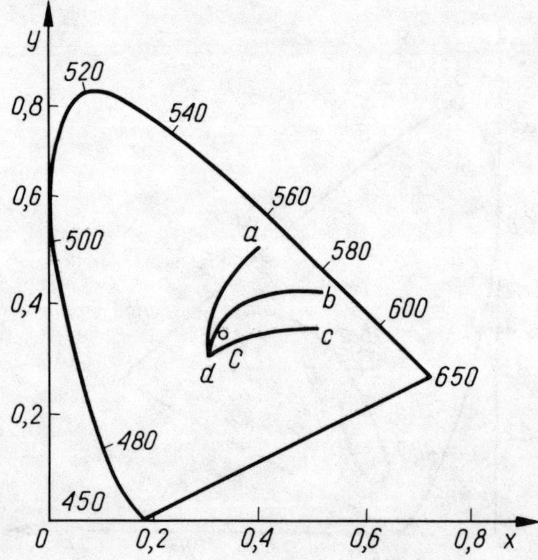

Fig. 18.24. Matching of viridian green (a) with cadmium orange (c) zinc yellow (b) with cadmium red (d) and of cadmium red with ultramarine (e), C—white point.

engaged in choosing the proper tanning method. However, it is a stochastic process and participation of functional groups in given kind of tanning may vary in particular parts of leather or in individual leather pieces.

Leather may as well contain tanning agents bound at one or many points, as well as other substances, which may possibly react with dyestuffs: in a vegetable leather these are: nontannins, i.e., sugars and phenols, and in mineral-tanned leather—complexes of chromium or other metals, as well as fats, surfactants, proteins and their destructs. There exist numerous analogies in behavior of collagen and other proteins like wool or silk. Among the specific features of leather dyeing as contrasted with textile fiber dyeing are:

(1) the necessity to dye textile fibers in bulk
(2) impossibility to use hot process for leather (boiling)
(3) own color of tanned leather, which makes leather different from raw fiber which is usually white.

Comparing, e.g., the content of acidic and basic groups, which may be determined by titration (Table 18.1) one may see that the number of these groups in collagen and in wool differs but slightly. Naturally, this value serves merely as general information.

More serious differences are due to amino acid composition of both proteins. A resultant of dissociation of functional groups at side chains is the surface potential. From the physical chemistry viewpoint it is analogous to the Donnan potential. The reaching of equilibrium equals to the Donnan equilibrium at the fiber surface, considered as membrane.

The isoelectric point of collagen lies in the range of 7.0-7.8 pH, that of chrome-tanned collagen ca 6.5 pH, of vegetable tanned one—ca 3.5, aldehyde, quinone and oil tanned—about 4.5 units. It is essential to chose dyestuffs so as to ensure opposite charges on both substances.

Modern dyestuffs technology and their intermediates are a highly developed branch of chemical industry. The knowledge of color properties of particular atom groups, of their ability to shift the color of other systems to red or blue is based on the knowledge of the behavior of the outer electron shells of atoms

Table 18.1.

Content of acidic and alkaline groups, as determined by titration in equivalents per 100 g of fibers

Fiber	Amount of equivalents	
	basic	acidic
Wool (keratin)	0.082	0.159
Silk (fibroin)	0.015	0.029
Collagen	0.094	0.124

and molecules. Dyestuffs used in leather dyeing in their great majority are aromatic organic compounds. Functional groups contained there are of chromophore character. These groups are shifting π-electrons in the rings thus forming some negatively or positively charged centers in an aromatic molecule.

Generally aromatic compounds are poorly, if at all, water-soluble. In this connection ionogenic groups are introduced into them such as: carboxylic, sulfonic, amino or hydroxylic groups.

Among the dyestuffs of industrial importance are distinguished the acid, basic and nonionic groups. Due to dissociation an acid dyestuff splits into a proton and an anion containing one or more chromophores. Basic dyestuffs will dissociate into chromophoric cation and an OH^- group. An example of an acid dyestuff may be fast red A

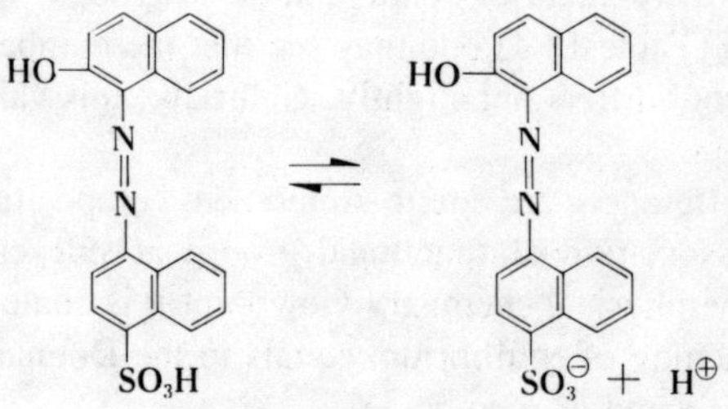

an example of a basic dye-chrysoidine

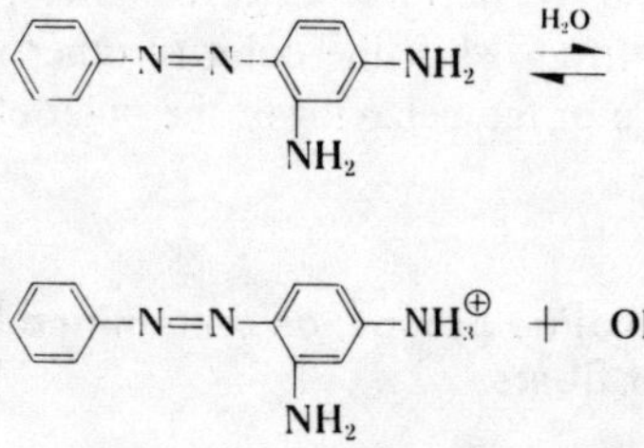

When the charges of leather and dyestuff after dissociation are equal, the dyestuff is bound very weakly, as only the secondary forces (van der Waals, induced dipoles) are participating in the bonding, and the dyestuff molecules easily penetrate into the leather. If the charges are different, the dyestuff will quickly and strongly be bound to collagen; its penetration is poor, however. This is utilized in surface dyeing.

Level dyeing with ionic dyestuffs is only possible if the differences between charges are not too great. Leather charge depends on the pH and on the isoelectric point (IP). If pH is lower than IP, leather is positively charged; if it is higher—the charge is negative. Capability for dyestuff binding by leather can be controlled by pH adjustment or by shifting the IP by retannage. Based on this, one can get information which dyestuff can be applied after particular tanning procedures. Chrome-tanned leather may be dyed with acid dyestuffs, whereas vegetable and syntan tanned leathers have greater affinity for the basic dyes. The bonds between a tanning agent and a dyestuff have to be taken into consideration as well.

Nonionic dyestuffs (e.g., azoic ones) are considered to be valuable models in the dyeing reaction as with their use it is possible to check which ionic system is necessary for obtaining colorings under definite conditions. The dyeing time is increased in the case of nonionic dyestuffs (ca. 15-20 times). Chrome-tanned leather binds nonionic dyestuffs much better than pelt, and the binding occurs at very broad pH range. Non-ionic dyestuffs are resistant to washing and abrasion. However, surfactants of glycols of polyether type do remove them, probably due to the formation of new hydrogen bonds. These dyestuffs may be totally washed out from pelt by acetone; in chrome-tanned leather, however, $\frac{1}{3}$ of them will remain. One may conclude that in the binding of dyestuffs to chromium complexes strong bonds participate; probably some parts of the dyestuff molecule are built in as ligands into the chromium complexes already bound to collagen. Complexes of other metals used as tanning agents (zirconium, aluminum) may have the same property. Behavior of a dyestuff in solution depends primarily on dissociation of its functional groups being responsible for its solubility. The overall dissociation constant can be estimated by potentiometric titration. This, however, is uncertain, as in this way it cannot be estimated, whether, e.g., it is only one sulfonic group that dissociates, giving the total proton pool, or also several carboxyls. Result of titration may be identical, and behavior of the substance entirely different.

The effect of individual substituents end groups on the pH has for long been known. Among the essential rules the following may be mentioned:
(1) Sulfonic group is strong dissociated, when attached to aromatic ring or in the presence of amino groups; its pK is 0.5;
(2) Carboxylic group is more dissociated when attached to benzene ring, than when linked to an aliphatic chain. Attaching of other groups to the ring changes the degree of dissociation of the group itself; direction of this change depends on the kind of group and on its position; nitro group in ortho position

increases dissociation the strongest. At this position it decreases the pK by the value of 2, at other positions—less. A second carboxylic group increases dissociation of the first one; methyl groups in meta and para positions and hydroxylic group in para position decrease the degree of dissociation of carboxylic group; amino group increases the pH. Groups: sulfonic, chloro and hydroxylic in ortho and meta positions do increase somewhat the dissociation of the carboxylic group;

(3) Hydroxylic group, which dissociates very weakly (pK = 10), is markedly affected by other substituents; nitro groups and halogens in phenols increase significantly the dissociation degree;

(4) Amino group dissociates to a small extent, other substituents do influence it like, e.g., hydroxyl.

Many dyestuffs in solutions tend to aggregate, forming systems of 10 or more molecules. This is a result of formation of hydrogen bonds or of dipole interactions between functional groups of the molecules. However, no correlation exists between the aggregate formation from dyestuff molecules and their ability to penetrate leather. Aggregating dyestuffs which penetrate leather very well despite their formation of colloidal solutions are known, as well as dyestuffs which form molecular solutions and do not penetrate leather due to a repellent action of their functional groups.

Two groups of dyestuffs of very specific way of action have gained in recent time importance for leather dyeing: metal complex and reactive dyestuffs.

Metal complex dyestuffs

In metal complex dyestuffs the use is made of ability of some metals to form coordinative compounds. Dyestuff residues are therefore introduced as ligands into complexes, having metal atom as a nucleus.

Chromium dyestuff complexes are stable, when at least four coordination valencies are saturated with dyestuff residues. An example of a dyestuff containing chromium atom is Eriochrome Black R, of the formula

In this compound two dyestuff molecules do saturate six coordinative chromium valencies. Less stable although much more reactive are the chromium-containing dyestuffs, having salicylic acid derivatives as their basic structural unit. Such a complex, containing still two water molecules as ligands, may crosslink collagen, e.g., compound

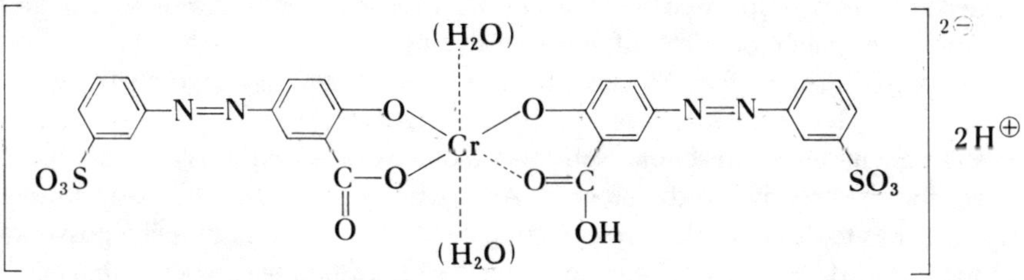

having charges positioned at both ends of a long molecule. It is stable only in weak acidic medium; at pH below 3 it decomposes.

Among the most stable metalcomplex dyestuffs are the copper complexes, e.g., phthalocyanins. In these compounds copper has coordination valency four. When as their ligands benzene rings occur, they may be very valuable anionic dyestuffs for leather, if sulfo groups are attached to the rings. When the metals have coordination number six and tridentate ligands, the complexes may be formed, containing one dyestuff molecule (1:1 metal complex dyestuffs) or two (1:2 dyestuffs). As a rule, the 1:1 dyestuffs are formed at pH 4, 1:2 at higher pH values. Formation and decomposition of dyestuffs occur according to general rules:

1) in acidic medium

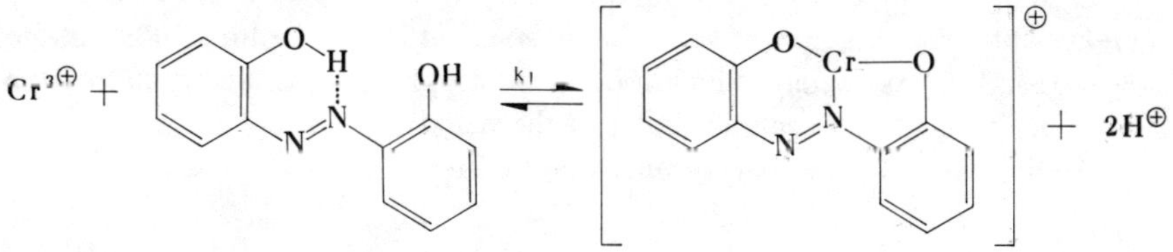

2) in weakly acidic or neutral medium

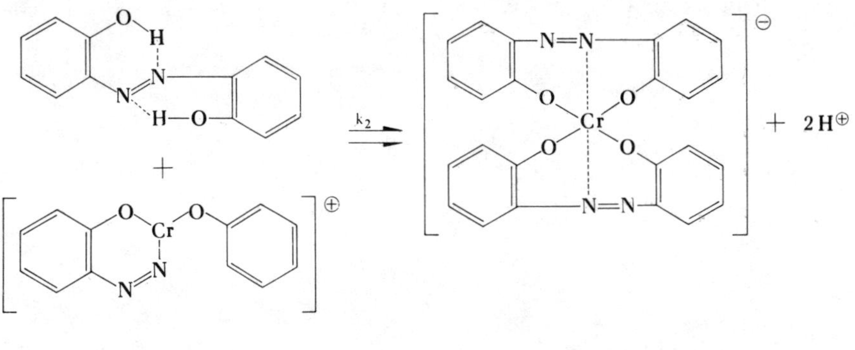

The already mentioned Eriochrome Black R and a vast number of industrial dyestuffs are in this group.

Reactive dyestuffs

Reactive dyestuffs form covalent bonds with the substrate as well as a metal complex. A major problem of reactive dyeing lies in a parallel run of two reactions: binding or forming covalent link to substrate, and a competitive reaction of the dyestuff hydrolysis. The hydrolyzed dyestuff formed is nonreactive in subsequent binding reaction. Control of the rate of competition is an important condition in carrying out the process. Also affinity of the tanning agent to the dyestuff has to be controlled under dyeing conditions, in order to make possible to the dyestuff to penetrate into fiber, and to be washed out, when unbonded. Binding between reactive dyestuff and fiber is particularly stable, which makes the resistance to washing out very high. Typical structure of reactive dyestuff is the following:

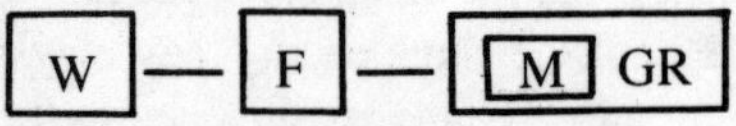

$$\boxed{W} - \boxed{F} - \boxed{\boxed{M}\ GR}$$

where GR is the reactive group, M—connecting (bridging) member, F— chromophore and W—is the group making the dyestuff water-soluble.

In the case of dyestuffs for leather or wool, their amino, hydroxyl, carboxyl or nucleophilic groups serve as the connecting members, whereas the center of the reactive group is of electrophilic character.

Reactive dyestuffs may be assembled as follows, according to their behavior:
(1) dyestuffs reacting according to the mechanism of nucleophilic substitution:
   —specific, base-catalyzed addition of nucleophilic functional group of the substrate to nucleophilic center of the reactive group;
   —elimination of the leaving nucleophilic group

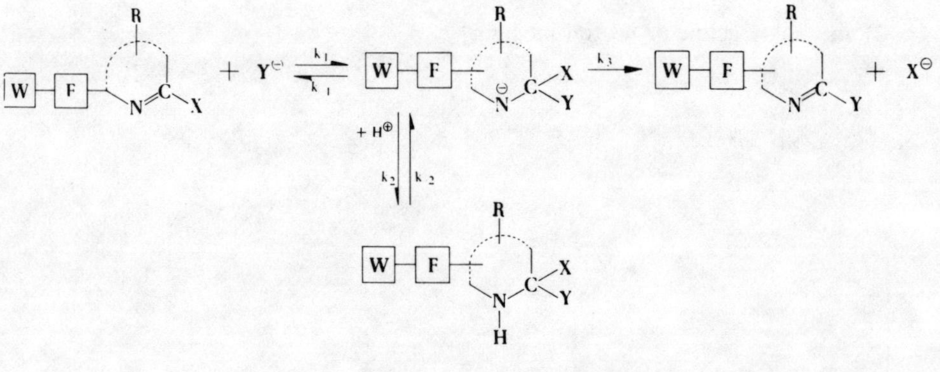

Y^- is a nucleophilic functional group of substrate. In commonly used dyestuffs frequently $X = Y = Cl$ or $X = Cl$, $Y = F$; sulfonic groups or quanternary ammonium bases are the leaving groups;

(2) dyestuffs reacting according to mechanism of nucleophilic addition. In this group frequently at first the reaction of elimination of nucleofuge or leaving group occurs, which is alkali-catalyzed, then the addition of the substrate functional group also alkali-catalyzed

$$Z-CH_2-CH_2-X \underset{k_{-1}}{\overset{k_1}{\rightleftharpoons}} -Z-CH=CH_2 \begin{array}{c} \overset{+HY \ \ k_2}{\underset{+H_2O \ \ k_{-2}}{\rightleftharpoons}} -Z-CH_2-CH_2-Y \\ \\ \overset{-H_2O \ \ k_{-3}}{\underset{+H_2O \ \ k_3}{\rightleftharpoons}} -Z-CH_2-CH_2-OH \end{array}$$

Z means the bridging member. HY is nucleophilic group of the substrate. A substantial difference between both reactions is that in the first case the substrate functional group participates in addition, as well as in elimination, whereas in the second one, the elimination step is independent of this group. Thus in the second group it is possible to come to optimal rate of formation of the reactive compound and to its diffusion by unchanged pH. This may be achieved by change in concentration and proper choice of the buffer;

(3) dyestuffs, reacting multistepwise by addition and elimination

$$-\underset{Z}{\overset{|}{C}}-\underset{X}{\overset{|}{C}}H- \overset{-HX}{\rightleftharpoons} \underset{Z}{\overset{|}{C}}=\underset{H}{\overset{|}{C}} \overset{+HY}{\rightleftharpoons} -\underset{Z}{\overset{|}{C}}-\underset{Y}{\overset{\overset{H}{|}}{C}}-H \overset{-HZ}{\rightleftharpoons}$$

$$\overset{-HZ}{\rightleftharpoons} \underset{Y}{\overset{|}{C}}=\overset{|}{C} + H_2O \longrightarrow -\underset{HO}{\overset{|}{C}}-\underset{Y}{\overset{|}{C}}H$$

HY means nucleophilic substrate group;

(4) Systems, acting as multifunctional components, connecting the dyestuff residue to substrate. The bridging member M defines reactivity of the system, its selectivity and the dyeing stability. Chromophore may have various chemical character.

In general to yellow, orange and red dyestuffs simple mono components are introduced, to violet, purple and blue—mono and diazocomponents containing copper, to the blues—antraquinone derivatives.

Influence of chromophore on dyestuff reactivity and on selectivity of its binding has been discussed by Rys [7]. Sulfonic groups are used in an amount of 1-4 per molecule of dyestuff as hydrophiles, making it possible to dissolve in water.

18.3. Chemistry of dyeing

A huge number of dyestuffs is currently available for leather, and still new brands increasingly appear. It is beyond the scope of this work to give to the reader the indications how to use or evaluate them. The author's intention is to give a generalized classification of leather dyes according to their mode of action and to say how their functional groups do affect their appearance or, more exactly, their visible spectrum. Apparently the most reasonable classification is that due to Griffiths [8]:
(1) n → π* chromogens
(2) donor-acceptor chromogens
(3) acyclic and cyclic polyene chromogens
(4) cyanine-type chromogens.
As chromogen a compound is defined which forms an unsaturated system that is either colored, or can become colored by attaching certain substituents. A chromophore is an unsaturated, colorless grouping of atoms.
(1) To the class of n → π* chromogens belong chemical grouping, in which n → π* absorption band lies in the visible region; compounds, having nitroso group are in this class (on n → π* transitions cf. sect. 7.2).
(2) Chromogens in this group contain an electron donor group linked to a conjugated π electron system. The site of the lone pair electrons must make it possible to be delocalized into the π-electron system. The visible absorption band then corresponds to this migration. There may be one or several atoms showing an increase of electron density, which are included in the π system, where electron density increases (complex system). Such an example is 1-aminoantraquinone which amino groups as donor, and the antraquinone residue as acceptor

This group of chromogens is the largest one.

(3) A molecule, having an alternating sequence of single and double bonds, forming a chain, ring, or both, may be regarded as a collection of sp^2-hybridized atoms. Complete overlap of all the p orbitals occurs then, giving a conjugated π electron system with as many electrons as there are p orbitals. An example of such systems are polyolefines $CH_3(CH=CH)_nCH_3$, or antraquinone.

(4) Alternant molecules containing an odd number of carbon atoms must have an odd number of molecular orbitals. The 'unpaired' electron lies on a 'nonbonding' orbital. If the terminal carbon atoms of such a system are replaced by heteroatoms, the electronic symmetry will not be significantly disturbed, and the system will retain certain properties of the hydrocarbon anion, among them a low-energy first electronic transition. This system was met first in cyanines; thus all colored systems of this type, containing the same π-electron system are called cyanine-type chromogens.

$$R_2N-(-CH=CH)_4CH=\overset{+}{N}R_2 \rightarrow R_2\overset{-}{N}=CH(CH=CH)_4-NR_2$$

The significance of particular substituents in the dyestuff systems is theoretically explained in terms of the LCAO (linear combination of atomic orbitals) method and perturbation theory with regard to the nonbonding molecular orbital. Starting with two positions of the nonbonding orbital, starred (active) and unstarred (inactive)*, the following rules (Dewar rules) will be mentioned, which may be helpful in the prediction of the dyestuff characteristics (Fig. 18.25)

(a) Increasing the electronegativity at an unstarred position gives a bathochromic shift;

(b) Decreasing the electronegativity at an unstarred position gives a hypsochromic shift;

(c) Increasing the electronegativity at a starred position gives a hypsochromic shift;

(d) Decreasing the electronegativity at a starred positions gives a bathochromic shift;

(e) Extending a conjugation with a neutral unsaturated group produces a bathochromic shift, irrespective of the point of attachment.

The last system (e), having the similar unformity of bonding to the odd alternant anions, is probably the best system for application of the rules. Also for other systems the Dewar rules work remarkably well. Based on these rules and knowing the chemical formulas of dyestuffs, it is easier to make a proper choice.

Considering the coloring process from the practical standpoint, it is to be kept in mind that except coloring with pigments, all other colorings are the diffusion

---

*Explanation of the conception of 'alternant,' and 'nonalternant' hydrocarbons and of 'starred' and 'nonstarred' atom, derived from Hückel's Molecular Orbital Theory the interested reader may find by Griffiths (ref. [10] ch. 7).

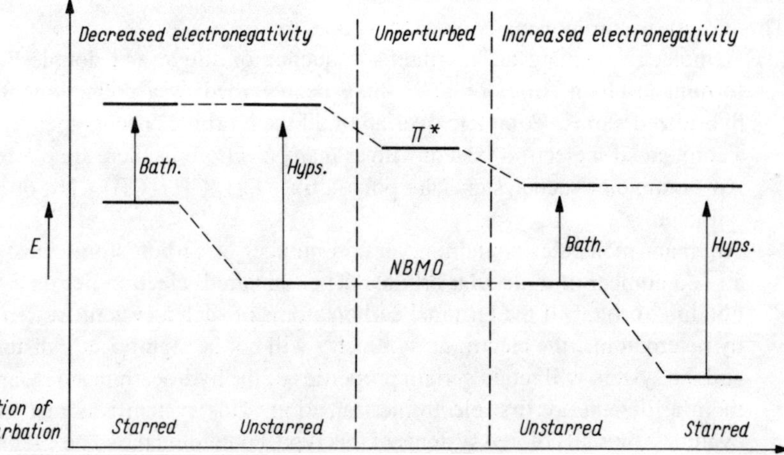

Fig. 18.25. The effect of replacing a carbon atom in an odd-alternant by an atom of greater or lower electronegativity on the orbitals and the energy of the first electronic transition. According to [10] Ch. 7. with permission

and sorption processes, most frequently accompanied by chemical reactions. Physicochemical approach to these processes on a basis of thermodynamics, quantum and statistical mechanics may be inaccurate. Models used in this sense are not exactly depicting the reality. They have half-quantitative sense only. The dyeing (coloring) process may also be defined as sorption which means partition between at least two phases. From the thermodynamic standpoint these processes are as irreversible as they are spontaneous, however they may be carried out in the opposite directions. Statistical dyeing reactions, i.e., equilibrium processes of sorption and desorption are reversible.

The dyeing process may be described by sorption isotherm. Parameters, influencing sorption, are not always determinable, as many processes have to be considered at the same time. Various kinds of bonds participate in it:

(1) electrostatic forces, which may be determined on a basis of the knowledge of charges of the dyestuff and the substrate molecule. Coulomb's attraction and other forces of this kind, e.g., dipole-dipole interaction, are decreasing proportionally to square distance of the centers of these charges; they are formally described by the Langmuir's isoterm;

(2) van der Waals forces, decreasing proportionally to the sixth power of the distance between centers of charges occurring when great amount of dyestuff molecules may be maximally approached to the centers of the substrate charges. Occurrence of these forces also depends on steric relations in the dyestuff-substrate system;

(3) hydrogen bridges discussed in sect. 2.2, as related to collagen—may be formed between certain atoms of the dyestuff and substrate;

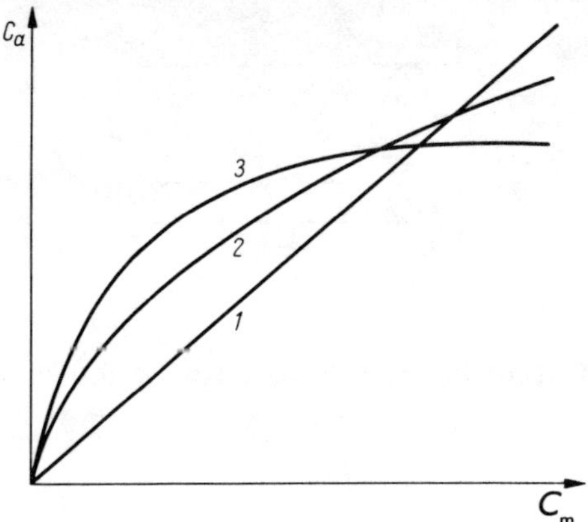

Fig. 18.26. Dyestuff distribution isotherms: 1—Nernst's isotherm, 2—Freundlich's isotherm, 3—Langmuir's isotherm, C_m—concentration in bath, C_a—concentration in fiber.

(4) hydrophobic bonds, formed between side chains of collagen and dyestuff. In their neighborhood water molecules become immobilized. Hydrophobic elements of the system tend to form greater aggregates. Distribution isoterms for dyestuffs based on experimental results have usually the shapes as shown in Fig. 18.26.

18.4. Dyeing kinetics

The curve that characterizes the passing of the dyestuff from the float to the substrate has several elements; it is mostly of a parabolic shape, in which the directrix is the y-axis. Usually it may be estimated only experimentally.

Dyeing may be described as based on the thermodynamics of irreversible processes. This description may not be used in interpretation of practical operations, because some quantities are hardly, or not, measurable. A rather clear model based on statistical mechanics, proposed by Rys-Zollinger [9], is of some use. This is shown in Fig. 18.27, where a substrate pore is depicted. Supposing the exchange is quick enough, i.e., mass transport in float is efficient, the dyeing process may include as a rule the following steps:

(1) diffusion (a) of substance 1 to substrate;
(2) immobilizing (b) of substance 1 (adsorption, or chemical reaction);
(3) mobilization (c) of substance 2 (e.g., desorption);
(4) diffusion (d) of the substance 2 from substrate.

If diffusion of moving substances 1 and 2 (steps 1 and 4) is considered to be a

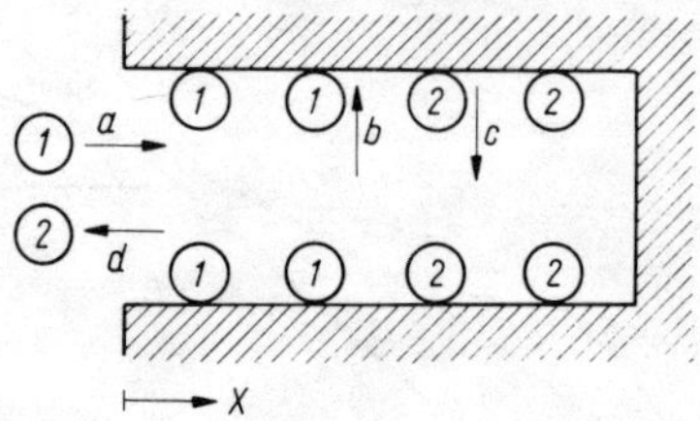

Fig. 18.27. Pore of fibrous substrate in dyeing process: 1, 2—the substances absorbed, where 1 displaces 2.

statistical motion (random walk), the rate is expressed by the 1st Fick's law

$$F = D\,\frac{dc_m}{dx_4}$$

where F is mass flow. It is proportional to concentration gradient $\dfrac{dc_m}{dx_4}$ of mobile molecules. Proportionality factor D (diffusion coefficient or diffusivity) is a measure of mobility of molecules in a medium: D is a property of the medium and takes values, which are greater, the higher is the average path of a diffusing molecule. The average path depends on total concentration of all molecules present in the medium. In diffusion processes in liquids D is constant, as concentration of liquid is usually constant. In Fig. 18.28 one may see D-values (or more exactly, D ranges) which may be expected for various media; e.g., in aqueous solution D takes the values of $10^{-6} - 10^{-5}$ cm²/s. In a homogeneous medium D is close to the real value—D_{real}. In a heterogenous medium one may initially presume that the dyestuff is diffusing to the pores, which are filled with water. In this case the effective volume of diffusivity is

$$D_e = D_{real}\,\frac{p}{b}$$

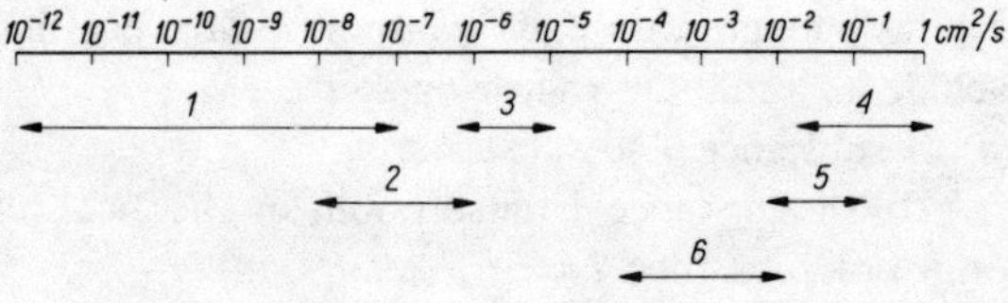

Fig. 18.28. Range of D_{real}: 1—solid, 2—solid electrolyte, 3—liquid, 4—free gas, 5—gas in large pores, 6—gas in small pores.

where p means porosity and b, assumed in textile industry as $= \sqrt{3}$ is a pores tortuosity factor.

Experimentally the diffusivity may be determined in two ways: by the method of stationary or nonstationary conditions. Now we will discuss the former: the latter is much more difficult from theoretical standpoint, and may give some problems in carrying out the measurements. It bases on the 2nd Fick's law. The first method based on the 1st Fick's law assumes that the substrate is a membrane. Principle of calculation is shown in Fig. 18.29. Taking into consideration that the concentration of the dyestuff in liquid c^{σ} is proportional to the concentration in pores c^{ϕ}

$$c^{\phi} = nc^{\sigma}$$

dc_m in the 1st Fick's law equation may be replaced by $n(c_1\text{-}c_2)$, and dx, by the membrane thickness L. Knowing n and P—D_{real} can be calculated from the formula for the mass flow. Usually one takes n $= 1$. In fact one has still to consider adsorption in the membrane, which usually gives for dc_m a value $n[c_1^{\sigma} - (c^{\phi}\text{-}c_2^{\sigma})]$. If we take leather as a layer, it will be $c_2^{\sigma} = 0$.

The processes shown in Fig. 18.27 mutually affect each other. Substance 1 increases the rate of transport of substance 2, however if its real diffusivity is greater than that of substance 2. This acceleration in turn slows down the transport rate of substance 1. In the first approach this mutual influence may be neglected, if molecular weight of the dyestuff does not exceed 800. However, the influence of processes of immobilization and mobilization on diffusion rate may not be neglected as it is visible from the formula

$$D_{real} = \frac{kT}{6\pi\eta\tau}$$

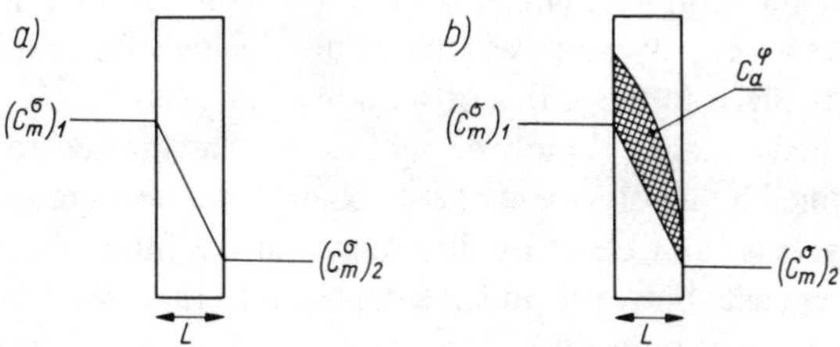

Fig. 18.29. Diffusion measurement by the stationary (membrane) method: (a) in the absence of adsorption, (b) with adsorption, C_m—concentration of dyestuff in float, C_a—concentration of adsorbed dyestuff, L—membrane thickness. According to [9]

where k is the Boltzmann constant, η—viscosity of the medium, τ—radius of dyestuff molecule, proportional to $\sqrt[3]{MW}$ (molecular weight). This is true if substance 1 has greater diffusivity than substance 2.

The rate at which the equilibrium is reached affects immobilization of dyestuff on the surface. Frequently adsorption may be complicated by the use of two dyestuffs and their immobilization due to chemical reaction. Another complication is sorption (or reaction) inside the fiber, rather than on its surface. As a rule, the processes are to be considered for a given case that during the process the temperature, pH and float concentration are changing. Specific factor, significant for the dyeing process is the tanning agent previously used. In dependence on this quite different color effects are obtained, independent on color addition rules, e.g., due to the color of vegetable tannins.

The pK values of typical vegetable tannins are 8-9, those of syntans are alike. These tanning agents have various nonionic groups which may participate in the binding of dyestuffs. Accordingly, multiple weak interactions can be expected. In fact such interactions are observed. If both agents contain ionic groups, their repelling or attraction depends on their charges. Aggregates of opposite-charge molecules are precipitating as it occurs in a case of a vegetable tannin and a cationic dyestuff. Such aggregates still in solution attach well to leather although they give unlevel colors which are dull and nonresistant to rubbing. However, using vegetable tannins and dye extracts (logwood, fiset wood) the old masters could achieve very nice and durable effects.

Strong and various interactions occur between mineral tanning agents and dyestuffs. Due to these interactions compounds of metal complex dyestuff are formed—compounds of this type have been found in leather. Their formation is a reverse function of dissociation: the weaker is the dissociation of the acid residue, the more stable are the compounds with anionic dyestuff residues. Adsorption of dyestuff by chromium salts goes through chromium coordinative bonds. Many salts are known to mask chromium complexes which expel dyestuffs from their complexes with chromium. It is also known, however, that, e.g., formate, nitrophthalate and citrate residues do not replace Brown A or Pluton black BL in chromium complexes. In turn Pluton black BL may replace Brown A in its chromium complex. This leads to the conclusion that in order to know how effective an ionic dyestuff will be in the dyeing of chromium salts—one has to know its place in the series of masking ligands.

Chrome-tanned collagen is more easily colored than the vegetable-tanned one. It can be explained by involvement of side-chain functional groups in the reaction with tanning agents, as well as by differences in the filling out of interfibrillar spaces. In either case, however, independent of the argumentation it is necessary to adopt technological procedure. Measurement of isoelectric point in a leather batch may give the best information about applying a given dyestuff. The most accepted method is the potentiometric titration. In tanning literature a concept of surface charge is widely recognized. It is incorrect, as it concerns the charge, which is formed on functional groups of the collagen side chains when concentration of hydrogen ion changes in solution in which collagen is immersed. Obviously, decrease of the pH decreases dissociation of carboxyls and appearing of positive charge and vice versa—the pH increase decreases dissociation of amino groups and appearance of a positive charge. A conception of surface charge in solution (as resultant of particular charges at single points on the surface of a membrane) occurs in biochemistry. Its value, however, depends on pH and ionic strength of the solvent used: it decreases when ionic strength increases and/or when pH of the solution approaches the isoelectric point of the membrane. It may be measured by electrophoresis under strongly constant conditions. Such a measurement is hardly possible on leather during its processing as the results would depend on above mentioned conditions, not on leather itself (cf. p. 468). Its numerical value is questionable. Measurements were done more than 30 years ago using the techniques, which are no longer applied [10]. Later on Gustavson [11] treated these values more marginally, trying in the quoted paper to establish collagen isoelectric point in dependence on conditions and tanning agents used. Practical sense of 'leather surface charge' may be reduced to a statement that, a tested sample of leather (or pelt) has a pH lower or higher than its isoelectric point. This information may be needed for predicting, if a dyestuff or impregnant which has to be applied will be bound strong enough. It may be obtained in about three ways: by applying a drop of dyestuff solution onto leather it may change its color by the effect of positive or negative charge, electrophoretically or potentiometrically. This latter method means to find the isoelectric point, e.g., by titration. Then it has to be compared with the pH value of a 0.1m KCl solution added to sample. In the practicisers' opinion the value of 'surface charge,' which consists of ionic charge and of coordinational interactions, gives more valuable

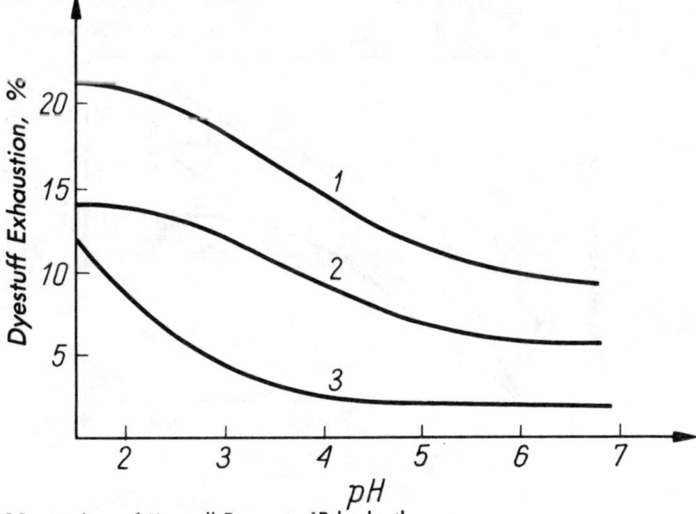

Fig. 18.30. Binding of Krystall Ponceau 6R by leather.

information about leather properties than the isoelectric point. Thus without neglecting this index one has to just remember that this definition is incorrect, suggesting, e.g., the possibility of leading it out by an electric cable, like the electrostatic charge. In fact this 'surface charge' shall be rather called 'internal charge' of leather. In Fig. 18.30 dyeing of chrome-tanned leather (1) collagen (2) and binding of dyestuff by $Cr(OH)_3$ (3) has been shown. (dyestuff Crystall-ponceau).

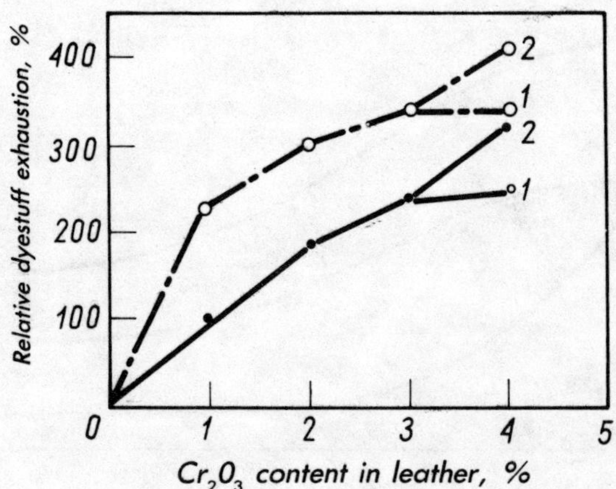

From this Figure it can be seen that the curve of binding of dyestuff by chrome-tanned leather is roughly a sum of both the remaining curves. In Fig. 18.31 binding of Diaminblack BH by chrome-tanned leather is roughly a sum of both the remaining curves. In Fig. 18.31 binding of Diaminblack BH by chrome-tanned, aluminum-retanned leather is shown. When tanning is done several times, with neutralization after every time (broken line) the amount of dyestuff bound will be increased. Aluminum retannage increases very much the binding ability of an anionic dyestuff.

Drying of leather between the tanning and dyeing also affects the dyestuff binding: chrome-tanned leather, dried after tanning, binds less dyestuff (Fig. 18.32) and then the dependence of the amount of dyestuff bound on the pH of

Fig. 18.31. Dyestuff binding by leather: 1—chrome tanned, 2—aluminum retanned.

float will be still more pronounced. From reactions shown it is accepted that binding of dyestuff by leather is a resultant of many forces and bonds.

Summing up the information concerning leather dyeing, a scheme of the reactions carried out can be set up.

The isoelectric point of leather, freshly chrome-tanned with basic sulfates, lies at pH = 7. This is why a fresh chrome-tanned leather is well dyed by anionic dyestuffs. The pH of such leather is about 3-4. It is a value very distant from the isoelectric point. The greater the difference is, however, the higher is the positive collagen charge, and the quicker and more vigorous is then the binding of anionic dyestuffs. For many tanning agents it is even too quick, owing to whether the dyeing is unlevel, and the leather has to be neutralized beforehand.

Control of the rate of the dyeing process is very important. Decreasing the rate of dyestuff attachment by decreasing of the difference between pH and the isoelectric point of leather, one will have at the same time a weaker binding of dyestuff, and also a better penetration. Reverse carrying out of the process is also possible, if one intends very shallow, highly intense dyeing. Experimentally, very many regularities have been found in the drum-dyeing of leather. The typical ones are the following:

(1) The better the dyestuffs penetrate into leather, the smaller is the float ratio; in floats of higher ratio they bound intensely to leather surface;

(2) A direct dependence exists between the binding rate of the dyestuff and float

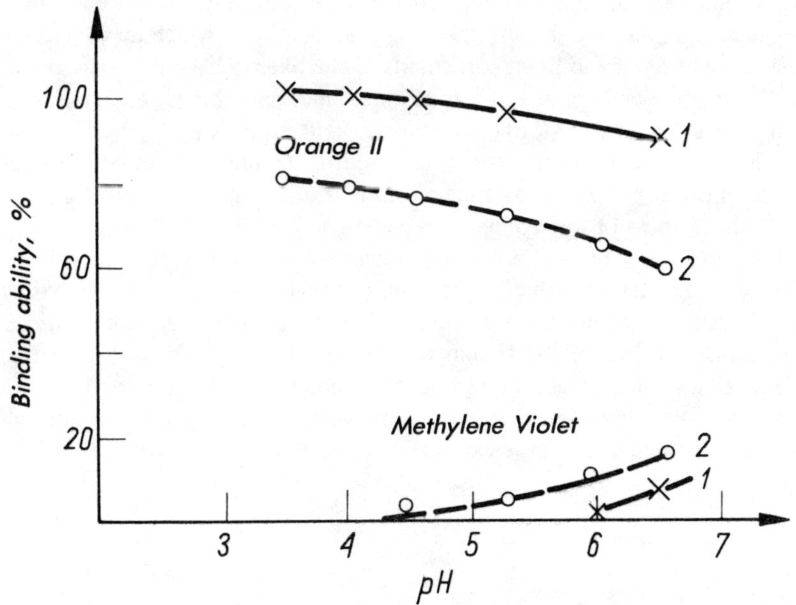

Fig. 18.32. Binding of dyestuff by: 1—freshly tanned leather, 2—leather dried after tanning.

temperature. Slow dyeing, however, is more level and deeper, whereas dyestuff will be bound more weakly;

(3) Dyestuff penetrates leather easier, the greater its concentration;

(4) Anionic dyestuffs of smaller molecule do penetrate more easily and more deeply, but they form weaker bonds;

(5) The stronger the tanning float is masked, and the longer the leather is stored before dyeing, the easier but at the same time the less strongly it will be dyed.

18.5. Light fastness of dyeing

A review of the problems connected with fastness of leather dyeings to light has been given by Heidemann [12]. These problems are of special importance, because of the requirements which contemporary leather for shoes and clothing has to meet. A starting point for the technique has to be dyestuff characteristics, as given by the producers. It is known that metal complex dyestuffs are of special fastness to light. Generally they have a rather complicated and great molecule, so they are more difficult to synthesize and thereby are more expensive.

How to have the best light fastness is not only a problem of the use of proper dyestuffs. It lies also with substances introduced into leather before dyeing, e.g., in syntans or in vegetable tannins, in which the quinoid groups favor oxidation of dyestuffs. It is known, e.g., that vegetable pretanned East India goats have colors not fast enough to light. This can be prevented by washing out the vegetable tannins, and then by choosing tanning agents, which may increase the resistance of dyeing to light, particularly if one likes to have pale, pastel colors. Beside tanning and tanning agents of importance are the fats, e.g., use of halogen derivatives of hydrocarbons increases light fastness of dyeing.

Also the masking agents which are used for tanning should be considered. Some of them may decompose the metal-dyestuff complexes, and thus decrease the light fastness of the dyeing with metal complex dyestuffs.

As a rule, the lightening (bleaching, fading) of nonproteinous fibers is believed to be due to oxidation, whereas proteinous fibers do lose their color by oxidation or reduction. Whether the bleaching occurs by oxidation or reduction can be found on the basis of the Hammett equation. This equation is a quantitative description of the effect of substituents introduced on the reactivity of groups, already present in the ring by making use of constants, specific for the substituents. Hammett equation is expressed by the formula:

$$\log \frac{K}{K_0} = \rho\sigma$$

or

$$\log K - \log K_0 = \rho\sigma$$

where K is constant of the reaction rate of a compound with substituted functional group, K_0 is a constant, when substituent X = H, σ—is the constant of the substituent and ρ defines the sensitivity of the reaction in question on changes in the electron density around the reacting atom. For a given reaction also ρ is a constant. For decomposition of a dyestuff the Hammett equation may be simplified into the form:

$$\log K_r = f(\sigma)$$

Depicting this dependence in a graph with coordinates log K and σ a straight line is obtained whose slope is equal to ρ. In oxidation reactions ρ is negative, in the reduction—positive. This is connected with the values of changes in density of electron clouds at single centers of the ring. This value may be positive or negative. If the decomposition of a dyestuff is a result of oxidation, it is always accompanied by fiber deterioration, which in turn decreases their tensile strength. Then probably the transfer of electrons from the dyestuff to substrate occurs. These residues are for sure participating in the reaction.

The above considerations give the reason for the use of various dyestuffs for nonprotein and protein fibers. In the case of nonprotein—e.g., cellulose fibers, i.e., fibers where dyeing agents are decomposed by oxidation, it is recommended to use dyestuffs having substituents in meta- and para positions with a very high value of the Hammett constant. The reverse situation is in a case of wool or leather. A particular sense in organic dyestuffs has the $n \rightarrow \pi^*$ transition excited by photon in groups N=O or C=O. In the carbonyl group in this case the bond is split in the following way:

$$\underset{2p}{\overset{2s}{>}C=O} \quad \xrightarrow{h\nu\,(n\cdot\pi)} \quad >\overset{*}{C}-\overset{2s}{O}{}_{*}$$

This photochemical reaction of the carboxyl groups is of special importance for dyestuffs. This chromophore influences in an essential way the electronic spectrum of a compound.

The $n \rightarrow \pi^*$ transition for the carbonyl group structure means a decrease of density of electrons on the oxygen atom. Oxygen is in this case more electrophilic than in the ground state, which makes the electron delocalization easier.

An example of group of compounds, exactly tested on their sensibility to light, are the aryl diammonium salts. By substituting in a compound:

$$\underset{R''}{\overset{R'}{>}}N-\!\!\left\langle\!\!\bigcirc\!\!\right\rangle\!\!-N_2^{\oplus} \quad Cl^{\ominus}\cdot\ 0,5\ ZnCl_2$$

for R' and R'' various groups (-H, -CH$_3$, $-C_2H_5$, phenyl) various degrees of fastness are obtained which may be estimated by testing of absorption of a compound in a range of 300-400 nm. The same amount of light gives to a salt substituted with one ethyl group a fourfold greater effect, than to the one substituted with two phenyls. In such a case delocalization of charges will occur, by occurrence or otherwise of a positive charge on the para carbon.

Light fastness of dyeing may be still influenced by other factors, e.g., humidity, particle size and composition of the atmosphere.

Influence of humidity is, as a rule, discussed together with that of temperature. Temperature increase by 10°C decreases fastness on average by 10%. Influence of atmospheric humidity is less clear; it depends on the kind of substrate and of dyestuff.

It should also be remembered that the capability of dyestuff to oxidation depends on the wavelength of the radiation absorbed. It is known that this capability is different in various dyestuffs and at various wavelengths, as it results from the laws of quantum mechanics.

Influence of the atmosphere does not need much explanation. It is obvious that the greater is the concentration of ozone, sulfur dioxide, nitrogen oxides and other aggressive products of the human activity in it, the smaller is the resistance not only of dyeing, but of the leather itself. Kinetics of the color change by slow light absorption is believed to be a reaction of the 1st order, as the absorption rate increases, the reaction goes to the zeroth order. Thus actually bleaching is a reaction of kinetics oscillating between these two extremal cases. It may be checked by drawing a bleaching curve, as time logarithm vs. concentration logarithm. They are known in literature as characteristic fading order (CFO). The curve of the reaction of the 1st order is a straight line, parallel to the x-axis the curve of reaction of 0-order-a straight line, sloped to x-axis. Determination of the CFO curves also gives information concerning the grinding of a dyestuff. Light fastness increases with particle size of the dyestuff [13]. The dyestuffs are not distributing themselves equally on fibers, rather they form clusters. This may be due to intermolecular interaction. This way of distribution is confirmed by the use of methods of X-ray diffractometry and of electron microscopy.

REFERENCES

1. see ref. 1 ch. 12
2. Kubelka, P., Munk, F. Z. Techn. Physik *12*, 593 (1931)
3. Kortüm, G., Schreyer, Z. Z. Angew. Chem., *75*, 658 (1955)
4. Rózycka, D. Prace IPS *15*, 61 1971 in Polish
5. Troy, D. J. SATRA Conference Blackpool 1971
6. MacAdam, D. L. J. Opt. Soc. Am., *33*, 675 (1943)
7. Rys, P. Textilveredlung *2*, 95 (1967)
8. see ref. 10 ch. 7
9. Rys, P., Zollinger, H. Leitfaden der Farbstoffchemie, Chemie-Verl. Weinheim 1970
10. Neale, S. M., Peters, L. Trans. Faraday Soc., *42*, 478 (1946)
11. Gustavson, K. H. J. Am. Leath. Chem. Assoc., *47*, 425 (1952)
12. Heidemann, E. Leder *23*, 253 (1972)
13. Baxter, G., Giles, C. M., McKee, M. M., McDanlay, M. J. Soc. Dyers Colourists *71*, 218 (1965)

19.

FATTING OF LEATHER

Degreasing and bating of hides and skins remove from them almost all their native fatty substances; in fatty skins, like, e.g., sheep, goat or pig, further degreasing is necessary for technological reasons. A secondary result of this is that on drying of the skin it becomes a stiffer, horny substance. The aim of the fat introduced into the leather later on is to assure it an appropriate friction coefficient between its fibers. If the Batzer's theory of tanning would be accepted, the fatting has to be considered as its integral part. A secondary problem, although important as well, is to impart more water-resistance to leather, and to ensure that fat remains in it during its use.

Fats may be attached to leather chiefly by ionic or polar bonds. They may be introduced into leather by four different ways, viz.

(1) directly, in the dry process, in solid or liquid form (e.g., oil tannage, hot stuffing);
(2) In the form of water-in-oil emulsion (see below);
(3) In the form of oil-in-water emulsion;
(4) In solution (impregnation).

The most used methods are (2) and (3); impregnation is considered as a separate process (Ch. 20).

Distribution of fat in leather

Proper distribution of fat in the leather layers is of importance, as from simple bending of a sample one may already conclude that these are differences and changes in leather shape, external layers have to shrink and extend, just the central one remains in this experiment unchanged. But from this only one may see the necessity of stronger fatting of external layers. This is from the beginning influenced by the thickness of the fibers from grain to the flesh side. The problem becomes more complicated if one considers unequal thickness of leather in various points of its topography; e.g., leather properly fatted in bellies may be unsatisfactorily fatted in shoulder and butt, and so it may crack in more stiff places. Overfatting is avoided, as a grease stain may arise, as well as fatty spews, which are exudations of crystals of fatty acids or of glycerides, occurring when fatty acids content in external layers exceeds ca. 10%.

Stratigraphic distribution of fats in leather is influenced by the kind and properties of fat and by the way it is introduced. General rules may give in this case only approximate orientation. For example in Fig. 19.1 the curves for distribution of fat in particular layers are shown. A typical example of commonly used fat was depicted there [1].

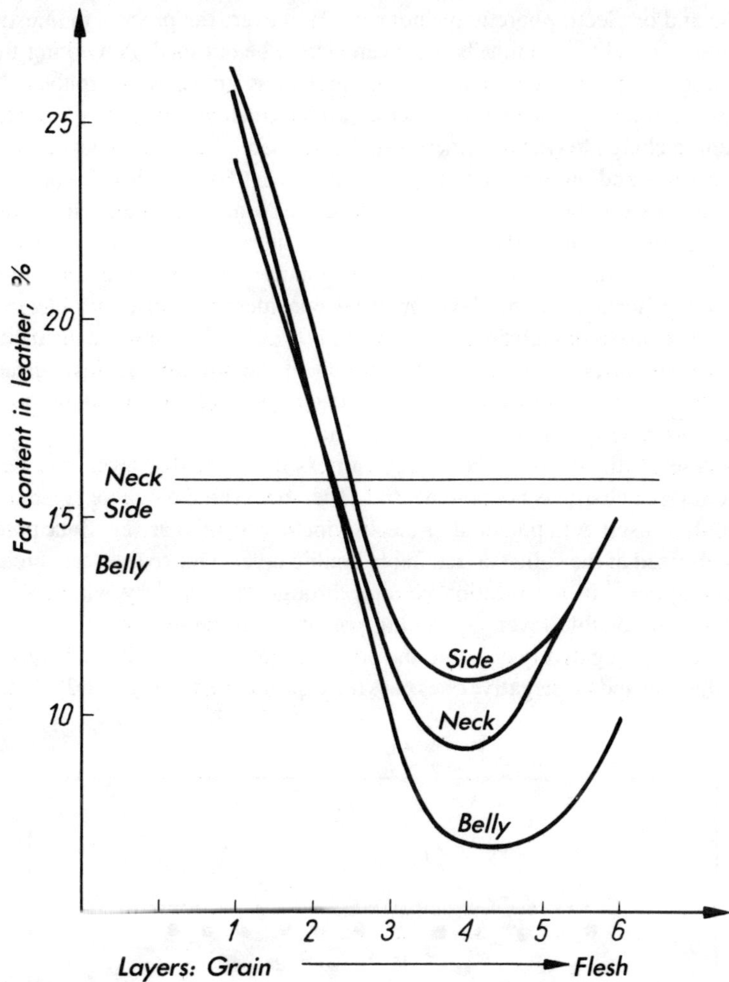

Fig. 19.1. Penetration of fat into cattle hide. Extraction with hexane.

Data much like those may be found in other papers, e.g., given by Stather [2] or more recently—in a very extensive and detailed study of Poré [3], where more examples may be found.

A question arises about the ways the fatliquoring emulsion penetrates inside the leather and how the fatliquoring operation is influenced by the preceding operations, as well as by the choice of fats and surfactants used, temperature, drumming speed, etc. It was concluded rightly [4] that zeta (ζ) potential may be the controlling factor. When this point of view can be presumed, the above questions could be answered by regarding the fatliquoring process as an electro-osmotic and/or electrophoretic phenomena. However, the process in question is very complex and for the time being it can merely be outlined. A working theory correlating practical observations on fatliquoring is given by Sharphouse [5].

On the surface of a protein or other colloidal molecule or particle situated in a solvent, a charged layer is formed, usually defined as the double layer, because of the cations and anions (or in the reversed order) fixed in it (cf. ch. 5). The 1ixed ion sites on solid extend into liquid as an 'ion atmosphere' in which the ions of opposite sign predominate. At greater distances from the solid the ion atmosphere becomes less dense and the net charge per unit value tends to zero.

The outer surface of this layer may be considered as discrete, electrically neutral structure (although diffused). It may, however, change when disturbances in the surroundings occur due to the motion of the solution against surface or due to the external electrostatic field. Collagen molecule is considered as such a surface as having a double layer around it.

In a case of disturbances the charge carriers in the double layer may be split off; the surface charge is no more neutral. Then between the double layer surface and solution itself zeta potential or electrokinetic potential arises. Zeta potential is immobilized at the solid surface in the double layer. Due to this, the streaming potential appears in the solution passing through the capillary wide enough in relation to the double layer. Streaming potential can be measured.

If the wall is negatively charged, the positive ions present in the solution move in its direction and the negative ones pass the capillary faster (Fig. 19.2). Potential

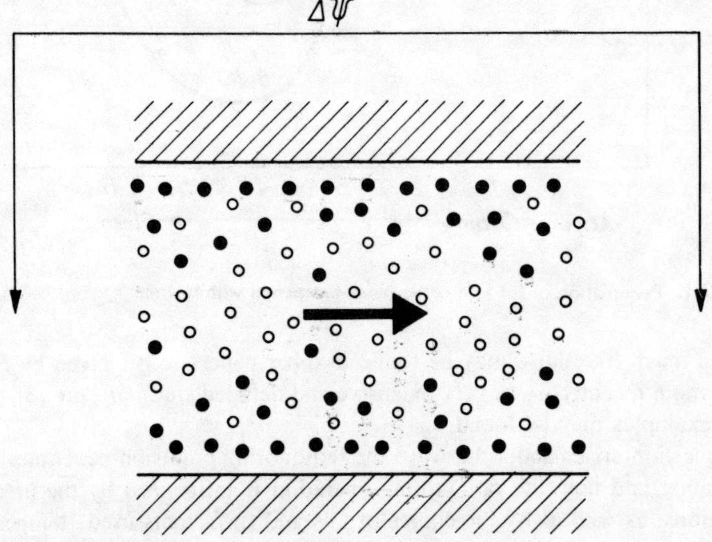

Fig. 19.2. Passing of colloidal, charged particles of an emulsion along the capillary. Potential difference between capillary ends equals Δψ.

differences arise between the capillary ends and the aqueous solution traveling along.

A different kind of potential—diffusion potential—arises when we have a solution of ions with colloidal particles suspended in it. Its source is unequal solubility of the organic and inorganic ions in the oil phase of the emulsion. Diffusion potential remains at equilibrium between the oil and water phase and is governed by the surfactants, if added. The zeta potential thus refers to the potential difference across the plane of motion between phases and has no necessary relation to the total potential between the interiors of the phases. Just as stated above, on the surface of protein exposed to water a fixed layer of water molecules would be embedded, in which there are ion sites. Some of the sites would be beneath the fixed water layer.

When the ion sites are located in cracks and crevices on the surface of a protein molecule or a particle—part of the ion double layer associated with the hidden sites would move along with the mobile particle. Thus the mobility of the protein molecule depends on the ionic strength in the electrophoretic process. When the solution, *not* the protein particle, is moving—electro-osmotic process takes place. The zeta potential cannot be measured and it is not quite strictly defined, as the double layer is not exactly limited. However, it is of value in the discussion of the electrophoretic theory, in the theory of streaming potential and of electro-osmosis as well. The theoretical values of zeta potential for uni-univalent electrolytes are about 50 mV/25°C. For more exact physical treatment of electrokinetic potential biophysical chemistry handbooks should be consulted [e.g., 6].

The situation is extremely complicated when we realize that neither tanned collagen nor the emulsion has unified charges (all negative or all positive). Certain indications, however, without referring them to the potential charges present are given in the paper of Poré [7] where the author was able to show the differences in penetration of various fat components into leather, working on standardized leather and emulsion samples which made the experiment easier to interpret. Perhaps the most consistent theory of the potential across the fixed double layer, as well as of the diffuse double layer was given by Stern [8]. His theory is of limited value as it is generally impossible to calculate the potential of a fixed ion layer. However, the conception of zeta potential can be related to stability of suspensions, the electroviscous effect and ion binding. The electro-osmosis phenomenon, no doubt participating in fat-into-leather penetration is considered to be a reason for water flow across the biological membranes. The conclusions from these biophysical considerations, maybe, extend Strakhov's point of view on the significance of zeta potential [4] and may throw some light on the fatliquoring operation. The smaller the potential difference between the double layer on collagen fibers and particles of fatliquoring emulsion, the deeper this emulsion penetrates into leather. Using anionic emulsion immediately after tanning with cationic chromium complexes a quick binding of fats is observed—however, in the surface leather layers only.

Neutralization of leather after tannage decreases its acidity and positive charge, which results in a deeper penetration and more equilibrated fat distribution across the leather. The degree of neutralization is of importance.

In order to have deeper penetration of fat it is recommended to fatliquor the leather near to its isoelectric point. However, this point depends on the operations done after tanning, particularly on retanning and it may be shifted towards lower or higher pH values. Vegetable or syntan retanning as well as dyeing with anionic dyestuffs give anionic character to the leather fibers resulting in deep penetration of anionic fatliquoring emulsions. Although we do not know much about reactions between vegetable tannins and fats, certain interaction between sulfato groups of syntans and fatliquors can be expected.

The factors influencing the result of fatliquoring are: fat and surfactant character, drumming intensity, temperature, horsing and drying. Drumming and heating influence the penetration speed for the above-discussed reasons, forced flow, mobility of water molecules due to heating and changes in pore width, thus the distances between double layers. Drying process changes the distances in pores (see ch. 22) by water removing and the equilibrium in double layers due to ionic strength increase. No exact data based on calculation of potentials and other factors in question have been available to the author.

Kinds and properties of fats applied in tannery

The number of kinds of fats, which are used in leather industry, is vast and the possibilities of combinations unlimited. Not only fats, i.e., acylglycerides or esters of glycerol and fatty acids, may be classed as substances, having ability of impregnation and of action as interfibrillar greasing factor, but also soaps, waxes, halogen derivatives and sulfo derivatives of fatty acids, amines, hydrocarbons, silicones, etc. Commercially today fats have almost completely come out of use *in substantia*, formerly used for hot stuffing.

Now for introduction of fats saturation with emulsions is done. From the electrochemical viewpoint the emulsions may be anionic, cationic or nonionic. This depends on the dissociating group contained in the fat used, and always has to be defined, and as a result the method of using emulsion depends strictly on its negative or positive charge. The knowledge of emulsions is a domain of the physical chemistry of macromolecular systems. The fat should be introduced into leather in a way permitting a hydrophilic part of the fat molecule (e.g., carboxyl or sulfo group) to be bound to collagen, while its hydrophobic part would stand out. This makes the collagen fibers water-repellent and decreases the value of internal friction coefficient.

Emulsions as disperse systems; emulsifiers

In emulsions two phases are distinguished: dispersed and dispersing ones. Usually the particles of the dispersed phase have a spherical shape, and a diameter from 0.5 to a few microns. Depending on which component is the dispersed phase, and which the dispersing one, we differentiate between oil-in-water (O/W) and water-in-oil (W/O) type emulsions. Low value of surface tension on the phase boundary makes the emulsion formation easier. Emulsifier or a surface active substance is used to decrease the electric charge in the double layer on the phase boundary. This layer controls the emulsion stability, making particle coalescence more difficult. This is responsible for whether the emulsion is of the O/W or W/O type.

Emulsifiers may be divided into four classes:

(1) Anionic, in which the active group is negatively charged;

(2) Cationic, having positively charged active groupings;

(3) Amphoteric, having both kinds of active groups; their charge depends on the pH of the medium;

(4) Nonionic, which are nondissociating (containing hydroxylic or ethoxylic groups).

The emulsifier may be characterized by determining its HLB (hydrophile-lipophile balance). This value can be calculated by several ways from semiempirical formulas, like, e.g., for polyhydroxylic esters of fatty acids from the formula

$$\text{HLB} = 20 \left(1 - \frac{S}{A} \right)$$

where S is its saponification value, and A—the acidity of the residue, or from formula.

$$\text{HLB} = 0.42Q + 7.5$$

where Q is the hydration heat. This formula is valid for liquid hydrophilic surfactants.

For esters of fatty acids and polyalcohols $\text{HLB} = \dfrac{E + P}{5}$ where E is weight content of ethylene oxide, P—of alcohol (in %).

Influence of HLB values on emulsifier behavior has been shown in Table 19.1.

A simple way to check HLB is the cloud point. This is a temperature at which 5% solution of water in fat shows a clear change in its cloudiness. In Fig. 19.3 is shown the dependence of HLB on cloud point according to Sherman [8] for nonionic systems. The process of turning of one kind of emulsion into another is called inversion of the emulsion. Layer of the dispersing phase around the drop consists of immobilized molecules. Emulsifier stabilizes emulsion and determines its type. Depending on its kind O/W or W/O emulsions is obtained.

Table 19.1.

The HLB values and emulsifying action of surfactants

HLB-range	Water solubility	Use
1-4	insoluble	
3-6	poor	emulsifier W/O
6-8	milky dispersion during mixing	wetting agent
8-10	stable, milky dispersion	" " emulsifier O/W
10-13	dispersion transparent or color-less	emulsifier O/W
13	solution, colorless	" O/W, solubilizer

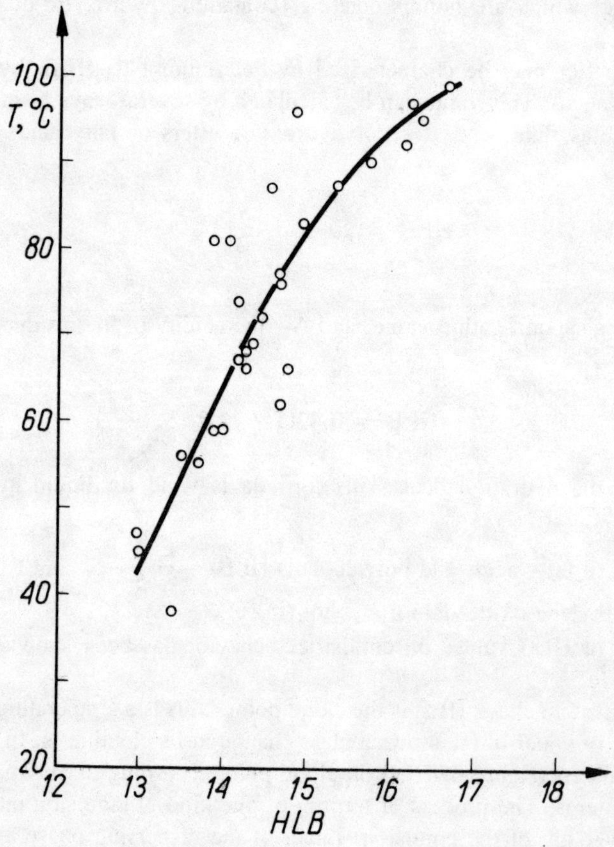

Fig. 19.3. Cloud point as a function of HLB, 5% solutions of water, heated to cloud point.

Which type of emulsion is formed depends on the type of emulsifier, its concentration, the way of mixing, temperature, volume ratio, etc. The amount of emulsifier needed depends on the type of fats used, and may be ca. 2-4%.

The significance of the double layer, which is a shell around the drop, lies in orientation of the emulsifier molecules, with polar groups pointing to the water phase (outwards in O/W emulsion, inwards in W/O emulsion). Nonpolar groups are immersed in nonpolar liquid, i.e., in drops in the O/W emulsion, in the surrounding phase—in W/O emulsion. The structure of such a layer is shown in Fig. 19.4. It shows some similarity with biological membrane of the cell wall type.

When discussing the interaction between collagen and fat molecules it can be concluded that the chances for bond forming are equal for hydrophilic polar functional groups and polar parts of the fat molecule, as between its nonpolar part and nonpolar side chain, e.g., leucine or valine. To chose a proper emulsifier as a fatting agent, it should be known which functional groups of collagen are already blocked by the tanning agent.

Most typical examples of anionic emulsifiers are soaps, in which hydrocarbon residue is a chain of at least 12 carbon atoms. Anion is a carboxylic group, and cation may be an alkali metal or ammonium ion. In the second group are the substances where cation is calcium group metal, aluminum or chromium. Such soaps have whole molecule hydrophobic, because metal ion is shielded by two or three hydrocarbon chains, being a steric hindrance. As hydrophobic substances the metal soaps favor the formation of W/O emulsion, however to much less extent than alkaline soaps favor the formation of O/W emulsions. The change of charge of droplets and thus breaking of O/W emulsion occurs when to an

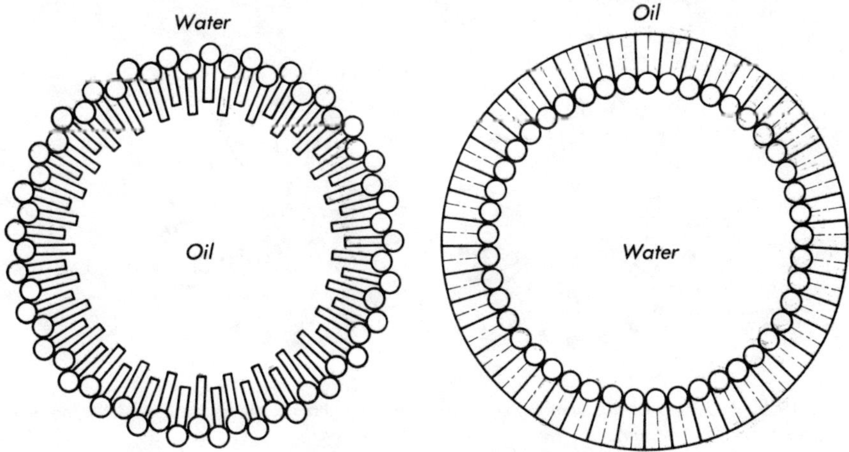

Fig. 19.4. Orientation of molecules in layer, immobilized around the drop (Figure from Thorstensen).

alkaline soap solution calcium salts will be added. In this case the electric charges on droplets of the dispersed phase become equilibrated by changes of ions in water. These ions may form 'bridges' between droplets, and thus cause the fats to precipitate. This is of some importance in degreasing of skins; chromium and aluminum soaps bind themselves very well to leather. Dissolved in organic solvents (alcohols), they subsequently hydrolize and form hydrophobic coat on collagen fibers becoming thus an impregnant.

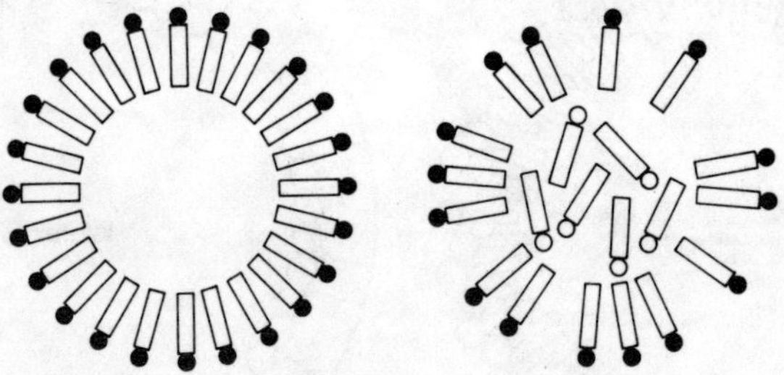

At low pH values ionization of carboxylic group becomes suppressed, part of molecules from double layer passes into the oil phase (Fig. 19.5), causing a destabilization of the emulsion. Decreasing pH may destabilize emulsion and break it up.

Sulfated derivatives of hydrocarbons (i.e., having the $-C-O-SO_2H$ group) are in the same group of anionic emulsifiers. Soaps at pH = 4, i.e., under conditions applied in the fatting of chrome-tanned leather, are too close to their flocculation point. In this case the groups: sulfonic (i.e., having the $C-SO_3H$ group) or sulfite, are much better emulsifiers. Stability of sulfated and sulfited

Fig. 19.5. Breaking of the immobilizing layer coat by change of the charge of the medium (Figure from Thorstensen).

oils at low pH (about 3) is still satisfactory: it depends on the kind of oil and degree of its sulfonation. It is the most abundant group of substances used for fatting. In this group several subgroups may be distinguished, e.g., Turkey red oil, which is a bifunctional compound of a type:

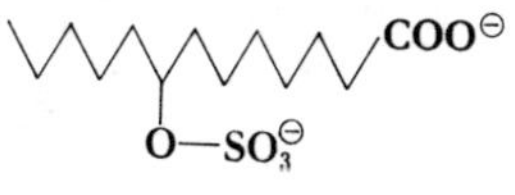

$$\text{(chain)}\diagdown COO^{\ominus}$$
$$O-SO_3^{\ominus}$$

triacylglycerol sulfates

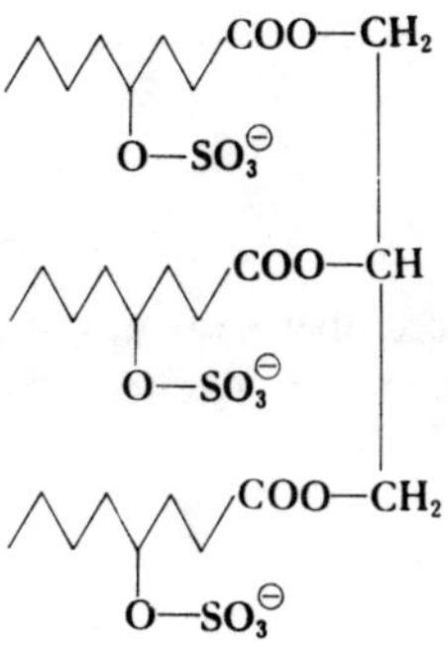

$$\text{(chain)}COO-CH_2$$
$$O-SO_3^{\ominus}$$
$$\text{(chain)}COO-CH$$
$$O-SO_3^{\ominus}$$
$$\text{(chain)}COO-CH_2$$
$$O-SO_3^{\ominus}$$

sulfated esters of monovalent alcohols

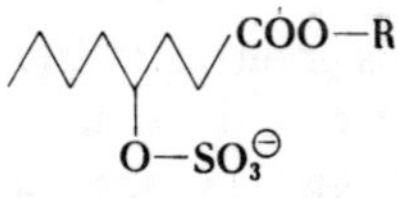

$$\text{(chain)}COO-R$$
$$O-SO_3^{\ominus}$$

sulfated derivatives of aliphatic alcohols

$$\text{(chain)}O-SO_3^{\ominus}$$

sulfated hydrocarbons

$$O-SO_3^{\ominus}$$

condensation products of fatty acids and sulfuric acid

$$COO-SO_3^{\ominus}$$

and alkylaromatic sulfonates

$$SO_3^{\ominus}$$

Sulfoderivative emulsifiers are also known, coming as ligands into chromium complexes. Cationic surfactants, or inverted soaps dissociate according to the formula

$$\left[\begin{array}{c} R' \quad R'' \\ N \\ R \quad R''' \end{array}\right] X \rightleftharpoons \left[\begin{array}{c} R' \quad R'' \\ N \\ R \quad R''' \end{array}\right]^{\oplus} + X^{\ominus}$$

In this formula R means long hydrocarbon chain, R', R'' and R''' organic radicals and X—an acid residue. To this group ammonium salts of all orders also belong, esters of phosphoric acid, esters and amides of organic acids (citric, maleic, succinic, adipic). These compounds may be characterized by bonds forming by their cations with negatively charged functional groups of the collagen side chains. Thus compounds in question are attaching poorly to chrome-tanned leathers, in which carboxyls of the collagen side chains are built into chromium complexes, as well as to those of aluminum. However, they are attaching well to vegetable and synthetic-tanned leathers. Because of this they remain in the outer layer of vegetable-tanned leather, making it hydrophobic. Methods of use of these emulsifiers after previous fatting with anionic components are known as well.

Nonionic emulsifiers are the substances containing semipolar bonds. They also affect orientation of molecules of the system. These substances contain hydrophobic backbone of the molecule, as, e.g., hydrocarbon chain, and hydrophilic groups, e.g., hydroxylic or amino groups. Being lined up in the system with their hydrophobic part towards oil droplet, and with their hydrophilic moiety towards collagen molecule, they cause attraction between fat and protein and certain fixation of the system. Fixation of collagen is a result of a decrease of interface tension, thus it is a result of operation of small forces, and this is why it is highly dependent on temperature and on the molecule shape. This has several consequences:

(1) stability of the system is limited to the range of pH, close to the isoelectric point; greater deviations do not fit into equilibrium range;

(2) non-ionic substances penetrate leather easily, as their binding is low and weak;

(3) these compounds may be mixed with other compounds either cationic or anionic in character;

(4) fatting with nonionic substances alone is not very efficient, as they are easily washed out for the most part;

(5) in their presence O/W emulsions form, containing oil droplets rather large in size.

Among typical examples of these emulsifiers are alkylphenols.

Fat is introduced into leather by formation of dispersion systems, where hydrophilic and hydrophobic components coexist. They shall coexist permanently, i.e., water should not wash out fat or come as liquid into leather, whereas the system should bind, give back and transmit water vapor. Stability of the emulsion substantially affects both the formation of the system and its maintainance in use. Problems of emulsion rheology are important, primarily for the formation of the system, as in a great majority the leathers are fatted in emulsion now, i.e., fatliquored. Thus some attention has to be paid to these two problems: stability and rheology of emulsions.

Stability of emulsion. Emulsions are systems consisting of two liquids insoluble in each other. One of them is dispersed in another, hence the name of dispersion. As diameter of the dispersed particles arbitrarily dimensions of 0.1 to 100 mμ are generally accepted. Emulsions are never permanently stable, as interface tension energy decreases with lowered common surface area, when two drops do coalesce. Thus coalescence of drops is a spontaneous process from a thermodynamical viewpoint, as then free energy of the system decreases. The reverse process requires work to be done. So-called spontaneous emulsifying may occur only if two phases coming in contact are unsaturated with each other. The work needed to increase their interface area is then done at the expense of free energy coming from mass transfer. Examples of such spontaneous emulsification are shown in Fig. 19.6. In the first case (a) deep funnels are formed at the interface, from which the drops are detaching. In (b) one may see what happens, if to the water column from the bottom alcoholic toluene solution is introduced. In Fig. c at a definite point the negative surface tension can be

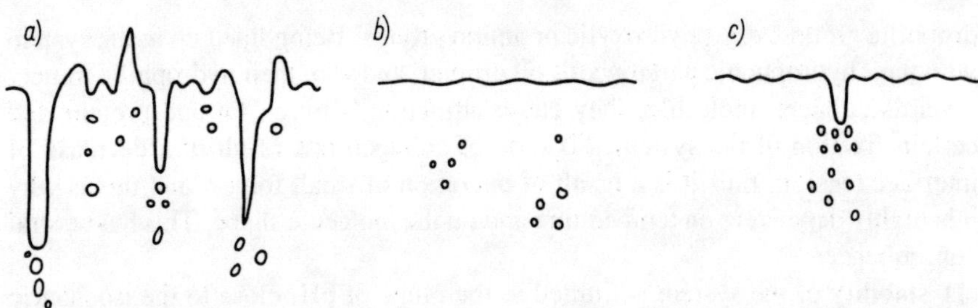

Fig. 19.6. Spontaneous emulsification under various conditions: (a) formation of drops due to vigorous stirring, (b) solution penetrates into the other phase and mixes with it, leaving drops of nondissolved substance, (c) nonuniform absorption from interphase surface.

observed; the surface area (interface) increases, forming drops. This case may be seen on the interface of ethyl alcohol in toluene and dodecyl sulfate in water.

Permanent solutions, containing colloid micells, cannot be considered as emulsion, because the particles contained in these cannot exist as independent phases.

Separation of the emulsion into components may occur by breaking, creaming of flocculation. Differences between these destabilization processes are shown in Fig. 19.7. Creaming can be regularly observed in emulsion, where the components are of nonequal density. During creaming, as well as during flocculation, droplets are not coalescing with themselves. No phase separation occurs in that case because slight stirring allows them to come back to their previous state. As a rule, coalescence occurs by breaking of the emulsion.

For emulsion stability the time and external conditions are of vital importance. There is no unambiguous stability index available, i.e., stabilized oil-in-water emulsion (hydrosol) separates at room temperature in seconds or minutes, unless mixed. The breaking process consists of collision of droplets due to Brownian motion, convection and increase in size of greater drops formed due to coalescence. Gentle stirring makes breaking faster, as collisions are more frequent. This process is called orthokinetic flocculation. Stirring at a high speed is a reason for splitting of large drops due to their elongation and shearing. The stirring speed can be fixed on a level which makes it possible to establish dynamic equilibrium of the medium. This speed depends on concentration, temperature, the form of stirrer and on the mean drop size.

Stabilized emulsion can be stable for a long time, if there is no contact between the contents of the drops. It is a result of action of the emulsifier, forming 'molecular barriers' on the surface of the droplets. Such a barrier (double layer) may resist against some dynamic or static pressure.

During centrifuging or evaporation the 'cream' drops contacting among themselves form polyhedra. Polyhedra of the same volume have a greater surface than spheres, so this process leads to the extension of interface area, and, as a consequence, may break it. Stability of emulsion may be influenced by the following processes:

(1) Dissolving rate of smallest droplets. This makes the average droplet size increasing during emulsion aging. This is more easily observable in foams (aerosols) than in emulsions (hydrosols). Influence of the droplet size on

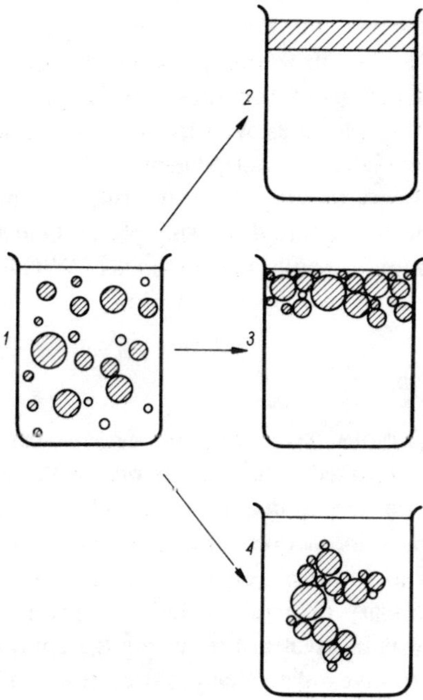

Fig. 19.7. Various ways of emulsion breaking: (1) primary dispersion, (2) breaking, (3) creaming, (4) flocculation.

solubility is given by formula

$$\ln \frac{S_1}{S_2} = \frac{\gamma V}{RT} \left(\frac{1}{r_1} - \frac{1}{r_2} \right)$$

where S_1 and S_2 denote solubility of particles of radii $= r_1$ and r_2, γ is interface tension and V is molar volume of the liquid in drops;

(2) freezing out. Ice crystals growing at the expense of water press the drops off the dispersed phase, then concentration increases and polyhedra are formed. Only some kinds of emulsion, having very thick 'skin' like, e.g., milk, may stay stable at freezing, however, to a limited extent as well, as then, e.g., crystallization in dispersed phase may occur;

(3) bacterial action, changing the medium;

(4) ionic strength of the medium. Particles of the dispersed phase are the carriers of electrostatic charges, dispersing phase has to contain an equal charge of opposite sign. Balance of these charges may change, it is a function of charge density in dispersing phase as well as in the dispersed one. One of these variables has to be chosen as independent.

There is no general theory as yet, giving the rules of emulsion behavior, in a case of change of ionic strength in dispersing phase. One may just presume that in a medium of great ionic strength, in emulsions of W/O type layers are formed, having character similar to two-layer lipid films.

Emulsion rheology

Some rheological equations have been touched in sect. 1.2, when the hide behavior model was discussed. Rheology is one of the new topics in physics, and cannot be discussed here in detail, among others because it requires rather extended mathematical introduction. Rheology deals with deformation or flow of bodies caused by known forces, or on the contrary—it answers the question, which forces are necessary to obtain a definite degree of deformation or the flow. This knowledge is of great use in testing the emulsion behavior. First of all, it is the best way to investigate changes occurring in physical characteristics of emulsion, e.g., during aging, transportation or changes of temperature. The simplest determination which almost always can be done is of the emulsion fluidity.

Elastic body does not flow, i.e., after removing of force applied it comes back to its primary shape. Viscoelastic body behaves alike, if, however, the force applied does not exceed a definite critical value. If this value is overcome, so it flows like fluid and if the applied force is removed it does not come back to the initial shape.

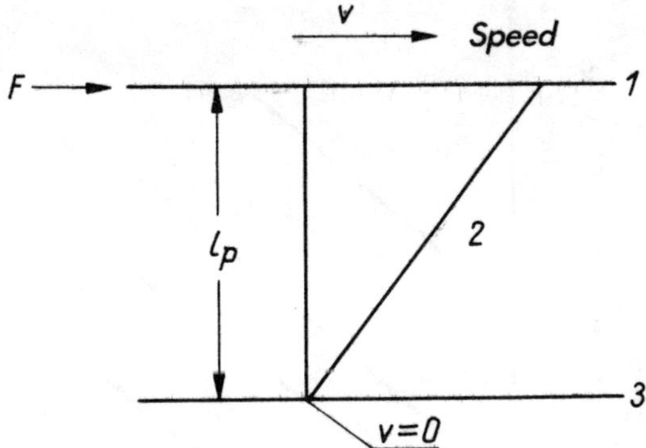

Fig. 19.8. Newtonian flow: (1) immobile plane v = 0, (2) velocity gradient $\frac{dv}{dx}$, (3) mobile plane.

Very dilute emulsions behave like Newtonian fluids, i.e., they obey the Newtonian flow law. In more concentrated solutions the interaction between drops starts; by further concentration increase flocculi are formed, having properties of a viscoelastic body. In Fig. 19.8 Newtonian flow is shown. In the model shown the fluid is contained between two planes of surface a cm², distant 1_p cm. Upper plane is moving by force F with a speed v. Flow speed in proximity to the upper plane is equal to v; it drops down to zero in proximity to another plane. Change in plane speed, i.e., average shearing rate V is proportional to the distance from upper plane and is $\frac{V}{l_p}$. At a point, lying in plane x to the upper plane the value V is a differential $\frac{dv}{dx}$. Force, coming on a unit of surface $\frac{F}{a}$ is shearing tension S. Fluid viscosity η is defined by quotient $\frac{S}{V}$ and can be determined by measuring S and V, as η does not depend on the force applied.

The majority of emulsions do not behave like that. Such substances are called non-Newtonian fluids. Behavior of several kinds of such fluids is shown in Fig. 19.9, where for comparison a Newtonian fluid is shown, too. Shearing tension is shown there as a function of v. In some cases v increase causes at start a quick S increase until the point, at which quotient $\frac{S}{v}$ becomes constant. Initially the viscosity η decreases, then it becomes directly proportional to the shearing speed. This kind of flow occurs, when the structure is destroyed by shearing. Pseudoplastics are emulsions of the same type, but with no threshold value. They start to flow immediately when the force is applied. Emulsions for which a certain threshold value is needed (i.e., some shearing force applied) are called plastic (real plastic flow). In such a case v increases more slowly than S. Thus η initially increases, then $\frac{S}{v}$ quotient becomes constant. Behavior of the body

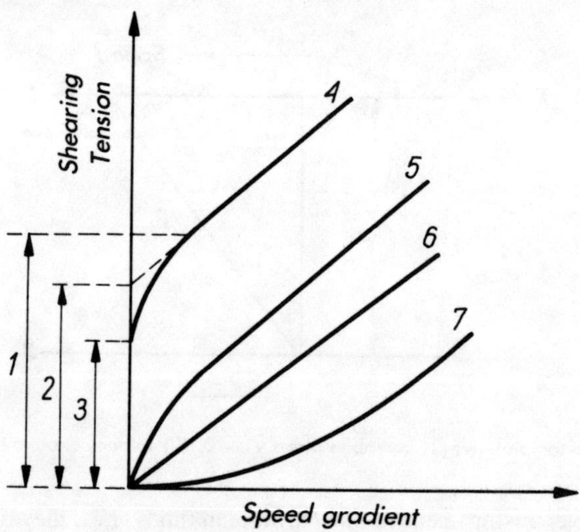

Fig. 19.9. Kinds of flow: (1) upper threshold value, (2) extrapolated threshold value, (3) lower threshold value, (4) plastic flow, (5) pseudoplastic flow, (6) Newtonian flow, (7) flow of dilatant substance.

is opposite to the pseudoplastic substance, if there is no threshold value, and to the plastic one, if such a value exists. Such behavior is shown by some loose dry goods and by close packed (in molecular sense) dispersions. During shearing the package will be loosened due to initial increase of volume, which is required to make possible the transitions of the particles. Such substances are called dilatant.

Viscosities of pseudoplastic, plastic and dilatant substances are changing with shearing value; determinations of viscosity, done at one shearing rate are of limited value. Particularly if flow characteristics of two emulsions are to be compared, measurements are to be taken in possibly wide range of shearing rates. Are the measurements done on non-Newtonian fluids so the ranges are always to be given. Gradient $\dfrac{S}{v}$ of linear sector for flow curves is sometimes called 'apparent viscosity.'

Frequently the change of viscosity with time can be observed whereas the experiment parameters remain unchanged. After some time it stabilizes. Time lapse decreases, if shearing rate increases. After removing the force applied and some time later the initial values are restored. This is explained by changes in the fluid structure and can be defined as tixothropy, despite that this definition was for the first time introduced for isothermal sol ⇌ gel transition. A dependence of viscosity on time may be defined as

$$\eta_t = \eta_f + (\eta_1 - \eta_f)e^{-\frac{t}{\lambda}}$$

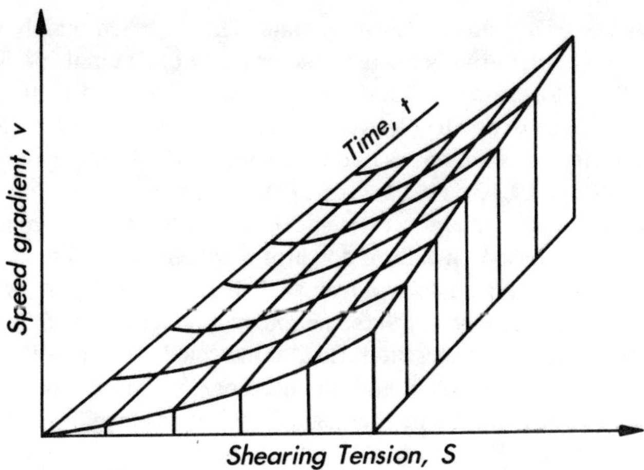

Fig. 19.10. v-S-t graph.

η_i is viscosity at start, η_f is a constant viscosity after infinite time, and λ relaxation time. In order to define graphically the behavior of tixotropic fluids, a space arrangement v-S-t was introduced (Fig. 19.10).

Hysteresis is also observed in tixotropic fluids. This occurs, when the shearing force does not come to its initial (threshold) value. This value not necessarily is observed there. In Fig. 19.11 a hysteresis loop is shown, where the shearing force comes back to its initial value (2) or otherwise, changing then the threshold value 1.

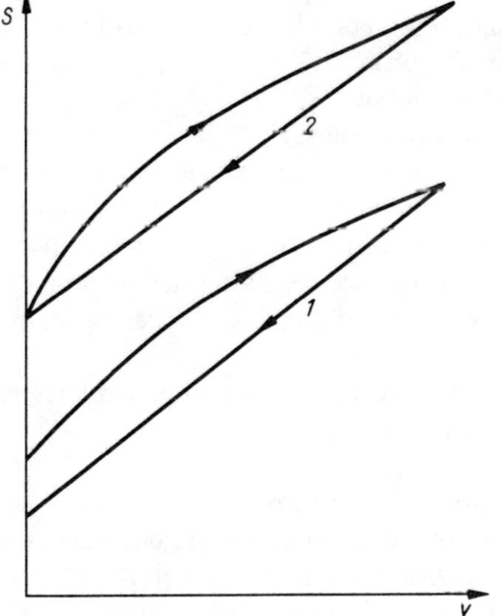

Fig. 19.11. Influence of hysteresis on the threshold value. (1)—changed, (2)—unchanged threshold value.

Characteristics of a fluid (inclusive of emulsion) are based mainly on investigation of its viscosity. Another substantial property of an emulsion is stability which is difficult to estimate. Some attention has to be paid to the viscosity measurements, i.e., viscometry. Viscometers are applied in rheology and in the physical chemistry of solutions. Viscometers used in industry may be classed in four groups (Fig. 19.12): (1) capillary, (2) axial, where a rotating spindle is positioned in a coaxial container, (3) conical, in which a rotating cone is based on a plate, and (4) based on falling down of a sphere in a calibrated tube. A review of viscometer types, their advantages and disadvantages, is given by Sherman [9]. Use can be made of the same source by calculation of the threshold value for a particular viscometer kind. He also indicated the suitability of various kinds of viscometers for testing liquids or emulsions of a given kind considering the influence of various factors on rheological properties of emulsion that should be taken into account. These are:

(1) in the dispersed phase:
   —concentration, hydrodynamic interaction between droplets, flocculation, leading to the formation of aggregates of droplets,
   —viscosity, deformation of droplets by shearing,
   —diameter, distribution of droplets, way the emulsion is prepared, interface tension interactions among droplets, and between them and the dispersing phase,
   —chemical composition;
(2) in the dispersing phase:
   —viscosity, other rheological properties,
   —chemical composition, polarity, pH, energy of interaction among droplets through dispersing phase,
   —concentration of electrolytes;
(3) factors, concerning emulsifiers:
   —chemical composition, potential of mutual droplet interaction,
   —concentration, solubility in both phases, emulsion type, inversion capability, solubilization of dispersing phase in micells,
   —thickness of 'skin' (double layer), formed on droplets and its rheological properties, deformation of droplets by shearing, motions of fluid in droplets,
   —electroviscometric effect (influence of electrolytes in medium on emulsifier, its ionization, etc.),
(4) additional stabilizing agents:
   —pigments, hydrocolloids, hydroxides,
   —influence on rheological properties and on interface properties.

Practical aspect of leather fatting. In the leather industry fatliquors are used. The smaller the particles of dispersed phase, the more easily they come over to the interfibrillar spaces of collagen. However, this statement is a rough approx-

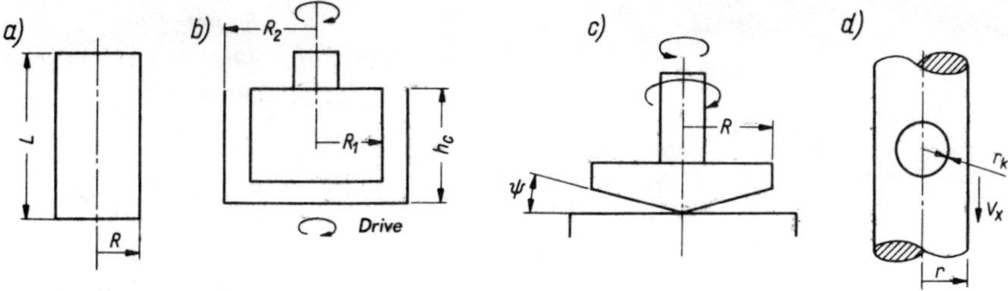

Fig. 19.12. Acting principle of the viscometers: (a) capillary viscometer—characteristic features: length, capillary radius R,

(b) axial (rotational) viscometer—characteristic features: radius of rotator R_1, radius of container R_2, its height h_c and revolutions/unit time,

(c) conical viscometer—characteristic features: cone radius R, angle of its generatrix with base ψ, and revolution number in unit time,

(d) falling sphere viscometer—characteristic features: tube radius r_k, sphere radius r_c, its density σ, falling rate v_x. According to [8]

imation to the rule. In fact the local charges in leathers control the penetration ability as well as rheological properties of fluid and other, already named, factors.

In approaching the fatliquoring mechanism we may point out two items—one is, what happens to the emulsion surrounding the leather and eventually penetrating it during technological process, and the other—what occurs when the molecules of fatliquoring substance (presuming one component for simplicity) have already approached the tanned collagen molecule in solution.

The first item may be best visualized in Fig. 19.13. As can be seen there the O/W emulsion concentrating itself on the leather surface becomes converted into W/O emulsion, and in the last phase it becomes separated—fat molecules are deposited in leather, aqueous phase (in practice the exhausted liquor, which means but waste water) remains outside.

In the second point we have to consider why leather collagen, being itself a hydrophilic substance, keeps the fat molecules attached and therefore less water is usually bonded to it. As it has been explained (sect. 2.2), the collagen molecule consists of hydrophilic and hydrophobic clusters of amino acid residues. The sequences of these clusters are now quite well-recognized and the charges occurring to them computed, although nobody still has worked out the computer program allowing it to predict what happens if a fat molecule having one end charged (or otherwise like, e.g., in mineral oil) approaches the protein. As we remember, the collagen molecule is surrounded by water molecules arranged as a shell around it.

The hydrophobic bonds which may arise between the hydrophobic collagen amino acid residues and nonpolar ends of fat molecules are of different character

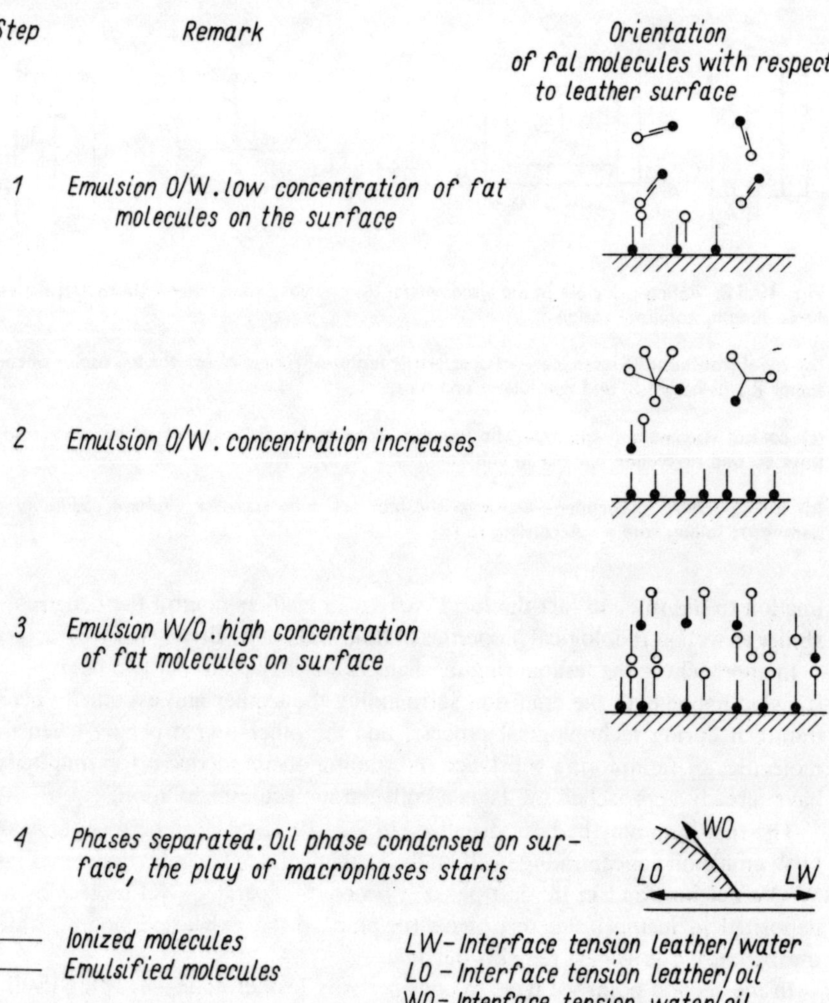

Fig. 19.13. Fatliquoring phases. According to [3] with permission

and strength. These bonds are very weak, and their action radius decreases very rapidly, proportional to r^{-6}, where r is the distance between molecules. If a surfactant (say amphoteric) is used, it may attach both kinds of molecules, 'pulling' an oil molecule along. Thus the oil molecule may become 'pressed' to the hydrophobic molecule part by the surrounding water molecules and fixed in this position. This is a way to increase the collagen 'affinity' to oil (expression very distant from real definition of affinity). The necessary condition is that oil molecules have to have an access to collagen molecules, which means either the surrounding amount of water is not very great (sammyed leather) or the leathers are in vigorous motion in drum, or else the bound water becomes more mobile (heating). This reaction may still proceed even better during leather drying. It cannot be expressed by figures, as it is too difficult, and too many independent variables are working at the same time. Experimental proof of this author's point of view was found by Sharphouse [10]*, who observed the expelling of some water from leather stuffed in drum after sammying. Of course, ionic groups when contained in fat molecules make the reaction easier, as they are anchored by them to the active groups in the side chains; however, after this anchoring they must not remain in a position, perpendicular to the collagen molecule axis, forming a 'hedgehog,' but they have to align to the molecule, and then probably the predicted effect occurs.

Usually the chrome-tanned leather has a positive charge. Emulsifiers contained in a fatliquor have thus to form negatively charged molecules, in order to connect to the charged sites of leather when penetrating into it, and thus making the fixation of fat easier there. Of course, of importance are the charge value and distribution in leather as well, thus the degree of masking of chromium floats and the kind of anions introduced into leather in this process are to be established, with regard to fatliquoring process and its effectiveness. As a rule, fat will be introduced to nappa, soft, glacé and waterproof leathers through the whole thickness. This requires strong masking of float components or exact neutralization before fatliquoring. Other sorts of leather are fatted only on their surface. An example is box calf, where no fatting through the whole thickness is done, or floats of lower masking extent are used.

As fatting substances animal fats are used: neatsfoot oil, lanoline, yolk oil, cod-liver oil, fish, plant and mineral oils, etc. As a rule, they are used as appropriate preparations not in substance. Much data concerning properties, chemical composition of fats, etc., are given by Poré [3].

Extractable and nonextractable fats

It is impossible by available methods to check if a given substance has not formed bonds with collagen, or has formed such ones of an induced dipole type, or hydrophobic bonds (sect. 2.2). Sherman considers the term *'hydrophobic bonds'* as incorrect; he proposes *'hydrophobic association'* because this association is related to water molecules and the bonds between them through hydrophobic groups. Extractable fat (or fatty matter) is a definition used for a long time for fat, which may be extracted by an organic solvent. Nonextractable fat is the fatty matter, which may be detached from collagen by very drastic means,

---

*An essential difference between hydrophobic interaction and hydrogen bond from thermodynamical viewpoint is that the first is exothermic whereas the other one endothermic. Both effects are slight.

e.g., by a hydrochloric acid hydrolysis. According to solubility of the fatting agents used, not only different amounts may become extracted from leather, but different fractions as well. Development of gas chromatography as well as of infrared spectroscopy made the problems of fat extraction more familiar, and better known [7]. One may demonstrate how deep and how fast the particular fatliquor components are penetrating into the leather.

Leberfinger and Matschkal [11] have tested a balance of fatting agents, using various solvents and alkaline hydrolysis, allowing removal of the bound fat from the leather. However, hydrolysis only allows one to split soaps and sulfited fats from collagen. Method of separation of the emulsion components into neutral fat and nonsaponifiable matter (waxes) and to sulfited and emulsifiable parts has been worked out by Panzer and Niebuer [12]. In this method a 50% water and alcohol mixture is used for extraction, and an organic solvent (white spirit), as a second phase. Some fatting agents cannot be removed from leather by methylene chloride extraction; nevertheless, it is considered to be the best extracting agent for fats. Also the use of several extracting solvents, like, e.g., methylene chloride, n-butanol, and n-butanol ammonia (alcalized) one after another, still gives incomplete extraction. Leberfinger and Matschkal have tested the content of fat in liquor before and after fatliquoring, as well as the extraction and hydrolysis of leather. Data from this paper contained in Table 19.2 may confirm this view.

Due to alkaline hydrolysis a bond between collagen and fat is broken, and hydrophilic groups are thus split off. In limited cases they may be washed out with ether from the hydrolizate. This concerns sulfated codliver oil, quaternary ammonium salts and ethoxylated fatty acids. Sufonic groups are much more hydrolysis-resistant. Further investigation concerned the use of n-butanol as a factor making it possible to extract strongly polar fatting substances. In this way about 90% of fat was extracted, bonded to the leather, except the one, fatted by

Table 19.2.

Fat balance of chrome leathers (%) by the use of CCl_2H_2 as extracting solvent

Fatting agent	Mass, extractable with CCl_2H_2	Mass, extractable with ethylene ether after hydrolysis	Extractable together	Fat uptake by leather
Alkyl sulfochloride	0.80	1.30	2.10	7.90
Alkyl sulfonate	2.30	3.20	5.50	10.00
Sulfonated codliver oil	6.45	0.35	6.80	10.00
Sulfated codliver oil	6.60	3.10	9.70	11.20
Cationactive agent	4.25	2.05	6.30	6.20
Nonionic agent	10.60	0.20	10.80	12.10

alkyl sulfochloride, where the recovery was 29%, alkyl sulfonate (68% recovery) and for cationic fatliquor (75% recovery). Attention was drawn to the fact that the amount of nonextractable fat is about equal to the amount of the emulsifier in the mixture. According to the authors, as a result of the polar solvent extraction of fatliquored leather, functional groups of side chains are restored in a nonionized form. They also found that methylene chloride extracts acylglycerides and hydrocarbons, butano-sulfated and sulfonated compounds and nonionic substances containing ester groups. These extracts do not contain proteins, as it can be seen from their IR-spectrum. According to Leberfinger and Matschkal—between collagen and fats coordinative bonds are formed due to a delocalization of free electron pairs on oxygen and chlorine and of π-electrons of the double bonds. This kind of affinity is not maintained during methylene chloride extraction; however, it stops fat migration in leather. This paper predicts an important step in contemporary technology of leather fatting. However, further investigations are required, concerning the selection of substances that connect to leather very strongly. This would permit one to limit the amount of fats and at the same time to influence the leather quality.

An inter-laboratory experiment initiated by the Commission of Chemical Analysis of IULTCS (Prof. Dr. E. Heidemann) is going on now. The aim of this experiment is to check the conditions of fat extraction from leather. According to the results obtained in 14 laboratories, the amount of fat extracted by the official CH_2Cl_2 method makes 64-89% of free fat only. Two-step extraction with n-hexane first, followed by the use of a chloroform-methanol mixture gives better results. The influence of additional factors like binders is visible and—which is a novelty in fat determination—has to be jointly considered. This experiment is carried out in the same way as Hyp determination [13].

Heidemann et al. [14] applying gas chromatography for investigation of bonds between alkylsulfonates and leather have shown that the longer their carbon chain, the more strongly they are bound to collagen. This is still more expressed, if syntans, like Basyntan BN are used for chrome leather retannage. Permanent binding of fat is a very important factor for clothing leather quality, as these leathers are expected to resist the cleaning process, and of upper leathers used for shoes with vulcanized soles. A certain amount of fat should remain in leather, as well as some water.

This effect can be obtained in various ways, as, e.g., by introducting bifunctional fats into leather (as dicarboxylic acids), and then fixing them by retanning. In this way one of the carboxylic residues will be built into the chromium complex, which is by one point connected to the leather, whereas the second group may attach a now-introduced complex. Thus an aliphatic chain becomes connected to chromium on both its sides.

A very old way to make vegetable-tanned leather waterproof was its incomplete tanning with a thin slice of nontanned hide left inside. Fibers of this layer, swelling by water, closed the access of water to the inside of the shoe. This kind of shoe could be worn for a shorter time, because of putrefaction of this layer; however, such leather was waterproof.

REFERENCES

1. Durande-Ayme, B., Dussaillant, R. Bull. AFCIC., *26*, 110 (1964)
2. see ref. 8 ch. 8
3. Poré, J. La nourriture du cuir Le Cuir, Paris 1974
4. see ref. 42 ch. 15
5. see ref. 25 ch. 12
6. see ref. 3 ch. 6
7. see ref. 2 ch. 11
8. Stern, J. Z. f. Elektrochem. *30*, 508 (1924)
9. Sherman, P. Emulsion Science Acad. Press., London 1968
10. Sharphouse, J. H. private communication
11. Leberfinger, R., Matschkal, H. XIII IULCS Congress, Vienna 1973
12. Panzer, A., Niebuer, W. Leder *3*, 219 (1952)
13. Heidemann, E., Moldehn, R., Coop, J. XIII IULCS Congress, Vienna 1973
14. IUC-Commission Report, XVII JULTCS Congress, Buenos Aires 1981

20.

LEATHER IMPREGNATION

Leather impregnation is closely connected to fatliquoring problems, as the impregnations should have an effect like that of fats. To make the shoe leather water-repellent, they have had to be heavily fatted in the past. This method, however, had several disadvantages. First, it was not possible to finish in a modern way a leather which contains over 8-10% of fat, which is the fat amount required to make it waterproof. Such leather has different adhesion characteristics, and it is difficult to dye it. Besides, such heavily fatted leather cannot be used for shoes produced in machines for cementing or vulcanizing soles.

Modern chemistry offers many preparations, permitting one to avoid these disadvantages, e.g., there are substances like silicones, preferably dimethylpolysiloxanes

$$\left[\begin{array}{ccc} CH_3 & & CH_3 \\ | & & | \\ -Si & -O- & Si- \\ | & & | \\ CH_3 & & CH_3 \end{array} \right]_n$$

which are chemically inactive, resistant to high and low temperatures, and preventing rubbing off and staining of leather. Another group of impregnants are alkylated derivatives of succinic acid

$$\begin{array}{c} R_2 \\ | \\ R_1-CH=CH-CH-CH-COOH \\ | \\ CH_2COOH \end{array}$$

These compounds are attaching to leather, forming bonds other than fats. They may be quantitatively extracted from leather by organic solvents.

Recently fluorine hydrocarbon derivatives are used in impregnating systems. They have a particularly low surface tension having thus highly desirable surfactant properties. A widely used type is perfluorooctanol $CF_3(CF_2)_6 - CH_2OH$, a compound very reactive, forming, e.g., with organic acid esters, like with acrylic acid which is

$$CF_3(CF_2)_6-CH_2-O-\overset{\overset{\displaystyle O}{\|}}{C}-CH=CH_2$$

commercially used.

These substances are not water-repellent, but they bind water to form W/O type emulsions. Accordingly, this is an impregnation with nonhydrophobic compounds. According to Neher and Gallegher [1] a bond is formed between succinyl carboxyl and amino groups of the side chains. Among numerous derivatives of the type considered only liquid substances are as efficient as water-resistance makers for leather. Based on their own observations, the authors assume that W/O emulsion formed at the water/impregnant interface is a barrier in interfibrillar spaces for penetration of further amounts of water. It is questionable whether it is right to speak about impregnation, if hydrophobization does not occur.

The amount of alkylated derivative of succinyl acid which is absorbed by leather changes according to temperature is smaller at 25°C and greater at 0°C. The sorption isoterm for both temperatures is a straight line, as if coordinate system one will take logarithm of molar concentration in leather as a function of logarithm of concentration in both. When investigating the extracts from leather, however, one may conclude that only 1-2% of alkylated derivatives are leather-bound. Any further amounts are then stored in the leather just filling out the empty spaces, without binding. Interface tension between water and impregnant is of an order of 1-3 dynes/cm (10^{-3}N/m). Emulsifying ability and capability of making leather water-resistant is different for derivatives of various carbon chain length: C_{17} and C_{21} derivatives are good emulsifiers and may make leather water-resistant, whereas the C_{18} ones are not. According to the quoted authors' conclusion, substances forming the W/O emulsion are good waterproofing agents.

Chromium^{3+} and aluminum stearates, used for impregnatives, have been described by Heyden [2]. These complexes, and the chromium complex of perfluoroacetic acid, which has similar properties, are used as alcohol solutions. Chromium soaps, formed on leather, have strong water repellency. However, the use of chromium soaps requires some technological caution, because they easily form precipitates with inorganic salts present in the solution. This is the reason for some complications.

Complexes of chromium and fatty acids have been studied by Küntzel and Mischnik from the viewpoint of the tanning chemistry [3]. For impregnation of waterproof type leather, containing 17-23% fat, special fatliquoring mixtures are used, containing, among others, animal fats—tallow and lanolin; these compounds are now less used in tanneries because of their price. A condition of introduction of such a great amount of fat into leather is its use in the form of W/O emulsion; usually this process is carried out in two steps: fatliquoring and greasing with paste. These operations do not increase, however, the amount of fat fixed (bond) by the leather; it remains on the same level, as in chrome leathers, fatted as usual. The rest of the fat remains unbound and is exudated slowly during the wearing.

Investigations on various W/O and O/W types of emulsions have shown [4] that resistance against water is to a great extent dependent on leather motion during the fatting. This motion is a reason for 'disintegration of emulsion drops, smearing out' of them on fibers, and thus the closing of the pores. When observing inversion of emulsion on the materials used for fatting, the authors have found that such inversion can be multiple-repeated; as it is repeated if preparation comes through the isoelectric point. The O/W type emulsion exhausts poorly from the bath; however, it goes easily through the leather. It becomes unstable under the influence of acids; fat is then separated on the fibers in the form of droplets (cf. also the remarks on p. 485).

REFERENCES

1. Neher, M. B., Gallegher, N. D. J. Am. Leath. Chem. Assoc., *59*, 403 (1964)
2. Heyden, R. Leder *20*, 4 (1969)
3. Küntzel, A., Mischnik, O. Leder *5*, 285 (1954)
4. Heidemann, E., Wolframm, C. P. XII IULCS Congress Prague 1971

21.

FINISHING COATS

Visual and palpable evaluation of leather depends on its surface. As a rule, this surface is treated in such a way that it emphasizes its natural beauty and conceals its minor defects. Development of polymer chemistry and technology has given us the possibility of changing the leather surface in many ways. In this chapter only principles of coat formation and the method of their applying are given, because it is impossible to describe here all the kinds of coats and their properties. The interested reader is referred to special monographs devoted to these problems. Independent of the kind of coat and the way it is applied, it only then meets the requirements if it is stable and uniform enough. Apart from chemical stability, i.e., resistance to damaging factors like air oxygen, water and its impurities, it has to be firmly attached to the substrate. The surface tension existing at the groundcoat interface during applying, i.e., when the coat is liquid, affects the effectiveness of coat applying. The capability of joining the two substances (adhesion) and internal coherence of the coat components (cohesion) are responsible for the stability of the coat, together with reactions occurring between the coat layers.

Surface tension

Increase of the liquid surface requires work, counteracting intermolecular forces, as the molecules from the inside of the liquid then come to the surface. The work needed for the surface increase is usually expressed in erg/cm^2 or in dyne/cm (Joule or Newton/sq.m). Surface tension between any two nonmixing liquids is called interface (interphase) tension. Methods of its measurement are typical in physical chemistry: measuring the liquid level in a capillary tube, counting the falling drops (stallagmometric) or by the de Nuoy method of ring disconnection.

Surface tension under a constant pressure and at a constant temperature is equal to the work needed to increase the surface of one unit, so it is equal to a free energy increase during the carrying out of the process. Surface tension (γ) decreases with temperature increase. It is an empirical value, difficult to express in figures. One of the proposed equations is

$$\gamma = C(\rho - \rho_o)^n$$

where ρ is density of liquid, ρ_o—density of vapor, C—constant, specific for every liquid as well as the exponent value n. Another equation expressing surface tension as a function of temperature is:

$$(M_o)^{\frac{2}{3}}\gamma = K(T_c - T - 6)$$

where T_c—critical temperature of liquid, K—Eötvös constant of a value about

2 for nonassociated liquids, M_o is molar volume of the liquid, $M_o^{\frac{2}{3}}$ is a surface of one wall of a cube containing 1 mole of fluid. Eötvös constant value may reach 4, as in the liquid fatty acids—i.e., in liquids where the molecules do orient themsevles on the surface.

Cohesion and adhesion

The work needed for separation of a column of liquid of 1 cm² cross-section area equals doubled surface tension values (2γ). This work is overcoming the cohesion forces. Thus water cohesion at 25°C is 143.9 erg (10^{-7}J). The work equivalent to adhesion W_a is done if we have to separate two different substances, having common surface of 1 cm². The work of overcoming adhesion between liquid and solid substance is given by formula:

$$W_a = \gamma_L + \gamma_{SV} - \gamma_{SL}$$

where γ_L is surface tension of liquid, γ_{SV} surface tension of solid, and γ_{SL} is interfacial tension between liquid and solid. However, this is a simplifying expression. By separation of the liquid and solid the latter remains liquid-saturated. Only substances having small internal energy, e.g., liquid paraffins, are not wetted and the actual surface tension of a solid surface is then γ_{SV}. More detailed considerations on this subject can be found, e.g., in a monograph of Kaelble [1]. Equation for the work may be taken from equations:
(1) for solid in vacuum

$$\pi_e = \gamma_{s_0} - \gamma_{sv_0}$$

(2) for the solid-liquid-liquid-vapor system

$$W_a = \gamma_{LV_0} (1 + \cos\Theta) + \pi_e$$
$$W_a = \gamma_{s_0} + \gamma_{LV_0} - \gamma_{SL}$$
$$W_a' = \gamma_{LV_0} (1 + \cos\Theta)$$
$$W_a' = W_a - \pi_e$$

where π_e is the value of degrease of free energy of the solid, when immersed in saturated vapor of the liquid. In these equations an angle Θ occurs, known as the wetting angle (see Fig. 21.1). The force acting on the contact line between

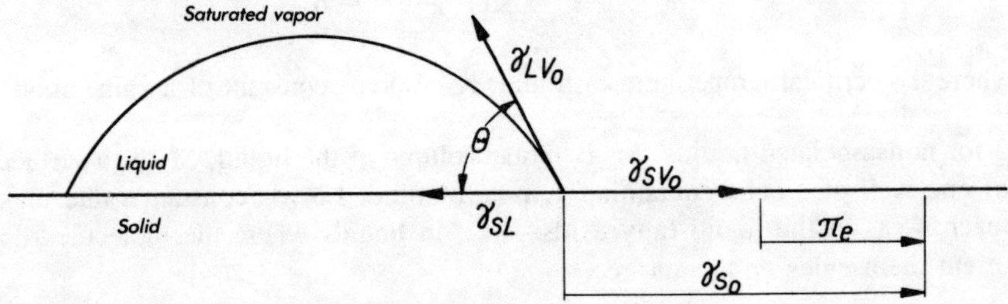

Fig. 21.1. The wetting angle components.

a liquid drop and a solid surface is expressed by formula:

$$\gamma_{SV_o} - \gamma_{SL} = \gamma_{LV_o} \cos \Theta$$

For the actual evaluation of the work of overcoming adhesion forces for non-smooth surface a roughness index is introduced

$$r = \frac{\cos \Theta_r}{\cos \Theta} = \frac{\text{actual surface}}{\text{apparent (smooth) surface}}$$

then

$$(\gamma_{SL})_r = r\,\gamma_{SL}$$
$$(\gamma_{S_o})_r = r\,\gamma_{S_o}$$
$$(W_a)_r = W_a + (r - 1)\,(\gamma_{S_o} - \gamma_{SL})$$

Surfaces for which $r = 1$ do not exist in practice, thus, in order to check the actual value of overcoming the adhesion forces, r has to be found out. This value is more than 1.5 for solids. It could be expected that rough surface would become smooth, if it would be possible to remove limitations in molecular motion due to viscosity, elasticity, and to tendency to minimalization of γ_{S_o} value. Pores, cracks and capillary holes in a solid surface are function of surface preparation as well as of the range, in which a spontaneous minimalization is possible for free energy acting on that surface. Smooth surface of fresh melted glass or of fresh poured liquid is an example of a surface energy minimalization process. Smooth surface of carefully split mica is an example of surface formed in a process without initial roughness. Special cast technique, together with centrifuging, allows the smoothest surfaces of soluble or melting polymers.

Leather surface after tanning is far from smooth; on the contrary, it is complicated to a very great extent. In Fig. 21.2 can be seen the cross sections of reticular layer of leather obtained by Stirtz with a raster microscope. Visible

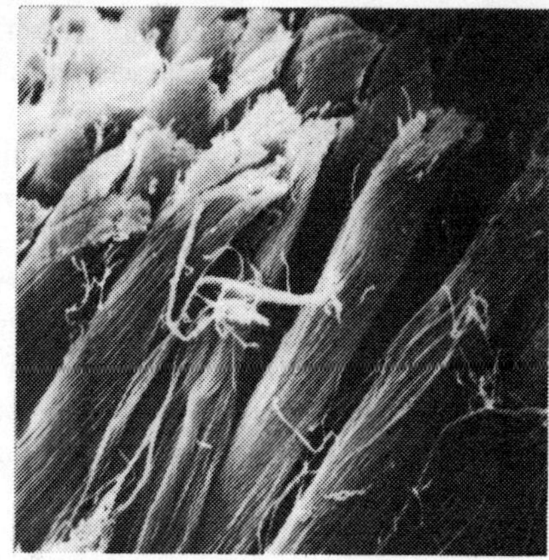

Fig. 21.2. Leather cuts. A scanning electron microscope photograph. Courtesy Dr. Theo Stirtz.

differences in the length of the fibers are a result of decay of tensions in them caused by partial network demolition.

21.1. Application of coats onto leather

Coats are to protect leather against damage, increase its stability, when leather is in immediate contact with mud, humidity and sharp edges, and to give it a desirable look, consisting of color, gloss and texture. However, they should not make the leather tight and impermeable for gases. Suede leathers are impregnated with invisibly thin impregnating coats only. The coat on leathers obtained from splits and imitating grain leather is thick and deformed by an iron press or cylinders. Leathers of other destination than shoes and clothes, as, e.g., for belts, or sole leathers, may have no coat.

Producers of chemical preparations for leather industry are offering huge amounts of coat components, or of ready to use compositions. Very many of them are mixtures of components, being producers' secret. It would be impossible to discuss single kinds of coats (finishes) and preparations used for this. In this chapter only factors will be discussed, influencing all finishes to a greater or smaller extent: strength of connection of the ground coat and dependence of leather appearance on coat structure.

The ATR (attenuated total reflectance) method is of great assistance in investigating the composition of coatings, when applied in the IR region. This work has been started by Nursten and Pearson [2]. Later these studies have been continued for identification as well as for research [3]. IR-spectrum of cross-linked, ordered polymers is not an additive property of a sum of the components due to limitations in vibration freedom by structure. The chemist dealing with problems of coat adhesion to leather has to remember that all the following factors are significant: characteristics of the ground surface, preparation of it, choice of substance, binding coat to the ground, way of binding of coat by binder, coat applying methods and the methods of its testing. As a rule, the surface of the ground is absorbent enough, sometimes it has to be decreased (sealed), lest the components applied would penetrate too deeply and their consumption would be too high. For this purpose resins of medium molecular weight are used. Their polymerization process will be completed in the leather. Such sealing is one of the steps of surface preparation. Other steps are: retanning, equilibration of charges (leveling), fatting, dyeing, buffing, smoothening, and other mechanical operations, whose purpose is to make softer or stiffer, smoother, or a more stretchable surface.

The choice of the binder is connected with choice of coat and leather properties. Its task is to connect in a stable way the coat and ground, without chemical or physical changes in them with time lapse, maintaining of the bond despite the work applied, temperature changes, water influence and, in case of special requirements, influence of organic solvents. Gases (air, CO_2, water vapor) should be able to come through. Methods of coat application depend on its type, and they belong to technology, as for the majority of kinds of coats, several kinds of applications can be used. Every one of them has its advantages and disadvantages.

Testing methods are subject of standards, settled by national and international organizations. Most of all—if we are speaking about physical testing—dynamic methods are used, which imitate the wearing processes: multiple bending (flexing) stitch resistance, tear resistance, lastometer (dome) testing, and fastness testing, i.e., testing of resistance to light in various and changing climatic conditions, to rubbing, washing, etc.

Coats on leather consist of several layers. The aim of the first coat is to 'anchor' it on to the leather and to form ground for the next one, of appropriate adherence. Anchorage in leather is a problem of penetration of the components into the leather in a way to form insertions that penetrate leather. This is like the connection between epidermis and dermis (sect. 1.1). Thus this layer should change its condition with time, coming from a mobile, easy penetrating liquid into a sticky, keeping its once-given shape, and elastic yet to some extent solid. By adhesive force the next layer is attached to the first. This layer has another significance of its own, like, e.g., mechanical protection or reflection and dispersion of the incident light. Connection of these layers occurs on the same basis, as cementing of macromolecular substances. Cementing process from a theoretical viewpoint is worked out for very many connections. In the case of coats on leather, as a rule one has to deal with a mixture of many materials, and really there is no way to test them by typical methods, as the coats are too thin and too weak to be tested by determination of the force needed for separation of two layers by stripping off. This is, however, the process, which occurs in wearing, when the layers peel off.

According to Kaelble *(loc. cit.)* the following diagram is applicable, if two bodies are separated and one of them is stiff, and another flexible (Fig. 21.3). In the diagram we can see

(1) separation of the layers (tearing) should occur in the connecting layer, rather than on its limit; thus, important for proper separation is the cohesion of the flexible layer. In Fig. 21.3 they are separated: connecting (2), flexible (1); on leather these may be one or two coats;

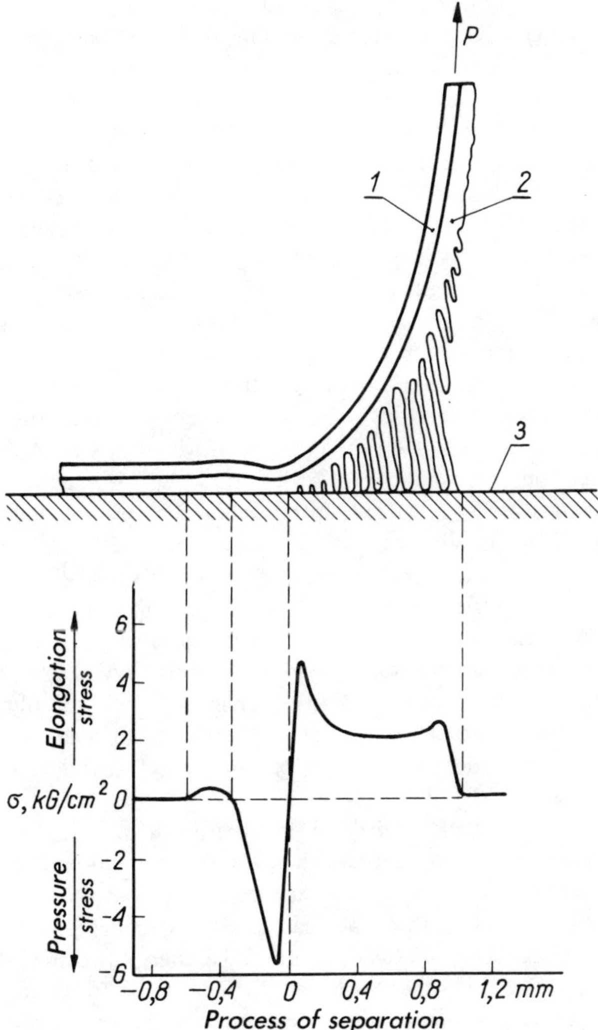

Fig. 21.3. Separation of two cemented bodies: stiff and flexible. A graph of forces acting in the system.

(2) forces during separation are not equally distributed: in the 'head' the flexible layer has increased thickness, then it is decreased and finally action of the force is a reason for forming holes in the connecting layer. Stress is the greatest where the holes are originally formed, then it decreases once more, to increase again just before the layers come apart. At a constant value of the disconnecting force P and the rate of disconnecting r in this picture distribution of stresses and reaction of solids are shown on viscoelastic deformation in the connecting layer and on micromechanism of tearing.

In peeling off a coat other components participate, in addition to stretching in a direction perpendicular to surface. Among them are: shearing, which acts parallel to the solid surface, multidirectional stressing (like in lastometer test), and multidirectional external lesion. As a rule, flexing testing is applied, as a practical test for evaluation of firmness of coat-leather connection.

The viscoelasticity mechanisms, acting at a point where two polymeres are connected, may be described by rheological models (sect. 1.2). There are two classical determination methods for these values: static and dynamic. In the first method deformation is taken as independent variable, and the stress measured, as the dependent one. According to the second method, the rheological properties are analyzed, taking into consideration the history of stresses (fatigue), and deformation, as independent variable. In the first method, rheological properties are tested in relation to the relaxation spectrum, in the second—to retardation spectrum. If the problem is to be defined in terms of the criterion already mentioned the former type shows the system of Maxwell's elements in a row, in which stress on every component (piston or spring) is equal to total stress, and deformations are summarized. In the latter type pistons and springs are settled parallel, deformation of every element is identical, and total stress is distributed between particular elements (Kelvin-Voigt's model, sect. 1.2). Both systems are too simplified to be a behavior model for polymer layer on leather, particularly because these layers are filled with pigments, plasticizers and pre-servatives. No theory as yet has been worked out which would allow us to predict the behavior of polymer coats on leather. Thus frequently the films tested are obtained by evaporation or coagulation of the finish. In such cases testing of relaxation and retardation behavior of polymer may provide information ex-pected.

Relaxation time τ_R is determined in an experiment in which deformation is a constant value (stressing to definite elongation). Retardation time is measured by determining the creeping under constant load or by determination of reversion time for sample, set free from the load. Kelvin's model illustrates well the properties of viscoelastic substance. Part of the energy becomes stored in the spring; another part shifts the piston, which is a reason for deformation retar-dation. By removing the load, the energy stored in the spring will be used for reversion to the starting position. Such a reversion is delayed by the action of the piston. The time required for coming to equilibrium is the retardation time τ_J. In this model, creeping occurs and deformation is expressed by the formula:

$$J = J_o \left(1 - e^{-\frac{t}{\tau_J}} \right)$$

where J is susceptibility to creep in time t, and J_o—susceptibility to creep in time t = 0. Susceptibilities are the reverse values of moduli $\left(J = \dfrac{1}{G} \right)$.

When a body is behaving according to the Maxwell model it will be deformed up to a constant value, in time t = 0 only the spring is stressed, and the total energy equal to the energy stored. With time the piston, under the influence of the reversing spring, starts to 'flow' and the energy from the spring comes over to the piston, where it is used for overcoming friction. The stress decreases slowly approaching zero. This is described by equation

$$G_t = G_o e^{-\frac{t}{\tau_R}}$$

The modulus decreases to ½ of its initial value, when $\tau_R = t$. According to Debye's equations, the J and G values for cyclic deformations, one obtains, taking ω (frequency) at reverse value of time:

for Kelvin's model
energy gain

$$J'(\omega) = J_o \frac{1}{1 + \omega^2 \tau_J^2}$$

for energy loss

$$J''(\omega) = J_o \frac{\omega \tau_J}{1 + \omega^2 \tau_J^2}$$

for Maxwell's model
energy grain

$$G'(\omega) = G_o \frac{\omega^2 \tau_R^2}{1 + \omega^2 \tau_R^2}$$

energy loss

$$G''(\omega) = G_o \frac{\omega \tau_R}{1 + \omega^2 \tau_R^2}$$

If from these formulas the curves for J and G as a function of $\ln \dfrac{t}{\tau_J}$ and $\ln \dfrac{t}{\tau_R}$ are plotted, the graphs are obtained (Fig. 21.4) which define the shape of functions and hysteresis loops. These curves being very close to a mirror reflection allow us to describe the material storage and some of its properties as the deformations and stress during wearing are alike.

More extended discussion of rheological properties of plastics may be found in Kaelble's textbook (loc. cit.) or in other sources, e.g., in Lenk's monograph [4]. There are no monographs published dealing with finishing coats only. Many data based on experiment are described in technical literature.

Investigations on rheological properties of coating materials are frequently done, mostly in a simplified way, like, e.g., by Toth and Posa [5], or Pector et al. [6]. In the first of these papers, deformation and stress have been tested for coats, in dependence on their casein and pigment content. In Fig. 21.5 the dependence is shown for deformation on the casein amount in mixture. It can be seen there that deformation work increases when the casein amount increases. When the curves for high casein content are in their first part they are steeper than they are level, which shows the decreasing elasticity of mixture with increase of the casein content. With increasing caseine content the losses due to hysteresis are increased and the speed of recovery of the initial shape after removal of the load decreases. The authors think the properties have some connection with flexing properties of films, as one may see from the results shown in Table 21.1. It is an obvious difference between properties of the films obtained by pouring a composition on a smooth surface, and the film on leather; however, this method

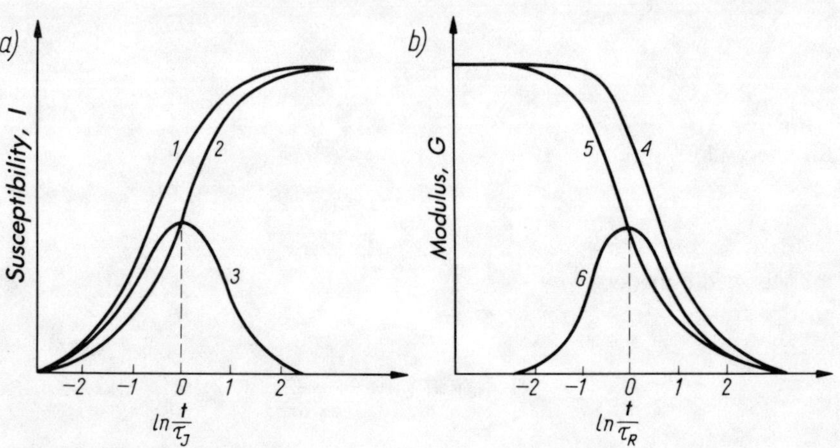

Fig. 21.4. Behavior of bodies:
(a) according to Kelvin's model and
(b) according to Maxwell's model.
(1) Complex susceptibility,
(2) "stored" susceptibility,
(3) lost susceptibility,
(4) complex modulus,
(5) "storage" modulus,
(6) lost modulus.

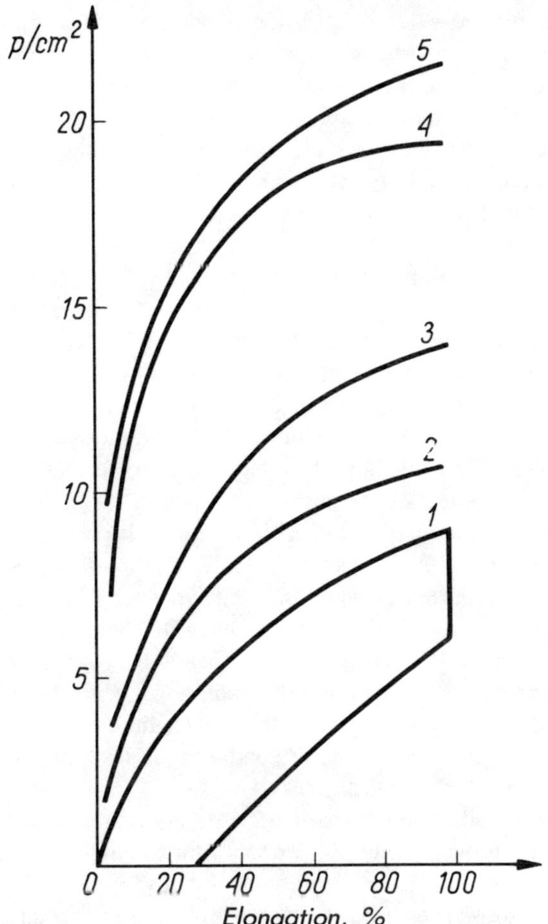

Fig. 21.5. The effect of casein addition to the binder on tensile strength of the film obtained (1) 0% casein in dry substance, (2) 15%, (3)30%, (4) 50%, (5) 70%. According to [5]

Table 21.1.

Dependence of flexing resistance on coat elasticity

Content of 30% casein in top	Hysteresis loss after 5 min of stressing	Rate of deformation decay $mm\ s^{-1}$	Flexing, kilocycles needed for obtaining visible damage
—	69	6.5	40
10	87	4.5	10
15	93	3.0	6
35	96	—	0.2

may provide much more exact information about them, then observation of leather behavior with the investigated coat on it. Most useful for obtaining films are Teflon® plates, because the films do not stick to them.

In the Pector work, authors were looking for correlation between simple parameters obtained by testing films and solutions on one hand and properties of ready coats on the other. They have found some dependences, however, for in the majority of parameters tested no correlations could be found. The following parameters correlate:

(1) Tensile strength of a film is approximately inversely proportional to elongation at break;

(2) Plasticity of a film (persistent deformation by stress) and its strength in second stressing cycle are dependent as follows: films having a strength of 40 kg/cm² (4 MPa) and over are of high and medium plasticity; of a strength of 15-40 kg/cm² (1.5-4.0 MPa) medium or minimal; those of strength below 15 kg/cm² (1.5 MPa)—medium plasticity. Similar plasticity minimum is observed when investigating the correlation plasticity vs. elongation at break. Films of great elongation are sticky. Authors have also observed reasonably good correlation between film strength and elongation modulus;

(3) Flexing resistance decreases when plasticity increases. Tested substances having plastic component of elongation over 0.4 have weak or medium resistance; substance of plastic component below 0.2—good or very good. For the medium values one may introduce additional dependence: the greater the force needed for film elongation, the greater the flexing resistance;

(4) Film strength correlates well with dry and wet rub fastness: the greater the resistance, the greater the rub fastness.

According to Pector, other properties of films do not correlate with properties of the coat formed by them. Over 400 various film-forming substances (called binders in tanning industry) have been used in his investigations.

Among modern substances used in compositions are the products of drastic solubilization of collagen from trimmings or damaged hides. The investigations have shown great stability of these compositions. They are more stable than casein coats, used for a long time for shoe upper. Collagen imparts to coating compositions very good plasticity [7]. However, the leather, finished with the use of a collagen containing agent is not air-permeable which is disadvantageous from the wearing viewpoint. Probably this is because under the conditions applied collagen becomes disordered and thus more gelatin-like *(parent gelatin)*.

The basic esthetic asset of upper leathers is their natural look. It is mainly obtained by the aniline effect.

21.2. Aniline effect of leather finishing

Among modern finishes of leathers in first place mention should be made of the aniline ones, as they do not spoil the natural appearance of leather, which is always highly appreciated. It is, however, not always possible to apply: if the leather grain is damaged, the 'semianiline' or 'aniline-look' finishes are made. There are many ways to obtain aniline effect. Accordingly, definition of aniline leather is not always connected with definite way of finishing. Also it is not necessary for the grain to be undamaged, as it may be buffed slightly. Classical aniline is a just colored leather. However, because some fastness is required, top coats and some hydrophibization are applied. In Fig. 21.6 a typical aniline finish is shown. It consists of a coat with shiny layer 1, then binding layer 2, containing particles of organic pigment suspended, of diameter below 500 nm. Diameter of pigment particles should be below the wavelength of visible light; thus they are transparent. Optimal diameter of pigment particle should be about 200 nm. Then incident light will be dispersed merely to a slight extent. Reflection occurs on the leather surface or on the ground applied on it. If the pigment ground is lighter than the coat, the 'aniline effect' is nicer and more distinct and the leather is brightening during stressing. This effect is particularly characteristic when the coat is of pure, bright color. Too great color differences are unfavorable. It is undesirable to have the organic pigments of large particle size. High dispersion level has to be maintained in the coating layer, formation of greater clusters (aggregates) during coating, and fixing of this layer would disturb its

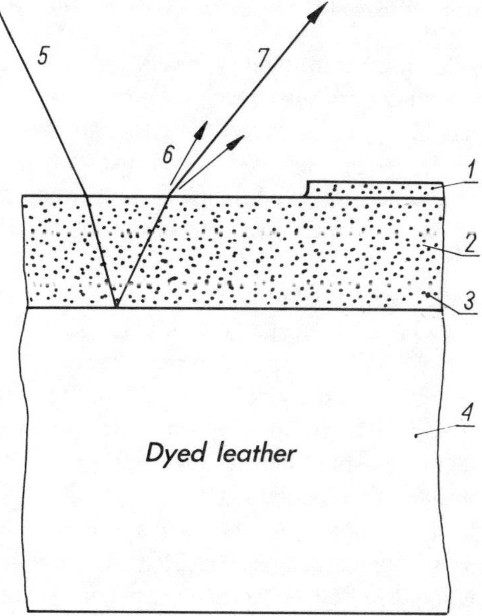

Fig. 21.6. Principle of aniline finishing of colored leather: (1) shiny layer, (2) layer, binding the shiny with pigmented layers, (3) pigment grain, (4) leather, colored, (5) incident light, (6) dispersion (slight) of reflected light, (7) apparent visible surface.

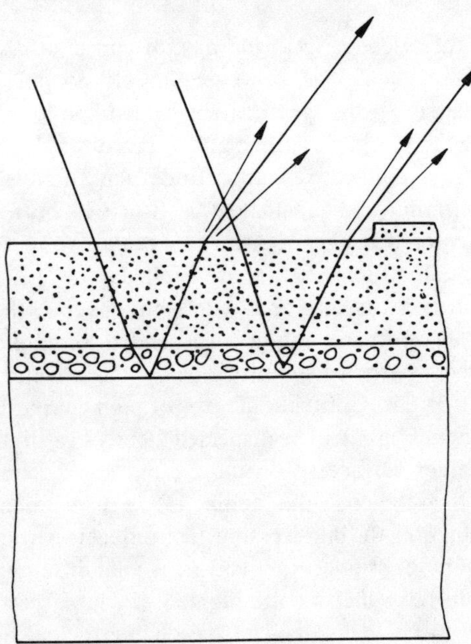

Fig. 21.7. Principle of semianiline finish. Refractive index in coating layer is higher than in the shiny layer; dispersion is somewhat greater than in aniline leather.

transparency. Thus the dispersion stability is very important, as well as the ratio of dry substance of pigment/dry substance of the binder which forms this layer. The more transparent the pigment layer, the easier to impart a natural aniline look to the leather. If the aniline leather is finished with a coat containing dispersed organic pigments, its rub fastness is improved, as well as its light fastness. Thus it is more durable, and the possibilities of its preservation are better.

Semianiline leather, a schematic cross section of which is shown in Fig. 21.7, has two coats on it. One of them has characteristics like the coat on aniline leather. It is rather thick, when compared to the other one that contains pigment particles, dispersing light, of a diameter over 800 nm. As a rule, these pigments are inorganic. The second layer, in which light reflection occurs, covers the leather surface, making its small defects invisible. In this way an effect very close to the aniline one may be obtained.

'Aniline-look' leathers differ from the semianiline ones by the thickness of layer containing coarse pigment grains (Fig. 21.8). Leather surface is completely covered with them, and the effect of 'aniline' dispersion is obtained by reflection in coat, breaking and dispersion in transparent layer. This finishing may also be called 'aniline-type finishing.'

Pigment finishing, shown in Fig. 21.9, does not contain a layer pigmented

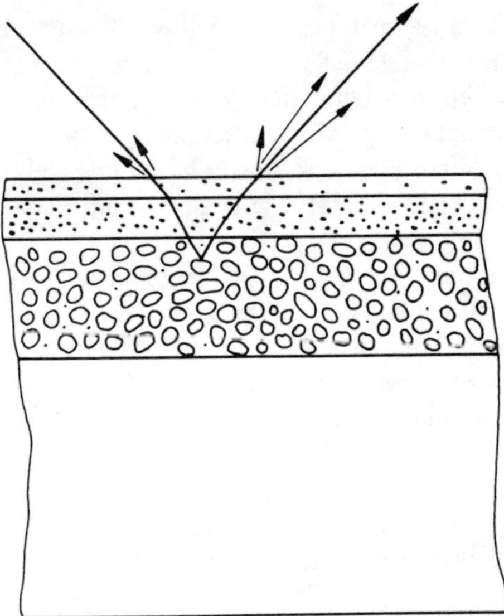

Fig. 21.8. Principle of the 'aniline look' finishing of noncolored leather; pigment grains are of greater diameter in bottom layer.

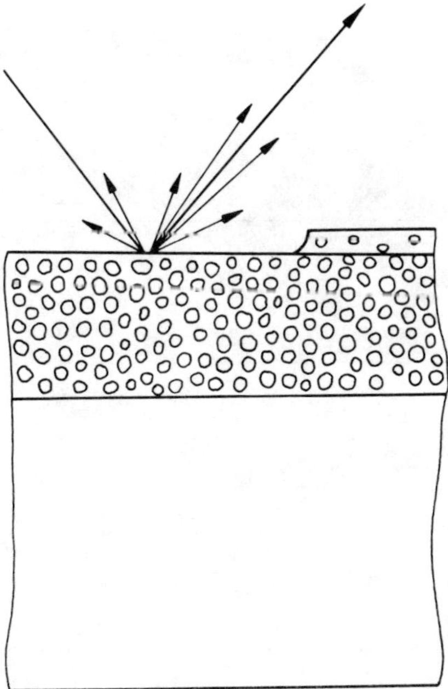

Fig. 21.9. Principle of pigment finishing of noncolored leather; very high dispersion.

with organic pigments of small particles (below 500 nm). The thick coating layer reflects and disperses visible radiation, whereas the thin shiny layer almost does not participate in light reflecting and dispersing. The coating layer can reflect or absorb incident light, depending on its color. White color reflects the light as do the majority of pastel colors. Binder layers give yellowish tones on white leather; so they are usually slightly pigmented, and the coating layer is supplied with some optical brightener.

Pastel hues of aniline-look leather can be obtained, if a thin, light-pigmented shiny layer (Fig. 21.10) lies on a heavy pigmented covering layer.

The foregoing review of methods of aniline-look and of 'aniline-type' finishing shows that this most typical for leather kind of finishing is an effect of two layers: covering and transmitting. The ratio of their thickness and the size of the pigment particles contained in them are the factors controlling the efficiency of aniline effect.

21.3. Easy care leathers

Modern industry, having in view the consumer's interest as its ultimate target, recommends a special method of finishing called 'wipe and wear.' Leather, finished in that way, does not require special care of brushing and using of shoe

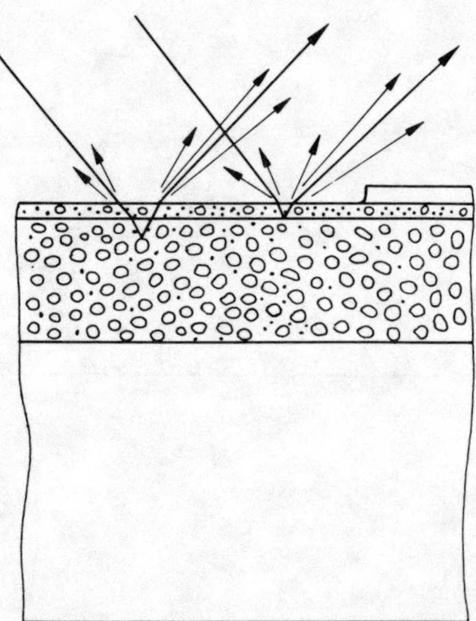

Fig. 21.10. Principle of pastel 'anilinlook' finish of noncolored leather. Shine slight matted by the last coat.

polish for removing dust and mud. Finishing by coating with polyvinyl resins in organic solvent, with polyacrylic resin, also in organic solvent, with an addition of a hardener, are in this group. Finishing with such polyacrylic resin is very resistant to mechanic damage; it resists rubbing off and is not easily scratched. A disadvantage of polyacrylic or polyvinyl resin finishes is their poor water vapor and air transmittance.

A typical easy care finish is the polyurethane finish. Polyurethane chains may be modified in various ways: extended, crosslinked, and stiffened by some segments built in, as, e.g., benzene rings with several functional groups. Polyurethane resins are expensive, but they provide significant stability, resistance and decreased sensitivity to external factors. If monocomponent polyurethane finish is applied, water vapor from the air is a crosslinking agent for it.

Polyurethanes obtained by polyaddition reaction are of great importance as substances of unusually valuable wearing properties. They are discussed and described in many monographs and textbooks.

REFERENCES

1. Kaelble, D. H. Physical Chemistry of Adhesion, Wiley N.Y. 1971
2. Nursten, H., Pearson, C. J. Soc. Leath. Tr. Chem., *48*, 456 (1964)
3. Tancous, J. J. Leather and Shoes *169*, 34 (1976)
4. Lenk, R. S. Rheologie der Kunststoffe, Hanser Munich 1971
5. Tóth, G., Pósa, W. XIII IULCS Congress Vienna 1973
6. Pektor, V., Dederle, T., Ambroz, M. ibid.
7. Pietrzykowski, W., Banaszewska, B., Wieczorek, J. Prace IPS *17*, 65 (1973) in Polish

22.

LEATHER DRYING

Tanned, dyed and fatted leathers are dried. Substances introduced into leather in the preceding operations are already in contact with collagen. They may still remain in water medium as solutions or emulsions. This state is changing during the drying, thus this process cannot be considered only as removing humidity from the leather. Its other aim is to trigger chemical reactions between leather and chemicals by concentration of solution. These two aims make drying a very important moment in leather processing.

22.1. Course of drying

A graph in Fig. 22.1 shows the content of water vapor in the air in dependence on temperature and relative humidity. The drying process may be controlled significantly by these two parameters of substantial importance. Among other things, technologically important variables are: air motion, kind and structure of material and speed of acquiring a defined humidity level. Under structure the size, distribution and shape of pores should be understood.

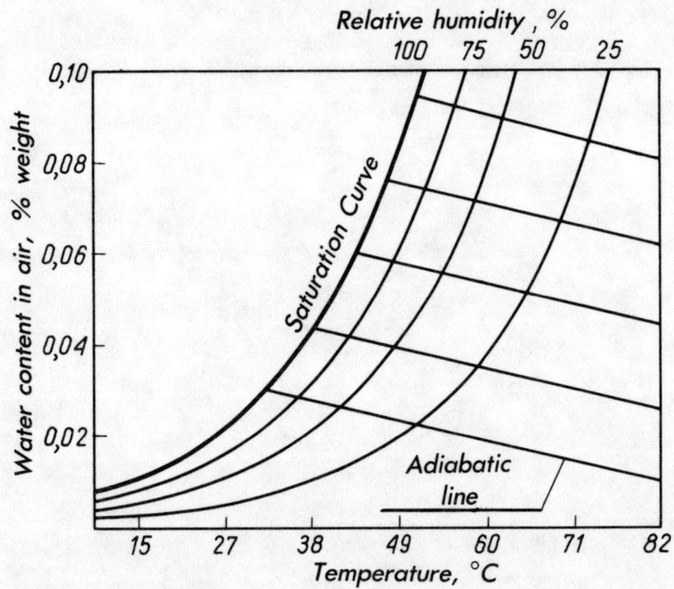

Fig. 22.1. Water vapor content in air versus temperature and relative humidity (simplified).

Temperature decrease and humidity increase are in regular dependence. Due to this the adiabatic line is straight—its slope is constant. By extending it to intersect with the saturation curve, the wet bulb thermometer temperature in hygrometer is obtained. Comparing the temperature shown by both thermometers, air humidity can be calculated or found in the tables. If air drying is introduced, having the humidity of 100% and temperature of 20°C, then it is heated in a dryer to 100°C, and the temperature at the exit, indicated by the wet bulb thermometer, is 80°C, then this difference (20°C) defines saturation degree with the vapor of the air coming out. The differences in temperature of the supplied and leaving air, and in its vapor content, depend on the atmospheric conditions, as in tanneries these parameters for the air supplied are not controlled.

22.2. Water content in leather

To analyze the drying operation the values are required, characterizing the water content in leather, the strength of bonds between water and leather and their character. These problems are discussed in sect. 5.4. For technological purposes they are treated jointly.

The water content in chrome- and vegetable-tanned leather, shown as a function of percent relative moisture, is of characteristic sigmoidal shape, shown in Fig. 22.2.

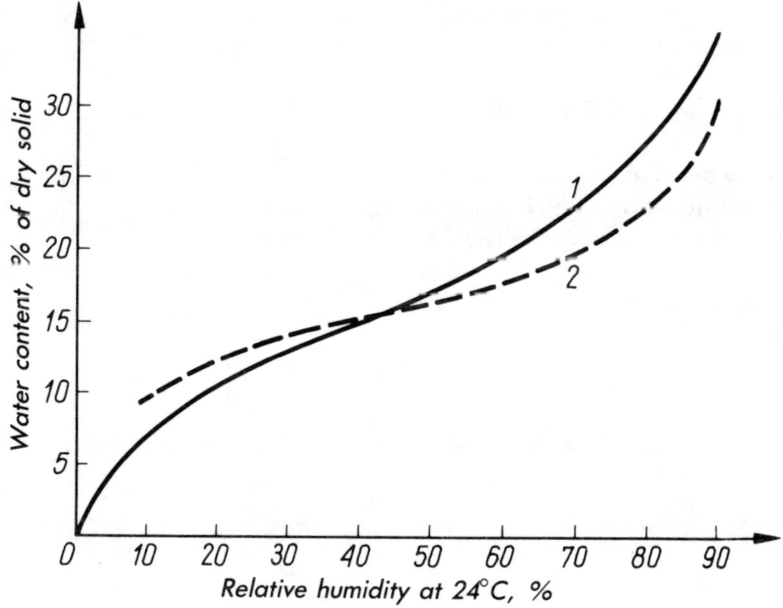

Fig. 22.2. Water content in leather in relation to air humidity: 1—chrome-tanned leather, 2—sole leather, vegetable tanned.

This sigmoidal shape differs in details, in dependence on the type and the method of tanning and all operations of the process are of importance. Leather characteristics change in dependence on the moisture content. With moisture increase suppleness increases and the surface as well. If the humidity is too great, suppleness converts into flaccidity.

By leather drying the equilibrium water content, i.e., the content by which no exchange with atmosphere of a constant humidity occurs, is of crucial importance (see sect. 5.4). Determination of this moisture can be done under static or dynamic conditions; measurements may be made in a number of different ways. This value is of particular importance for drying, as it is the limiting value for a given humidity and temperature of the air supplied. Forcing beyond this limit in the direction of lower values is dangerous, as it results in deep and irreversible changes in the structure of the leather collagen fibers. In Table 22.1 the influence of leather moisture content on its basic physical properties is shown. As can be seen from the Table, coefficients for dry leather, accepted as 100, increase markedly as leather moisture increases. This is more probable for chrome-tanned leather. Thus the necessity of conditioning the samples in an atmosphere of a constant, definite humidity before physical testing is obvious. In most countries this is 65% relative humidity; in the U.S.A. it is 50%. A constant temperature of conditioning is required as well because the relative moisture of the atmosphere changes significantly with changing temperatures. This is of particular importance if in the humidity control a saturated solution of inorganic salt is used.

22.3. Theory of drying of solids

During the drying of solids two basic processes simultaneously occur:
(1) transfer of heat which makes the liquid evaporate;
(2) transfer of mass of the liquid to be evaporated.
The factors controlling all these processes determine the effectiveness of the drying operation.

Table 22.1.

Leather humidity effect on its strength and extensibility

| | Relative humidity, % | | | | | | |
| | 0 | | 33 | | 52 | | 97 | |
Kind of leather	tensile strength	extens-ibility	tensile strength	extens-ibility	tensile strength	extens-ibility	tensile strength	extens-ibility
Vegetable tanned	100	100	113	113	113	113	98	120
Chromium tanned	100	100	144	135	153	135	171	147

Industrial drying uses convection, conductance or radiation of heat or their combinations. Independent of the heat transfer mechanism, it comes first to the surface of a solid, and then—to its interior. An exception is high-frequency drying, which produces internal heat. In such case the inside temperature of a drying solid is higher than the external one. This in turn results in heat flow from inside to the surface of a drying solid.

In drying mass will be transferred: as liquid and/or vapor inside a substance, and as vapor, from its surface. Concentration gradient of liquid is dependent on its flow inside a solid.

Investigation of the way of drying of a solid can be based on a study of flow mechanism of liquid or on the influence of external conditions on the internal ones. This influence is exerted by: temperature, humidity and the amount of air flowing. The former testing method is seldom used, despite the fact that it gives results, theoretically fully explained. Internal liquid flow may have various mechanisms; which of them is working in a given case depends on the solid structure. Among these mechanisms are:
(1) diffusion in the solid of compact structure,
(2) capillary flow in granulated and porous substance,
(3) flow due to gradients of shrinkage and pressure,
(4) flow by gravitation,
(5) flow due to sequences of evaporation and condensation processes.

The mechanisms mentioned do not exclude each other. It would be justified, however, to speak about predominance of one of them under definite conditions. The prevalent mechanism may change as well in dependence on the drying step.

One may differentiate the mechanisms of liquid flow, e.g., by observation of moisture gradients in dependence on the distance from the surface. In Fig. 22.3a the difference is shown between capillary and diffusion flow. Capillary flow (Fig. a) gives a curve of a sigmoid shape, having two curvatures and an inflexion point. Diffusion flow, however, (Fig. b) gives a curve of one curvature, concave upwards. The diffusion rate has no constant value; it decreases with rising moisture content. In Fig. b the dashed line is appropriate for a system where constant diffusion is presumed; the continuous line is the experimental one.

Among the phenomena observed in leather drying are: capillary flow and diffusion of vapor and liquid. The course of capillary flow, as considered in drying technology, is as follows: water in solid is a liquid, covering its surface and filling out the pores. In sect. 5.4 it has been shown that in fact the matter is more complicated: state of water in part most closely bound to surface of collagen fibers is an intermediate one between solid and liquid. Water undergoes gravitational motion. In the drying process the liquid flow at a humidity greater than the saturation point (equilibrium with the surroundings) occurs in fabrics, paper, leather, etc.

Diffusion is also a reason for the coming of vapor through a solid if a temperature gradient due to heating exists there. Vapor diffusion and evaporation

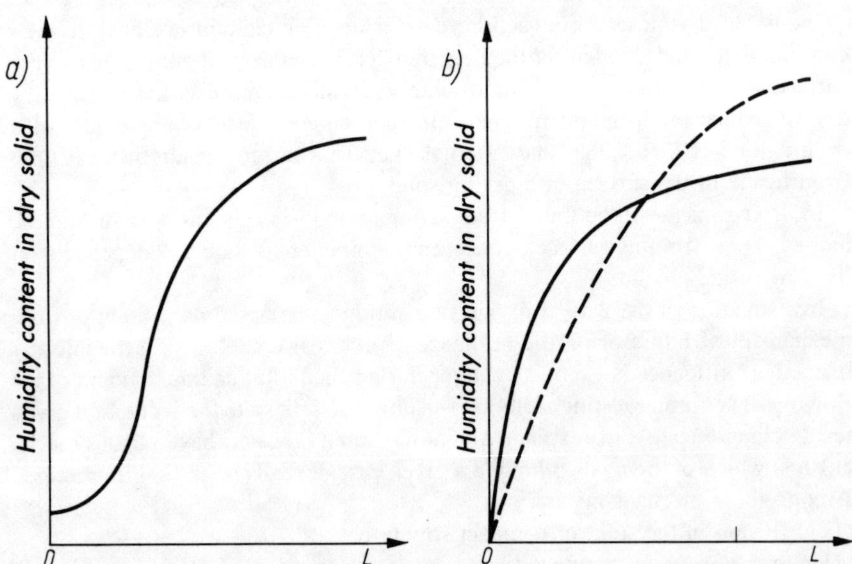

Fig. 22.3. Moisture flow inside solid: (a) capillary, (b) diffusion; L—distance from surface.

always occurs if a solid is heated on its one side and it dries on the other, as it occurs by leather vacuum drying, or if the liquid clusters are separated by structure of solid.

Diffusional motion of liquid is limited to a moisture content below the atmosphere saturation point. The last stages of periodical drying belong to that type.

External drying conditions are a typical problem in chemical engineering. Optimal conditions for drying materials are to be found. As basic external values, one may mention air temperature, humidity, and flow, condition of the material to be dried, that is, motion and contact surface between hot plates and a moist solid.

Results of experimental drying of solids are usually expressed as dependence of humidity on time, and more correctly, as the moisture content W versus time t (Fig. 22.4). This curve shows a general case, when a moist body loses its water, starting from a free, wet surface. Then this surface decreases, and the evaporation process gradually comes into deeper layers. Dependence of the drying rate on time (Fig. a) may be demonstrated more clearly if we differentiate the curve of Fig. a and plot $\dfrac{dW}{dt}$ vs. humidity W or time t (Figures b and c).

From these pictures it can be seen that drying is in fact not a unit process, controlled by action of one mechanism. Plotting it as in Fig. c has the advantage of showing at which time every part occurs. Segments of the curves have a following physical sense: segment AB corresponds to the heating period; segment

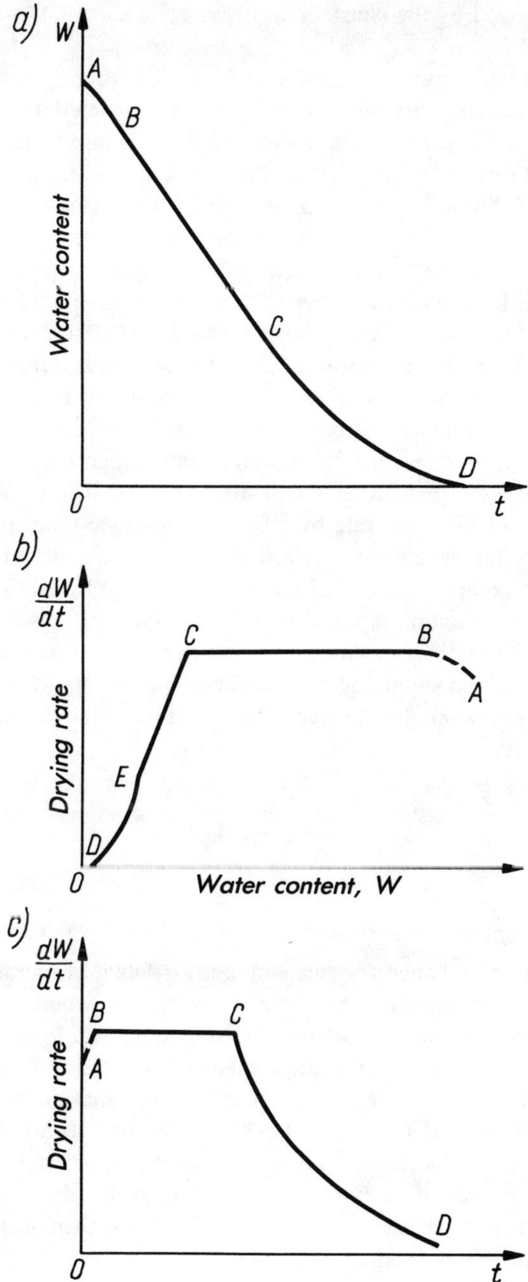

Fig. 22.4. Methods of characterization of the drying process: (a) water content as a function of time; (b) drying rate vs. water content, (c) drying rate vs. time.

BC—is characterized by the constant drying rate, segment CD—by decreasing the drying rate.

Point C, at which a transition between BC and CD occurs, is defined as the critical moisture content. Drying in the BC period is characterized by evaporation of free water from the solid surface. Evaporation rate is in principle independent of the solid, and drying is equivalent to water evaporation from its free surface. The drying rate is limited by the diffusion rate of water vapor to the air stream flowing through the immobile layer of air, adherent to the solid. The constant evaporation rate on the surface of a solid maintains it at a constant temperature. This temperature if no additional factors come in question is close to the psychrometer wet bulb temperature. It is somewhat lower if heat needed for evaporation reaches the evaporation surface by radiation or transfer and convection. If the heat comes to the wet body by transfer through hot surfaces and convection is not very great, the temperature is closer to the boiling point than to that of the wet bulb. In such a case the drying rate is much higher than in convection drying at the same temperature as the heating surfaces. The result of radiation is an increase of evaporation rate by increase of surface temperature. Usually it is a secondary factor against convection or transfer; however sometimes it becomes a basic component, e.g., by evaporation by infrared radiation. If the heat needed for evaporation in a constant drying rate period is supplied by hot gas (usually air), a state of dynamic equilibrium condition sets up between the rate of heat-to-material supplying and the removing of vapor from the surface. This equilibrium between the transfer of mass and heat may be expressed by the following formula:

$$\frac{dW}{dt} = \frac{h_\tau \, A\Delta\tau}{\lambda} \; k_g A \; \Delta p$$

In this formula $\dfrac{dW}{dt}$ is the drying rate, h_τ is total coefficient of heat transfer, A—surface of heat exchange and evaporation, λ—latent evaporation heat at T_s temperature, k_g mass transfer coefficient; $\Delta\tau = \tau_a - \tau_s$ where τ_a is air temperature, and τ_s—temperature of surface of evaporation, $\Delta p = p_s - p_a$ where p_s = vapor pressure of water at τ_s i.e. surface temperature and p_a = partial pressure of water vapor in the air. From this formula it results that the evaporation rate at a constant rate period (BC sector) depends on three factors: mass or heat transfer coefficient, surface contacting with the drying agent, and the temperature difference between the air flowing and the surface of the dried solid. All these factors are external variables: internal liquid flow mechanism does not affect the drying rate when it is constant.

Drying in the falling-rate period (sector CD) starts at a critical moisture content C. If the corresponding value is below the humidity required for the dried material, the process will be finished at a constant rate. Segment CD has two zones: a zone of unsaturated surface drying, and a zone controlled by internal liquid flow. The first zone follows immediately after the critical point. The decrease in the rate of drying in this zone is caused by a decrease in the wetted surface area of the material, i.e., the surface is no longer wetted as a whole. Dry elements of the solid which are 'immersed' in adherent air layer decrease the part of the surface of evaporation in the whole surface. Frequently it is presumed that the effective wetted surface in this zone is a linear function of water content. Due to this the drying rate is changing directly with average water content (sector CE in Fig. 22.4b).

The drying mechanism is in principle the same as in drying at a constant rate. Thus the influence of external factors is in principle the same if one keeps in mind the decreasing wetted surface area. The second zone is of decisive importance for determining the drying rate if it has to go down to the humidities below the critical point. In this zone several various, independent controlling mechanisms are operative. Most important of them are: diffusion, capillary system and gradient of tensions due to shrinkage.

Drying rate in period CD may be expressed by the following equation:

$$\frac{dW}{dt} = \frac{\pi^2 D}{4L^2} (W - W_e)$$

where $\frac{dW}{dt}$ is the drying rate, D is diffusion coefficient, L—half of layer thickness liquid is diffusing through. If evaporation goes through one surface only, so L is equal to the whole thickness of the solid. W is average value of humidity after drying time τ, W_e—equilibrium humidity. This equation is true for gelatin, leather, glue, starch, fibers, fabrics, clay and other materials, if

$$\frac{W - W_e}{W_o - W_e} < 0.6$$

(W_o is humidity at point C). By integrating equation for the drying rate, we obtain the drying time in the falling-rate period

$$\tau = \frac{4L^2}{D\pi^2} \ln \frac{(W_c - W_e)}{(W - W_e)}$$

In order to obtain value τ one has to know the humidity at point C—i.e., W_c. This value may be obtained experimentally by finding out C-point. The C-value is given in tables for various materials; if it has to be really exact, it has to be confirmed by experiment. If the drying rate is high, so is the temperature gradient inside the substance high, and the average value is much higher than on its surface. Critical moisture content, as an average for material, increases with the drying rate and thickness of the layer. For chrome-tanned leather it is 125% water, calculated on dry weight, for vegetable-tanned—90%, for average wool—31%, for worsted yarn—8%, for gelatin—13%. Values for diffusion coefficient D are best found graphically.

Sense of shape and size of capillaries. To show the forces acting inside leather during drying, let us consider the sense of the shape of surface and the size of capillaries. These considerations are concerning idealized cases, as the real fiber network contains all variants.

On a molecule of a liquid located on free surface, various forces may act in dependence of the shape of the liquid and surface, whether it is convex, plane or concave. The molecules on concave surface are attached stronger to the liquid phase than if the 'surface' of the liquid is plane. Still weaker they are attracted if the surface is convex. Vapor pressure over the concave surface is smaller, and over the convex greater than over the plane one. Vapor pressure over smaller liquid drops is smaller than over the larger ones.

Relative humidity is a function of force needed to maintain the liquid level in capillaries, due to the pressure over liquid surface. Suction may be defined as a measure of affinity of water to the porous material. One may define it as well as theoretical height of the liquid column in a capillary, which corresponds to the force needed to keep water in the porous material structure. In an ideal capillary, the surface tension decides the height of the liquid column in the capillary and of the meniscus shape. This is given by dependence

$$gh = \frac{2\gamma}{dr}$$

where g is gravitational constant, h—liquid column height, γ—surface tension between water and air, d—water density and r—curvature radius of the meniscus.

Vapor pressure over the meniscus P depends on the surface curvature. It differs from vapor pressure over plane surface p_s—the difference is a value equivalent to the mass of the vapor column of a height h. Values h and r are bound together, so the pressure over curved surface may be related to curvature radius. This is determined by Kelvin's equation

$$\ln \frac{P}{p_s} = \frac{2\gamma M}{rdRT}$$

where M is molecular weight of water. From both equations above we obtain

$$h = \frac{RT}{M_g} \ln \frac{P}{p_s}$$

this is true for real h, as well as for the hypothetical, and connects the water column height with the curvature radius of the water surface in the pores of the material. This value can also be defined as suction S. It is a real index of changes in proximity of the saturation point with water.

In Table 22.2 some relations are given between ratio $\frac{P}{p_s}$ (in %) and water column height h (in mm) (or capillary rise), and value pF, being log S. This value is better to operate than value S.

Several methods of suction measurements and diagnostic value of this index are discussed by Penner [1]. It is suitable for measuring humidity of clay, plaster, fibers and other materials, leather inclusive.

Kelvin's equation, however, cannot be used for leather drying without additional conditions, as curvature radius of water surface in leather capillaries may be of a different value. It depends not only on the leather origin, i.e., animal species, but on the hide processing as well. In practical chemical engineering a substitute radius is accepted, which means the value at which experimental data fit to the formula. Water adsorption layers on the capillary walls have a concave surface; as water wets the collagen fibers and the saturated vapor tension over them is lower than over a plane surface. Condensation in these curved layers requires lower tensions than p_s of the substance adsorbed. Earlier condensation in capillaries, occurring before the reacting saturation in the vapor layer over the plane surface, is called capillary condensation.

Leather contains pores of various shapes and especially there is a difference between grain and flesh side. Introduction of theoretical values, valid just for one type of capillaries, does not provide a basis for conclusions concerning the behavior of material during drying.

Table 22.2.

Vapor pressure over matter in capillaries of various diameter

$\dfrac{P}{p_s}$	Curvature radius, nm	Capillary rise h, cm	pF
	0.147	10^7	7
0.08	1.47	10^6	6
49.0	14.7	10^5	5
93.0	147	10^4	4
99.3	1470	10^3	3
99.9	147000	10^1	1
99.999			

22.4. Industrial methods of leather drying

Papers devoted to leather drying contain primarily technological and machinery information. It is almost an impossibility to assume leather is a homogenous material. The general laws given above are explaining to some extent the drying process.

Leather is dried usually at a rather low temperature in order to prevent structural and chemical changes. The initial process temperature is lower (usually 40-50°C for chrome leathers, 35-45°C for vegetable-tanned ones). Then, as the water content decreases, temperature increases, however never higher than 8-10°C below the shrinkage temperature. It is done so to avoid local overheating and deterioration of components introduced in tanning and finishing.

Before drying, leathers are stacked and sammyed. Drying starts at 40-60% humidity (below point C). The constant-rate period may occur locally only due to nonequilibrated moisture content, or through the use of aqueous glue solution in pasting.

A typical process now is the two-step drying. In this process leathers are dried first from initial humidity 40-60% to 8-14%. Then they are stored in rooms of about 85% humidity until their water content becomes 18-22% and still moisturized in an atmosphere of a humidity close to 100% until they reach 26-32%. Such moist leather is mechanically finished and dried again. Traditional humidification is done by sawdusting, i.e., piling of leathers powdered with wet sawdust.

This way of humidity control is mentioned not because it is of special value, as modern climatization chambers make it much better, without contaminating leathers but to show how the important industrial processings are developed empirically. In the two-step process one obtains leathers of well equilibrated humidity and of proper constitution for mechanical operations, not breaking, soft and resistant. A more modern solution is the drying of leathers to attain exactly to the humidity needed in industrial operations. Condition *sine qua non* of this method is interoperational control, sometimes by hygrometers built into the machinery. Piling is still necessary after drying according to this system in order to equilibrate water distribution in various leather parts.

Among industrial drying methods are:
(1) air-drying without energy supply;
(2) air-drying with energy supply
   —with circulation,
   —tunnels, channels,
   —chambers,
   —pasting on glass plates,
   —tunnel drying on perforated metal plates or frames (stretched flat under tension by toggles);

(3) drying with water flow
   —Secotherm, a commercial pasting process in which the reverse side of
   plate comes into contact with warm and hot water;
(4) drying with infrared radiation;
(5) vacuum drying;
(6) high-frequency drying.

Air drying (method 1) is applied only in some special cases; methods 2-5 are based on the heat supply from outside, and removal of the leaving vapor; method 6 is based on the forced heat production inside material dried due to quick reorientation of water molecules resulting from quick changes in electric field directions. Reorientation energy will be transformed into heat. The advantages of this novelty just gaining in acceptance are: remarkable drying rate and automatic process control. The control follows by loss of reorientation possibility at definite frequency of changes in the field direction. If the amount of water molecules in material is decreased to a certain limit, their motion freedom becomes limited. Reviewing the data concerning the drying method by pasting may be found in the paper of Dolley [2]—those concerning vacuum drying—by Sharp [3].

Leather is dried either stretched, as in pasting, or toggled on frame, or pressed to a table in a vacuum dryer, or without tension, hanging free. Before drying the chrome-tanned leather may contain about 100-120% of moisture. This is in contradiction to the above-said (initial moisture 40-60%); the starting point of the dyeing, however, depends on introductory operations (sammying) and on the technique applied.

In such leathers stretching tensions may arise of about 0.54 kg/mm^2 (5.4 MPa). Plastic component of the deformation after removing of the load is about 90%. When dried without stretching, the elastic deformation relaxes, elasticity of leather increases significantly, its surface area, however, decreases. Rehumidification, stretching and drying, then drying on frame does not restore the initial surface to leather. In processing as a whole leather surface changes about several percent, sometimes even more than 10%. These changes are measured and registered only at last step of the finishing operations, as this surface is a base for store calculation. The time and the method of measuring are controlled by technological regulations. In order to demonstrate the differences resulting from single operations Skelham's observations [4] can be used. Taking a surface of a tanned, chrome grain leather (sides) for 100, he found after toggling, 108%, after drying—102%, after taking off the frame 101%, after conditioning 102%, after staking (Molissa) 99%, after vacuum drying 97%, after finishing 96.5%. The figures for leather with corrected grain are similar.

During drying by pasting several physical processes are running simultaneously. Plastic component of the deformation decreases; in the final stage it is about 50%. If an action is undertaken whose purpose is to decrease the plastic component (use of springs), the tension in leather increases to 0.9 kG/mm² (9 MPa). With this increase a shrinkage of capillaries occurs, due to which either the leather surface is decreased or, if it is impossible (toggles), the tension will still become higher. In capillary tension the surface tension of liquid is a significant component. In the last drying stage the collagen cohesive forces become operative (although this definition is not very precise). These forces fix the structure deformation.

At the same time, when the deforming forces decrease, relaxation in leather occurs which lowers the stresses. Relaxation is favored by high moisture of leather in its initial drying stage and at a high temperature of the medium. Thus the internal stresses, small at the initial drying stage, subsequently are increasing.

In the pasting drying process, which is now very common, of high importance is the kind of the glue used. The adhesive should hold stresses of an order of 0.5-0.7 kG/mm² (5-7 MPa). Thus for proper operation of the pasting drying it is necessary to choose plates of adequate surface and the glue to bind leather with appropriate strength. For reasonable drying it is necessary to know deformations occurring in leather of a given kind. Choice of the moment of cementing leather to plates is of fundamental importance (e.g., moisture content); one more thing to remember is to know whether the material has to be more or less plastic.

The drying process has also a chemical aspect, indicated at the start of this chapter: wet leather contains solution of substances previously introduced into it. This solution is concentrating during drying. Simultaneously, with decrease of the volume of the system, the collagen fibers are approaching each other as well as the substances introduced. Accordingly, e.g., to the tanning agent bound at one point a functional group of another chain may approach and further crosslinking of collagen may occur. Other reactions, like those between mole-

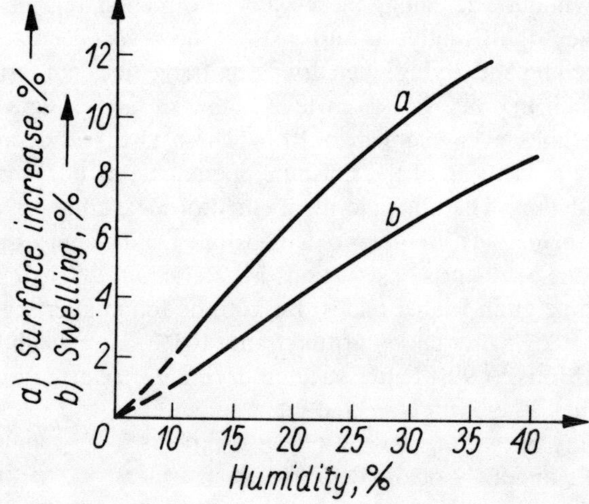

Fig. 22.5. Humidity and surface increase (a) and thickness increase (b). According to [5] with permission

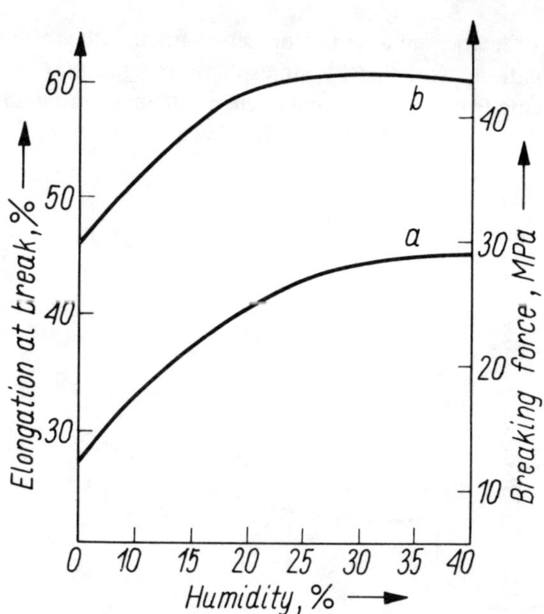

Fig. 22.6. Humidity and break load (a) and elongation at break (b). According to [5] with permission

cules of vegetable tannins, aggregate formation, etc., may occur as well.

The relations between humidity and surface, density, stress-strain ratio and dielectric parameters are shown in Fig. 22.5., 22.6. and 22.7., between humidity and heat conductivity or length and sorption isotherms, all for chrome-tanned leather are shown in Fig. 22.8., 22.9. and 22.10. [5]. As one may see in these summarizing figures, humidity is a very important factor for all properties in question and significance of these aspects of drying must not be neglected.

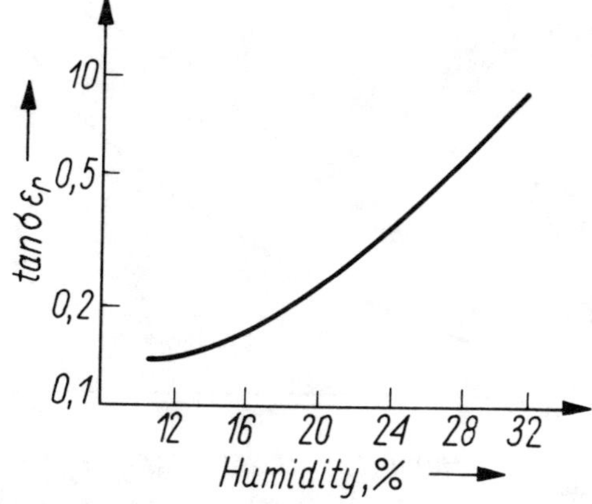

Fig. 22.7. Humidity and quotient tan $\delta \varepsilon_r$, δ loss angle, ε dielectric conductivity. According to [5] with permission

As the pores of leather are after this operation filled with air, beside reactions occurring in solution, oxidation by atmospheric oxygen may also take place. Thus leather drying from the standpoint of chemical engineering differs not much from the drying of other porous materials. However, it has some peculiarities.

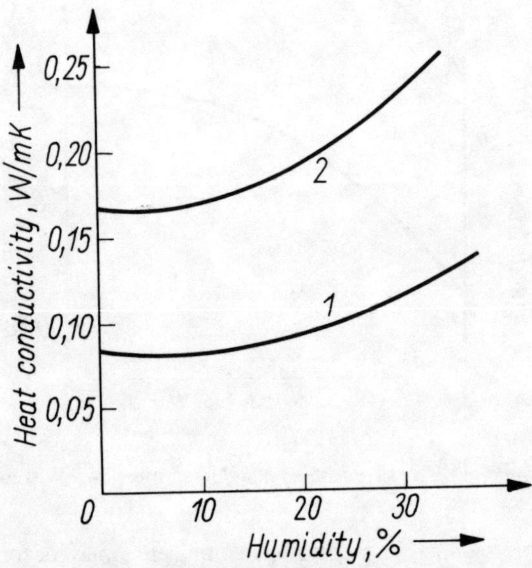

Fig. 22.8. Humidity and heat conductivity (1) chrome leather, (2) veg. tanned leather. According to [5] with permission

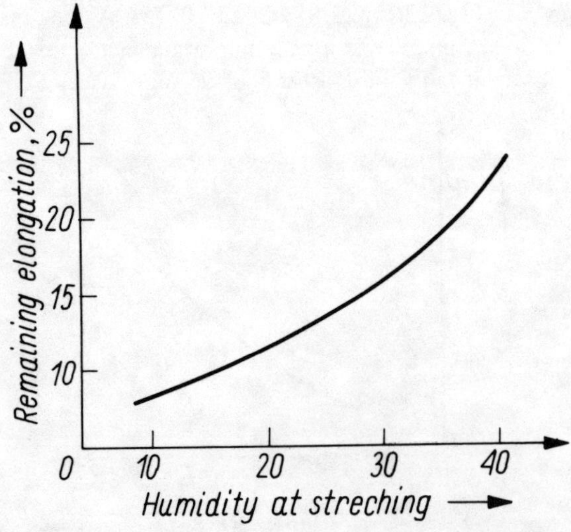

Fig. 22.9. Humidity and persisting elongation (initial elongation 30%). According to [5] with permmission

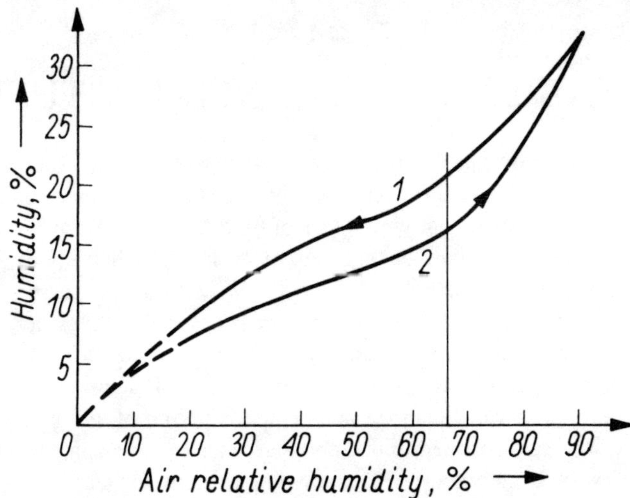

Fig. 22.10. Isotherm of sorption 1. sorption, 2. desorption at 20°C. According to [5] with permission

REFERENCES

1. Penner, E. in Humidity and Moisture (ed. Winn., P. J.), vol. 4 Reinhold N.Y. 1965
2. Dooley, W. E. Leather Manufacturer *1*, 25 (1973)
3. Sharp, B. ibid., *1*, 28 (1973)
4. Skelham, R. W. T. Satra Bull., *17*, 298 (1976)
5. Werner, W. (ed) Ledertechnik, Fachbuchverl. Leipzig 1979

23.

STABILITY OF TANNED LEATHER

Ancient leather items can be encountered today still in good condition. Stored under proper conditions, they may last for ages and for thousands of years without losing their shape but with a loss of their firmness. Many archaeological museums in their collections have goods made from leather: shoes, harnesses, gloves, armor parts, bags, bolts and other items, mostly of personal use. Old books used to be bound between wooden boards covered with leather, their pages are of parchment. Documents of historical importance are still now bound primarily in leather. In the Royal Castle (in Cracov, Poland) walls covered in XVIIth century with Cordovan leather (which is a kind of upholstery leather of sheepskins, vegetable-tanned and embossed) can be appreciated. These walls still have a very attractive appearance. Some reasons for leather stability have been explained previously, among them are the resistance against proteolytic enzymes, especially when dry, crosslinking bonds and specific helical structure, stiffened with iminoacid rings.

The factors responsible for leather instability are more frequently classed as follows:

(1) atmosphere pollution,
(2) climate factors, acting on stored leather,
(3) perspiration,
(4) molds and bacteria or more exactly enzymes of these organisms.

To the factors of atmosphere pollution radiation of various wavelengths should be included. Now more and more substances produced or obtained and accumulated by mankind do influence the biosphere by ionization of its gases. They may also have a strong influence by hitting leather with high-energy elementary particles.

Influence of atmosphere pollution. In vegetable-tanned leather stored in dry air, a very slow binding of the tanning agents to collagen proceeds as the amount of washable substances decreases. According to Stather [1], this is observed to a greater extent in leather tanned with quick-reacting tannins (chestnut, quebracho, mimosa) and to a smaller extent—when slow working tannins are applied (fir bark). However, even Faraday observed that the products of the gas-burning light are damaging bookbinding, upholstery leather and trunks. Fibers of leather become brittle and easy to pulverize, giving the so-called red rot. From analyzing this powder, a very high level of sulfur and acids can be found. The backs of books kept on shelves and exposed to the atmosphere deteriorate much faster than the remaining parts of the covers. Bookbinding leather may take up sulfur from the air in the amounts corresponding to as much as 8% of the sulfuric acid of their mass. Lifetime of a bookbinding may be extended many times if air exchange in storage rooms is restricted, and they are protected against the sunlight. Observations of Innes [2] allowed him to say that bookbinding leathers obtained by tanning with hydrolyzing tannins are more stable than those tanned by condensing tannins (of catechol type). Hydrolyzing tannins contain more pyrogallol-derivatives.

Some data concerning deterioration of leathers as it was observed during more than ten years in a library, were given in the paper of Beebe et al. [3] and are shown in Table 23.1. Comparison of leather aging in a library with the aging during 12 weeks in a chamber where SO_2 is generated does not give unequivocal results, as well as a comparison of the tanning methods used. Aluminum retannage has advantages as seen from Table 23.1.

Sulfur binding does not directly correlate with the sample strength, which may be considered one of the most important aging criterion. The presence of traces of ionic iron catalyzes the deterioration of vegetable-tanned leather, as may be seen on old iron studded upholstery. Sharphouse (private communication) reports that the presence of suitably high concentrations of iron-chelating ions (oxalate, lactate or EDTA) can greatly minimize this catalytic deterioration.

The authors of [3] emphasize that soluble nitrogen content does not increase significantly in leathers which have lost a great deal of their strength. It is rather surprising that some correlation between these two indexes could be expected. They are, however, more complicated.

The effect of ionizing radiation on collagen is a subject of protein chemistry interest. Because atmosphere even highly polluted contains just small amounts of radioisotopes and the problems of action of this radiation are rather the field of special services—they are not discussed here. Some related data can be found in Cooper's review [4].

Table 23.1.

Comparison of the aging degree of various leathers used for bookbinding. Leathers stored over 10 years in library rooms

Tanning	Leather	Loss of strength %	Sulfur binding calc. on H_2SO_4. %
Veg-chromium	cattle	6.1-36.8	1.9-2.5
Veg-aluminum	sheep	7.9-22.6	0.7-1.5
Chromium	sheep	19.0-58.3	1.5-2.2
"	cattle	20.8	0.9
"	goat	21.8-53.9	1.5-2.8
Veg-chromium	sheep	25.5-65.9	1.7-2.1
" "	goat	35.5-70.3	1.5-2.3
Vegetable	shark	61.9	3.1

Influence of the climate and wearing has been elucidated in a work done by Bowes et al. [5]. This extensive report covers several topics, among them those referred to above. Beside this, a particular interest presents the point concerning interaction between tanning agents. The work of Bowes, to which an interested reader is referred, may be only briefly summarized here.

The main cause of deterioration is the moist heat. All types of leather investigated were affected when exposed to increased temperature and high relative humidity of the atmosphere. Chrome- and aldehyde-tanned leathers were less influenced; the vegetable-tanned ones become dark and cracky; their strength decreased significantly. The most damaged were the chrome vegetable-tanned leathers, the slightest—glutaraldehyde-tanned ones. These results agree well with those obtained in Germany during preparation of World War II [6], and with those achieved in the U.S.A. during World War II, as there was a need to evaluate storage possibilities for military shoes, etc., in tropical climates.

In the work quoted [5] Bowes discusses the forms of protein degradation. It may take two forms—breakdown of polypeptide chain and disturbance of the molecular structure. Hydrolytic breakdown of polypeptide chain occurs in collagen stored in a warm and moist climate and is increased when acidity, temperature and humidity increase. Thus, when pH is below 3.5 it increases sharply. Probably the oxidative degradation occurs under the same conditions; however, the author, very cautious in drawing conclusions, has not found substantial proof of it.

Hydrolytic breakdown is reduced to some extent by tanning agents; even relatively small amounts of chromium salts and aldehydes are effective. Vegetable tans are distinctly less efficient, as they tend to oxidize and polymerize, thus changing their chemical activity and probably decreasing the amount of hydrogen bonds between them and collagen.

No significant changes in chromium complexes were found in leather, at least not of the kind which may be of influence on collagen deterioration. Damages resulting from wearing are essentially due to moist heat. They are accelerated—or mitigated—by the absorption of perspiration and by mechanical action. The most important factors in this case are detannage, pH rise, and accumulation of salts in the leather. Generally—the rise of pH reduces hydrolytic breakdown, while salt accumulation accelerates it.

The composition of perspiration changes; as an average it contains [7]:

NaCl	1.0-5.7 g/l
K^+-ions	0.04-0.6 g/l
Lactic acid	0.4-4.0 g/l
Urea	0.12-0.4 g/l
NH_4^+	0.05-0.09 g/l

Chrome leather is primarily damaged by perspiration due to detannage by lactic acid and by the rise of pH. The chief cause of damage of vegetable leathers is a pH increase and oxidation of vegetable tannins. Aldehyde-tanned leathers appear to be unaffected by perspiration except that the accumulation of salts affects the physical properties and water retention.

Ideally for maximum resistance to moist heat and perspiration a leather should have a pH of 5.0-6.0 in order to reduce hydrolytic breakdown of protein to a minimum and be crosslinked with a compound, forming stable covalent bounds. The importance of stable crosslinking bonds is sharply emphasized in Bowes's report. Glutaraldehyde was found to be particularly efficient in this case, as well as some cyanuric chloride derivatives, which was confirmed by Bieńkiewicz and Orszulski [8]. However because the use of cyanuric chloride for tanning is technologically difficult, the glutaraldehyde may be considered at present as the most important anti-deterioration factor.

As a result of hydrolytic breakdown, the scission of polypeptide chain with release of α-amino and carboxyl group occurs. This could be confirmed by Sanger's method of α-amino group determination [9] with fluorodinitrobenzene. Such hydrolysis is insignificant at pH values above 4.0 and at low relative humidities. It increases sharply with pH fall below 3.5 and at high relative humidities. Under extreme conditions (pH = 2.5, 40°C and 100% relative humidity) the average chain molecular weight becomes about 9000 and collagen is almost completely water-soluble. Tanning causes certain reduction in hydrolytic breakdown, particularly chrome and aldehyde tanning, as it reduces the molecular deorganization. Multiple hydrogen bonds caused by vegetable tannins are less effective. This was checked by investigation of heat resistance, which decreases more slowly in tanned leathers than in collagen itself.

Oxidative breakdown by free radical mechanism could not be proved by Bowes [5] due to the lack of a suitably sensitive method although this mechanism is supposed to act in the following way:

$$\text{-CH-CO-NH-CH-CO} \rightarrow \text{-CH-CO-NH}_2 + {}^-\text{O}-\overset{\displaystyle O}{\underset{|}{\text{C}}}-\text{O}-$$
$$\underset{R}{|} \qquad \underset{R'}{|} \qquad \underset{R}{|} \qquad \underset{R'}{|}$$

The breakdown of hydrolytic type is probably acting during deterioration of bookbinding and upholstery leathers. In this case the moisture content is very low, the times of exposure very long, and the pH value is continuously maintained at a low value by the absorption of acids from the atmosphere.

Interaction between vegetable tannins and chromium was suggested as the cause of poor stability of semichrome and combination-tanned leathers to moist heat. This view is supported by decrease of pH of combination-tanned leathers during storage [10]. Loss of chromium or prolonged vegetable retannage of chrome-tanned leather [11] and the fall of T_s in combination-tanned leathers exposed to moist heat [5] suggest that the vegetable tannins in the leather may displace collagen carboxyl groups from chromium complexes, thus causing detannage. This view was confirmed in Bowes's report by spectroscopic method, demonstrating significant changes in the absorption curves of the tanning agents in question when brought together and heated.

It is not possible to give a complete explanation of the reaction that occurs as yet, however there is no doubt that such a reaction takes place. It was shown as well that the presence of chromium salts increases the oxygen uptake by vegetable tannins, particularly by sulfited quebracho. These reactions are observed if the concentration of both components is sufficiently high, which is a rule in leather, where water content is low.

Influence of microorganisms on leather. Action of mold and bacteria is first of all a matter of moist and dry tropical climate. However, moist leather left in an isolated place or container will develop molds in a moderate climate as well. This is why polyethylene sheets or other plastic packages are harmful and dangerous for leather.

Molds growing on leathers are primarily *Penicillia*, *Aspergilla* and *Trichoderma*. They decrease leather strength and fat content in it. According to a certain opinion, decrease of the strength due to the mold action is a result of fat consumption by the mold, with a resultant internal deterioration of greasing. We do not know much about the mechanisms of changes in leather due to molds. The information is sometimes contradictory. Discolorations and spots due to mold action are difficult to get rid of, and mold will appear again in these places. The only reasonable procedure is to use fungicides.

Development of the bacteria on leather is a secondary process, occurring after hydrolysis, long-term action of mold, etc.

Action of enzymes has been discussed in sect. 9.3. Tanned leather is *ex definitione* resistant to the action of proteolytic enzymes. Only leather previously deteriorated due to hydrolysis or detannage is sensitive to it.

REFERENCES

1. see ref. 8 ch. 8
2. Innes, R. F. J. Soc. Leath. Tr. Chem., *17*, 725 (1933)
3. Beebe, C. W., Frey, R. W., Harningan, M. V. J. Am. Leath. Chem. Assoc., *51*, 20 (1956)
4. Cooper, D. R. in Biophysical Properties of the Skin, (H. R. Elden ed.) Wiley—Intersc. N.Y. 1971
5. Bowes, J. H. A Fundamental Study of the Mechanism of the Deterioration of Leather Fibres, BLMRA Egham 1963
6. Stather, F., Herfeld, H. Collegium *877*, 210 (1943)
7. Robinson, S., Robinson, A. H. Physiol. Revs., *34*, 202 (1954)
8. Bienkiewicz, K. J., Orszulski, L. XV IULTCS Congress Hamburg 1977
9. Sanger, F. Biochem. J., *39*, 507 (1945)
10. Williams-Wynn, D. A., Sykes, R. L. J. Soc. Leath. Tr. Chem., *45*, 214 (1961)
11. Feher, S., Kerese, S. Rev. Techn. Ind. Cuir *51*, 95 (1959)

Index

531